TAILINGS AND MINE WASTE '01

PROCEEDINGS OF THE EIGHTH INTERNATIONAL CONFERENCE ON TAILINGS AND MINE WASTE '01 / FORT COLLINS / COLORADO / USA / 16 – 19 JANUARY 2001

Tailings and Mine Waste '01

A.A. BALKEMA / ROTTERDAM / BROOKFIELD / 2001

Organized by Geotechnical Engineering Program, Department of Civil Engineering, Colorado State University

ORGANIZING COMMITTEE

John D. Nelson	Conference Chair, Colorado State University, Fort Collins, Colorado
Debora Cave	HSI Geotrans, Inc., Westminster, Colorado
William A. Cincilla	Golder Associates, Lakewood, Colorado
Cary L. Foulk	M.F.G., Boulder, Colorado
Linda L. Hinshaw	Colorado State University, Fort Collins, Colorado
Robert L. Medlock	Environmental Consultant, Fort Collins, Colorado
Debora J. Miller	ESA Consultants, Fort Collins, Colorado
Sean C. Muller	SRK Consulting, Fort Collins, Colorado
Daniel D. Overton	Shepherd Miller, Inc., Fort Collins, Colorado
Charles D. Shackelford	Colorado State University, Fort Collins, Colorado
William P. Van Liew	US National Park Service, Fort Collins, Colorado
Dirk J.A. Van Zyl	Mackay School of Mines, Reno, Nevada

The texts of the various papers in this volume were set individually by typists under the supervision of each of the authors concerned.

Published by
A.A. Balkema, P.O. Box 1675, 3000 BR Rotterdam, Netherlands
Fax: +31.10.413.5947; E-mail: balkema@balkema.nl; Internet site: www.balkema.nl
A.A. Balkema Publishers, 2252 Ridge Road, Brookfield, VT 05036-9704, USA
Fax: 802.276.3837; E-mail: info@ashgate.com

ISBN 90 5809 182 1

Printed in the Netherlands

Tailings and Mine Waste '01 © 2001 Balkema, Rotterdam, ISBN 90 5809 182 1

Table of contents

Geotechnical considerations

Liners, covers and barriers

Tailings and Mine Waste '01 © 2001 Balkema, Rotterdam, ISBN 90 5809 182 1

Preface

This is the eighth annual Tailings and Mine Waste Conference held at Colorado State University in Fort Collins, Colorado. The purpose of these conferences is to provide a forum for discussion and establishment of dialogue among all people in the mining industry and environmental community regarding tailings and mine waste. Previous conferences have been successful in providing opportunities for formal and informal discussion, exhibits by equipment and instrumentation companies, technical exhibits, and general social interaction.

This year's conference has over 50 papers. These papers address the important issues faced by the mining industry today. These proceedings will provide a record of the discussions at the conference that will remain of value for many years.

Site characterization

Tailings and Mine Waste '01 © 2001 Balkema, Rotterdam, ISBN 90 5809 182 1

Identification of mine wastes using the CASI and SFSI hyperspectral imagers at the Bauer Mill tailings site, Stockton, Utah

E.D. Dillenbeck, L.G. Closs & K. Lee – *Department of Geology and Geological Engineering, Colorado School of Mines, Golden, Colo., USA*

D.C. Peters – *Peters Geosciences, Golden, Colo., USA*

P.L. Hauff – *Spectral International Incorporated, Arvada, Colo., USA*

G.A. Borstad – *G.A. Borstad and Associates Limited, Sidney, B.C., Canada*

ABSTRACT: Stockton, Utah (40 miles southwest of Salt Lake City), was a center of lead, silver and zinc mining from 1864 to 1970. Many small processing facilities were located in and near the town, with several homes in the town built on tailings. The Bauer Mill and tailings site is located one mile north of Stockton and was the largest non-smelter mineral processing facility in the immediate area. The Bauer Mill was active between 1900 and 1974, and arsenic-rich unreclaimed tailings cover at 160 acres. The primary environmental concern identified by the U.S. Environmental Protection Agency (EPA) and Utah Dept. of Environmental Quality (UDEQ) at Bauer is the transportation of tailings from the site by the prevailing southerly winds. Agricultural, residential, and light industrial development has been and is being considered for area near and potentially downwind of the Bauer site. Hyperspectral airborne remote sensing data were collected using the SWIR Full Spectrum Imager (SFSI) and Compact Airborne Spectrographic imager (CASI), with the goal of mapping the distribution of tailings and characterizing area of mineral processing activity at the Bauer Mill site and near the town of Stockton. Classification and mineral mapping images of both short-wave infrared (SWIR)-active and iron-minerals maps were produced in an attempt to define the limits of waste impacts. SWIR-active minerals, including smectite and gypsum, were found in samples collected at the Bauer Mill site (analyzed with the PIMA™ II spectrometer). This work was completed as part of the NASA EOCAP program in association with Spectral International/Peters Geosciences/Borstad Assoc. and the Colorado School of Mines. Selection of the Bauer site was made in consideration to the EPA Region 8 Utah Abandoned Mine Lands Program.

1 SUMMARY

In the fall of 1998, Short-Wave Infrared (SWIR) Full Spectrum Imager (SFSI) and Compact Airborne Spectrographic Imager (CASI) (owned by Borstad Associates Ltd.) data were collected over portions of central Utah for use in evaluating abandoned and inactive mines and mineral processing facilities for potential environmental hazards. The data were acquired as part of NASA's Earth Observations Commercial Applications Program (EOCAP) and co-funded by a team led by Spectral International Inc., Peters Geosciences, Borstad, and HENDCO Services. Part of the team's investigations were in association with the EPA's (Region 8) Utah AML Watershed Hyperspectral Analysis program. The Bauer Mill site was included among this team's study areas and was selected for further study as part of a M.Sc. degree currently underway at the Colorado School of Mines.

The Bauer Mill tailings site (Figure 1) is an example of an unreclaimed mineral extraction facility where the potential for environmental damage caused by waste products has not been fully quantified. Bauer is typical of other abandoned or inactive mill and mine lands (AML) where fluvial, eolian, or human processes are redistributing mineral extraction waste products

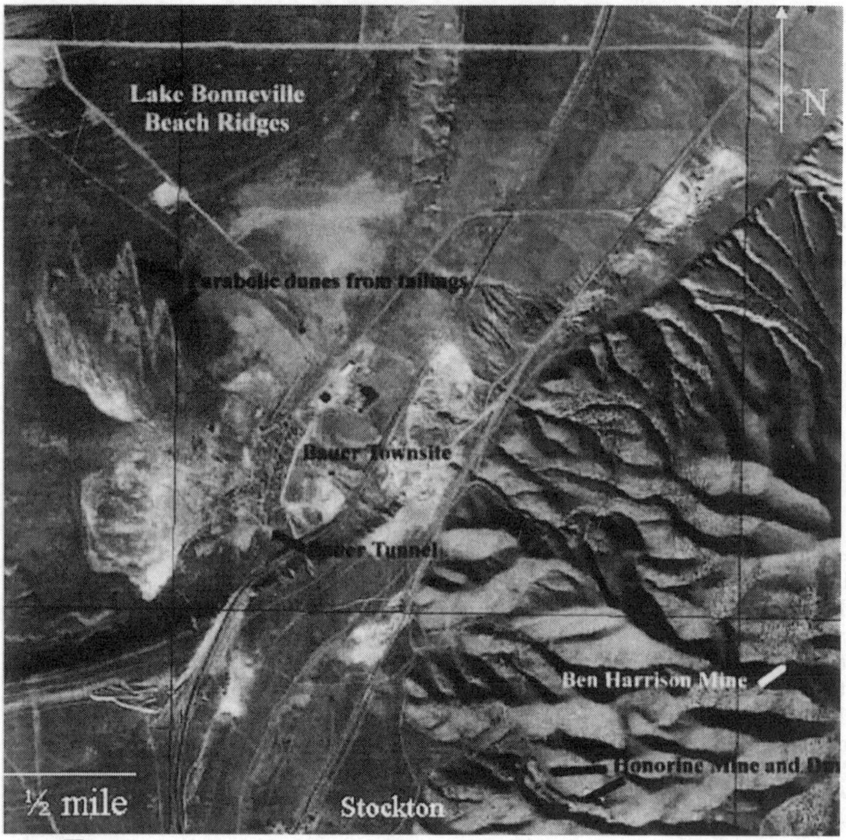

Figure 1. False-color CASI image of Bauer-Stockton area.

Figure 2. Windblown tailings at the Bauer site.

(Figure 2). There is a clear need for an efficient method to identify, characterize, and prioritize reclamation of these sites. It has been shown that research sensors such as NASA's Airborne Visible and InfraRed Imaging Spectrometer (AVIRIS) can identify potentially hazardous mine wastes (Swayze et al, 1996) and that this technology is at the point of commercialization (Hauff et al, 1999). The Spectral International – Peters – Borstad – HENDCO team's work at Bauer and elsewhere is intended to demonstrate that commercial hyperspectral surveys are reliable, reproducible, and efficient means of investigating environmental issues on mine lands.

2 REFERENCES/BIBLIOGRAPHY

Boardman, J.W., 1993, Automated Spectral Unmixing of AVIRIS Data Using Convex Geometry Concepts, in Summaries of the Fourth Annual JPL Airborne Geosciences Workshop, October 23-26, 1993, JPL Publication 93-26, v. 1, p. 11-14.

Chavez, L.D. 1999, Bauer Dump and Tailings/Blackhawk Resin Expanded Site Inspection Analytical Results Report, UTD980635528.

Dillenbeck, E.D., D.C. Peters, P.L. Hauff, G. A. Borstad, L.G. Closs and K. Lee, in press, Identification of Mineral Extraction Waste Products Using the SFSI and CASI Hyperspectral Imagers at the Bauer Mill Site, Stockton, Utah, in Proceedings of the Fourteenth International Conference on Applied Geologic Remote Sensing, Las Vegas, Nevada,

Garcia, Z.P., and others, 1976, Brief History of Stockton, Utah: Stockton Bicentennial Committee.

Green, A.A., M. Berman, P. Switzer and M.D. Craig, 1988, A Transformation for Ordering Multispectral Data in Terms of Image Quality with Implications for Noise Removal: IEEE Transactions on Geoscience and Remote Sensing, v. 26, no. 1, p. 65-74.

Hauff, P.L., N. Lindsay, D.C. Peters, G.A. Borstad, W. Peppin, L. Costick and R. Glanzman, 1999, Hyperspectral Evaluation of Mine Waste and Abandoned Mine Lands NASA and EPA Sponsored Projects in Idaho, in Summaries of the Eighth JPL Airborne Earth Sciences Workshop February 9-11, R.O. Green (ed.), JPL Publication 99-17 p. 229-238.

Krahulec, K.A., 1999, Road log from Salt Lake City to the Mercur gold mine, Tooele county, Utah: Geological Society of Nevada Special Publication No. 30.

Swayze, G.A., R.N. Clark, R.M. Pearson and K.E. Livo, 1996, Mapping Acid-Generating Minerals at the California Gulch Superfund Site in Leadville, Colorado using Imaging Spectroscopy, Summaries of the Sixth Annual JPL Airborne Earth Sciences Workshop, March 4-8, R.O. Green (ed.), JPL Publication 96-4.

Thiriot, S., April 13, 2000, Regional manager UDEQ, Personal conversation concerning the Bauer Mill site.

Tailings and Mine Waste '01 © 2001 Balkema, Rotterdam, ISBN 90 5809 182 1

Waste identification and characterization for the International Smelter area using CASI/SFSI hyperspectral remote sensing data

D.C. Peters – *Peters Geosciences, Golden, Colo., USA*
P.L. Hauff, W.A. Peppin & E.C. Prosh – *Spectral International Incorporated, Arvada, Colo., USA*
G.A. Borstad – *G.A. Borstad and Associates Limited, Sidney, B.C., Canada*
F.B. Henderson III – *HENDCO Services, Nathrop, Colo., USA*
E.D. Dillenbeck, L.G. Closs & K. Lee – *Colorado School of Mines, Golden, Colo., USA*
J.L. Thiros – *Utah Department of Environmental Quality, Salt Lake City, Utah, USA*
D.W. Coulter – *Newmont Exploration Limited, Englewood, Colo., USA*
D.A. Robbins – *ASARCO Incorporated, Salt Lake City, Utah, USA*
K.W. Wangerud – *US Environmental Protection Agency, Denver, Colo., USA*

ABSTRACT: The International Smelter, 3 km east of Tooele, Utah, operated from 1910 to 1971 and consisted of copper and lead smelters, concentrators, and various other processing facilities as well as a large tailings disposal area. The site was reactivated for tailings disposal by the Carr Fork copper mine and mill from 1979 to 1981. Reclamation and revegetation of the smelter complex and combined tailings facilities was completed to the satisfaction of State regulatory agencies in 1989 (Braxton & Buck 1989). The site was converted to a wildlife reservation following reclamation. The smelter site is upstream of farming and residential areas near Tooele, and the town of Tooele gradually is growing east toward the tailings area.

However, areas of tailings were missed or not addressed by reclamation activities. Other areas within the smelter-concentrator complex are showing chemical resurgence from sulfidic wastes and metal-rich wastes that were only covered up by a layer of gravel during reclamation. The degree to which revegetation has been successful at the site and areas of missed or resurgent wastes were the primary topics of interest for a remote sensing analysis of the area.

Hyperspectral remote sensing was used at this site to identify mineralogical and vegetation differences that were indicative of unreclaimed and resurgent wastes. Specifically, the CASI (Compact Airborne Spectrographic Imager) and SFSI (SWIR (short-wave infrared) Full Spectrum Imager) sensors were flown together to provide reflectance data in the visible, near-infrared, and short-wave infrared ranges of the spectrum. Accuracy of image processing results (mineralogical classification) were confirmed through ground sampling and analysis with portable spectrometers. The results of this site characterization process and implications for environmental impacts and further remediation of the site were the focus of this presentation.

1 INTRODUCTION

This paper is a brief description of remote sensing work performed for the International Smelter site (also called the "Carr Fork Property") near Tooele , Utah, as part of studies for a National Aeronautics and Space Administration (NASA) Earth Observations Commercial Applications Program (EOCAP) project on "Application of Airborne Hyperspectral Data to Characterization of Mined Lands and Analysis of Associated Watersheds and Impacts for Environmental Management" (Contract #NAS-13-99004). We are unable to cover all the details of this project and related work in the space available, so please see section 4 (Additional Information) on how to access web sites that can provide you with more information.

The EOCAP in general is designed to demonstrate remote sensing technologies that are on the cusp of commercialization and to advance the respective fields of these technologies through identification of advantages and pitfalls that need to be addressed before full operational use can be achieved. Our project in particular was designed to prove the applicability of airborne hyperspectral data to characterization of mine and mill sites and related

Figure 1. Oblique aerial view of the International Smelter complex (except for tailings impoundments) in the late 1960s, looking to the south. Photograph from the Wilbur H. Smith Papers (MS563), Special Collections, J. Willard Marriott Library, University of Utah.

wastes. The environmental industry in general has made little use of the capabilities of hyperspectral sensors and data, so the goal was to prove that critical and detailed characterization of sites and wastes can be performed reliably and more thoroughly using hyperspectral remote sensing than can be achieved using standard ground surveys alone.

In this paper, we will touch upon some of the methodology and results for the International Smelter area (Figure 1). Studies on one of the other Utah sites, the Bauer Mill, are discussed briefly in Dillenbeck and others (2001) in this book. Unfortunately, we are unable in this paper to show color images that are a critical part of the remote sensing technology and investigations for the International Smelter. Such images are available through the web sites discussed in section 4.

2 DATA ACQUISITION AND ANALYSIS

The primary hyperspectral sensors used were the Compact Airborne Spectrographic Imager (CASI) and Short-Wave Infrared (SWIR) Full Spectrum Imager (SFSI) airborne systems (Table 1), which are flown together in a light aircraft, at an altitude of approximately 3,500 m above

Table 1. Sensor specifications for CASI and SFSI and comparison with AVIRIS. "VNIR" refers to the visible and near-infrared range of the electromagnetic spectrum.

Sensor	SFSI-2	CASI	AVRIS
Spectral Range	1220-2320 nm (SWIR)	403-914nm (VNIR)	400-2500 nm
Bandwidth	5 nm	>2 nm	10 nm
Swath Width	2 km	2 km	10.5 km
Number of Bands	240	15 (selectable out of 100)	224
Ground Pixel Size	4 m	4 m	17 m

mean terrain. G.A. Borstad Associates performed all CASI and SFSI airborne data collection for this project. In addition to the data flown for use by our team for this NASA project, the entire Oquirrh Mountains area was flown in August 1998 with the Airborne Visible-InfraRed Imaging Spectrometer (AVIRIS), which is described in Table 1 for comparison with CASI and SFSI. These data were available to our NASA EOCAP team as part of a parallel study known as the "U.S. Environmental Protection Agency (EPA) Region 8 Utah Abandoned Mine Lands (AML)-Watershed Hyperspectral Analysis Project", which involved demonstration of hyperspectral technology for EPA environmental analysis and monitoring purposes.

CASI and SFSI data for the International Smelter were acquired in October 1998. Initial calibration and rectification of these data were performed by G.A. Borstad Associates as part of their work on the EOCAP project. Further analysis of the CASI data was performed by Borstad Associates to produce a vegetation classification of the study area. This vegetation classification proved to be especially valuable for full assessment of the International Smelter area because of the revegetated nature of much of the site.

Processing of the AVIRIS and SFSI data, and selected CASI data, for mineralogical analyses was performed by Spectral International and aided by review and other input and GIS work by Peters Geosciences. Other team participants had technical input to the site analysis process and/or background information and the GIS database for the study areas. ENVIO was used for most image processing work and ARC/INFOO and ArcViewO were used for all GIS work, with use of Adobe PhotoshopO and miscellaneous other software packages for other project-related work.

With respect to field work, the International Smelter area was visited and samples were collected to provide a representative set of mineralogical data both for the overall study area and for specific targets within the study area. Exhaustive sampling was not attempted, but major differences in mineral classifications (from the hyperspectral data) and physical features of study areas were examined within the limits available for the project. This resulted in approximately 100 sample sites being visited to date in the International Smelter area. The number of samples per sample site varied depending on the type of site and materials present. Only surficial materials were sampled. Samples are stored in sealable plastic bags to limit changes and prevent contamination of the samples. Ground photographs were taken at or near most sample sites to provide a visual record of site conditions that may be important for analysis of the hyperspectral data. GPS points were collected for sample locations for 1999 and 2000 field visits and subsequently corrected for system-based errors to provide locations that were reliable within the limits of the 4-m ground resolutions of the CASI and SFSI data.

Spectral analyses of the samples were performed with a PIMA II portable spectrometer (spectral coverage = 1200 - 2500 nm; internal light source) and an Analytical Spectral Devices "Hand-Held" portable spectrometer (spectral coverage = 300 - 1100 nm; solar or external light source). These devices provide detailed spectral coverage for most of the ranges covered by the airborne sensors used in this study. This allows detailed comparison of spectral mineralogy acquired both on the airborne scale and individual sample scale and the identification of significant mineralogical components that may be discriminable through processing of the hyperspectral data.

3 SELECTED RESULTS

The following are some bullets that represent significant results from the International Smelter study:

- Not all of the tailings have been cleaned up or covered up at the International Smelter site, because tailings were observed and spectrally confirmed in a number of samples and, equally as important, there is present acid drainage from the smelter/concentrator complex along the Dry Canyon drainage, and potentially from the Northwest Dump. These sample areas likewise appear as "tailings" and potential mineralogical anomalies in the hyperspectral images.

Figure 2. View of part of the International Smelter complex, looking to the northeast in May 1914, showing smelter emissions blowing to the east and southeast toward the adjacent Oquirrh Mountains. Photograph from the Wilbur H. Smith Papers (MS563), Special Collections, J. Willard Marriott Library, University of Utah.

Figure 3. View of east flank (in background) of Pine Canyon (Oquirrh Mountains) in August 1999, showing denuded lower and middle slopes that are believed to have been impacted by emissions from the International Smelter complex. The slag dump for the smelter is in the middleground at left.

- Although AVIRIS shows general distributions of minerals, the higher spectral and spatial resolution of the SFSI sensor provides the ability to better separate transitional, acid-drainage diagnostic mineral species that cannot be discerned as well in the AVIRIS images. Many of the failed revegetation areas and unreclaimed areas also are too small to be resolved adequately by the AVIRIS data.

- The most common infrared active minerals at the International Smelter Complex include smectite, illite, kaolinite, gypsum, calcite, and dolomite, with minor alunogen, jarosite, and selected copper sulfates. Aside from the exotic sulfates, these are the species most readily identified from the aircraft data.

- Acid drainage associated minerals at this site include alunogen, jarosite, gypsum, and transitional kaolinites.

- Gypsum can be a weathering product of sulfide, or SO_2 combined with carbonate, and is stable in a dry climate. Both input components are present at this site. The climate, however, is perhaps too humid to sustain extensive development of this mineral, except in areas of higher concentration of sulfide. This actually makes it a good pathfinder for the mine and mill wastes.

- Poorly crystalline kaolinite is the more common occurrence of kaolinite. It is also easier to explain its presence as a function of acidic ground water activity. Therefore, it would be expected to be present anywhere there is a possibility of sulfides or oxidized sulfides or SO_2 emissions. This appears to be the case as this material is particularly concentrated at the smelter complex where there are remnant sulfides, in drainages draining from the smelter complex, and on dumps and tailings piles.

- Impacts of smelter smokestack emissions (Figure 2) have been detected and preliminarily mapped with the hyperspectral images. This issue of smelter emissions previously has not been addressed at this site. It is felt to be very important relative to the impact on the Pine Canyon area vegetation (Figure 3) from the past H_2SO_4 generation at the site.

- An important key to understanding geologically oriented image processing is understanding the spectral variability that occurs with mineral species. If this is not addressed, then the image processing will produce inaccurate results.

- Certain applications may require only that the presence or absence of a mineral is verified and its detailed and accurate distribution may not be required to be known for a reconnaissance pass over an area that will be sampled in detail on the ground. In applications where the image data are tied to other information on the ground, such as distribution of toxic materials, the criticality of site specific spectral libraries and spectral variability is more apparent.

- Although the ground data may show the presence of a category of materials, there may not be enough of it in large enough concentrations and large enough areas to be detected from the air. This is a learning process to determine what the optimum parameters are for specific applications. This presents almost insurmountable problems with 17-m pixels (AVIRIS) and even significant problems with 4-m pixels (CASI and SFSI) (Table 1).

- Vegetation masks the soil signature and anything else with mineral spectral characteristics because the organic signature is strong. For the International Smelter area, where revegetation is extensive, the area must be studied in the negative sense (i.e., look for good mineral response of wastes as an indication of failure of revegetation).

4 ADDITIONAL INFORMATION

The full details and results of this study, and related results for other Utah study sites, are too voluminous to cover in this proceedings paper. Therefore, readers are directed to the web site *www.pimausa.com* for further information on the International Smelter study area and other work conducted under this NASA EOCAP project. Please click on the NASA icon at this web address for project information and how to obtain the reports for the NASA project. Further information on the NASA EOCAP program, including commercial partners, can be obtained from web site *http://www.crsp.ssc.nasa.gov/hyperspectral/hypermain.htm*.

5 REFERENCES

Braxton, L.P. & B.W. Buck 1989. Reclamation of the Carr Fork Property, Tooele, Utah. In G.E. Cordy (ed.), *Geology and hydrology of hazardous-waste, mining-waste, waste-water, and repository sites in Utah*: 115-120. Utah Geological Association, Publication 17.

Dillenbeck, E.D., D.C. Peters, P.L. Hauff, L.G. Closs, K. Lee & G.A. Borstad 2001. Identification of potentially hazardous wastes using the SFSI and CASI imaging spectrometers at the Bauer Mill tailings site and the Stockton mining district, Utah. In *Tailings & Mine Waste '01*: in press. Rotterdam: Balkema.

Tailings and Mine Waste '01 © *2001 Balkema, Rotterdam, ISBN 90 5809 182 1*

Characterization and control of alkaline seepage at closed tailings ponds, Arkansas

L.N. French
MFG Incorporated, Austin, Tex., USA (Currently: URS Corporation, Austin, Tex., USA)
B.L. McCulley
MFG Incorporated, Boulder, Colo., USA

ABSTRACT: Tailings ponds at a mining and ore processing facility in central Arkansas managed alkaline liquor and residual solids for over 40 years. During reclamation, a french drain was installed to intercept seepage from tailings materials. The aggregate flow ranged between 40 and 100 gallons per minute (gpm) of alkaline (pH = 13.5) fluid.

The study evaluated whether the flow represented simple gravity drainage of mostly residual liquor or a combination of residual liquor and infiltrated rainwater. A primary goal was to evaluate what the likely steady-state porewater levels would be and to what degree the porewater levels would continue to feed seepage. Three types of porewaters were observed: (1) porewaters with seasonally fluctuating levels; (2) porewaters with decreasing levels reflecting the dissipation of excess pore pressure and/or depletion of alkaline liquor; and (3) porewaters with high amplitude and/or variable level changes. The seasonal pattern of porewater fluctuations suggests the french drain flow is likely to persist indefinitely. Based on the study, extensions to the french drain system were added to facilitate drainage and reduce elevated porewater heads.

1 AREA DESCRIPTION

The process water lakes at a mining and ore processing facility contained alkaline liquor and tailings materials beginning in 1954. Drainage, closure, and reclamation of the lakes began in the early 1980s and final reclamation was accomplished in 1997 and 1998. Figure 1 shows these former lakes. The Stockpile is flat high area in the south center portion of the lake complex formerly occupied by impounded alkaline liquor, but later filled and raised with apparently dredged or otherwise deposited tailings. Lake 1 is a bowl-like structure that contained tens of feet of alkaline liquor, which was drained in 1997 and 1998. Remnants of the original dikes enclose this area. Pond 1 is a low area north of the Stockpile and Lake 1 that formerly contained blends of alkaline liquor and other plant waters. A feature termed the "mud wave" occurs in the north-central portion of the pond and consists of a roughly 300-foot diameter dome (25 feet high) of residual wet tailings that were pushed up in the final stages of pond reclamation.

The areas of focused seepage were observed to occur in a central swale within and upstream (west) of the Pond 1, and a french drain system on the northern boundary of the Stockpile.

The central swale is on the southern margin of former Pond 1 and conveys runoff east from the topographically higher area west of Pond 1 to Pond 2. The swale is characterized by diffuse outbreaks of alkaline liquor seepage, heaving sand and sand boils, as well as numerous small flowing seeps. Seepage flow measurements in 1999 ranged from approximately 4 to 14 gpm.

A french drain system was installed in mid-1998 to intercept and control seepage encountered north of the Stockpile. The drain is apparently 5 to 10 feet below the surface and intercepts a variety of materials, including lithified and fractured tailings materials, possibly native soils, and some soft tailings residues. Alkaline liquid flowed from fractured material near the eastern por-

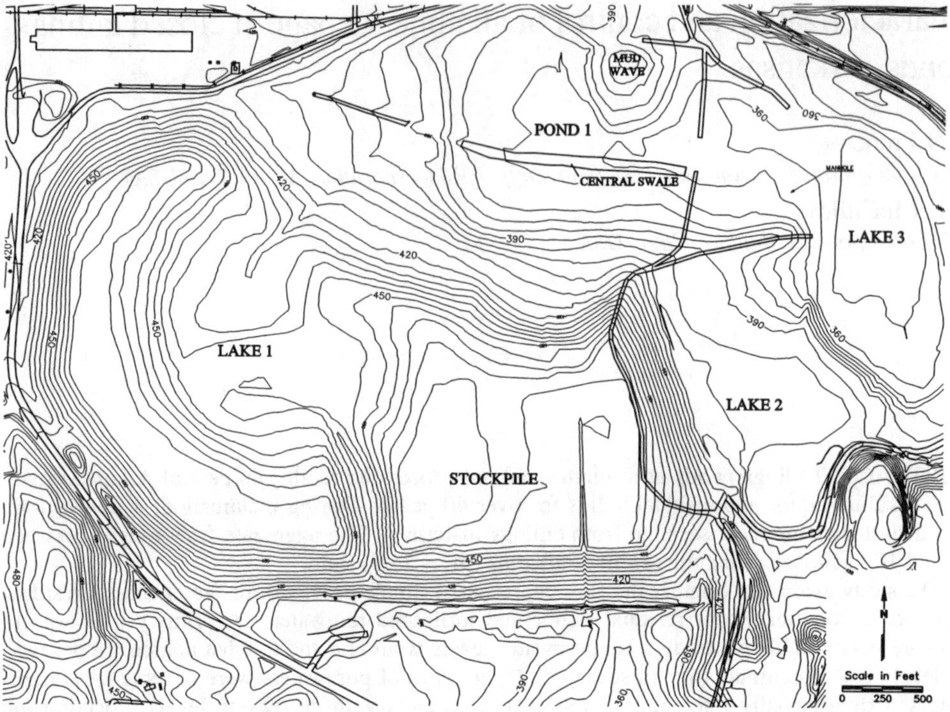

Figure 1. Topographic map of the closed tailings ponds.

tion of the central swale prior to the construction of the drain. Between May 1998 and March 2000 the discharge varied widely, from about 40 gallons per minute (gpm) during dry periods to over 100 gpm after periods of heavy rainfall.

2 OCCURRENCE OF POREWATER

2.1 *Potentiometric conditions*

At various depths below ground, the tailings materials are highly alkaline, saturated, and able to yield water to piezometers. Results of drilling, CPT soundings, pore-pressure readings during CPT soundings, pore-pressure dissipation tests, and multiple water level measurements support the concept of a complex, vertically discontinuous porewater occurrence. Figure 2 is a potentiometric surface contour map that illustrates the elevation of the uppermost occurrence of porewater in the lake complex in January 2000. The degree of hydraulic continuity (if any) between different process lakes is unknown

The depth to porewater varied from several feet (initially) above the ground surface at Pond 1, to tens of feet below ground at the Stockpile. Porewater in Lake 1 and the Stockpile occurs under unconfined (water table) conditions. Porewater levels at Pond 1 were initially above the ground surface, but have declined since December 1998 so that the potentiometric surface is near the ground level. The CPT soundings, particularly at the Stockpile, suggest multiple occurrences of porewater in isolated and discontinuous zones. Pore-pressure dissipation tests, and some of the piezometers illustrate that there are multiple piezometric levels, and hydraulic gradients that vary in magnitude and direction.

2.2 *Seeps and french drain discharge*

Table 1 summarizes the 1999 flow and field water quality measurements, and Figure 3 illustrates the fluctuations of the french drain flow and monthly rainfall totals. Discharge rates from

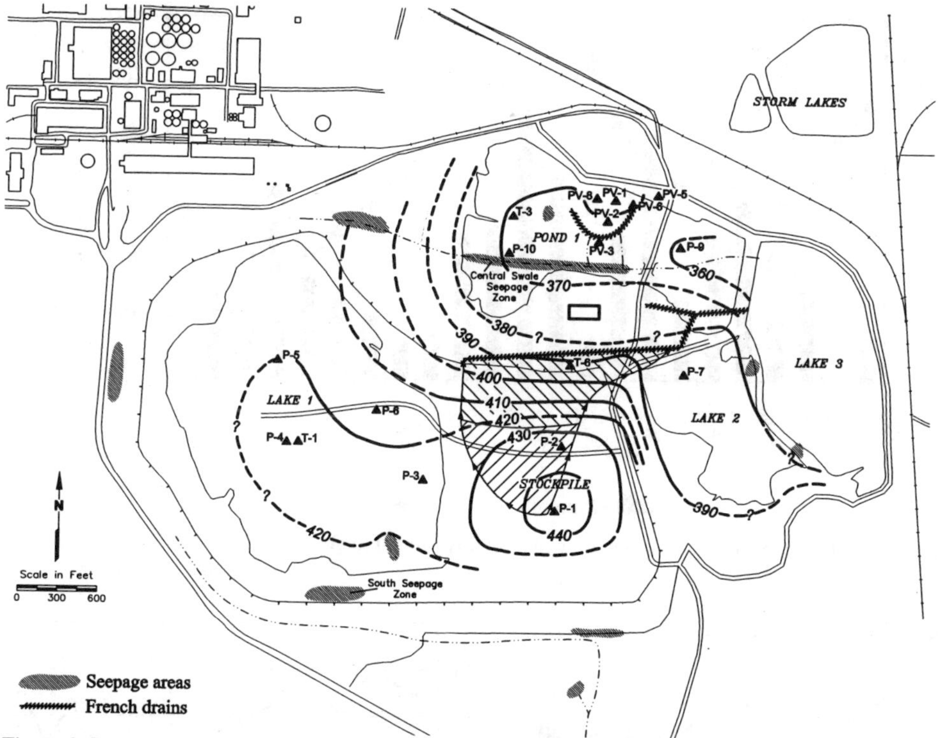

Figure 2. Potentiometric surface map and seepage areas of the closed tailings ponds.

Table 1. Range of Seep Flow and Water Quality Measurements.

Location	Flow (gpm)	Conductivity	pH	Alkalinity (g/L)
French Drain	40 – 106	>50,000	13.1 – 13.5	18.8 – 20.4
Mud Wave drains	1.3 – 1.5	---	10.6 – 13.1	3.1 – 13.1
Central Swale seeps	4.6 – 14	---	13.1 – 13.2	14.6 – 18.9
Upper Central Swale seeps	1.9 – 18.2	39-50,000	12.3 – 13.3	2.7 – 16.7
Lake 2 Seep	0.76 – 0.85	3,600	10.2 – 10.6	1.9 – 2.3
South Seep Collection	1.5 – 25.8	32-42,000	10.4 – 12.7	1.7 – 12.5

Note: Locations of seepage areas shown on Figure 2.

the french drain have ranged between about 40 and 106 gpm since May 1998. The discharge from the drain fluctuates seasonally and roughly in accordance with rainfall amounts. Field measurements of alkalinity and pH have shown virtually no change whether in a dry or rainy weather pattern. This observation suggests that the porewater continues to be generated by infiltration of rainfall through the tailings materials and then discharged through the drain (or other seeps). Had the increase in flow been due to shallow percolation of rain directly into the drain, then some amount of "dilution" (e.g., lower pH, alkalinity) should have been seen during rainy periods. A dilution effect (mixing with runoff) was observed in some smaller seeps.

The amount of alkaline seepage emerging in the Central Swale in the form of diffuse seepage and in micro seeps and boils at the southern boundary of Pond 1 has been difficult to quantify. In addition to the visible flow, it is suspected that underflow in the swale channel is a likely mechanism for transport of seepage water downstream. Measurements of flow in this area since April 1999 indicate fluctuations between no observable flow to 14 gpm. The highest flow rate (18.2 gpm) in the channel was on December 9, 1999, but was influenced by runoff judging by the much lower alkalinity value (2.7 g/L).

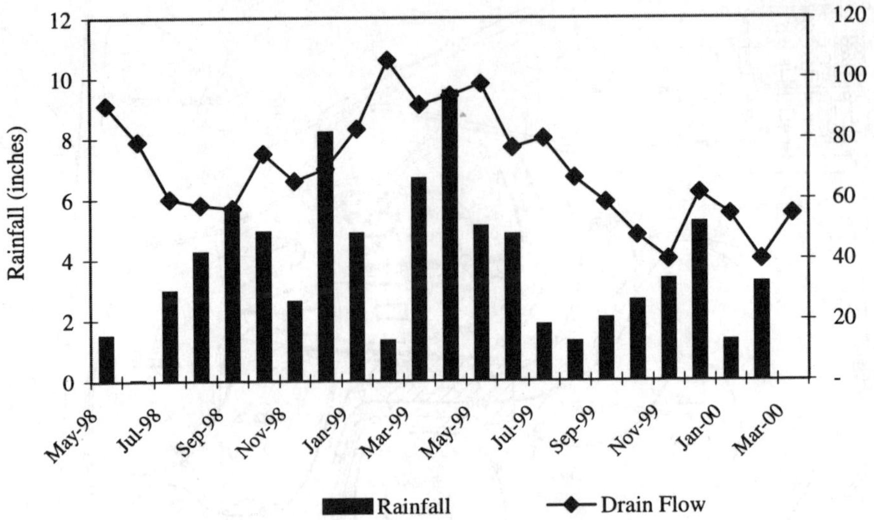

Figure 3. Discharge rate of the french drain with monthly rainfall totals.

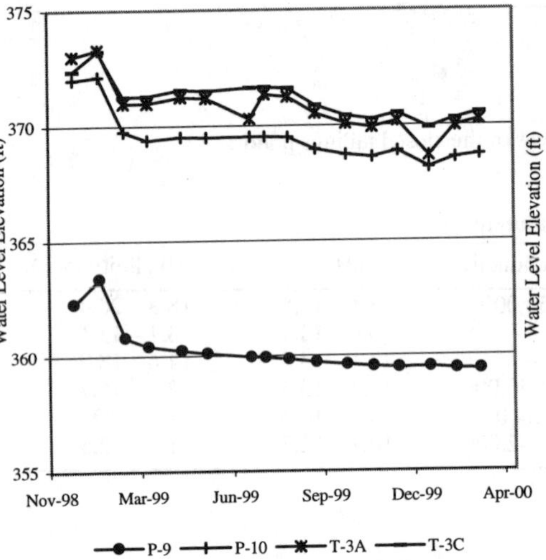

Figure 4. Hydrograph of Pond 1 piezometers.

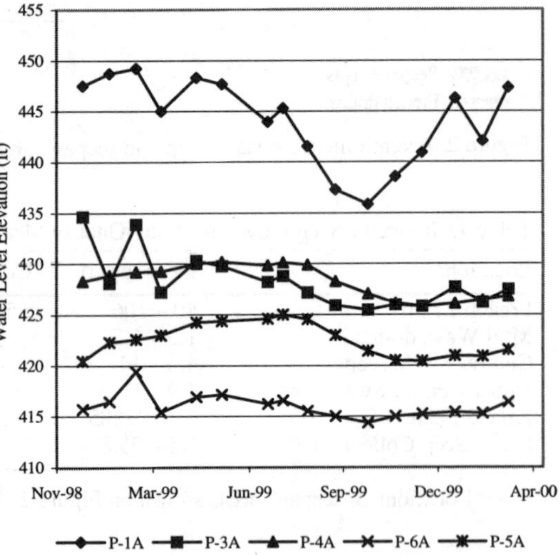

Figure 5. Hydrograph of Stockpile and Lake 1 piezometers

2.3 *Porewater fluctuations*

Figures 4 and 5 are hydrographs of selected piezometers that illustrate the porewater fluctuations at the lake complex Nearly all piezometers had decreasing porewater levels, with most of the declines in the 1 to 2 foot range. The greatest declines in porewater levels occurred in the Pond 1 piezometers (Figure 4). All piezometers in the flat area of the Pond 1 and the Central Swale showed decreases of between 3 and 4 feet during the period. However in many wells there are intraannual fluctuations indicative of recharge. Fluid levels within the Stockpile (Figure 5) did not behave in a manner characteristic of simple gravity drainage of a system hydraulically isolated from recharge processes.

There appear to be several groups of water level trends:

1. Relatively steady or declining levels, primarily at the Pond 1 piezometers (Figure 4). Most of the porewater level declines in these locations appear to be related to the drainage of excess porewater and reduction of pore pressures in soft tailings. For example, four of the six piezometers completed in the mud wave registered declines and did not resemble the pattern displayed by piezometers in the Lake 1 or Stockpile.
2. Porewater levels that suggest a seasonal response to rainfall by increasing levels in the winter through early summer, then steadily decreasing in the summer and fall months during drier conditions. These locations are shown on Figure 5 and are mostly at the Lake 1. Comparing these fluctuations to the rainfall hydrograph (Figure 3), the increased rainfall amounts correlate to moderately rising porewater levels and drier conditions correspond to falling porewater levels;
3. Wells/piezometers exhibiting variable trends with no obvious increasing or decreasing trends, or correlation with rainfall patterns. Piezometers with these characteristics appear to be grouped at the Stockpile and displayed widely variable porewater levels, possibly due to a combination of factors such as irregular drainage of the tailings materials at depth, consolidation of tailings, delayed recharge, etc.

Groundwater levels in strata underlying the lake complex are much lower than the porewater potentiometric surface in the tailings material itself.

There are several possible explanations for these trends:

- Zones with seasonally fluctuating water levels do have hydraulic interconnection with recharge/discharge processes. For those wells/piezometers located beneath the clay cap, the rising water levels reflect deep percolation of infiltrating rainfall through the cap system. The wells/piezometers at the Lake 1, portions of which are low and tend to collect runoff, appear to exhibit these characteristics.
- Zones with decreasing water levels, even during the wet winter season, may reflect the consolidation of tailings and the dissipation of excess pore pressure created by the surficial placement of material during construction. This zone occurs primarily in the Pond 1, and Central Swale, where artesian conditions initially occurred. As water levels continue to decline, some seepage areas in the Central Swale may dry up.
- Zones with high amplitude and/or variable porewater level changes that do not have significant, direct hydraulic interconnection with recharge/discharge processes and/or represent zones subject to a combination of rapid drainage or recharge conditions, along with some zones of pore fluid with relatively static hydraulic pressure. Potentiometric conditions in these zones appear to be variable and difficult to predict, although the general trend appears to be toward decreasing porewater levels. Piezometers at the Stockpile appear to exhibit these characteristics.
- The observed trends in water level fluctuations suggest that the dewatering of the tailings materials would take decades. However, this statement is based on the unlikely situation of a linear decline in porewater levels, when it is more likely that the decline is asymptotic. For example, even in a dry year such as 1999, the roughly 1-foot average annual decline of porewater levels extended linearly would translate into a period of about 40 years for the porewaters in the Lake 1 to dissipate and about 70 years for the porewaters in the Stockpile to dissipate. Furthermore, this simple analysis ignores the contribution of infiltrated rainwater to sustaining porewater levels in the tailings.

3 EVALUATION OF RESULTS

3.1 *French drain flow*

The flow to the french drain originates from a combination of discharge of infiltrated rainwater (based on the observed relationship between volume of drain flow and precipitation patterns) and drainage of residual process water. Based on potentiometric conditions and inferred flow lines for isotropic permeable media, the area contributing flow to the french drain is in primarily the northern part of the Lake 1 (shaded area on Figure 2). An accurate definition of the area contributing flow is not possible, given the sparse distribution of piezometers in the lake complex and the fact that preferred flow paths that may exist (desiccation cracks, fractures, buried or abandoned piping, etc.). For example, when the ponds were active, seepage in focused, discrete areas was observed, which suggests that preferential flow paths likely do exist.

Figure 6 is a conceptual model illustrating how the french drain may receive porewater flow originating at the Stockpile. Although the construction details of the french drain are not well understood along its length, the conceptual model suggests that the french drain is probably incapable of completely intercepting and diverting porewater flow originating from the Stockpile, but instead captures flow only in the upper saturated zone. Underflow at the french drain would proceed downgradient to the Central Swale Area. However, as discussed earlier, this conceptual model does not consider potential preferential flow paths. It is possible that the french drain collects a larger or smaller component of flow than one would infer assuming homogeneous, isotropic permeable media.

To develop an estimate of the volume of rainwater infiltrating through the clay/vegetated cap that could be discharging at the french drain, the relatively flat areas of the Stockpile (and possibly areas of the Lake 1) are considered to be where infiltration would be greatest (refer to Figure 1). These areas have lower slopes than the other parts of the lake complex, and several poorly drained areas of possibly closed depressions that likely promote infiltration of rainwater.

It is unclear how much of the infiltration on top of the Stockpile would migrate northward towards the french drain and how much would migrate southward, due to the uncertainties in the potentiometric surface (sparse piezometer control) and the role of preferential flow paths. The relatively flat area of the Stockpile shown in Figure 1 is about ten acres. To approximate infiltration conditions that may exist at the Stockpile, a previously prepared HELP model of Pond 1 was used to estimate of the range of infiltration rates that may occur within the upper part of the Stockpile (Table 2). The associated flow rates that could be provided by ten acres of cap is estimated to range up to about 9 gpm, depending on a variety of factors, including the actual vertical hydraulic conductivity of the cap.

During dry periods of this study, the french drain flow stabilized at about 50 gpm, so that it is reasonable that the flow greater than about 50 gpm is primarily attributable to the discharge of rainwater infiltration on a "seasonal" scale. (However, to produce an average instantaneous flow of 50 gpm, it would take a two-foot drop in porewater levels over an area of about 130 acres (assuming a specific yield of 30%). However, Lake 1 and the Stockpile encompass only about 50 acres.) The average annual french drain flow rate during the study period was 73 gpm, and thus the average "seasonal" infiltration is estimated at approximately 73 minus 50, or 23 gpm, expressed on a instantaneous basis. However, it should be expected that some portion of

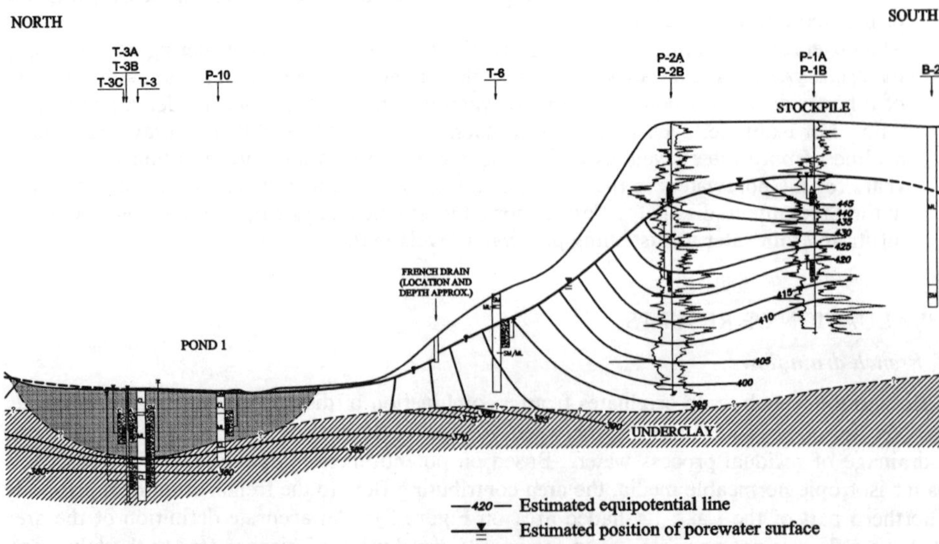

Figure 6. Conceptual hydrogeologic cross-section model of the Stockpile and Pond 1.

the 50 gpm flow is not seasonal in occurrence, and probably includes a long-term component of infiltrated rainwater that is sustained over the years. (This would be similar to considering the water table – a subtle expression of the topography – that persists over the long term regardless of the short-term or seasonal inputs of rainfall.) Therefore infiltration over an area as small as 30 acres (note that the combined flat areas of the Lake 1 and the Stockpile represent about 50 acres) could be providing the estimated component of french drain flow attributable to seasonal recharge. Alternatively, of course, lesser rates of infiltration over a much larger area could account for an equivalent flow rate.

There are complicating factors that could act to increase and decrease the magnitude of this estimate. These include:

- The contributing areas of flow to the french drain are not defined. Probably not all of the relatively flat areas described above contribute equally to the french drain, and infiltration that may occur on slopes could contribute some flow to the french drain.
- Buried pipes, desiccation cracks, and variabilities in permeability of tailings materials confound efforts to accurately predict subsurface flow regimes and estimate future flow to the french drain.
- Based on the configuration of the potentiometric surface (Figure 2), a portion of the infiltrated water flows to the south seeps or other seeps such as the area west of the Pond 1.
- Some of the infiltrated water could flow below the french drain structure and ultimately discharge to the central swale of the Pond 1.

Some of the infiltrated water could migrate to shallow groundwater zones away from the lake complex and not be expressed in local seeps.

Regardless of the magnitude of the above factors which act to increase or decrease the infiltration available to the french drain, it appears that a combination of gravity drainage and rainfall infiltration will sustain flow from the french drain for many decades. The fact that the Stockpile area was drained of standing process water approximately 20 years ago provides additional support to a rainwater infiltration source for water discharging from the french drain.

3.2 *Addressing Future Seepage Conditions at Pond 1*

Even though the potentiometric surface has declined substantially in the Pond 1 area, it appears more likely than not that the potentiometric surface will eventually stabilize at a level at or just below the land surface. As a topographically low area, the Pond 1 is hydraulically downgradient of most of the lake complex and pore water will tend to migrate toward and accumulate in this area. In addition, much of the internal surface drainage within the lake complex and also areas west of the lake complex is directed through the Central Swale. The flat and slowly draining surface of the Pond 1 will promote infiltration of water during winter and spring months. These conditions should act to sustain the continuing high level of the potentiometric surface. Therefore, it is more likely than not that the surface of the Pond 1 will experience zones of alkaline pore water breakthrough, possibly threatening the growth and stability of the vegetative cover. As noted earlier, active seepage has occurred along much of the length of the central swale and in at least one area just west of the mud wave. These areas would more likely than not have persistent or seasonal seepage. Because of the likelihood of persistent alkaline

Table 2. Simulated Infiltration Rates and Resultant Expected Flow Rates at the French Drain

Simulated Permeability of Clay/Vegetated Cap	Simulated Infiltration Through Cap				Contributing Area of Stockpile	
	Low Range		High Range		Low Range	High Range
	%	Inches	%	Inches	gpm	gpm
1.00E-07	--		1	0.5	--	0.5
6.80E-07	7	3.5	9	4.5	1.8	2.3
1.20E-06	13	6.5	16	8.0	3.4	4.1
3.60E-06	25	12.5	31	15.5	6.5	8.0
1.90E-05	31	15.5	35	17.5	8.0	9.0

Note: Average annual rainfall is about 50 inches.

seepage zones at Pond 1 is a concern to a long-term reclamation program, additional french drain(s) at Pond 1 are being manifolded into the existing french drain system to mitigate high potentiometric levels and dissipate excess hydraulic head.

4 CONCLUSIONS

The purpose of this paper has been to address the possible sources of alkaline seepage at a reclaimed tailings pond complex and evaluate the future flow rate and duration of seepage features. The principal conclusions included:

1. Alkaline porewaters in the tailings ponds occur as both unconfined and discontinuous zones and under flowing artesian conditions depending on the age and topography of the ponds. Pronounced downward vertical hydraulic gradients have developed in areas such as the Stockpile, whereas very low gradients were observed in Pond 1.

2. Discharge from a french drain that receives flow from the Stockpile varies seasonally, with a greater than 100% increase in flow (from about 40 to 100 gpm) in rainy periods.

3. Porewater fluctuations observed in multiple piezometers installed in the tailings materials displayed either steady decreasing trends or fluctuations that appeared to vary directly with rainfall.

4. Evaluation of possible flow paths and porewater behavior suggest that the french drain is fed by both gravity drainage of residual liquor and direct infiltration of rainwater in contact with alkaline tailings. As flow continues, a greater portion will be attributable to infiltration as the residual liquor is depleted. Nevertheless, the porewater fluctuations and measurements of french drain discharge over a 2-year period suggest that seepage will continue indefinitely.

5. The potentiometric surface of alkaline porewaters in Pond 1 will likely not decrease much below the land surface, and will represent a potential threat to the integrity of the vegetated cap. To mitigate the elevated porewater levels, an expanded french drain system has been installed in Pond 1.

Tailings and Mine Waste '01 © 2001 Balkema, Rotterdam, ISBN 90 5809 182 1

In situ investigation of solute transport in a spent heap leach pad

G. Zhan
Hydrology Department, Barrick Goldstrike Mines Incorporated, Elko, Nev., USA
G. Danko
Mining Engineering Department, Mackay School of Mines, University of Nevada-Reno, Nev., USA

ABSTRACT: Hydraulic and solute transport characteristics of a spent leach pad were investigated in the Carlin Trend, Nevada. 61 suction lysimeters were installed into 13 lysimeter holes at an approximately 30 x 30 x 30 m pad domain. Under steady water flow ($3*10^{-4}$ cm/s) applied by sprinkler irrigation, a pulse of 49 ppm bromide was injected for 82 hours. The downward tracer transport was sampled by the suction lysimeters for 60 days. The concentration breakthrough curves (BTCs) of the lysimeters were obtained using ion chromatography analyses. Strong variability of solute movement between different lysimeters was found. In order to obtain the statistical hydrologic properties, individual BTCs were grouped to form area-average BTCs which were then simulated applying a dual-porosity model available in CXTFIT. The simulated results indicated that dispersivity of the test pad approximately equaled the test scale and the differences between the dual-porous model and the convection-dispersion model were insignificant.

1 INTRODUCTION

Heap leaching has become the primary method of extraction for low-grade precious metal ores in recent decades, and cyanide is the main leaching agent. Rinsing is currently a common method of removing cyanide from a leach pad using fresh water or solutions that have been chemically treated to destroy the cyanide.

Even though laboratory column tests on leach materials have been conducted by numerous researchers (McClelland 1991; Dix et al. 1992; Decker and Tyler 1998), field hydrological study on the leach pad has not been done as yet. McClelland (1991) conducted laboratory column tests in his cyanide leaching and rinsing studies. The research focused on the rinsing time and required volume of rinsing water. Their tests indicated that cyanide concentration fell below 0.2 ppm, the standard required by the government agency, after 15 pore volume. Dix et al. (1992) performed laboratory chloride tracer tests on different leach pad materials from four mines in Nevada, USA. To closely represent the field rinsing situation, Decker and Tyler (1998) carried out a large scale laboratory test using a 6 m height column filled with materials from the same pad of our field study. The breakthrough data from Dix et al., as well as Decker and Tyler's tests were analyzed using both the convection-dispersion equation (CDE) and the dual-porosity mobile/immobile (MIM) models (Decker and Tyler 1998). It was found that the MIM model could represent observed breakthrough curves (BTCs) slightly better than the CDE model for the most cases. The results from the 6 m column test indicated that ratio of mobile water content to total water content was about 0.75; and dispersivity was about 1 m.

In addition to these laboratory tests, some field tests on the leach pads were also performed (Schafer and van Zyl 1991; Comba et al. 1992). These field investigations focused exclusively on the cyanide degradation processes and did not study the hydrologic mechanism which affected the degradation processes.

Preferential flows caused by pad heterogeneity are expected to occur during rinsing. Eriksson et al. (1997) conducted bromide tracer tests in both laboratory columns and field weighting lysimeters to investigate preferential flow and transport behavior of mining waste rocks. In their tests, no indication of significant preferential flow was found in the column tests. However, in the field, water flowed preferentially. It is believed that the waste rocks and leach materials share some similarities.

To comprehend the hydrological mechanism of cyanide rinsing, a large scale field tracer test on a spent leach pad was conducted. The flow and transport characteristics of the leach pad were evaluated using the numerical code – CXTFIT (Toride et al. 1995).

2 FIELD HYDROLOGIC TEST

2.1 *Site Description*

The field study was conducted at a spent leach pad located in the Carlin Trend of Nevada, USA. The climate of the test area is semi-arid. The area of the whole pad is approximately $9*10^5$ m^2 with a height between 45.7 and 76.2 m. Ore mined from both open pit and underground was hauled to the leach pad. Depending on the grade of the ore, it was either placed directly on the leach pad or crushed and agglomerated before placement on the pad.

2.2 *Experimental Setup*

A total number of 61 suction lysimeters were installed into 13 boreholes within an approximate 30 x 30 x 30 m pad domain. Each hole contained one lysimeter assembly, which consisted of 4 or 5 lysimeters. During the test, some lysimeters did not yield measurable water samples. The failures of these lysimeters could be due to inappropriate couplings with pad materials or the lysimeters were located at zones isolated from flow domains. There were 34 active samplers while 27 lysimeters failed to yield measurable amounts of samples.

Figure 1 shows the top view of the drilling hole pattern. Initially, the drilling pattern was designed to have a regular grid, but because of the drilling failures, some holes were relocated. The lysimeters are numbered with three digits where the first two digits represent hole number and the last digit represents the depth. Table 1 details the depths for all lysimeters.

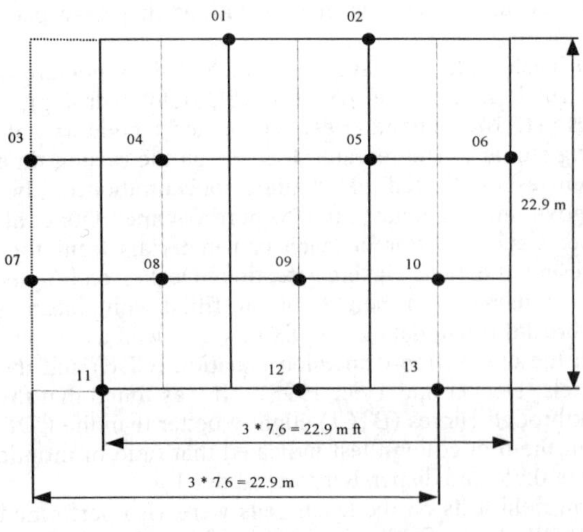

Figure 1. A planar view of the drilling hole pattern.

Table 1. Installation depths of the lysimeter assemblies.

Hole number	Lysimeter depth (m)				
	1	2	3	4	5
01	28.8*	22.7	16.6*	10.5	4.4*
02	26.7	20.6*	14.5	8.4	2.3
03	18.3*	12.2	6.1		
04	30.8*	24.7*	18.6*	12.5*	6.4*
05	28.4*	22.3*	16.2	10.1	4.0*
06	27.4*	21.3	15.2	9.1	3.0*
07	30.8*	24.7*	18.6	12.5	6.4
08	30.8	24.7	18.6	12.5	6.4
09	30.7*	24.6	18.5	12.4	6.3
10	21.3*	15.2	9.1*	3.0*	
11	26.8	19.2	11.6	4.0*	
12	29.3*	23.2*	17.1*	11.0*	4.9
13	25.2	19.1*	13.0	6.9	0.8

Note: * the lysimeter is inactive.

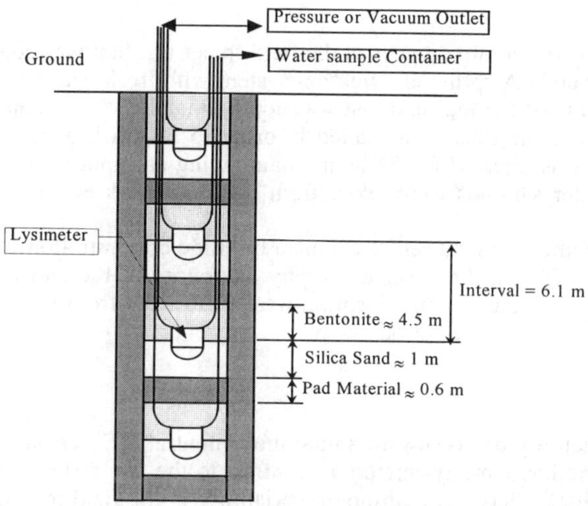

Figure 2. Schematic diagram of a lysimeter assembly installation (not to scale).

Figure 2 shows the schematic diagram of a lysimeter assembly installation in one hole. In order to maintain connections between the ceramic cups of the lysimeters and the pad materials, either silica flour or silica sand was backfilled around the cups. The effect of preferential pathways that may be caused by the drill holes was minimized using low-permeability bentonite backfill between neighboring lysimeters.

In order to generate a hydraulic gradient around the ceramic cups of the lysimeters and obtain water samples, a certain vacuum level was built up and maintained in the lysimeter reservoirs. The tests indicated that five hours were needed to get a measurable amount of water for most lysimeters. Following the vacuuming procedure, pressurized air was delivered to purge the samples out of the lysimeter reservoirs and deliver them into the sampling tubes. Each lysimeter was pressurized individually; otherwise, air would bypass through the least resistant flow path.

The sampling process was controlled by solid state relays (SSR) that were automated, and controlled through a computer program. Figure 3 exhibits the automatic sampling station.

Figure 3. Controlling center of automatic sampling station.

2.3 *Field Tracer Test*

Potassium bromide (KBr) was used as a conservative tracer in the field tracer test that was conducted in 1996 from August to October and. A sprinkler irrigation system with fresh water was utilized for the test. The irrigation rate used for the field test was approximately $3*10^{-4}$ cm/s. The pad was irrigated for 30 days before the tracer was added in order to establish a quasi-steady-state flow situation. The tracer was injected for 82 hours maintaining an input concentration of approximately 49 ppm. Water samples were taken from the lysimeters every day during a period of 60 days.

Ion chromatography measurement method was chosen to evaluate bromide concentration for each sample. The evaluation was conducted in the chromatography laboratory of the Nevada Bureau of Mines and Geology. The BTC of each sample location was constructed from the individual sample concentrations over the 60 day time period.

2.4 *Field Observed BTCs*

The bromide BTCs indicated high variability of downward solute movement at different locations. Rapid solute breakthrough occurred at some lysimeters in contrast to the late arrival and slow passage of solute at others. The high degree of hydrologic variability in the field may be illustrated by examining the observed BTCs at certain depth segments. Table 2 divides the 34 active lysimeters into four groups. Each group includes either 8 to 9 lysimeters. The average depths of lysimeters are also listed in this table.

Figure 4 demonstrates the observed BTCs for both the shallowest and the deepest lysimeters in each group. As shown, a wide range of concentration variation exists within each group.

The variability between different lysimeters could be blended together using averaging. This method has been used to deal with field data to eliminate the effect of local heterogeneity (Butters et al. 1989; Butter and Jury 1989; Tseng and Jury 1994). Through this process, the leach pad hydraulics and transport properties can be characterized to describe area-averaged behavior.

Table 2. Lysimeter groups for all observed BTCs.

Group number	Lysimeter name	Averaged depth (m)
1	135 025 125 033 095 075 085 134	5.0
2	024 064 054 014 113 032 094 074	10.8
3	084 133 023 063 102 053 093 073 083	15.8
4	112 062 012 092 082 131 021 111 081	24.7

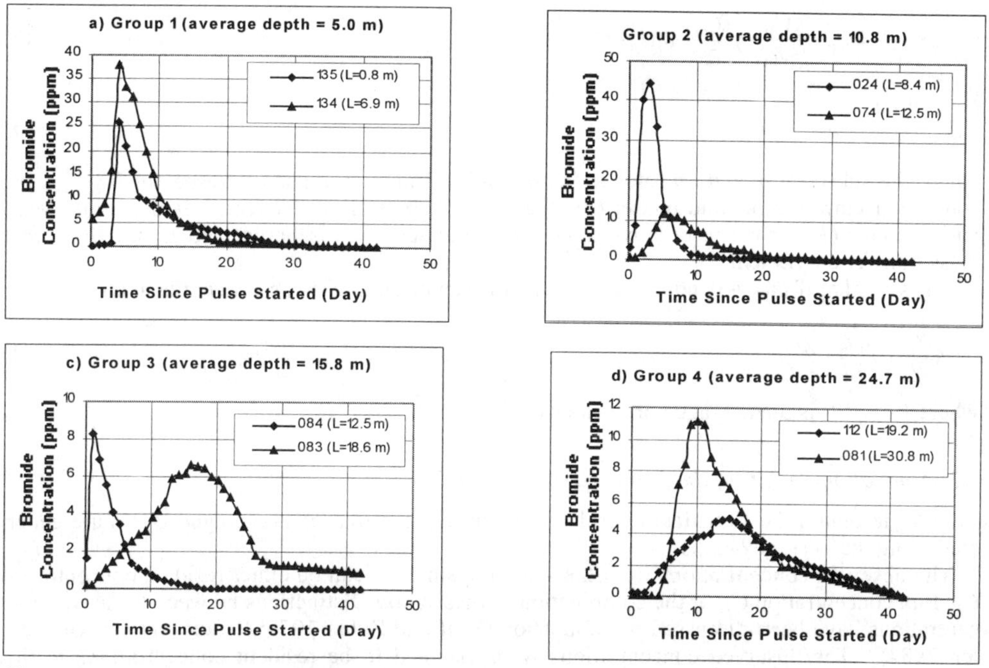

Figure 4. Observed BTCs at the deepest and the shallowest depths for each lysimeter group.

The four representatives BTCs obtained by arithmetic averaging are shown in Figure 5. All BTCs show long tailings. The peak concentration decreases while the breakthrough timing and spreading increases with depth.

There is no significant difference between the Group 3 and Group 4. It is possibly due to preferential flows that bypassed the lysimeters in Group 3.

3 NUMERICAL SIMULATION OF THE TEST RESULTS

Zhan and Danko (2000), and Danko (2000) have used CXTFIT (Toride et al. 1995) and MULTIFLUX (Danko 1999) respectively to evaluate the data set of 34 lysimeters. The simulations were completed to achieve low root-mean-square (RMS) error between the measured and calculated concentrations of all individual lysimeters. The simulated results indicated that even though the derived "average" hydrologic parameter set did not exactly match individual results, it followed the basic trend of the full data set and overall solute mass balances were achieved.

The purpose of the analysis in this paper is to generate "average" hydrologic behavior of the leach pad based on the area-averaged BTCs. The simulation is carried out using the time-honored CXTFIT code with third-type boundary conditions.

3.1 *Numerical Simulation Model*

CXTFIT MIM model assumes that the liquid phase can be partitioned into mobile and immobile phases. Solute exchange between the two phases is modeled as a first-order process. In the model, the total volumetric water content is the sum of mobile and immobile volumetric water contents, θ_m and θ_{im}:

$$\theta = \theta_m + \theta_{im} \tag{1}$$

For one-dimensional vertical flow and transport, assuming no sources/sinks for water and solute and no solute degradations, the MIM model is governed by:

$$\theta_m R\frac{\partial C_m}{\partial t}=\theta_m D\frac{\partial^2 C_m}{\partial z^2}-q\frac{\partial C_m}{\partial z}-\alpha(C_m-C_{im}) \quad (2)$$

$$\theta_{im}\frac{\partial C_{im}}{\partial t}=\alpha(C_m-C_{im}) \quad (3)$$

where the subscript *m* and *im* refer to the mobile and immobile phases, respectively; *C* is the solute concentration of fluid phase; *D* is the solute dispersion coefficient; *R* is the solute retardation factor (*R*=1 for conservative tracer); *q* is the specific discharge; and α is a first order mass transfer coefficient.

It is seen that if $\theta_{im} = 0$, equations (2) and (3) can be simplified as conventional CDE:

$$R\frac{\partial C}{\partial t}=D\frac{\partial^2 C}{\partial z^2}-V\frac{\partial C}{\partial z} \quad (4)$$

where $V = q/\theta$ is the average pore velocity.

3.2 *Simulation Results of the four area-averaged BTCs*

In order to obtain the statistical hydrologic properties, the four area-averaged BTCs are generated using the average depths.

The observed concentrations from the suction lysimeters can be either resident concentration C_r, flux concentration C_f, or the combination of these two. Distinctions between these two concentrations have been discussed in publication (Kref and Zuber 1978; Parker and van Genuchten 1984). The observed concentrations were assumed to be resident concentrations in this study. The third-type boundary condition was applied in the simulations because it conserves solute mass.

Because the average pore velocity *V* is one of the fitted parameters, the estimated value for *V* has to be in a reasonable range in order to achieve good solute mass balance. For the applied irrigation rate, the theoretical average pore velocity can be calculated from water retention curve (WRC) of the pad material, resulting in around 1.5 m/day. In reality, the pore velocity in the pad differs from this idealized velocity because of various reasons: 1) homogeneity doesn't exist in the field test; 2) irrigation rate on the pad surface can not be completely even; and 3) potential preferential flow patterns exist in the pad. In this simulation, the average pore velocity was fitted around the idealized velocity, i.e. between (1.3-1.7 m/day).

Three other parameters were optimized within relative large ranges. It was realized that CXTFIT doesn't produce unique results. The simulated results can vary when different initial values are applied. Different combinations of initial values were tried intensively. The minimum root-mean-square (RMS) error that could be achieved was 1.7 ppm with a mean error of 0.64 ppm. Considering the possible measurement errors made by the test units, these average errors are believed to be insignificant. The optimized parameters and goodness of the fitting are listed in Table 3.

In Table 3, *V* is pore water velocity, $V=q/\theta$; β is partitioning coefficient, $\beta=\theta_m/\theta$; ω is a dimensionless mass transfer coefficient, $\omega=\alpha L/(\theta V)=\alpha L/(\theta_m V_m)$; α_d is dispersivity, $\alpha_d=D/V$; Two parameters, dispersivity α_d and mass transfer coefficient α, are derived.

Table 3. Simulation results of the MIM model using the four area-averaged BTCs.

Fitted parameters	V	D	$\beta=\theta_m/\theta$	ω	α_d	$\alpha=\omega*(\theta V)/L$
	m/day	m^2/day			m	1/day
	1.37	41.86	0.99	100	30.5	1.8
Goodness	RMS	Mean error	Max error	Min error		
of fitting	ppm	ppm	ppm	ppm		
	1.70	0.64	6.86	-7.28		

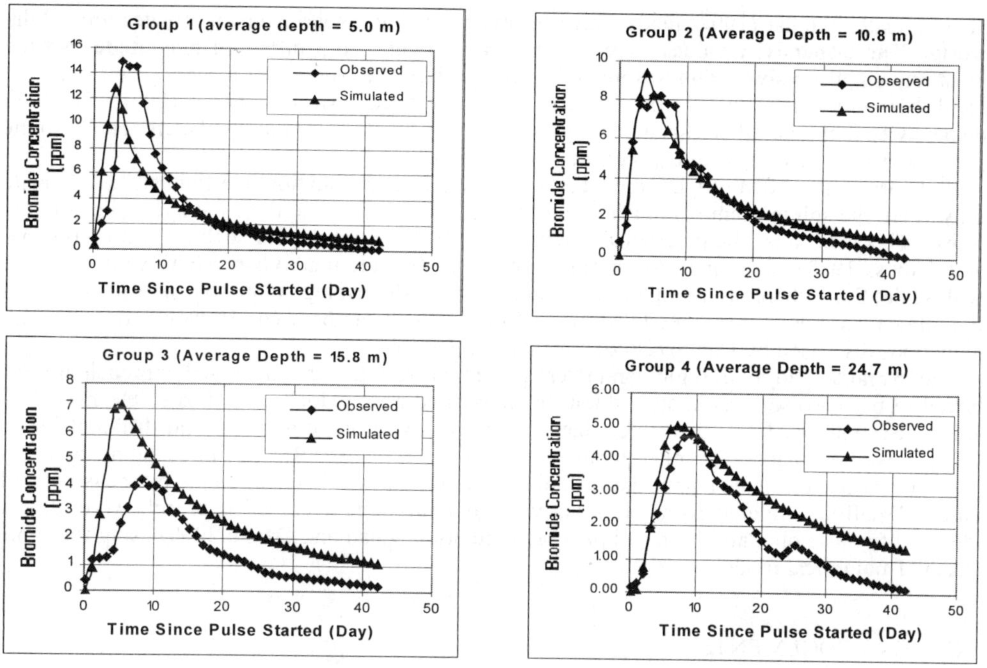

Figure 5. Observed and simulated area-averaged BTCs using the simulated parameters.

The characteristic length L is the averaged depth of the lysimeters group. A water content of 0.19 is estimated based on the irrigation rate and measured WRC of leach pad material in the laboratory. The simulation errors are defined as the differences between the simulated and the observed concentrations. Because the simulated mobile water content and mass transfer coefficient are high, this MIM model is closely equivalent to a CDE model with same boundary conditions. The simulated BTCs for these four average depths, shown in Figure 5, are derived from the "average" pad parameters listed in Table 3.

4 DISCUSSIONS

The field data imply a complex flow and transport system. While laboratory tests use mainly uniform materials and well controlled condition, field tests involve heterogeneity of the materials and variation of the test conditions. For example, it is impossible to maintain perfectly even irrigation rate over a time period of 60 days. Different wind speeds and directions, sprinkler working patterns would definitely generate an uneven irrigation field. Some lysimeter locations had more water and solute inputs than other locations. Meanwhile, heterogeneous properties, such as unsaturated hydraulic conductivity, water content, and preferential channels, exist in the pad. All these factors would affect the solute transport process and make lysimeters behave individually.

The purpose of this study is to generate the common hydrologic properties of the pad through the area-averaged lysimeters groups. The discrepancies of the individuals are blended using averaging within a group.

The simulated results demonstrated that dispersivity of the test pad is about 30.5 m; mobile water content is extremely high; mass transport between mobile and immobile water content is also significantly high. Since strong mass transfer exists between mobile and immobile phases, the MIM model can be simplified into regular CDE model. It is interesting to notice that the simulated dispersivity approximately equals the field test scale. This number seems to be relatively large because it is commonly believed that the dispersivity is usually about one tenth of the test scale. Gelhar (1992) made a critical review of field experiments at 59 sites around the world. The dispersivity values were found to range over 2 to 3 orders of magnitude, but the most reliable dispersivity values were at the lower end of the range. For example, he found that the dispersivity values varied from 1 to 200 m when test scale was only about 100 m. The high dispersivity found based on this modeling exercise may be a result of strong heterogeneity in the pad and/or the weak physical basis of the model approach.

In theory, using MIM or CDE model to solve the transport equation in a field site is adequate only when the solute plume is much larger than the correlation length of the hydraulic conductivity. This is because the macrodispersivity is nearly a constant under such a situation (Gelhar and Axness 1983). For small to intermediate test scales, the macrodispersivity values change with scale. Since there are no other better equations for describing solute transport in small and intermediate scales, it is assumed that the MIM model still can be used. However, the other explanations discussed by Danko (2000) may also be possible.

The averaged pad hydrologic parameters generated from the area-averaged approach are believed to be reasonable estimates. First, the simulated pore velocity satisfies solute mass balances. Second, the fitted parameters have been set in very large domains and different initial values have been tried. No other parameter set can gain a lower RMS. Third, the mean simulated errors are less than 1 ppm that is comparable with possible errors made by measurement units. Finally, simulated parameters derived from this area-averaged approach were almost identical to those obtained from full data set simulation (Zhan and Danko 2000) using the same CXTFIT numerical model.

ACKNOWLEDGEMENTS

This paper is part of the research program supported by Barrick Goldstrike Mines Inc. and the Generic Mineral Technology Center in Mine System Design and Ground Control, Office of Mineral Institutes, U.S. Bureau of Mines. Their financial assistance is gratefully acknowledged. Authors thank Dr. Paul Lechler from Nevada Bureau of Mines for providing chromatography laboratory support for the research.

REFERENCES

Butters, G.L., W.A. Jury & F.F. Ernst 1989. Field scale transport of bromide in an unsaturated soil 1. experimental methodology and results. *Water Resour. Res.* 25 (7): 1575-1581.

Butters, G.L. & W. A. Jury 1989. Field scale transport of bromide in an unsaturated soil 2. dispersion modeling. *Water Resour. Res.* 25 (7): 1583-1589.

Comba, P.G., R.B. Dix, & S.L. McGill 1992. A field study to assess natural degradation of cyanide species in an inactive leached ore heap. *Proceedings: Second International Conference on Environmental Issues and Management of waste in Energy and Mineral Production.* Calgary, Alberta, Canada. 2: 535-543.

Danko, G. 1999, *MULTIFLUX User Manual*, CRWMS M&O MOL.19990409.0096, DOE, Yucca Mountain Project Record Center, Nevada, USA.

Danko, G. 2000, Coupled convection-diffusion modeling with MULTIFLUX, *Proceedings of the International Symposium on Hydrology and the Environment.* Wuhan, Hubei, P.R.China.

Decker, D.L. & S.W. Tyler 1998. Evaluation of flow and solute transport parameters for heap leach recovery material. *J. of Environmental Quality.* 28 (2): 543-555.

Dix, R.B, P.G. Comba & S.L. McGill 1992. Laboratory rinsing behavior of spent ores. *Proceedings: Second International Conference on Environmental Issues and Management of Waste in Energy and Mineral Production.* Calgary, Alberta, Canada. 2: 917-924.

Er, C. & M.E. Henderson 1999. Hydrological modeling of heap deoxification(Abstract). *Closure, Remediation, & Management of Precious Metal Heap Leach Facilities Workshop*, Univ. of Nevada, Reno.

Eriksson, N., A. Gupta & G. Destouni 1997. Comparative analysis of laboratory and field tracer tests for investigating preferential flow and transport in mining waste rock. *J. of Hydrol.* 194: 143-163.

Gelhar, L. W. & C.L. Axness 1983. Three-dimensional stochastic analysis of macrodispersivity in aquifers. *Water Resour. Res.* 19 (1): 161-180.

Gelhar, L.W, C. Welty & K.R. Rehfeldt 1992. A critical review of data on field-scale dispersion in aquifer. *Water Resour. Res.* 28 (7): 1955-1974.

Kref, A. & A. Zuber 1978. On the physical meaning of the dispersion equation and its solutions for different initial and boundary conditions. *Chem. Eng. Sci.* 33: 1471-1480.

McClelland, G.E. 1991. Laboratory scale heap detoxification. *Heap and Dump leaching Newsletter*. 6: 1-7.

Parker, J.C. & M.Th. van Genuchten 1984. Determining transport parameters from laboratory and field tracer experiments. Bull. 84-3. *Va. Agric. Exp. Sta.*. Blacksburg.

Schafer, W.M. & D. van Zyl 1991. Cyanide degradation field study of spent heap-leach ore at the landusky Mine. *Heap and Dump leaching Newsletter*. 6: 1-7.

Tseng, P.H. & W.A. Jury, 1994, "Comparison of tracer function and deterministic modeling of area-averaged solute transport in a heterogeneous field," Water Resour. Res., Vol. 30, No. 7, pp2051-2063.

Toride, N, F.J. Leij & M.Th. van Genuchten 1995. The CXTFIT code for estimating transport parameters from laboratory or field tracer experiments, version 2.0. *Res. Rep. No. 137*. U.S. Salinity Laboratory, Riverside, California.

Zhan, G. & G. Danko 2000. Experimental and numerical study of in-situ leach pad hydrology. *Proceedings of the International Symposium on Hydrology and the Environment*. Wuhan, Hubei, P.R.China.

Design, operation, and disposal

Tailings and Mine Waste '01 © 2001 Balkema, Rotterdam, ISBN 90 5809 182 1

Evaporation from wet and "dry" beaches of tailings dams

G.E. Blight
Department of Civil Engineering, University of the Witwatersrand, Johannesburg, South Africa

ABSTRACT: Water lost by evaporation is a major component of most tailings dam water balances. In semi-arid, arid and desert climatic regions the replacement of water lost by evaporation can also represent an important component of production costs. In addition to representing large water losses, evaporation and seepage losses may be difficult to quantify separately. For decades an argument has continued as to whether evaporation losses from a wet beach are more than those from an apparently dry beach and if it is preferable to minimize the area of wet beach to minimize both seepage and evaporation losses, or whether evaporation and seepage losses from dry beaches are much the same as from wet beaches.

This paper throws some light on the problem by making a direct comparison of evaporation rates from wet and dry surfaces using the radiation balance method. It is shown that evaporation losses from wet and apparently dry surfaces are almost exactly the same.

1 INTRODUCTION

Hydraulic fill tailings impoundments are often operated on a seven to fourteen day cycle of deposition with roughly one seventh to one fourteenth of the hydraulic fill beach area being deposited on each day. This area is visibly wet for a day or so, while the remaining area usually appears dry although it is wet to a varying degree just below the surface. When calculating the water balance for the impoundment, the question often arises as to whether the evaporation rate for a wet beach surface differs from that for an apparently dry surface, and if so, by what proportion - is it 10 per cent more, or 50 per cent? The question is important, as one millimetre of evaporation per day represents one litre per square metre per day, or one million litres (1 000 m^3) of water per day per 100 ha of impoundment, and evaporation from a tailings surface can be more than 5 mm per day in warm dry weather.

To answer this question, two sets of parallel evaporation measurements have been carried out, one on a light-colored power-station fly ash surface, and the other on a dark-colored soil surface. The method used was the surface energy balance, and in the case of the ash surface, gravimetric water content measurements were also made.

2 THE SURFACE ENERGY BALANCE

The balance of radiation between space and the earth's surface can be described by the equation (Robinson 1966, Flohn 1969):

$$R_n = R_A(1 - \alpha)(1 - a) \qquad (1)$$

R_A is the direct solar radiation received at the outer limit of the earth's atmosphere, α is the planetary albedo, the proportion of solar radiation reflected back into space by the atmosphere, a is the albedo of the ground surface, and R_n is the net radiation (incoming less reflected) received at the ground surface. α depends on the amount of cloud cover and dust in the atmosphere, but averages about 0.4. At a given location, the albedo of the ground surface varies throughout the day with the inclination of the sun's rays and depends on the texture and color of the ground surface. It is defined by

$$a = R_o/R_i \tag{2}$$

where R_i is the incoming solar radiation and R_o is the reflected radiation.

The surface energy balance, i.e. the way R_n is converted to other forms of energy at the ground surface is given by:

$$R_n = G + H + L_e \tag{3}$$

where G is the soil heat flux, the energy consumed in heating the near-surface soil;
H is the sensible heat flux, the energy consumed in heating the air above the surface; and
L_e is the latent heat flux for evaporation, the heat consumed in evaporating water from the surface.

R_n can be measured directly, G can be estimated from changes in temperature of the near-surface soil and the specific heat capacity of the soil and H can be calculated in a similar way for air, or from temperature and relative humidity gradients in the near-surface air. The analysis will not be given in detail in this paper as it is available elsewhere, for example, Blight (1997).

Equation (3) refers only to solar energy. If energy from an additional source is present (e.g. wind energy) a term describing this energy must be included. A modified form of Equation (3) including wind energy would be:

$$R_n + W = G + H + L_e \tag{3a}$$

where W is the energy flux arising from the wind.

In practice, the sensible heat H is very much less than the soil heat G and can be neglected with little error.

Conservation of energy requires that the heat used for evaporation must equal that supplied, while conservation of mass requires that the rate at which water evaporates from the surface be equal to the rate at which it is dispersed into the atmosphere.

When using the surface energy balance analysis, it is simpler to evaluate W separately, from wind speed and relative humidity measurements. Penman's empirical expression (Penman 1965) for the evaporation caused by W is:

$$E_W = 2.6/D\ (0.5 + 0.15u)e_{sat}(1 - r) \text{ in mm.day}^{-1} \tag{4}$$

where u is the mean wind speed for the day of measurement in $km.h^{-1}$ at a height of 2m, e_{sat} is the saturation vapour pressure of the air in kPa, and r is the air's relative humidity.

Values of e_{sat} and D are related to temperature as shown in Table 1.

Table 1. Values of e_{sat} and D at various air temperatures

Temperature °C	e_{sat} kPa	D (Dimensionless)
O	0.6	1.5
10	1.2	2.3
20	2.3	3.3
30	4.2	4.8

3 THE EFFECT OF A WET OR APPARENTLY DRY BEACH SURFACE

Possible reasons for differing evaporation rates from wet and apparently dry beach surfaces are as follows:

- differing albedos for wet and dry surfaces, resulting in differing absorption of solar energy, and hence differing values of G and H;
- differing permeabilities of the immediate surface layers, with the lower permeability of the dry surface impeding the upward evaporation flow of water; and
- differing soil suctions or free energy of the pore water in the surface layers. If the suction of the apparently dry surface were very different to that of the wet surface, the near surface permeability would be reduced and the evaporative solar energy would have to supply the free energy of the soil water which would reduce the amount available to cause evaporation.

4 FIRST COMPARATIVE MEASUREMENTS OF EVAPORATION

Figure 1 compares two sets of simultaneous surface energy measurements, one made for a visibly wet surface, the other for an apparently dry surface. The measurements have been shown cumulatively. The material was a pale grey colored power station fly ash, the wet surface had been deposited on the previous day and the dry surface a week previously. The two sites were adjacent (200 m apart) and the effect of wind has been ignored for the time being as it would have been very similar at both sites.

The lower diagram shows that the albedo of the dry area was slightly higher than that of the wet area throughout the day, probably because of its lighter color. This resulted in ΣR_n for the wet surface being slightly larger than for the dry surface. The wet near-surface ash also heated up more and its ΣG was larger. This resulted in ΣL_e for the wet surface being less than for the dry surface. 1.7 mm of water evaporated from the dry surface as compared with 1.5 mm for the wet surface. The wind evaporation calculated from equation (4) was 1.0 mm and the calculated evaporation for the two surfaces ended up being 2.7 mm for the dry surface compared with 2.5 mm for the wet surface.

Suctions measured at a mean depth of 75 mm below the wet and dry surfaces by means of tensiometers were as follows:

42 kPa for the "dry", and
51 kPa for the "wet" area.

in practical terms there is no significant difference between these two values. In other words, at 50 to 100 mm deep, the ash underlying the "wet" and "dry" areas was, in effect, equally moist.

After the evaporation measurements had been conducted, samples of the ash were taken for water content determinations. The samples were taken from depths of 0 to 50, 50 to 100, 100 to 150, 150 to 200, 200 to 250 and 250 to 300 mm. The sampling was repeated at the same depths at 2, 3, 4, 6, 10, 12 and 16 days after deposition.

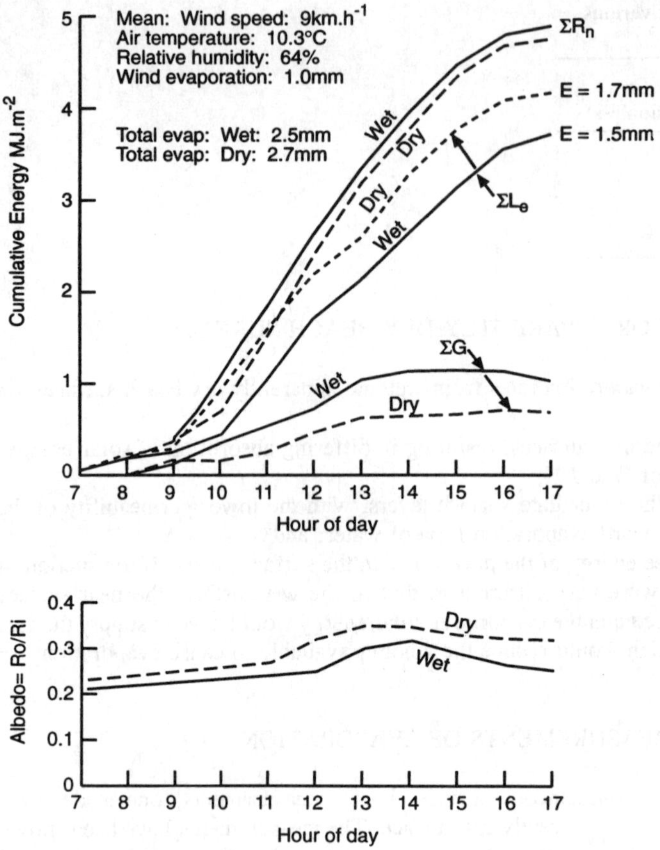

Figure 1. Comparative surface energy balance measurements made on wet and apparently dry areas of a fly ash dam beach.

The resultant water content profiles are shown in Figure 2a. The profiles were close to a series of vertical lines, with the water content at the end of the first day being about 41 per cent. The ash appeared to dry out or drain fairly uniformly over the top 300 mm, although there is a tendency for the measurements from 250 to 300 mm to be slightly wetter. This seems to show that the depth to which short-term drying or drainage extended was about 300 mm, probably with some downward percolation also occurring below this depth. The field moisture capacity of the ash was also measured at about 40 per cent which agrees with the data of Figure 2a. However, if deeper drainage into the ash occurred, a suction would be set up that would cause drainage to occur to below the field capacity, as measured conventionally.

Figure 2b shows the water content measurements for all depths plotted against the time in days since deposition. As this figure shows, the average gravimetric water content of the ash decreased almost linearly from about 41 per cent to about 24 per cent during the period of 16 days. Over the first six days, the water content reduced almost linearly with time, at about 2 per cent per day.

The water content has also been expressed in Figures 2a and 2b as mm of water stored in the top 300 mm of ash, on the assumption that the dry density of the ash was 1 000 kg/m^3. On this basis, the water stored declined from about 120 mm to just over 70 mm over the 16 day period, a loss of 50 mm.

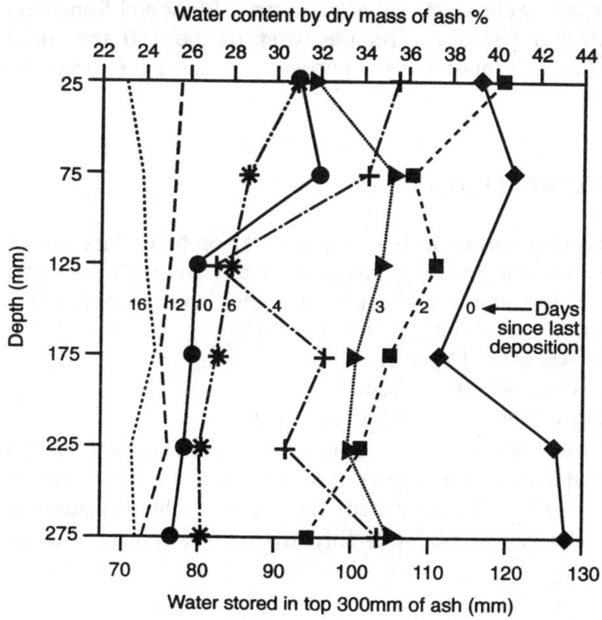

Figure 2a. Results of water content sampling of fly ash beach for 16 days following measurements on wet beach - water content profiles.

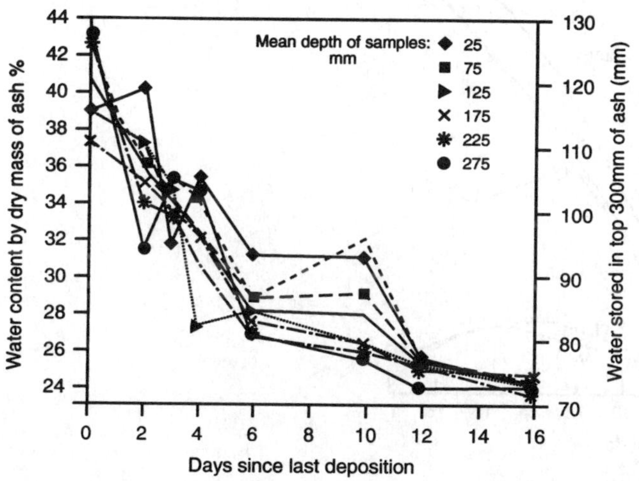

Figure 2b. Results of water content sampling of fly ash beach for 16 days following measurements on wet beach - decline of water content with time.

Over the first six days about 30 mm of water was lost, or 5 mm per day. The evaporation rate had been about 2.5 mm/day, hence about 2.5 mm/day was being lost by percolation into the deeper layers of the ash dump. Over the last 10 days of the water content sampling, 20 mm or 2 mm/day was lost, and this could well have been lost mainly by evaporation through the ash surface.

These conclusions are supported by the results of microlysimeter tests (Boast and Robertson 1982) which averaged a loss of 2.6 mm per day. As the bases of the 150 mm deep microlysimeters were sealed, all of this loss must have occurred by evaporation from the surface.

5 SECOND SET OF COMPARATIVE MEASUREMENTS

The second set of comparative evaporation measurements was performed on bare, dark brown colored natural soil surfaces, one of which was low-lying and permanently wet with the water table at about 100 mm below the surface. The other was about 1.5 m higher, and apparently dry. The water contents of the upper 50 mm of soil at the two sub-sites were:

wet area: 44 per cent
dry area: 30 per cent

The results of the energy balance measurements are shown in Figure 3.

The lower diagram shows that the albedo was almost the same for the two areas, with the dry site having a slightly higher albedo in the afternoon. The curves of ΣG are also very close to each other, and hence so are the curves of ΣL_e. The calculated evaporation for the two sites was almost the same, even though the "dry" area had a considerably lower water content than the wet area, in this case.

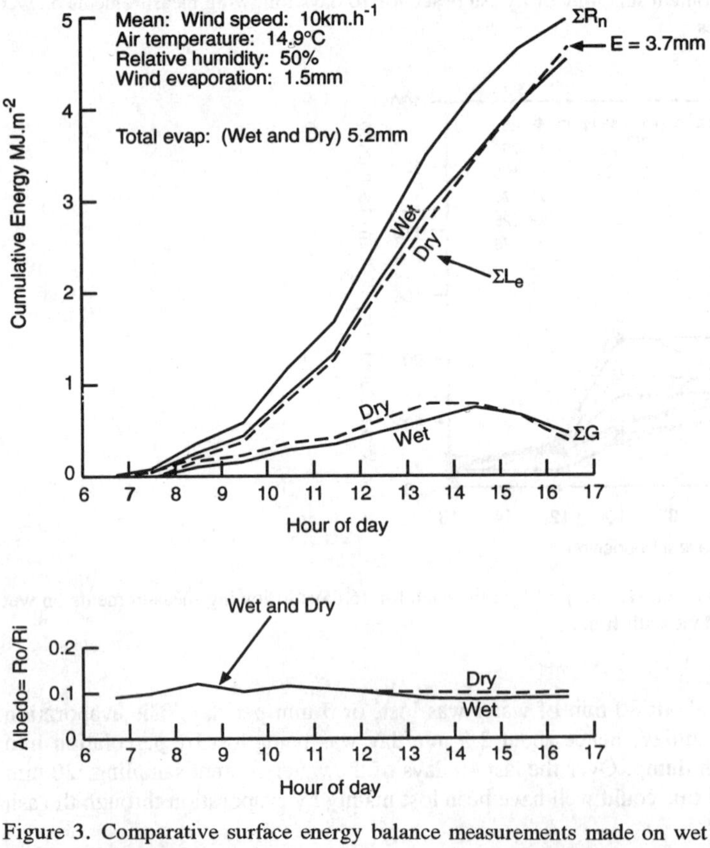

Figure 3. Comparative surface energy balance measurements made on wet and apparently dry areas of a natural soil surface.

6 CONCLUSIONS

The two sets of comparative measurements have shown that the rate of evaporation from a wet soil or tailings surface is almost the same as that from an apparently dry surface, even when the difference in water contents of the two areas is quite large. Figures 2a and 2b show that this conclusion is valid over quite a wide range of water contents. However, as the water content of the near surface layers of a dry beach reduces, there comes a time when water is no longer freely available to evaporate, and the rate of evaporation from the surface reduces.

REFERENCES

Blight, G.E. 1997. Interactions between the atmosphere and the earth. (37th Rankine Lecture) *Geotechnique*, London, U.K., 47(4): 715-767.

Boast, C.W. & Robertson, T.M. 1982. A microlysimeter method for determining evaporation from bare soil: Description and laboratory evaluation. *Journal of Soil Science Society of America*. 46: 689-696.

Flohn, H. 1969. *Climate and Weather*. London, U.K.: Weidenfeld and Nicolson.

Penman, H.L. 1963. *Vegetation and Hydrology*. Technical Communication No. 53. Commonwealth Bureau of Soils, Harpenden, U.K.

Robinson, N. 1966. *Solar Radiation*. Amsterdam, Netherlands: Elsevier.

Tailings and Mine Waste '01 © 2001 Balkema, Rotterdam, ISBN 90 5809 182 1

Sealing of penstock decant towers and pipelines on tailings dams in South Africa

D.J.Venter, R.J.Stuart & P.Labrum
Steffen, Robertson and Kirsten Consulting, Johannesburg, South Africa

ABSTRACT: The use of penstock decant towers and pipelines to remove supernatant free water from the top of tailings dams in South Africa is historical, dating back to the early 1930's. The development of decant towers as the preferred return water recovery methodology in South Africa has been documented in a paper entitled "Tailings dam penstock decants in the South African mining industry" prepared by the same authors. This new publication deals with procedures to effectively seal the decant towers and pipelines during normal tailings dam operations and in emergency situations.

The paper addresses the design aspects of simulated dam failure due to piping into or adjacent to deformed or collapsed decant towers and buried pipelines. Design considerations in regard to the extent and type of the concrete plugs required to create effective permanent seals are discussed in detail.

During the past three years the development of robust video camera equipment which can operate in wet, confined spaces and sometimes corrosive environments has opened up opportunities to effectively seal penstock towers at depths in excess of twenty metres by allowing ongoing visual inspection and control of the sealing activities.

The successful sealing of penstock decant towers at depth and pipeline outlets are documented as are various problems which must be avoided. The different plant and materials used in the various sealing processes are also discussed.

1 INTRODUCTION

During the latter 1920's improvements in the metallurgical processing of the gold bearing reef started to take place. This was partly due to the fact that the ore body was being ground to a finer product than in the past. Furthermore as underground mining methods were also being improved, this resulted in larger volumes of finer sands being produced at the plants, for disposal onto the sand dumps. This increase in production as well as the finer product being produced resulted in problems being experienced with the disposal via coco pans onto the sand dumps.

These coarser old gold dumps still contain some gold and this is why so many of the early gold sand dumps in South Africa are currently being re-mined and re-processed.

The problem of disposal of the finer gold waste material was overcome by the construction of slimes or tailings dams into which the fine material was hydraulically deposited. The fine material was pumped as a slurry from the plant to the impoundment. This method of slurry disposal resulted in supernatant or free water occurring, which needed to be removed from within the confines of the dam. The removal of the supernatant water was achieved through the construction of an inlet structure within the dam basin and an outlet pipe which discharged the water at the perimeter.

The development and improvement of these concrete shaft penstock inlet structures in South Africa with time is documented in a paper entitled *"Tailings Dam Penstock Decants in the South African Mining Industry"* prepared by the same authors.

The removal of the supernatant water from the tailings dam plays a major role in the safety and stability of the dam. The loss of either the penstock inlet shaft system or the penstock outlet pipe on any tailings dam will result in major implications to the mine. These implications could be so serious that shutting down of the operations of the mine for a period of time could result.

2 PENSTOCK SYSTEM FAILURES

The failure of a penstock system can occur as a result of the following:-

- Failure of the concrete shaft of the penstock inlet structure or structures due to poorly constructed concrete rings, poor construction of intake structure, excessive deformation or external forces.
- Failure of the penstock outlet pipe due to foundation settlement, overburden pressures, loss of support or corrosion of the steel pipe.

2.1 *Failure of the Penstock Concrete Shaft Inlet Structures*

Recent penstock concrete shaft inlet structure failures have occurred on various dams in South Africa. In all instances of failure, no major environmental damage or loss of life has occurred due to the quick response of both the tailings dam operators and the mine.

In all instances to date, the failures have been as a result of crushing or failure of the precast concrete penstock inlet rings forming the shaft.

In most instances the failures have occurred where the depth of the inlet structure is approximately 20 metres in depth. The International Commission on Large Dams (ICOLD) Bulletin 101 – 1995 *"Tailings Dam Transportation and Decantation"* recommend a maximum height of 20 metres for shaft penstock inlets using precast concrete rings.

It is important to note that the manufacture of concrete penstock inlet rings (Refer to Figure 1) by the various South African suppliers have been standardised since the implementation of a draft Code of Practice in the mid 1970's. However, the quality of the precast concrete rings being produced and installed have only been monitored effectively at tailings impoundments over the last 4 to 5 years.

Based on the specifications in the draft SABS Code of practice, the penstock rings should be able to achieve a depth of between 17 m and 20 m before crushing of the rings at the base of the structure occurs. This depth will depend on the saturated density of the materials being depos-

Table 2.1 Approximate Safe Depth of 510 mm dia. Precast Concrete Ring Penstock Intake Towers

Type of Material	S.G. of Material	Saturated Density kg/m^3	Approximate Friction Angle	Maximum Safe Depth of Construction		Ultimate condition	
				Unlined m	Lined m	Unlined m	Lined m
Gold	2,66	1578	33°	18	27	22	33
Ash	2,134	1232	33°	20	31	24	37
Chrome	3,55	1630	28°	18	27	22	33
Platinum UG2	3,63	1640	33°	18	27	21	33
Platinum Merensky	3,13	1544	33°	18	28	22	33
Platinum Platreef	3,00	1643	33°	17	26	21	32

ited around the inlet structure, the grind of the material, the downdrag fences and the rate of consolidation of the material. Table 2.1 indicates the theoretical maximum safe depth of intake tower based on the material characteristics, as well as the safe depth of intake tower if the towers are lined.

Investigations into the failures mentioned above have not been conclusive as to the cause of the failure. However, in all instances, at a current depth of tower of approximately 20 metres problems can be anticipated. The penstock inlet failures that have been investigated have all occurred on dams originally constructed in the mid 1970's, and the failures may be due to the manufacture and installation of "poorly" constructed precast concrete rings at that time.

It is important to note that in South Africa the penstock inlet shaft structures generally occur in the portion of the tailings dam where the finest particle size of material occurs. Furthermore, the materials in this area of a tailings dam are usually totally saturated. The possibility of flow from the centre of the dam via the damaged or failed penstocks cannot be overlooked.

At one specific tailings dam complex, blasting for open pit mining took place within 300 metres of a double compartment tailings dam. Of the three penstock inlet structures in the first compartment two failed within a short space of time.

At a large tailings impoundment at another mine five inlet structures failed simultaneously. Four of the inlet structures were grouped together (± 2 m apart) and the fifth inlet was spaced approximately 30 metres away. In this instance the penstock concrete rings at a depth of 3 m above the base crushed. An earth tremor / seismic shock was noted two days prior to the failure occurring.

2.2 *Failure of the Penstock Outlet Pipe*

The use of steel pipes, often uncoated, as a penstock outlet pipe during the early years of tailings dam construction, particularly in the gold mining industry often resulted in collapse due to corrosion failure. The result of this type of failure was often the development of sinkholes on the top and the side of the tailings dams.(Refer to Figure 2)

When failure occurred the dam had to be taken out of commission or compartmentalised so that a new elevated penstock could be installed.

As a better understanding of the long term performance of materials was obtained, designers changed over to the use of reinforced spigot and socket concrete pipes for use as buried penstock outfall pipelines below the tailings impoundments.

The availability of modern robust video camera equipment is now assisting tailings dam designers and operators to better understand the problems that occur with penstock inlet structures and outlet pipes. In the case of outlet pipes, however, it is currently only possible to inspect some 200 to 300 metres of the outlet pipe from the toe of the dam using a self propelled vehicle

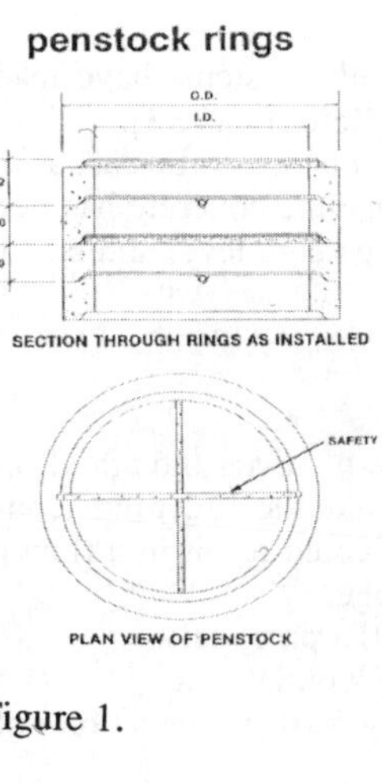

Nominal Diameter mm	I.D. mm	O.D. mm	Approx mass kg
500	510	660	35
750	750	915	61
900	900	1 066	74

Note: 1. Safety rods are supplied by others.
2. Penstock rings with recesses for safety rods are to be ordered as "Specials".

Figure 1.

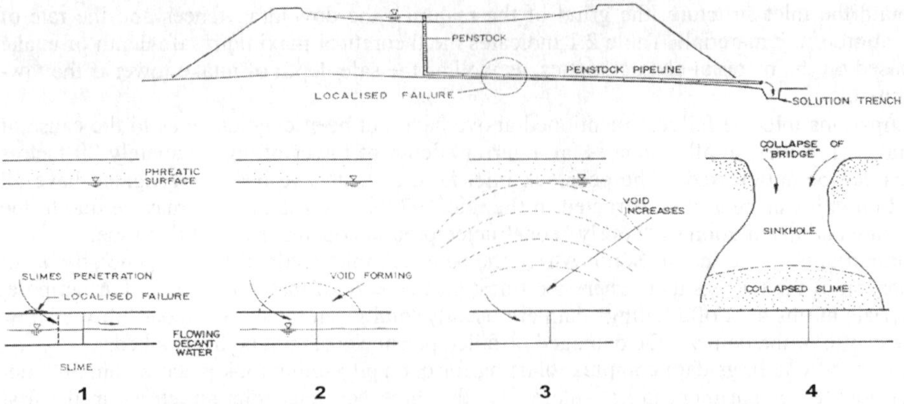

Figure 2. Formation of a sinkhole at a localised penstock pipeline failure

housing the video lens. With time this technology will probably improve so that it will be possible to inspect the full length (over 500 m) of a small bore penstock outlet pipe in future.

3 DESIGN REQUIREMENTS FOR SEALING THE PENSTOCK SYSTEM

The design requirements for the sealing of a penstock system will vary depending on the condition at the tailings dam at the time of sealing.

Penstock shaft inlet structures and outlet pipes should at closure not constitute a further source of problem. There should be no element of failure or risk to persons and the environment.

The sealing of the penstock system must ensure that no further uncontrolled release of water or tailings from the tailings dam can occur which may result in the loss of life or environmental damage.

Sealing of an element of the penstock system or the entire system may be required for the following reasons:-

- The penstock shaft inlet structures have reached their design height capacity.
- A failure of a penstock shaft inlet structure.
- A failure of the penstock outlet pipe.

3.1 *Penstock Shaft Inlet Structure Seal*

Recent designs of the concrete base structures for penstock shaft inlet systems have made provision for a system whereby the inlet structure can easily be sealed (Refer Figure 3).

The installation of the crossbars at the base of the structure permits concrete to be cast at the base of the shaft, thereby effectively sealing off the entire intake. Once the concrete has been cast, the remaining shaft can be filled with tailings to the current tailings dam level and no further placing of precast concrete rings takes place.

3.2 *Penstock Outlet Pipe Seal*

The sealing of the penstock outlet pipe is the area of greatest concern. Even though all the shaft inlet structures have been sealed, the ingress of water and tailings into the outlet pipe could occur if failure of the pipe at a later date took place. This later failure could result in a flow of tailings out of the dam with resultant loss of life or environmental damage.

The following should be considered when designing the penstock outlet pipe seal:-

- Final height of construction of the tailings dam above the penstock outlet pipe.
- The possibility of a slope failure above the penstock pipe during further deposition onto the tailings dam.

Figure 4 indicates a slope stability analysis carried out for a particular tailings dam height above the penstock outlet pipe. The phreatic surface within the tailings dam is adjusted in order to produce a slope failure, either as block failure through the underlying soils or as a slip failure. The extent of the failure zone is then determined indicating the minimum length of concrete seal required within the penstock pipe. Should a slope failure actually occur, sufficient concrete plug should still be present within the penstock pipe to prevent the release of tailings within the dam occurring.

4 CONSTRUCTION OF PENSTOCK SYSTEM SEALS

The method of construction of the seals for the various components of the penstock system will depend on the materials and plant available at the time of the sealing operation, as well as the urgency within which the seal needs to be accomplished.

Where no time constraints are applicable, penstock system seals could be installed as described below. During the installation of the seals it is important to ensure that the work is carried out correctly. The use of modern video camera technology can be used to ensure that the seals have been correctly installed.

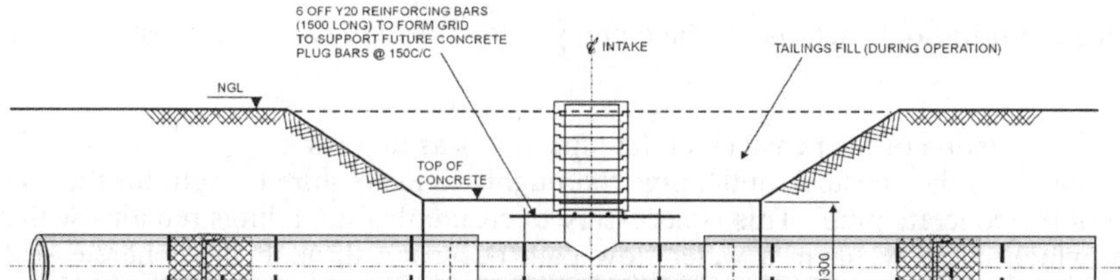

Figure 3. Typical base structure for a penstock intake tower

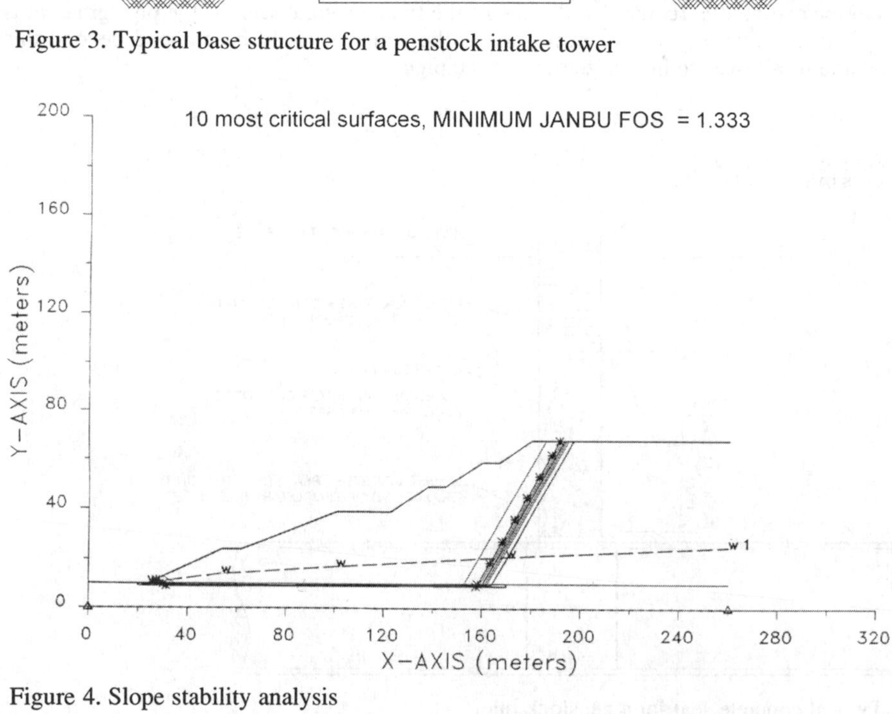

Figure 4. Slope stability analysis

4.1 *Penstock Shaft Inlet System Seal*

Figure 5 indicates the various materials to be employed to effectively seal the penstock inlet shaft.

The method of installation of the seal is as follows:

- The penstock shaft inlet structure must be raised above the tailings and water levels to ensure no further decantation can take place.
- The steel base plate and breather pipe if required is then installed into the base of the structure. The geofabric wrapped around the base plate prior to installation is to try and seal the base plate to the structure to prevent the loss of concrete into the penstock pipe.
- The first layer (± 1m in depth) of concrete or grout is then poured into the structure to form the initial seal at the base.
- After 24 hours, a second layer of ± 1m depth of concrete or grout is poured into the structure to further seal the intake and to ensure that any shrinkage zones that may have occurred in the first lift is sealed off.
- After a further 24 hours, the shaft intake is filled with dry tailings to the current tailings level on the dam, and any excess precast concrete rings removed.

4.2 *Penstock Outlet Pipe Seals*

Figure 6 indicates the various materials to be employed to effectively seal the penstock outlet pipe.

The method of installation of the penstock outlet pipe seal is as follows:-

- Clean the inside of the penstock outlet pipe thoroughly to the required length for the construction of the concrete plug. This is necessary to ensure that no tailings remains within the pipe, which could result in a water flow path occurring through the concrete seal. This can be done by means of high pressure cleaning equipment.
- Install the breather/pumping pipes required to the various depths in the penstock pipe.
- Install the rear shutter required at the end of the final concrete seal. If the pipe gradient is steep enough, this may not be necessary as the concrete could be pumped beyond this point and be allowed to lie horizontally in the pipe.

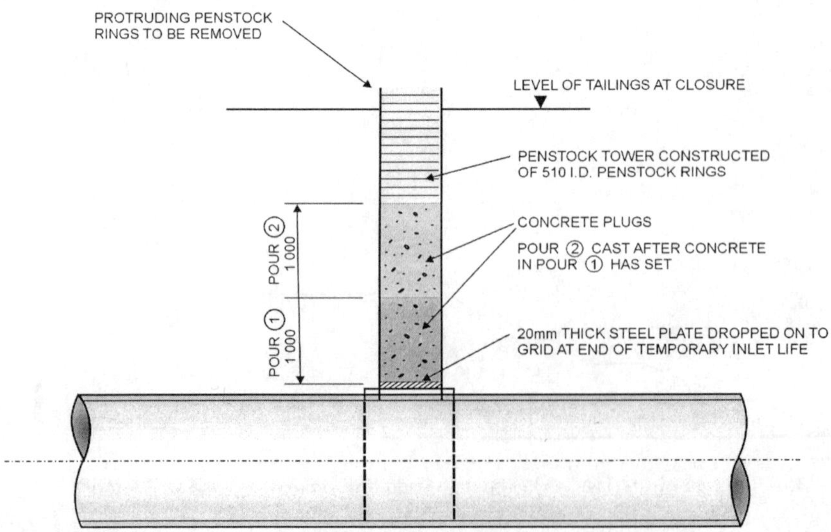

Figure 5. Typical concrete seal for a penstock inlet

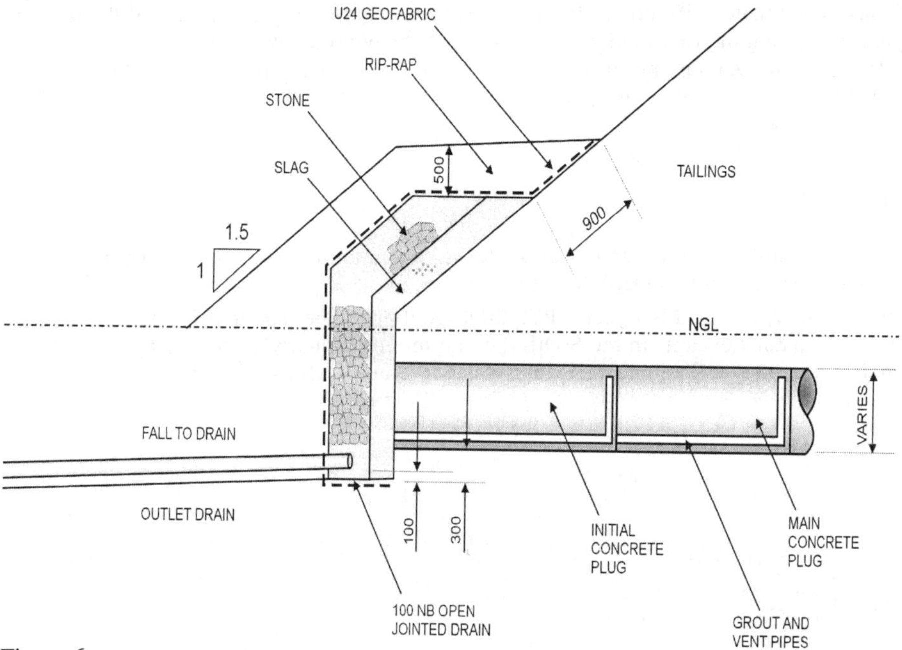

Figure 6.

- Install the shuttering required to install the initial grout plug. This plug is normally 10 – 15 m deep and acts as the initial plug to assist with the higher pressures encountered when pumping of the remainder of the concrete seal. Allow sufficient time for the initial concrete seal to gain strength before continuing.
- Once the initial seal has strengthened sufficiently, the remaining concrete to form the final seal can be pumped. It is important to note the volume of concrete pumped as well as the pump pressures to ensure that blockages are not occurring. The continuous venting of air through the breather pipe should also be noted.
- On completion of the main concrete seal placement, sufficient time for curing and drying of the concrete should once again be allowed. The vent pipe installed in the penstock pipe should then be sealed using high pressure grouting equipment.
- The remainder of the penstock pipe can now be filled with tailings pumped through the vent pipe installed in the shaft intake structure. A process of pumping tailings and removing excess water from the pipe will be required, until the penstock pipe and breather pipe have been filled with tailings. The filling of the remaining open length of penstock pipe is not critical and serves purely to prevent the formation of any future sinkholes from forming on the dam due to a loss of tailings into the pipe void.
- On completion of the concrete seal, a reverse filter and drain outlet system must be constructed at the penstock pipe outlet to ensure that the flow of seepage water along the barrel of the pipe is permitted to continue, without the loss of tailings fines occurring.

5 EMERGENCY SEALING METHODS

Materials used to date to carry out emergency sealing of the penstock shaft intakes, where the installation of a base plate has not been possible, have included:-

- Fetta Bags – This is a rubber/PVC bag which is lowered into the intake structure base and then inflated with air to form a seal, against which to cast the grout or concrete.
- Sand Bags – Bags filled with tailings have been lowered into the base of the penstock shaft to form a plug for the initial concrete seal.

- Coarse Aggregate – 19 mm or 25 mm stone has been used to fill the base of the penstock pipe as bedding in order to allow for concrete to be poured down the base.
- uPVC – A quick coupling uPVC pipe (sewer pipe) has been used successfully as an alternative for a vent pipe during an emergency sealing operation of the penstock shaft intake structure.

6 REFERENCES

International Commission on Large Dams (ICOLD) Bulletin 101 – 1995 "Tailings Dams, Transportation, Placement and Decantation".

Stuart, R.J., Venter, D.J. and Shepherd, P.J., 2000, Steffen, Robertson and Kirsten. "Report on Tailings dam penstock decants in the South African mining industry", Proceeding of Eighth International Conference on Tailings and Mine Waste '01, Fort Collins, January 2001.

Tailings and Mine Waste '01 © 2001 Balkema, Rotterdam, ISBN 90 5809 182 1

Tailings dam penstock decants in the South African mining industry

R.J.Stuart, D.J.Venter & P.J.Shepherd
SRK Consulting, Johannesburg, South Africa

ABSTRACT: The preferred and most often implemented method of removing free water from the top of operating tailings dams in South Africa is by penstock decants linked to a buried pipeline which in turn discharges at the outer edge of the tailings dam.

This paper initially documents the historical background to the original selection of this method of capturing and returning free water back to the plant. It then deals with the various types of penstock decants in use in South Africa from the standard single decant structure capable of decanting at the rate of approximately 0.30 m^3/sec, to special towers capable of decanting at a rate of 2.1 m^3/sec.

Operational issues are discussed including safety controls, access and monitoring. Also presented are the mechanics of past failures and the changes in design which have taken place over time to improve the safety and performance of these structures.

The paper concludes with the latest developments in the design, operation, monitoring and closure of penstock decants in Southern Africa. These new developments include the use of video camera inspections not only for monitoring but for the sealing of penstock decants under operational conditions.

1 INTRODUCTION

Methods used to decant free water from the top of operating tailings dams are varied and differ from country to country. Factors that influence the decision making process as to which decant method is most suitable for any given project include climatic area, local topography, safety considerations, economics, localised or owner preference, past performance and ease of operation.

In valley dam tailings impoundments the use of multiple pumps located on a floating barge is often the preferred option, although use is also made of large diameter towers located close to the outer wall. In this instance the decanted water is led directly to a tunnel which in turn discharges into pipelines or launders leading back to the plant. In perimeter paddock or ring dyke tailings dams the use of centrally located penstock decants linked to a buried outfall pipeline are often preferred. This is the method most often used in South Africa.

This paper deals with various types of tailings dam penstock decants used in the South African mining industry and specifically with small bore precast reinforced concrete ringed structures.

2 HISTORICAL BACKGROUND

The use of penstock decants on South African tailings dams has its origins in the development of the Witwatersrand Goldfields (Figure 1). Earlier development of gold mining in South Africa took place in the mountainous region in and around the Barberton area where tailings were

at best impounded in side hung dumps clinging precariously to the sides of mountains or in valley dams, or in the worst cases directly in streams and rivers.
The Witwatersrand Goldfields were developed on generally flat or gently sloping terrain requiring a different approach to tailings dam design. Initially tailings was deposited on sand dumps using coco pans but as gold metallurgical processing developed a finer tailings product was produced requiring a radical change in transportation and deposition methodology. It was at this time that the concepts for hydraulically depositing tailings in perimeter paddock type tailings dams with centralised penstock decants were introduced. The decants were linked to a buried pipeline which discharged at the lowest point on the edge of the tailings dam perimeter into a pump station or a return water facility. Later the same technique was modified as perimeter ring dyke impoundments for the platinum industry.

The first penstock decants were made of timber. Little evidence of these structures exists today but occasionally the remnants of one is found on a decommissioned dam. Second generation decants were made of brickwork, steel and timber (Figure 2), precast pipe sections or reinforced concrete. Some of these types of decant structures were three sided (U-Section) with the fourth side closed off progressively as the dam was raised by inserting wooden or precast concrete blocks.
Precast reinforced concrete ring sections, 500 mm ID, 100 mm high, with a wall thickness of 75 mm were then developed (Figure 3). As time went by improvements were made to the precast rings by introducing strict manufacturing specifications and increased strength parameters for the individual rings. Further refinements included the introduction of offset rings to bring tower structures made of individual rings back into the vertical plane. Construction and on-site quality controls were also introduced.

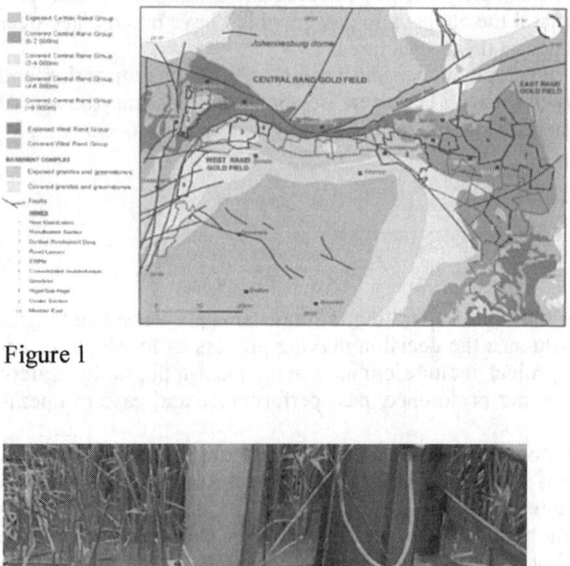

Figure 1

Figure 2

Figure 3

Figure 4

Penstock decants can be used in multiple configurations to match the required plant water recovery requirements, or the need to remove excess water captured on the tailings dam after a significant storm event.

Large bore decants in excess of 1.0 m diameter have been designed over the past 20 years for specific project requirements both in the gold and the platinum industry. Figure 4 shows the twin 5.5 m ID precast reinforced concrete decant towers at Impala Platinum Mine in the North Western Province of South Africa.

A similar decant tower is in operation at Ergo. This gold tailings dam impounds recycled gold tailings recovered from many of the old tailings dumps scattered all over the central Witwatersrand area.

3 FEATURES OF DESIGN

Penstock decants are essentially shafts constructed using different types of materials such as timber, brickwork, steel, precast concrete pipe sections or reinforced concrete. In Australia reinforced asbestos cement pipe sections have also been used but the decant methodology is significantly different to the South African systems. Hydraulic design characteristics vary according to the type of decant, for example, a timber, reinforced concrete or brickwork structure with a "U-shaped" form operates as a sharp crested weir. Precast reinforced concrete structures built using rings or in some instances pipe sections operate as a sharp-crested shaft or a "morning glory" spillway.

Discharge quantities over a sharp-crested weir or a circular shaft spillway are given by the equation:

where $Q = CLH^{3/2}$

Q is the discharge in (m^3/s)
L is the length of the crest (m);
H is the head of water above the crest(m);
C is a discharge co-efficient equal to $1.8m^{½}/s$

provided the overflowing water can fall freely down the shaft without converging to form a single jet down the shaft axis. In this case L becomes the perimeter of the shaft, 2 π. R, where R is the inside radius. The above condition applies providing the ratio H/R does not exceed 0.45.

If H/R exceeds 0.45 the same formula for Q can be used, but the value of C must be modified as follows:

H/R	C ($M^{½}S$-1)
0.45	1.8
0.70	1.6
1.00	1.0
1.30	0.9

The situation where H/R exceeds 0.45 should be avoided as it leads to pressure surges in the penstock that may result in the dislodgement of rings.

Other design considerations that must be taken into account are that the shaft or tower must be designed to resist both the lateral pressure of the residue and the frictional down – drag caused by settlement of the residue which surrounds the structure. It is further recommended that the design lateral pressure of the residue plus the water pressure, while the down-drag be taken as the tangent of the angle of shearing resistance times the effective lateral pressure. In seismic active areas, the effect of seismic forces in conjunction with liquefied tailings surrounding the shaft should be considered as recommended in ICOLD Bulletin 101.

It is important to note that the design of the buried pipeline should assume gravity flow rather than pressure flow. This is done to avoid pressure surges and is achieved by ensuring the buried pipeline is not the restricting element.

The number and size of penstocks is mainly dependent on the climatic conditions and legislative requirements of each country.

It is general practice when determining the number of penstocks required for a tailings dam to remove the rainwater from a 1:100 year event within 72 hours to reduce the risk of failure of the tailings dam. The head above the rings should not be higher than 200 mm to avoid pressure surges in the penstock which may result in the dislodgement of rings.

The highest penstock level also forms part of the new legal requirements of freeboard on a tailings dam. A brief history of the past or present freeboard requirements is described below.

In South Africa the National Water Act (Act 36 of 1998) and its regulation 704 superseded the Water Act (Act 54 of 1956) and its regulation 287. A schematic representation of the freeboard requirement is presented in Figure 5.

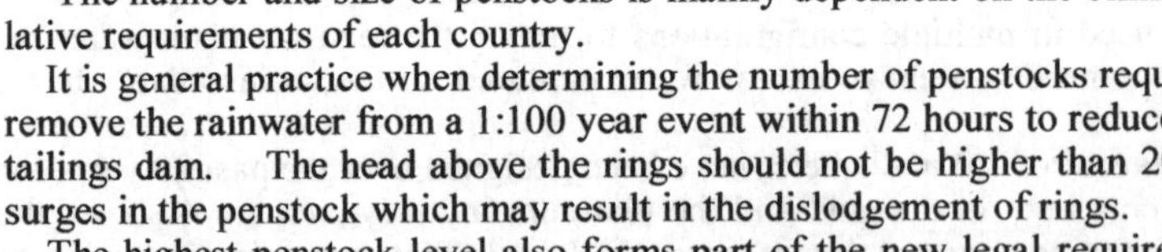

SCHEMATIC REPRESENTATION OF TAILINGS DAM FREEBOARD REQUIREMENT

Figure 5

The new regulations are also simpler to implement and takes away any hydrological based assumptions. However, it should be carefully checked when working with smaller surface area ring dyke impoundments. It is important to still calculate the 1:100 year storm event to ensure that the volume can be removed within the 72 hours as discussed earlier.

4 OPERATION OF PENSTOCK DECANTS

Penstock decants in the context of perimeter paddock or ring dyke tailings dam design are usually placed at selected intervals along the alignment of the buried discharge pipeline. The decants are sealed off as the deposited tailings increase in height and the operating pool moves back to the next intake. Eventually the operating pool is located around the last decant at the head of the discharge pipeline, and from this point onwards the recovery of free water from the top of the tailings dam is controlled at the centre of the dam. (Figure 6)
During the main operational phase of the tailings dam when the operating pool is controlled around the central decant access is achieved by building pool walls using the tailings or slimes being deposited on the dam. This method reduces the length of catwalk required to a short link only.

The construction of precast reinforced concrete penstock decants as the tailings dam is raised is simplistic in the sense that the tower is extended by stacking single rings one upon another.

The nature of the limited skills required to extend the decants can cause problems, however, if no checks are carried out to ensure that the tower is raised as near to vertical as possible. This can be achieved by the use of plumb bobs and the introduction of offset spacer rings. Poor control in this regard has resulted in the premature failing of decants. This in turn often results in significant capital expenditure required to install a replacement elevated penstock.

5 SAFETY ISSUES

Safe access ways and working platforms to and around single or multiple decant arrangements is an important issue in terms of operating decants. Access during the initial development of the dam is normally achieved by the construction of wooden walkways from the starter embankment. In South Africa these walkways are called "catwalks" perhaps to depict the need to walk over the often saturated tailings on what at first glance may be considered to be somewhat light structures. The "catwalks" are raised by extending the uprights and deck planking (Figure 7). The safety track record of these wooden structures is nevertheless very good. Where rare incidences of dam operator staff or equipment falling into the tailings or slimes have been recorded

Figure 6

Figure 7

Figure 8

the cause of the accident was found to be due to lack of maintenance, however, theft of safety ropes, planking etc. has become cause for serious concern resulting in the use of alternative materials such as long rolled wire and HDPE grating to build the catwalks and working platforms.

Decàntation of free water in small bore penstock decants, particularly precast ring type structures should not be carried out with more than two single rings removed. This is because the suction formed around the top of the decant shaft can be extremely dangerous to the personnel operating the dam when they are adding or removing new rings or sections.

Recorded fatalities due to operators being pulled down onto the open mouth of the decant tower have resulted in safe practice restrictions (max 2 rings off) combined with safety cages, signs and harness requirements being required under legislation in the South African mining industry (Figure 8).

One contracting company in South Africa have developed and patented a mechanical lifting device which allows the operator the facility to remove or replace individual units from a safe distance from the suction zone surrounding the top of the decant.

Taken into account the numerous tailings dams in South Africa the historical track record of penstock decants has been very good. Where failures have occurred they have been made safe immediately and in certain instances the penstocks have been given extended operational life spans without incurring excessive capital cost or plant downtime

6 PAST FAILURES

The mechanism of failure of small bore penstock decants are varied. Listed below are a selection of failures that have occurred in South Africa, the list is not exhaustive, however, as it records only those which the authors have identified by personal experience:

- Crushed rings within 3 metres of the bottom of decant towers in excess of 20 metres deep due to down drag forces.
- Collapsing of rings due to exposure and serious corrosion of precast ring reinforcement probably due to installing rings which may have had hairline cracks. Specifications and quality control systems in the early years of development of reinforced precast rings were not at the levels existing today in RSA.
- Poorly seated or aligned rings moving out of position under pressure.
- Penstock decants developing excessive distortion in the vertical plane. In one specific instance on an ash dam a decant of some 27 metres high was found to have developed a full 180 degrees spiral prior to failure.
- Rings crushing or distorting out of position due to a seismic event or an underground pressure burst.
- Multiple tower clusters made up of single units failed due to down drag forces on one tower leading to sympathetic failure of the other three towers. Loss of support due to the release of slimes at the base of the first tower to fail may have accelerated the failure mode to the other three towers.

In the majority of cases the penstock decants have been successfully operated up to a vertical height of between 20 to 25 metres in height. The average life of these towers was in the order of 10 to 15 years depending on the individual dams rate of rise.

7 LATEST DEVELOPMENTS

Two specific areas of improvement in performance in terms of operational life were identified some four to five years ago, these were:

i) Develop ways to extend the life of each decant tower to operational heights in excess of 20 to 25 metres.

ii) Develop designs that would allow final decant towers located at the centre of tailings dams to reach life of mine height requirements of between 40 to 45 metres.

In the first instance the development of robust video equipment has opened up significant opportunities for tailings dam designers and operators. On three separate occasions SRK tailings engineers have been involved with clients and tailings dam operators in the emergency closure of operating decants under severe stress and in failure mode. In each instance the use of video cameras and play-back equipment at the site during the placing and curing of the concrete plug played a very important role in successfully sealing the failing decant tower.

It is also now standard procedure to use video cameras to monitor the internal status and behaviour of penstock decants in excess of 18 metres deep on an annual basis.

To extend the operational life expectancy of penstock decants a construction programme is in place to internally sleeve the precast concrete loose ring structures with a 400 mm ID, 6 mm wall thickness steel pipe suitably protected internally against corrosion. The void between the concrete rings and the external wall of the steel sleeve is then epoxy grouted. Decants in excess of 20 metres deep sleeved in this way have, to date functioned well with no problems being recorded.

A new construction / operation programme is also currently being implemented to determine if internal steel sleeving of precast concreting towers should take place for every 2 to 3 metres of lift. The sleeving would be stopped within 0.5 m of the top of the slimes level so that the benefits of stacking or removing the lightweight individual rings is not nullified.

The second area of investigation to improve the performance of the decant towers was the need to extend the final height potential of the centrally located main decant structures. This particular problem was solved by introducing a two-staged approach to the construction of the overall tower.

The approach involved the initial construction of a standard decant to deal with the first 15 m of tailings lift. Alongside this decant a 15 m high precast reinforced concrete tower is built which only comes into operation as an extended ringed decant after the concrete tower is covered by tailings.

The two staged system allows centrally located final decants to be raised approximately 35 m to 40 m while still linked to the original outfall pipe, thus obviating the need for the additional capital expenditure involved in the installation of elevated penstock pipelines and decant systems. The use of penstock decants on tailings dams in South Africa has a long history. The authors have prepared this paper to document not only the varied types of decant structures, but also the important issues of operations and safety. Recorded failure modes are included to assist future designers as are the latest developments to extend the life of operational penstocks.

8 REFERENCES

ICOLD BULLETIN 101 – 1995 Tailings Dams, Transport, Placement and Decantation – Review and Recommendations.

Tailings and Mine Waste '01 © 2001 Balkema, Rotterdam, ISBN 90 5809 182 1

Fugitive particulate control at an iron mine tailings basin

K.G. Paterson & S.J. Vitton
Department of Civil and Environmental Engineering, Michigan Technological University, Houghton, Mich., USA

ABSTRACT: The objective of this study was to assess the effectiveness of the Superior Recycled Fiber, Inc. (SRFI) waste residue as a control strategy for minimizing dust generation from an iron mine tailings impoundment. The study investigated the effectiveness of four treatment methods to reduce fugitive particulate matter emissions from a closed tailings basin at the EvTac mine near Eveleth, Minnesota. Specific objectives included measurement of particulate matter emissions produced under controlled wind speed conditions during on-basin field tests, and comparison of field-scale fugitive particle emission reduction strategies. The method of analysis used a field deployable wind tunnel to produce constant wind speeds over the sections of the basin where various control applications had been applied, as well as over areas without control. Particulate matter less than 10 μm in diameter (PM_{10}) and meteorological variables were measured continuously throughout each of the fifteen test plot experiments. The field tests were conducted in late October 1999 during periods of evening freezing with day temperatures above freezing. Control strategies were assessed through comparison of test plot PM_{10} emissions to those for the test plot with no controls. Test results provide strong evidence for particular SRFI waste mitigation strategies on this basin.

1 STUDY OBJECTIVES

The objective of this study was to assess the effectiveness of the Superior Recycled Fiber, Inc. (SRFI) waste residue as a control strategy for minimizing fugitive airborne particulate matter generation from an iron mine tailings impoundment (Goff 1999). Airborne particles are associated with numerous environmental and health effects (Pope 1998; Seaton et al. 1995; U.S. EPA 1997). The study investigated the effectiveness of four treatment methods to reduce fugitive particulate matter emissions from a closed tailings basin at the EvTac mine near Eveleth, Minnesota. Specific objectives include:

- Measurement of particulate matter concentrations produced under controlled wind speed conditions during on-basin field tests
- Comparison of field scale fugitive particle emission reduction strategies

The method of analysis was to use a field-deployable wind tunnel to produce constant wind speeds over the sections of the basin where various control applications had been applied as well as over areas without control. The field tests were conducted in late October 1999 during periods of evening freezing with day temperatures above freezing.

2 METHODOLOGY

2.1 *Particulate Matter Sampling*

Field site testing of fugitive particle emission treatment methods was conducted using a portable wind tunnel (Figure 1) developed by the Civil & Environmental Engineering Department at Michigan Tech (Vasher 1999; Raupach and Leys 1990). Wind is generated by a fan mounted on a 5.5 m (18 ft) trailer along with a 70 kW (85 Hp) Detroit Diesel engine connected to a 1.5 m (5 ft) industrial fan. The engine is connected to the fan via belts which results a in a 2.6:1 engine to fan ratio and operated between 800 and 2600 RPM generating a maximum airflow volume of approximately 50,000 cfm at a top fan speed of 1000 rpm. This allows for wind tests to be conducted up to a maximum wind speed of 14.3 m/sec (32 mph) through the main working section of the wind tunnel. Prior to entering the working section the air is conditioned (straightened into a laminar flow) by passing the air through a transition zone, which starts at 1.8 m by 1.8 m (6 ft by 6 ft) section and drops approximately 0.5 m (1.5 ft) to the level of the ground where the air exits into the working section. Within this transition zone are two sets of plastic honeycomb panels that are used to straighten out the airflow prior to flowing into the working section. The working section is approximately 7.5 m long (25 ft), 1.2 m high by 1 m wide. The base of the working section is open to the ground for 6 m (20 feet).

Thirteen field plot tests were performed on October 27, 1999: three tests of compacted light paper sludge application, two tests of heavy paper sludge application, two tests of old hay mulch, two tests of light paper sludge application, two tests of no treatment (uncontrolled), and two ambient air tests. Each test followed a fifteen-minute air particle sampling protocol: one minute of ambient tunnel particle sampling, four minutes each sampling at 10, 20, and 30 mph sustained wind speeds, and a two minute quiescent ambient tunnel sampling.

The day of testing followed a night of freezing, resulting in some sublimation of the near surface moisture. Sublimation is suspected to cause a thin layer of readily transportable particles at the soil surface (Law and van Dijk 1994) . The morning was primarily clear followed by cloudy conditions in the late afternoon. Temperatures ranged from freezing in the morning to approximately 10°C at mid-afternoon. Periods of dusting occurred throughout the day increasing towards the late afternoon as winds picked up.

During the testing particulate matter less than 10 m in diameter (PM_{10}) was measured every 5 seconds over this 15 minute period using a model 8520 DustTrak Aerosol Monitor, manufactured by TSI Incorporated. The DustTrak is a small, transportable instrument that uses light scattering technology to determine the real-time mass concentration of an air sample. The sensing mechanism of the DustTrak contains a laser diode directed at the inlet steam. Light, scattered by the particles in the inlet stream, is collected on a lens located at a 90° angle from the sample stream and the laser diode. This light is then focused onto a photodetector, which converts the light to a voltage. The voltage is proportional to the amount of light scattered, and thus the mass concentration of the particulate matter in the sample stream. The measurement range of the DustTrak has a lower limit of 0.1μm for particle size and a mass concentration of 1 g/m^3. Sampling in relative humidity higher than 80% can cause particle concentrations to be biased on the high side of the actual concentration of dry particles. However, none of these limitations were a problem for this study, since ambient particle size and concentrations were one to five orders of magnitude greater and the relative humidity was low. The sample inlet to the DustTrak was located at the end of the portable wind tunnel, at a point approximately 2/3 of the available space above ground level (refer to Figure 1, the DustTrak unit is immediately right of the tunnel outlet).

A weather station manufactured by Met One Instruments was used to obtain meteorological conditions: wind direction, wind speed, temperature, relative humidity, and vertical wind speed. All sensors were positioned on a 10 ft tower, except the wind speed sensor, which was located at the terminal end of the wind tunnel (see Figure 1). The parameters measured by the met station were recorded every minute.

2.2 *Soil Sampling and Testing*

Samples of the control materials as well as the tailings were collected for moisture, density, and particle size analysis. Grab samples were conducted after the individual wind tunnel tests had been completed. The results of the moisture and density tests are reported in Table 1.

Moisture measurements were made using a standard oven drying technique in which the mass of water was determined and divided by the mass of solids. It is believed that the lower moisture contents, particularly in the paper waste, resulted form both surface drying as well as from sublimation effects due to freezing of pore water, which occurred the previous night.

Table 1. Moisture and density measurements for paper sludge, tailing and old hay mulch.

Treatment Method	Average Moisture Percent	Particle Density (g/cm^3)
Light paper sludge application	9.4%	1.917
Light paper sludge application with compaction	5.4%	1.930
Heavy paper sludge application	8.0%	1.929
Old hay mulch	13.3%	-
Tailings	10.2%	3.093

Figure 1. Mobile wind tunnel deployed on field plot with compacted light paper sludge application. The sampling equipment is set up in the foreground end of the tunnel.

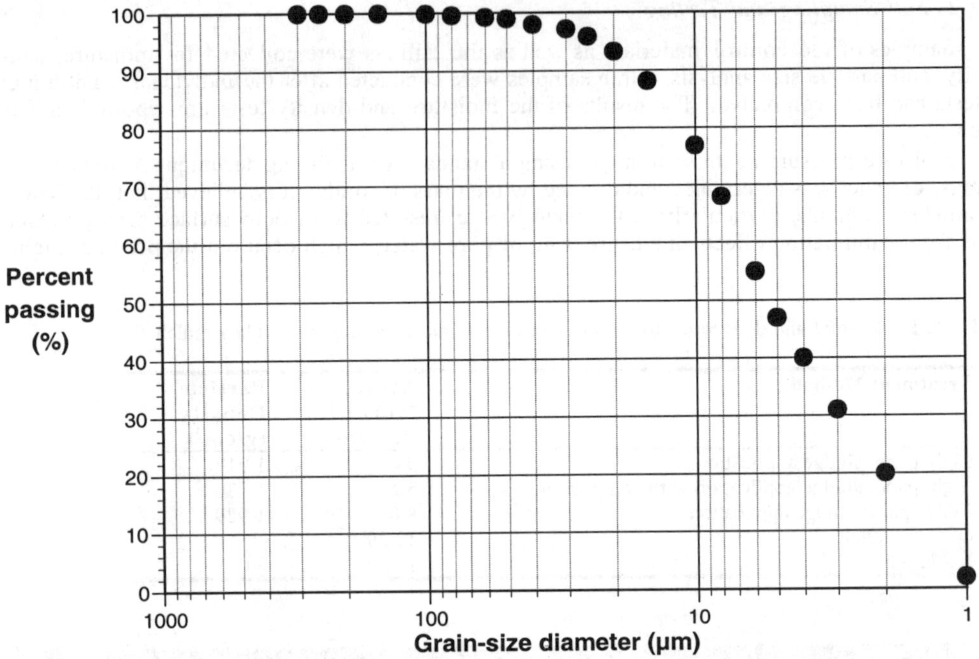

Figure 2. Average grain-size analysis of the tailings.

The particle density of the paper waste and tailings were determined using a Micromeritics 1360 helium pycnometer. This device uses helium gas, which can penetrate into the structure of the solid, to determine the volume of the solid portion of the particles. The grain-size of the tailings was determined using a Micromeritics 5100 Sedigraph. The 5100 Sedigraph uses X-Ray absorption to determine the particles mass over time in a suspension. The results of the grain-size analysis have been averaged together and provided in Figure 2.

3 RESULTS

To minimize data scatter and turbulence induced particle concentration variations, the five-second PM_{10} data was transformed to one minute running averages. An example of this transformation is provided in Figure 3. The smoothed data sets were used in all subsequent analyses. As suggested by Figure 3, it was consistently observed that the onset of higher wind speeds resulted in a transient spike in the PM_{10} concentrations due to liberation of particulate matter from the basin surface. This is the nature of fugitive particle emissions due to wind events. The transient concentration events begin with the increase in wind speed, rapidly increase, and more slowly decay, lasting approximately two minutes in total.

In order to compare the various field treatments, the background concentrations existing during each trial were removed from the data to ensure that variations in tailings (size, moisture content, deposition patterns) were normalized. For each of the two minute transient concentration spikes which occurred at the three wind speed conditions, the average concentration over the four-minute test period was removed, leaving the concentrations due to wind-induced liberation only. These concentrations were then averaged for the three wind speeds. Lastly, the three wind speed averages were averaged for each field treatment (e.g. the three heavy application 10 mph results were averaged) to yield more robust representation of the performance of each treatment methodology.

Table 2 shows overall average concentrations resulting from all field treatment tests at the three wind speeds (10, 20, 30 mph). The values in the table have been adjusted by removal of their background concentrations before averaging, as such these values can be considered as the particle contribution from the treated basin plots. As expected, particle concentrations (both average and maximum) increase with wind speed. The mean values are representative of performance over sustained wind conditions, whereas the maximum values suggest performance under gusts. Air quality violations are more likely to occur under sustained high wind conditions, as ambient air quality standards depend on average concentrations over an 8 or 24-hour period. In other words, while gusts may yield very high particle concentrations, these tend to be short-lived.

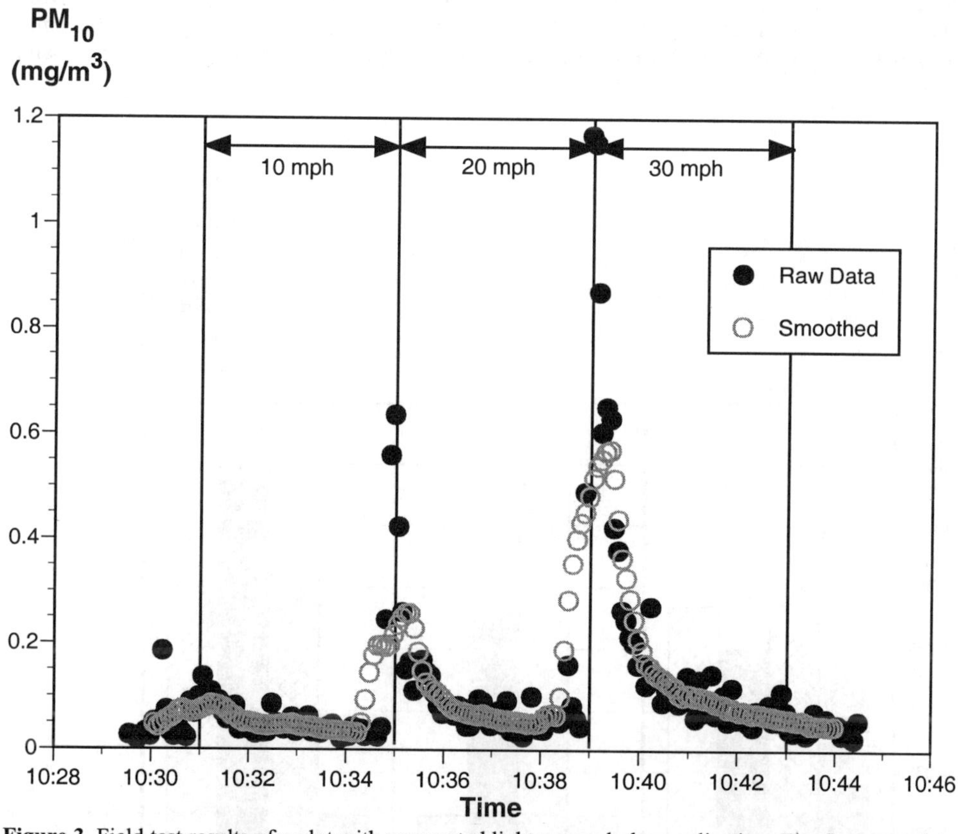

Figure 3. Field test results of a plot with compacted light paper sludge application. The five-second (raw) PM_{10} data are shown by solid circles, the one-minute running average transformed (smoothed) data are shown by hollow circles.

Table 2. Comparison of average mean and maximum PM_{10} concentrations produced at each wind speed during all treatment tests. Background concentrations have been removed.

Wind speed (mph)	Overall Average Mean PM_{10} Concentration (μg/m^3)	Overall Average Maximum PM_{10} Concentration (μg/m^3)
10	9.9	171.8
20	121.4	524.7
30	158.8	971.3

While Table 2 shows some overall trends in performance of the treatment methods, a comparison of the particle reductions provided by the different treatment methods is important for identification of the most effective treatment. Figures 4 and 5 show the reductions in average and maximum concentrations (or emissions) as compared to the uncontrolled field plots, resulting from the four treatment methodologies. Several findings are evident. All the treatment methods yield similar reduction (60-70%) of average emissions at the 10 mph tests. Based on the low average concentrations liberated at this wind speed, as shown in Table 2, this is not surprising. At 20 mph, the light application of paper sludge and old hay mulch provide the poorest reductions to the average PM_{10} emissions (about 30%), whereas the heavy paper sludge application and light application with compaction provided much better suppression (about 65% and 88% respectively). At the highest wind speed, 30 mph, the heavy application provides the best control of particle emissions (90%), followed by the light with compaction (72%), old hay mulch (69%), and light application. The emission reduction performance of the heavy paper sludge application improved with increasing wind speed.

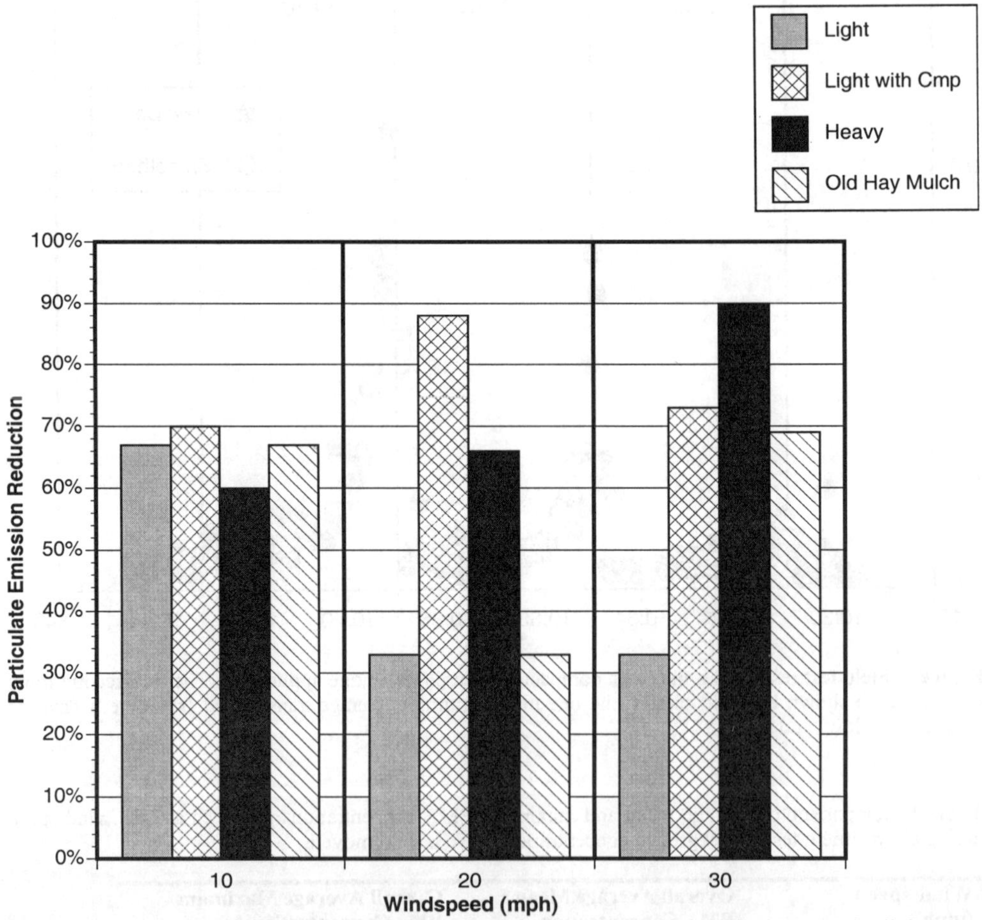

Figure 4. Average particulate emission reductions of four treatment methodologies: and (1) light paper sludge application, (2) light paper sludge application with compaction, (3) heavy paper sludge application, and (4) old hay mulch, at three wind speeds (10, 20, and 30 mph). EvTac Basin particulate emission testing, October 27, 1999, while reduction percentage is calculated using the uncontrolled (no treatment) field tests as the reference. All data is with the background PM_{10} concentration removed.

The maximum concentration plot (Figure 5) provides additional performance information. At 10 mph, the old hay mulch produces no control of maximum particle concentrations; in other words, it behaved similarly to the uncontrolled plot. This may be due to the enhanced surface roughness of the hay. At low wind speeds, the hay may result in the greatest turbulence and hence particle liberation. The light paper sludge with compaction results in the best control of maximum particle concentrations at low wind speed (10 mph). At 20 mph the light application with compaction provides superior emission control (nearly 90%) with the heavy application providing moderate control (55%). The old hay mulch and light application provide little control, about 13% and 9% respectively. At the 30 mph test level, the heavy paper sludge application provides the best control (85%), followed by the light application with compaction (65%), old hay mulch (50%) and the light application (22%). As was the case for the average emission control, the heavy paper sludge application provided improved maximum emission suppression with increasing wind speed.

To compare the overall performance of the treatment methods, their emission reductions (as provided in Figures 4 and 5) were averaged (Table 3). This method gives equal

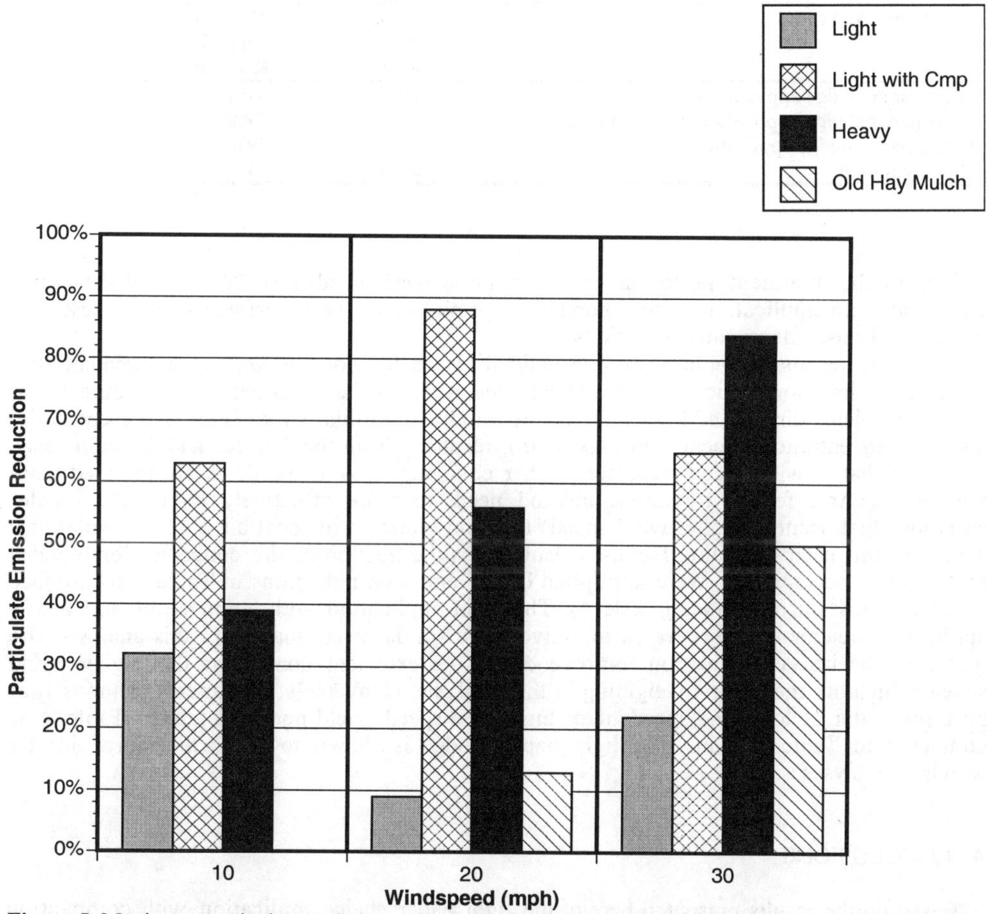

Figure 5. Maximum particulate emission reductions of four treatment methodologies: and (1) light paper sludge application, (2) light paper sludge application with compaction, (3) heavy paper sludge application, and (4) old hay mulch, at three wind speeds (10, 20, and 30 mph). EvTac Basin particulate emission testing, October 27, 1999, while reduction percentage is calculated using the uncontrolled (no treatment) field tests as the reference. All data is with the background PM_{10} concentration removed.

Table 3. Average particulate emission reduction of four treatment methodologies for all wind speed conditions. Reductions are based on average and maximum PM_{10} concentrations produced during testing. Reduction percentage is calculated using the uncontrolled (no treatment) field tests as the reference.

Treatment Method	Mean Emission Reduction	Maximum Emission Reduction
Light paper sludge application	44%	21%
Light paper sludge application with compaction	77%	72%
Heavy paper sludge application	72%	60%
Old hay mulch	56%	21%

Table 4. Total particulate emission reduction of four treatment methodologies for all wind speed conditions. Reductions are based on average and maximum PM_{10} concentrations produced during testing. Reduction percentage is calculated using the uncontrolled (no treatment) field tests as the reference and weighted by the relative concentrations provided in Table 2.

Treatment Method	Mean Emission Reduction	Maximum Emission Reduction
Light paper sludge application	19%	34%
Light paper sludge application with compaction	72%	79%
Heavy paper sludge application	70%	79%
Old hay mulch	33%	54%

waiting to the treatment performance at the three wind speeds (10, 20, 30 mph). On this basis, the light application with compaction provides the best performance to sustained (mean) and gust (maximum) conditions.

While the results in Table 3 are one valid measure for comparison, an assessment that weights the emission reduction performances according to the potential to contribute to air quality problems directly addresses the concern motivating this work. Table 3 presented the average concentrations (mean and maximum), resulting from the 10, 20, and 30 mph tests. It is clear that as wind speed increases, higher concentrations of particles are released from the surface. An effective treatment method needs to more effectively reduce the greater emissions that occur at higher wind speeds to avoid violation of local air quality regulations. To assess this need weighting factors, calculated as the fraction of the concentration divided by the total concentration, were multiplied by the emission reductions in Table 3 to produce weighted emission reductions (Table 4). The light application with compaction and heavy application treatments are revealed to deliver very similar performance. In this analysis, the merits of the heavy application improve due to its excellent control at the highest wind speed, which has the highest weighting in this analysis. Conversely, treatment methods that perform better at low wind speeds than high wind speeds yield poorer totals in Table 4 as compared to Table 3. Thus, the light paper sludge is shown to perform poorer in the weighted analyses.

4 CONCLUSIONS

Based on the results presented herein, the light paper sludge application with compaction and heavy paper sludge application provide similar fugitive particle control for sustained and gust wind conditions (approximately 70% and 80% reductions, respectively). The light paper sludge application and old hay mulch offer much poorer control.

As the particulate emission criteria are similar for the light paper sludge application with compaction and heavy paper sludge application, a final selection between these two methods should be based on other factors, such as cost and safety

5 REFERENCES

Goff, K. 1999. Fugitive Dust, *Erosion Control*, 6(2):86-95.

Law, J. and D. van Dijk. 1994. Sublimation as a Geomorphic Process: A Review, *Permafrost and Periglacial Processes*, 5:237-249.

Pope, III, C.A. 1998. Epidemology Investigations of the Health Effects of Particulate Air Pollution:Strengths and Limitations" *Applied Occupational Environmental Hygiene*. 13(6):356-363.

Raupach, M.R. and J.F. Leys. 1990. Aerodynamics of a Portable Wind Erosion Tunnel for Measuring Soil Erodibility by Wind, *Australian Journal of Soil Research*, 28:177-191.

Seaton, A., MacNee, W., Donalson, K. and D. Godden. 1995. Particulate Air Pollution and Acute Health Effects, The Lancet, 345:176-178.

U.S. EPA. 1997. *National Air Quality and Emissions Trends Report*, EPA-454/R-98-016. U.S. Environmental Protection Agency, Office of Air Quality. Planning and Standards..

Vasher, D.R. 1999. *A Technique for Assessing Fugitive Particulate Emission Potential from an Active Mine Tailings Basin*, M.S. Thesis, Michigan Technological University, 146 pp.

Tailings and Mine Waste '01 © 2001 Balkema, Rotterdam, ISBN 90 5809 182 1

A robust water-balance model for sizing heap leach solution ponds and reservoirs

J.R. Kunkel
Knight Piésold and Company, Denver, Colo., USA

ABSTRACT: This paper presents an overview of the water balance approach, a summary of how to manipulate climatological data, and an example design. An example, using only a single pass through a typical long-term climatological record versus the more robust approach, is presented. The results of the sizing of a heap leach pad solution pond and for predicting makeup water supply are presented for the robust approach.

1 INTRODUCTION

A water balance is a calculation procedure which provides for detailed analyses of the state of the system using climatological and operational variables. A water balance is an accounting of water using the general hydrologic equation (Thornthwaite and Mather, 1955, 1957)

$$P - Q \pm \Delta S_w - E \pm \Delta S_s - D = 0 \quad (1)$$

where P is precipitation, Q is runoff, ΔS_w is change in storage of water ponded on the surface, E is evapotranspiration (evaporation plus transpiration), ΔS_s is change in soil (ore) moisture storage, and D is deep percolation (unrecoverable by vegetation or evaporation).

Water-balance inputs typically include pond/reservoir characteristics, precipitation, evapotranspiration, inflows, seepage, water demands, and operating variables. For heap leach pads, the loading sequence of ore is an operating variable and must be taken into consideration in the water balance. One of the most sensitive times for water handling at a heap leach pad is at startup when large footprint areas are exposed but little ore is available to hold water. Another sensitive time is at shutdown or draindown during a power outage when water cannot be recirculated from storage ponds.

Water-balance outputs are the state of the system at any time for any variable of interest. For heap leach operations, these variables may include the volume of water in pregnant and barren storage ponds, makeup water demands, or flow rates which must be treated and released. Design usually involves predicting the state of the system for extreme events such as meteorological droughts or excessive precipitation. Therefore, the water balance is a design tool for sizing and operating the facilities. Water balances are generally more sensitive to operational variables than to climatological variables.

The purpose of a water balance, then, is to answer the questions "How do we know that the state of the system may not overtop the available storage?" or "How do we know that the state of the system may not be reduced to zero at a time of heavy water demand?" The following sections present the traditional approach to answering these questions and a more robust approach.

2 TRADITIONAL WATER-BALANCE APPROACH

The traditional approach to a water balance is to use the systematic climatological record with either averages or once-through operation. The approach using averages is to use only a 12-season model with one season representing one month. Thus, even if the climatological time series has 30 years of data, the average approach uses the average of the January values, the February values, and so on through December. An alternative approach for the traditional water balance is to use the full-time series of climatological data but only have a once-through simulation. In order to assess the sensitivity of the initial state of the system, for example an empty reservoir, a large number of once-through simulations must be run for each initial system state.

The fall back position of the traditional approach is to perform sensitivity analyses by assigning a wet or dry sequence within the average time series. Often the climatological variables are arbitrarily increased or decreased by some percentage in an attempt to simulate a wetter or dryer than average year. Another fall back of the traditional approach is to use a random time series of equally likely climatological variables. These variables, precipitation for example, are generated stochastically but used deterministically in the water balance.

There are two obvious drawbacks of the traditional water-balance approach. These drawbacks are (1) the approach does not analyze all possible combinations of climatological and operational variables, and (2) it uses an arbitrary sensitivity analysis. Some arbitrary sensitivity analyses used in the traditional water-balance approach include synthetic extreme wet years, synthetic extreme dry years, wet years followed by wet years, and dry years followed by dry years.

None of these traditional water-balance analyses can give a full assessment of running a water balance for all possible combinations of climatological and operational variables. Therefore, a more robust approach is needed to fully assess the impacts of a water balance on the system being analyzed.

3 ROBUST WATER-BALANCE APPROACH

The robust water-balance approach uses the systematic climatological record or a synthetically generated long-term time series. The difference between the traditional approach and the robust

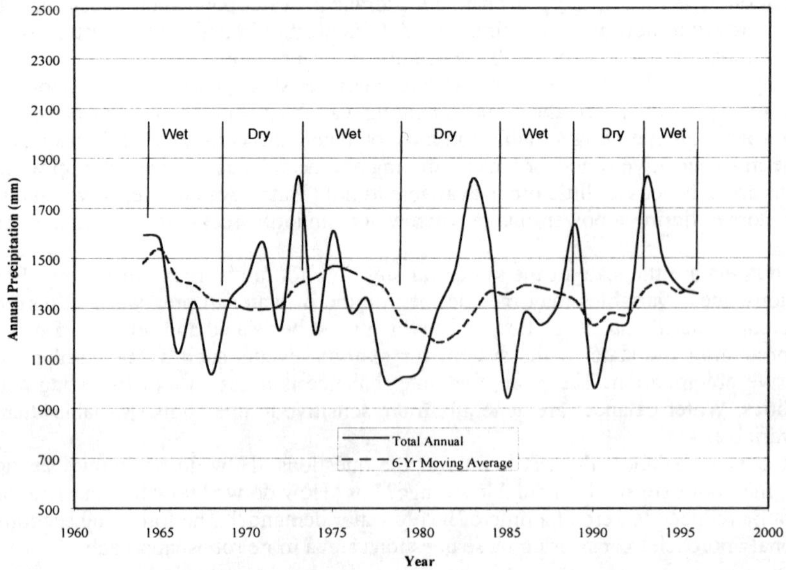

Figure 1. Total annual and six-year moving average precipitation showing wet and dry periods.

approach is that in the robust approach the operation begins in each year of the systematic climatological record and runs for the full life of the project.

For example, if a heap leach pad has an expected life of 12 years and there is a 20-year climatological record, the first 12 years of the pad life are first run using the first 12 years of the systematic record, the 12 years are run again starting in year 2 through year 13, and so on until year 20 is reached. Obviously, at year 10 there are not 12 years of climatological data remaining. In this case, year 1 is transposed to the end of the climatological record to always provide a full 12 years of climatological data. If the water-balance run begins in year 15, for example, the first six years of data are transposed to the end of the record to always provide a full 12 years of climatological data. In this way, all possible combinations of climatological record are experienced by the project operation.

Obviously, in a mining or water resources project, it is not possible to determine *a priori* if the project will come on line in a wet, dry, or average climatological year. Figure 1 shows a 33-year time series of monthly precipitation in Perú showing wet and dry periods based on a six-year moving average analysis. By analyzing the system using a robust water-balance method, the probability of having a successful project or finding areas of concern during system operation can be identified. The next section gives an example design for the volume of makeup water and makeup pond volume for a heap leach pad. The robust water-balance model also can be used to design the lixiviant pond volume, or the volume of water treatment required, as well as many other operational variables.

4 DESIGN EXAMPLE

The following example project demonstrates the use of the robust water-balance approach. This project is a 2.8 million tonnes per year (tpy) gold heap leach project in northern Perú. Previous designs had been done, and this project was an expansion to an existing project. The existing process ponds had been sized and constructed to store 182,500 cubic meters (m³) of fluid, and these ponds were to remain with the owner's desire to not construct either new ponds or expand the size of the existing ponds due to land space constraints.

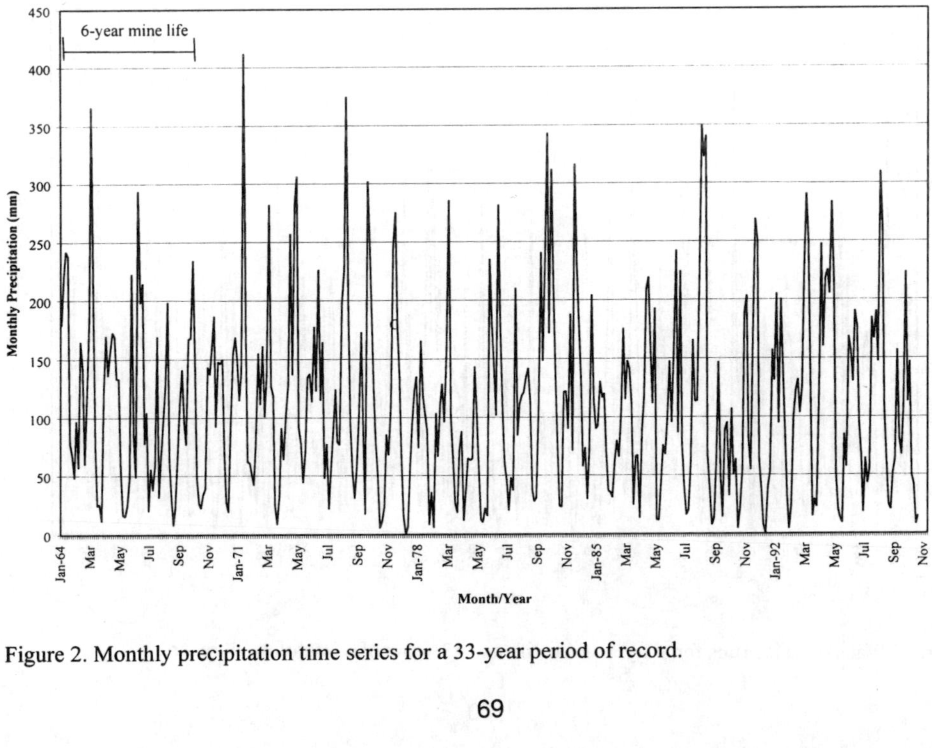

Figure 2. Monthly precipitation time series for a 33-year period of record.

4.1 *Design Procedure*

After developing a six-year loading plan for expansion of the heap leach pad, a water balance was done using the 33-year period of monthly precipitation from a nearby climatological station. Figure 2 shows the 33-year precipitation record in relation to the six-year project life.

The mine life was estimated by the owner to be approximately six years based on the proven ore reserves. The problem was to calculate the quantity of makeup water required to expand the heap leach pad to 2.8 million tpy.

Because it is unknown *a priori* which six years of the 33-year precipitation record would produce the largest quantity of makeup water, a systematic approach was used to assess this value. The first step was to perform a monthly water balance using the first six years of the precipitation record. The monthly values of required pond storage and makeup water were calculated from the water-balance equation (1). After the first six years of the precipitation record were used, a second water-balance run (second run on Figure 3) using years 2 through 7 of the precipitation was completed. Subsequent six-year water balances were run by stepping through the 33-year precipitation record by beginning each new water-balance run in each succeeding year as shown on Figure 3. For the last five water balance runs, climatological data from the beginning of the 33-year time series were transposed to the end of the existing time series so each run used a full six years of data.

4.2 *Interpretation of Results*

By beginning the six-year project water balance in each of the 33 years of climatological data, all possible combinations of operational (ore loading) and climatological possibilities were modeled. Using 33 years of the monthly data and six years per run meant that 198 values of required pond volume and makeup water volume were generated for each of the 12 months of the calendar. This total number of values was 2,376 for the 33 water-balance cases. Given so many values, 198 per month, or 2,376 total values, raises the question of how to interpret the meaning of these values.

Because engineers often design facilities for the largest or smallest values generated from a time series analysis, the robust method uses univariate statistics to summarize the 2,376 values

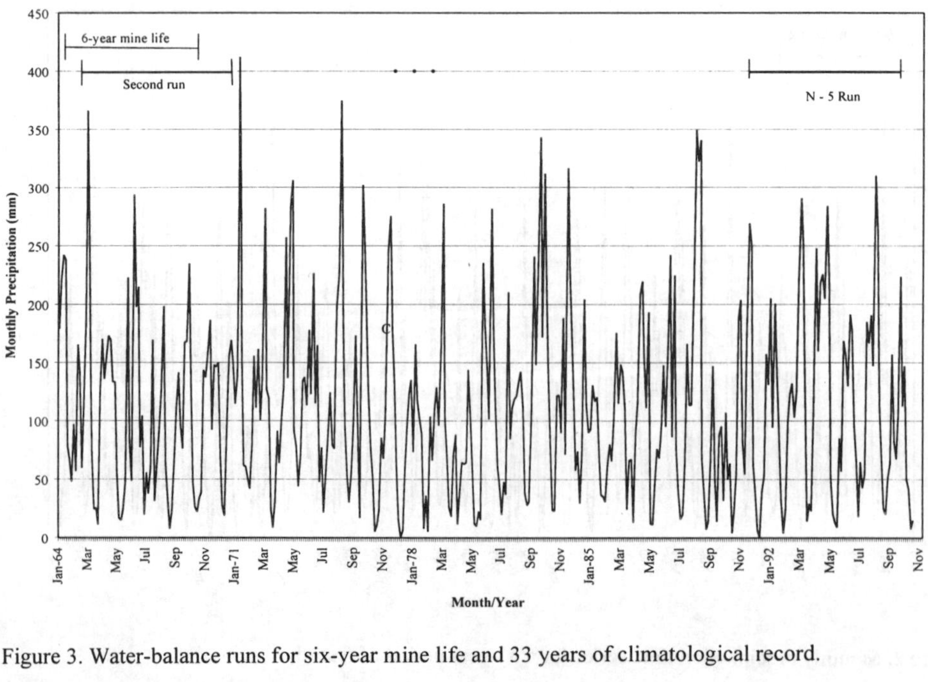

Figure 3. Water-balance runs for six-year mine life and 33 years of climatological record.

into usable design values. Univariate statistics are defined by the mean (average), variance (or standard deviation), maximum, and minimum values for a given set of events.

Figure 4 shows the required pond volume for six years of heap leach pad operation based on using all 33 years of monthly climatological record. Figure 4 shows the average, maximum, and minimum monthly values for the six-year project operation period. Figure 4 indicates that

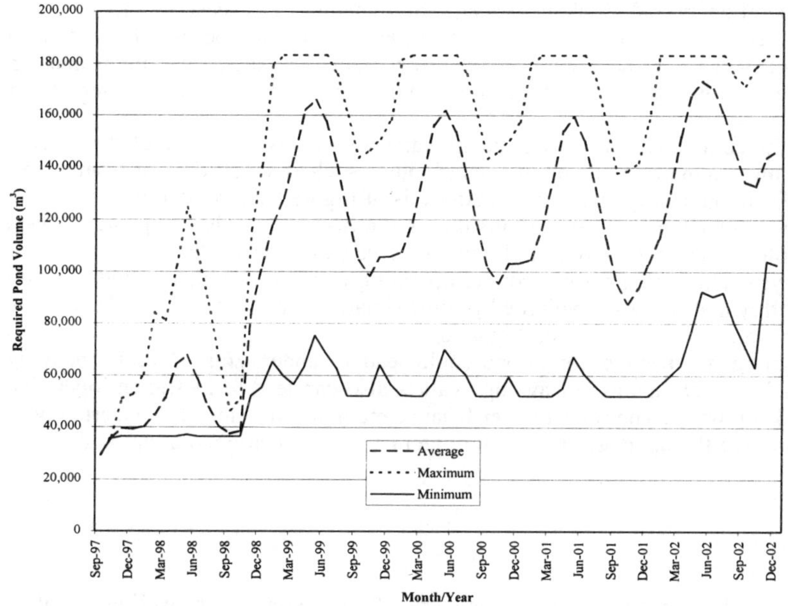

Figure 4. Average, minimum, and maximum required pond storage volume.

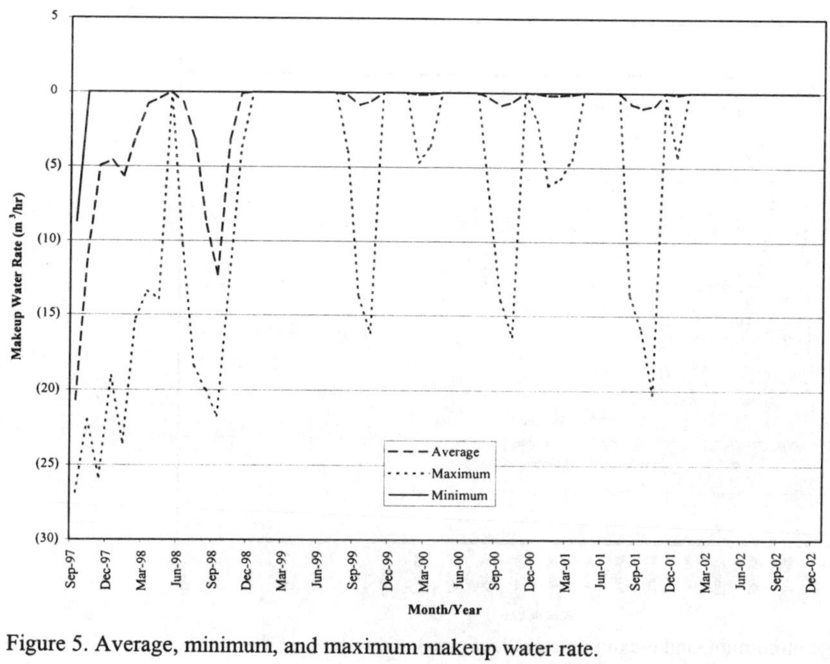

Figure 5. Average, minimum, and maximum makeup water rate.

the maximum storage does not exceed the existing pond capacity of the 182,500 m³ of fluid. Also, the water-balance model predicts that the minimum pond volume required is approximately 37,000 m³ at the beginning of the six-year operation and increases to over 100,000 m³ near the end of the operation period (Figure 4). Therefore, the water balance analysis helps set operational criteria for the operating pond(s) in order to minimize the makeup water requirements.

Figure 5 shows the monthly average, maximum, and minimum makeup water rate for the six-year project operation period. Makeup water requirements are largest in the first few months of the expansion project (over 25 m³/hr) because the ore must be initially wetted to bring its moisture content up to the leaching moisture content. The larger maximum makeup water requirements occur during the end of the Peruvian dry season in the months of September and October.

Figure 6 shows the cumulative monthly makeup water volume over the life of the project. This makeup water must come from either an outside source such as a well field or from a storage pond separate from the operational pond(s). Analysis of Figure 6 shows that the maximum cumulative makeup of 80,000 m³ of water occurs during the first six months of operation when operational pond volumes are generally low (Figure 4). This demonstrates that there is a combination of the 33-year climatological record and operating conditions which would increase the cumulative monthly makeup water volume by approximately four or five times the average cumulative makeup water volume shown on Figure 6.

Therefore, designing for average conditions could lead to undersizing of both operating ponds and other facilities. Designing for average values also can lead to inefficient operation and shortage of makeup water. The robust water-balance method can reduce the uncertainty of poor design and improve the understanding of the interaction between operational and climatological variables.

5 CONCLUSIONS

Using a robust water-balance method to size storage ponds and reservoirs as well as to calculate operational water demands is important to assess the interactions between operational and

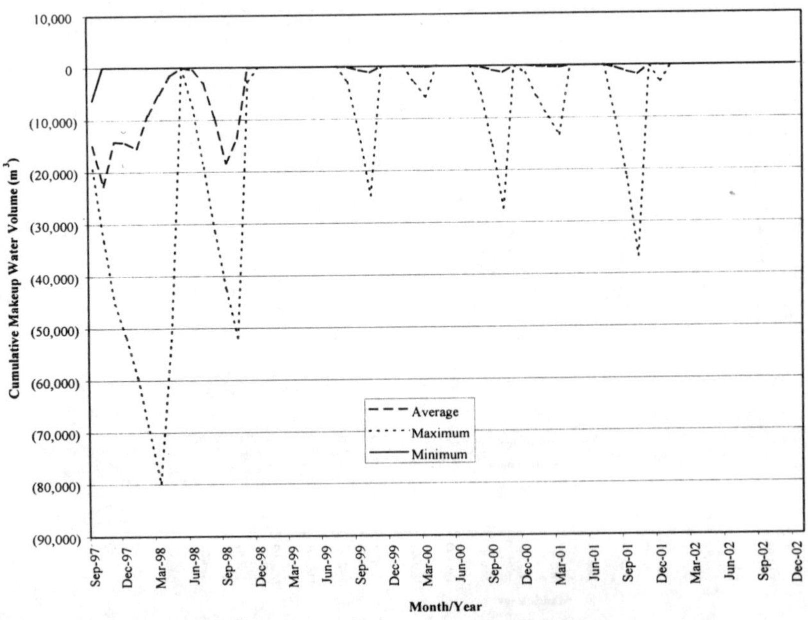

Figure 6. Average, minimum, and maximum cumulative makeup water volume.

climatological variables for the project. Typically, operational variables play a much larger role than climatological variables in design of the water management facilities.

In summary, the following conclusions are made related to use of the robust water-balance approach.

- Let the water-balance calculations begin in every year of the climatological record over the life of the project.
- Include operational variables in addition to climatological variables.
- Use univariate statistics to summarize the results and select design values.

REFERENCES

Thornthwaite, C.W. & J.R. Mather 1955. The water balance. Drexel Institute of Technology, Laboratory of Climatology, *Publications in Climatology*, volume 8, number 1, Centerton, New Jersey

Thornthwaite, C.W. & J.R. Mather 1957. Instructions and tables for computing potential evapotranspiration and the water balance. Institute of Technology, Laboratory of Climatology, *Publications in Climatology*, volume 10, number 3, Centerton, New Jersey

Tailings and Mine Waste '01 © 2001 Balkema, Rotterdam, ISBN 90 5809 182 1

Slope prediction for thickened tailings and pastes

F. Sofra & D.V. Boger
Particulate Fluids Processing Centre, University of Melbourne, Vic., Australia

ABSTRACT: Prediction of the slope formed by the deposition of dewatered tailings allows the residue disposal area to be sized to ensure maximum storage efficiency and stability. A small scale inclined plane apparatus was used to identify and understand the parameters affecting the deposition behaviour of dewatered mineral tailings. The scope of parameters examined included rheological properties such as yield stress and viscosity, operational parameters such as depositional flowrate, and the topography or slope of the underlying base. An industrial tailings slurry, a synthetic swelling clay suspension and a mineral slurry were used to develop a model relating the material properties and operating conditions to the depositional slope formed.

1 INTRODUCTION

Technological advancements in mineral processing have increased the feasibility of refining low grade ores. Because of this progression and continual depletion of high grade ore deposits, much larger volumes of waste material are produced. At the same time, increasing environmental legislation and cost competitiveness dictate that these wastes must be disposed of in a manner that is both environmentally acceptable and economically viable.

Conventional 'wet' methods of tailings disposal are problematic because of the ever present risk of ground water contamination and difficult rehabilitation. Furthermore, tailings dams are at risk failure due to leakage, instability and liquefaction. Since 1970, there have been 31 *major* tailings dam failures worldwide, resulting in at least 456 human deaths and untold environmental damage (WISE 2000). In addition to safety and environmental considerations, there is a distinct worldwide trend towards minimising water use and storage in order to reduce operational costs and increase the profitability of mine operations.

The means to dramatically reduce dam failures and the subsequent environmental impact is available in the form of dry disposal. Dry disposal techniques present significant safety advantages over conventional wet disposal.

Dry disposal requires that tailings be dewatered prior to disposal. Schemes such as dry stacking, thickened tailings disposal and paste fill all involve the deposition of dewatered tailings to improve water and reagent recovery and decrease tailings volumes and footprint. This reduction in the tailings footprint will increase the ease with which the site can be rehabilitated.

Dry stacking involves the progressive deposition of dewatered tailings onto sloped, underdrained drying beds. Once a drying bed is covered to the desired depth, the discharge point is moved to an adjacent bed to allow the freshly covered area to dry via evaporation and underdrainage (Glenister and Abbott, 1989). When all drying areas are covered, the process is repeated by returning to the original (now dry) bed and depositing a fresh layer of tailings. The dry stacking method allows the timing and depth of successive depositions to be varied to accommodate changes in climate and material characteristics (Chandler, 1987).

Thickened tailings disposal involves deposition of the dewatered waste material from a central disposal point to form a tailings stack of conical geometry (Robinsky, 1975). The formation of such a cone is advantageous as the need for high dam walls is eliminated and the stability of the deposit is enhanced through improved consolidation. In addition, problems associated with steep tailings slopes and the surface ponding of slimes are avoided. Land utilisation is improved as higher tailings densities are achieved compared with conventional disposal methods. The deposit is fully drained and consolidated at decommissioning so the reclamation procedure is simplified (Chakravatti et al., 1982).

Paste fill is used for backfilling underground mines in order to ensure adequate support properties for the overlying land and to reduce the surface placement of tailings. The thickened tailings that form the paste are often mixed with small amounts (3-5%) of cement to increase the paste's strength characteristics (Brackenbush and Shillabeer, 1998). The paste usually flows in a pipe by gravity down a mineshaft and then horizontally to the area being filled. Once deposited, the material must spread sufficiently to fill the void whilst quickly consolidating to provide the required strength to support overlying land and enable mining of adjacent areas.

Dry disposal represents a paradigm shift in engineering. The tailings are engineered to suit the disposal requirements, as opposed to engineering a facility to accommodate the tailings. It is therefore important to understand how the material properties and operational parameters can be modified to produce the desired tailings deposition characteristics.

The prudent design of a dry disposal system, whether it be dry stacking, thickened tailings disposal or paste fill, requires prediction of the slope formed by the deposited tailings. Prediction of the slope formed by the tailings allows the residue area to be sized to ensure maximum storage efficiency and stability. If the relationship between the tailings' rheology, operating conditions and the depositional behaviour is known, the desired slope of the deposit can be achieved by manipulating the operating or rheological parameters.

An inclined plane apparatus has been used to identify and understand the parameters affecting the deposition behaviour of dewatered mineral tailings. The scope of parameters examined include rheological properties such as yield stress and viscosity, operational parameters such as depositional flowrate, and the topography or slope of the underlying base. An industrial tailings slurry, a synthetic swelling clay suspension and a mineral slurry were used to develop a model relating the material properties and operating conditions to the depositional slope formed.

2 RHEOLOGICAL CHARACTERISATION OF MATERIALS

Most thickened tailings slurries and pastes exhibit non-Newtonian flow behaviour above a certain solids concentration. Thickened tailings display a yield stress which is the limiting stress below which irreversible deformation and flow does not occur. Above the yield stress, the viscosity of the tailings may not be constant with shear rate and/or time of shear. Pseudoplastic or shear thinning slurries show a reduction in the viscosity with increasing shear rate while for thixotropic materials, the viscosity and, more significantly, the yield stress decreases with increasing time of shear.

Shear thinning is attributed to the alignment of particles or flocs. An increase in the shear rate from rest results in the alignment of particles in the direction of shear, therefore providing a lower resistance to flow. Thixotropy is generally considered to be the result of shear induced destruction of a structural network of interparticle bonds.

To ensure that a wide range of material properties were available for slope prediction experiments, three suspensions with varying degrees of pseudoplasticity and thixotropy were used.

Rheological characterisation of both red mud and a titanium dioxide suspension was undertaken using a capillary viscometer. The Laponite suspension was characterised using a controlled stress cone and plate viscometer. The yield stress of all three materials was determined using the vane technique (Nguyen, 1985).

2.1 *Red Mud*

The red mud tailings were obtained from the Alcoa Kwinana alumina refinery in Western Australia. Samples were collected from the superthickener underflow and from the pipeline discharge to determine the extent of thixotropic breakdown occurring during pipeline transport. The yield stress of the underflow was found using the slump test to be 75.2Pa and this reduced to 45.3Pa when the material reached the final discharge point. The solids concentration of both samples was found to be identical. The pH of the red mud was 13 and the solids density was measured to be 3100 kg/m^3.

Capillary rheometry provided data to construct flow curves. In order to observe the thixotropic nature of the material, the runs were repeated several times with rapid mixing between each run. Figure 1 illustrates the typical shear thinning and thixotropic nature of red mud flow behaviour. Although the viscosity and, more significantly, the yield stress reduce with time of shear, the shape of the flow curves is relatively constant.

For each solids concentration, although the yield stress (and viscosity) reduce with the time of shear, the flow curves maintain a constant shape. As such, the measured yield stress can be used to vertically shift the flow curve so that it intersects the shear stress axis at the yield stress. This observation reduces time required for rheological characterisation. Figure 2 shows the yield stress - solids concentration relationship for red mud in the initial (unsheared) and the equilibrium (fully sheared) states.

The flow behaviour of red mud tailings is best described using a Herschel-Bulkley model, which contains a yield stress term followed by a power law term to describe the changing viscosity with shear rate. However, for the purposes of this work, the Bingham model provided an acceptable approximation to the flow behaviour. The Bingham model contains a yield stress term (τ_y) followed by a constant viscosity term (η) as shown in Equation 1. τ is the shear stress and γ is the shear rate.

$$\tau = \tau_y + \eta\gamma \qquad (\tau > \tau_y) \qquad (1)$$

During an inclined plane test, the tailings initially flow with a high velocity and slow before coming to rest. Because of this large variation in the velocity, it was necessary to use a constant, representative Bingham viscosity in analyses rather than a shear rate dependent viscosity.

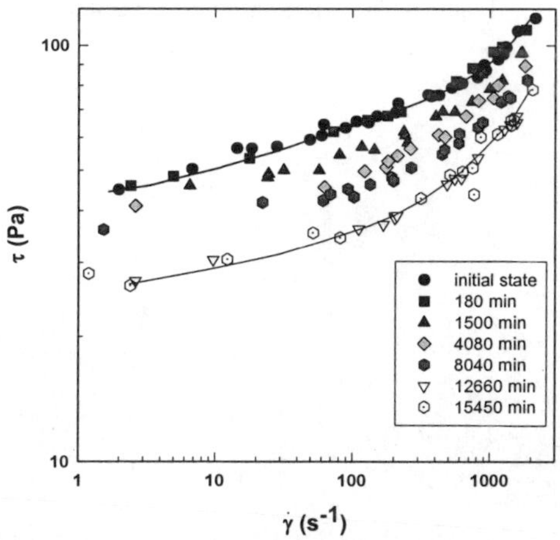

Figure 1. Flow curves for 47wt% solids red mud

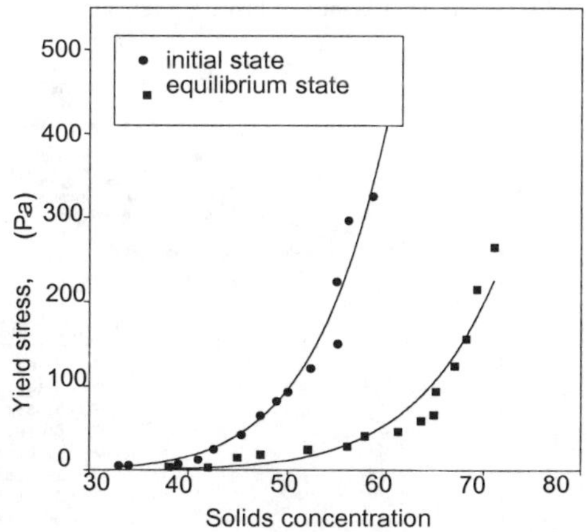

Figure 2. Red mud yield stress vs concentration

2.2 *Titanium Dioxide Suspension*

Titanium dioxide (TiO_2) is a pigment used in the paint industry and was used in this study to model a flocculated mineral suspension. The pigment used has a solids density of 4100kg/m^3. The flow behaviour of a 60wt% solids Titanium Dioxide suspension was determined and is illustrated in figure 3. The suspension was found to be a yield stress material and was shear thinning and non-thixotropic. The Bingham model provided an acceptable fit to the data. Figure 4 shows the yield stress - concentration profile.

2.3 *Laponite*

Laponite RD (Hydrous Sodium Lithium Magnesium Silicate) is a synthetic layered silicate that swells in water to give a transparent colloidal dispersion. It is used industrially as a thickener for toothpaste, paints and cosmetic formulations. Laponite was included in this study as it is very highly thixotropic and shear thinning and represents an extreme case of these rheological attributes. As shown in figure 5, thixotropic breakdown occurs on a very short time scale. Also observed, but not shown here, is the fast thixotropic recovery of Laponite to its original rheological state. As with red mud and TiO_2, the Bingham model provided an acceptable fit to the flow data.

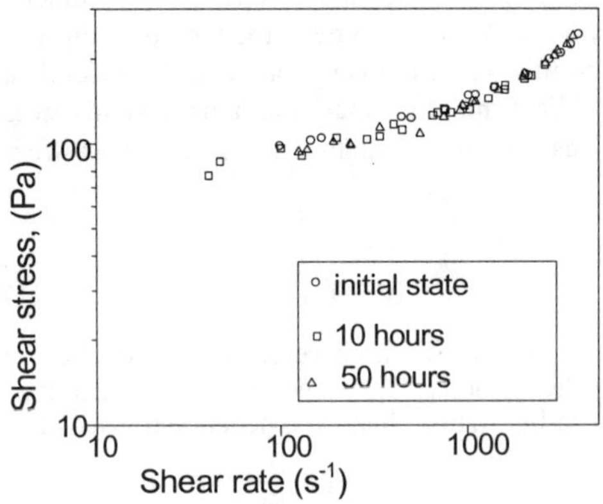

Figure 3. Flow behaviour of 60wt% TiO_2.

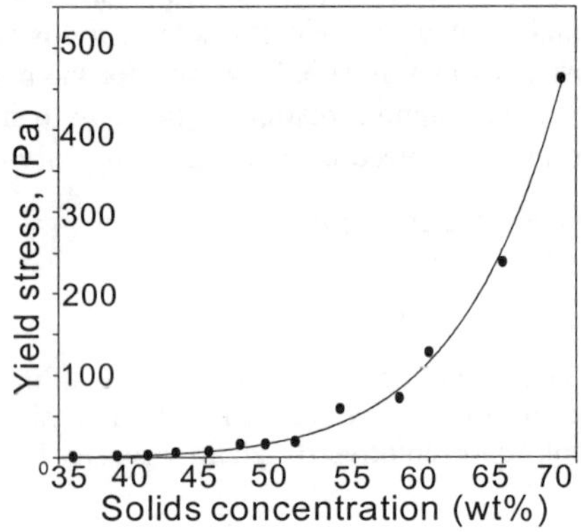

Figure 4 TiO_2 yield stress vs concentration

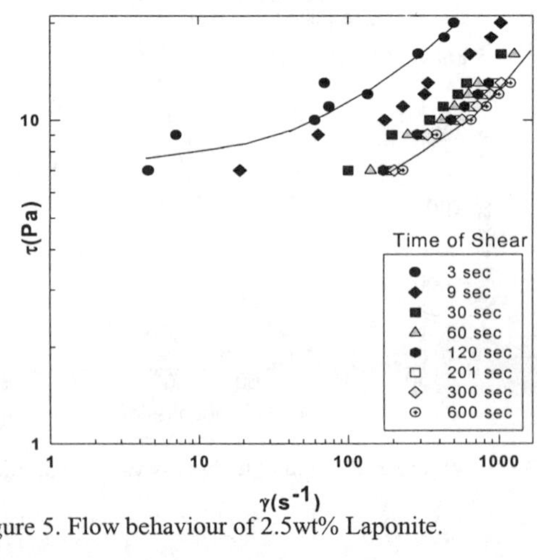

Figure 5. Flow behaviour of 2.5wt% Laponite.

3 INCLINED PLANE TESTS

3.1 *Apparatus and testing procedure*

Figure 6 shows a schematic diagram of the inclined plane apparatus used and the shape of the stationary fluid on the inclined plane. Prior to a run, the density of the material was determined. For inclined plane testing, a sample was placed in the fluid reservoir and the height of the material in the reservoir recorded. The angle of the base, θ is determined to the nearest 0.1° using a digital level. Once the gate is lifted, the material flows down the plane and comes to rest at a natural angle of repose, θ_r. Once flow had ceased, the angle of repose of the material (relative to the plane surface) is measured. The length of flow and the fluid height at the origin and the distal end (toe) are measured for determination of this angle as indicated in figure 6. A sample of the material is taken for immediate determination of the yield stress using the vane technique.

Tests were conducted for base angles ranging from 0 to 4 degrees. Materials with yield stresses in the range of 17 to 210Pa were used. The Bingham viscosities ranged from 0.006 to 0.035Pa.s.

3.2 *Inclined Plane Results*

Figure 7 shows the relationship between the angle of repose and the yield stress for a constant volume (constant initial height in the reservoir) of red mud deposited onto a flat plane. Both titanium dioxide and Laponite exhibited a similar linear relationship between the angle of repose and the yield stress, however, the gradients of the relationships for the three materials differed.

Inclined plane results presented in figure 8 show the angle of repose to be a consistent function of the base angle and the material yield stress for red mud. Similar results were obtained for the titanium dioxide and Laponite suspensions. As the base angle is increased for a given yield stress, the angle of repose decreases due to the greater apparent force of gravity. The initial volume of fluid in the reservoir was similar for all measurements shown.

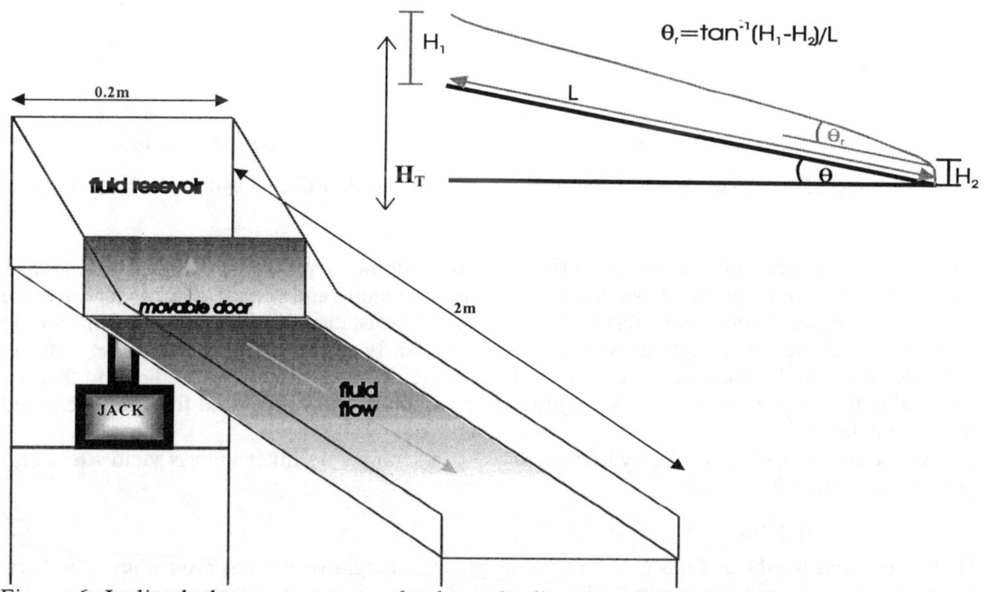

Figure 6. Inclined plane apparatus and schematic diagram of stationary fluid on the inclined plane.

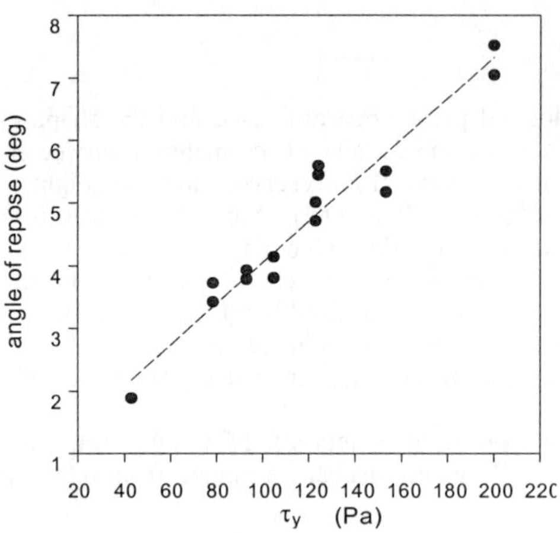

Figure 7. Angle of repose vs yield stress for red mud deposited onto a flat plane, const H_i.

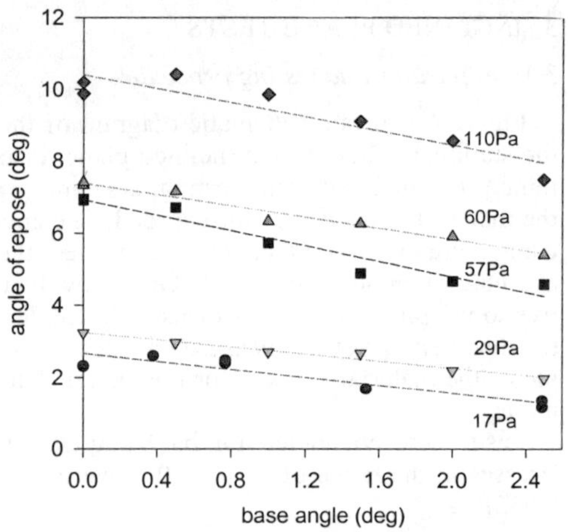

Figure 8. Angle of repose vs. base angle for red mud at various yield stresses, constant H_i.

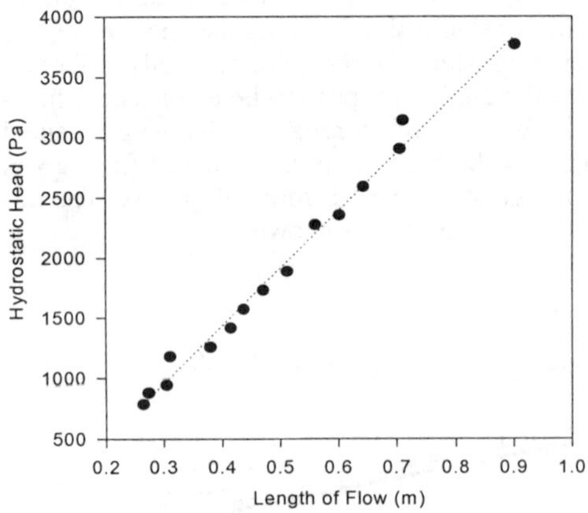

Figure 9. Effect of hydrostatic head on flow length.

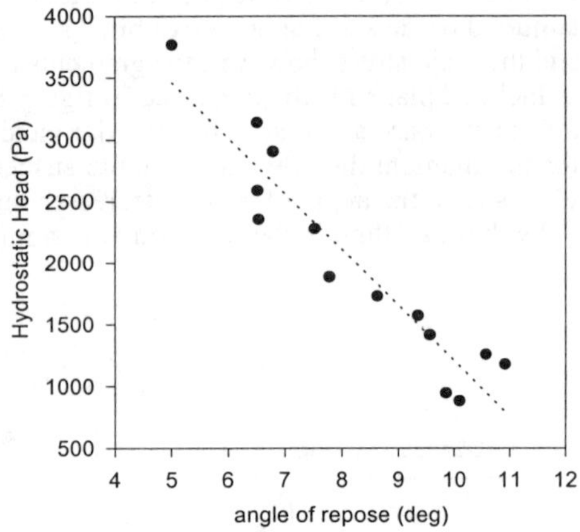

Figure 10. Effect of hydrostatic head on angle.

An interesting observation arises when the reservoir volume is varied by filling it to a greater height. If the angle of repose depended on only the base angle and the yield stress, an increase in the volume deposited would result in a greater flow distance (fig. 9), but the fluid would come to rest at the same angle of repose. Figure 10 clearly shows that this is not the case. As the volume of fluid deposited increases, the final angle of repose decreases, indicating that the material at the distal end of the flow is subjected to a greater 'pushing' force from the increased hydrostatic head.

In order to take the effect of the hydrostatic head into account, a dimensionless yield stress (τ_y') shown in equation 2 was used;

$$\tau_y' = \tau_y / \rho g H_T , \quad (2)$$

H_T is the initial height of fluid in the reservoir plus the height of the reservoir above the horizontal as shown in figure 6. ρ is the material density and g is gravitational acceleration. Figure 11 shows the variation in the angle of repose with dimensionless yield stress (τ_y') for two samples of red mud, titanium dioxide suspensions and Laponite suspensions. The dimensionless yield stress was varied by changing the sample yield stress, the volume of fluid used or the angle of the base.

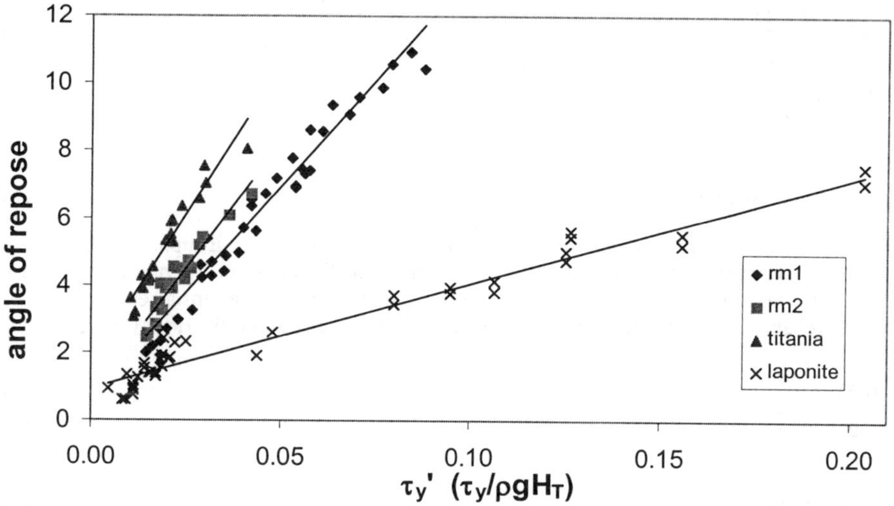

Figure 11. Angle of repose vs dimensionless yield stress for Red mud, Titanium dioxide and Laponite slurries

Table 1. Parameters for dimensional analysis

Parameter	Symbol	Units	Dimensions
Angle of repose	θ	degrees	1
Plane width	W	m	L
Flow Velocity	v	ms^{-1}	Lt^{-1}
Slurry density	ρ	kgm^{-3}	ML^{-3}
Yield stress	τ_y	Pa (Nm^{-2},$kgm^{-1}s^{-2}$)	$Mt^{-2}L^{-1}$
Viscosity	η	Pa.s ($kgm^{-1}s^{-1}$)	$ML^{-1}t^{-1}$
Gravitational accel.	g	ms^{-2}	Lt^{-2}

It is evident from figure 11 that the angle of repose is a unique function of the dimensionless yield stress for each of the materials. The relationship indicates that the forces experienced by the free surface of the flow determine the angle of repose for a given material. The free surface of the flow will come to rest at an angle which is governed by the balance between the force required for flow to occur (the yield stress) and the 'pushing force' determined by the hydrostatic head. However, this relationship differs for the three materials, indicating that there is a material property not accounted for by the dimensionless yield stress alone.

4 DIMENSIONAL ANALYSIS

A dimensional analysis was carried out in order to investigate the material and geometrical parameters important in the inclined plane tests. Dimensional analysis allows the results and trends from the small scale situation to be compared to industrial observations. The parameters used are listed in table 1.

The dimensional analysis indicated that the important dimensionless groups were the dimensionless yield stress (τ_y'), the Reynolds number (Re) and the Froude number (Fr), shown in equations 3,4 and 5.

$$\tau_y' = \frac{\tau_y}{v^2 \rho} \tag{3}$$

Note: This definition of the dimensionless yield stress is numerically identical to the alternative definition given in equation 2.

$$\mathrm{Re} = \frac{Wv\rho}{\mu} \tag{4}$$

$$Fr = \frac{v^2}{Wg} \tag{5}$$

so,

$$\theta = f\left(\tau_y^{'}, \mathrm{Re}, Fr\right) \tag{6}$$

The instantaneous initial velocity was used in all of the dimensionless groups. In reality, the velocity varies between this initial value and zero, as the fluid flows down the plane and comes to rest. The initial velocity was calculated by assuming that all of the potential energy (E_p) of the fluid in the reservoir is converted to kinetic energy (E_k) (ie negligable frictional losses).

$$E_p = E_k \tag{7}$$

$$\rho g H_t = \tfrac{1}{2}\rho v^2 \tag{8}$$

therefore,

$$v = \sqrt{2gH_T} \tag{9}$$

Where H_T = initial height of fluid in the reservoir relative to the horizontal ie the height of fluid in the reservoir added to the height of inclination of the plane.

Having determined the relevant dimensionless groups, inspection of the parameters included indicates the appropriate form of the relationship between these groups and the angle of repose as shown in equation 10.

$$\theta = f\left(\frac{\tau_y^{'} Fr}{\mathrm{Re}}\right), \quad = f\left(\frac{\tau_y \eta}{\rho^2 W^2 g v}\right) \tag{10}$$

Equation 10 shows that the angle of repose is governed by the balance between the parameters opposing flow (τ_y and η) and those contributing to flow. Application of the dimensional analysis to the inclined plane data in figure 12 shows that all of the data lie on a straight line with a correlation coefficient of approximately 0.95. The data points encompass a wide range of slurry densities, viscosities, yield stresses, flow velocities and base angles. The good fit indicates that the pertinent material and geometrical parameters have been included for the inclined plane case.

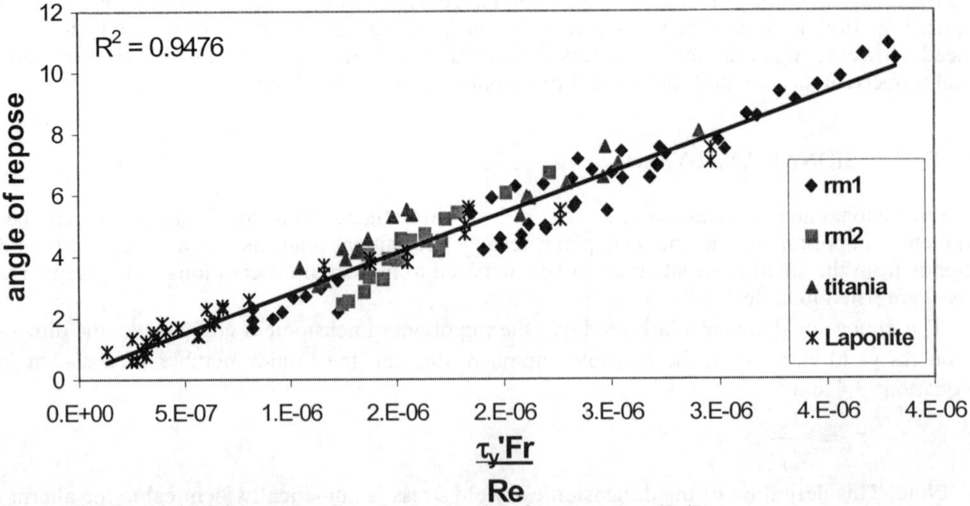

Figure 12. Results of dimensional analysis. Angle of repose vs τ_y'Fr/Re

The results presented here indicate that a knowledge of the material rheology, the slope of the underlying base and the depositional velocity allows the angle of repose to be predicted. However, for this analysis to be applied to a tailings disposal situation, the effect of lateral spreading needs investigation.

5 CONCLUSION

The angle of repose formed by thickened slurries deposited onto an inclined plane can be predicted based on a dimensional analysis. The angle of repose for materials of varying rheology and density has been found to be a function of the dimensionless yield stress, the Reynolds number and the Froude number. It is anticipated that the model will be a useful tool in the prediction of the angle of repose formed during the disposal of thickened tailings and pastes.

REFERENCES

Brackebusch, F. and Shillabeer, J. (1998) Use of paste for tailings disposal. *Minefill '98*. 6th Int. Symp. on mining with backfill. AusIMM Publication No 1/98 :53-58.

Chakravatti, J.L., LaRocque, E., Reades, D.W., and Robinsky, E.I. (1982) Thickened tailings experiment for close out of uranium mill tailings at Denison Mines Limited. *Management of wastes from uranium mining and milling*. Viena:IAEA. :417-428.

Chandler, J.L. (1987) The Stacking and Solar Drying Process for Disposal of Bauxite Tailing in Jamaica. *Bauxite Tailings - Proceedings of an International Conference*. Wagh, A.S. and Desai, P. (Eds) Kingston, Jamaica:101-105.

Glenister, D.J. and Abbott, T.M. (1989) Dewatering and dry disposal of fine bauxite residue. *Dewatering Technology and Practice Conference*; Aust. Inst. Min. Met.:105-110.

Nguyen, Q.D. and Boger, D.V. (1985) Direct Yield Stress Measurement with the Vane Method. *Journal of Rheology* 29(3):335-347.

Robinsky, E.I. (1975) Thickened Discharge-A new approach to tailings disposal. *Canadian Mining and Metallurgical Bulletin* 68:47-53.

WISE (2000) Chronology of major tailings dam failures. WISE World Information Service on Energy, Uranium project. Internet Publication.

Tailings and Mine Waste '01 © 2001 Balkema, Rotterdam, ISBN 90 5809 182 1

Tailings and disposal brine reduction – Design criteria for potash production in the 21st century

Henry A. M. Rauche & Dietrich Fulda
ERCOSPLAN Ingenieurgesellschaft Geotechnik und Bergbau mbH, Erfurt, Germany
Volker Schwalm
ERCOSPLAN Ingenieurbuero Anlagentechnik GmbH, Erfurt, Germany

ABSTRACT: Potash fertilizer production from underground extracted ore is compound with a relative high amount of solid and liquid waste materials and with surface subsidence above the mines. Commonly, the solid residues are stored on tailings piles at the surface and the disposal brines are injected in the underground by wells. The need for food supply for a growing world population will necessitate a further increase of fertilizer production in the future. That requires strategies for the reduction of these environmental impacts too. A reduction of both the surface subsidence and the disposal of residues at the surface is possible by backfill of the mine openings, as realised in some mining districts since the early years of potash production. In this connection the slurry backfill technology (in German: *Spülversatz*) posses remarkable advantages and may upgrade the economics of potash production.

1 INTRODUCTION

Since the first potash fertilizer production started in 1861 in the Stassfurt region (central Germany), the worldwide annual production has increased up to over 25 million tons of K_2O in the last three decades. The need for food supply for a growing world population will necessitate a further increase of fertilizer production in the future.

Unlike to other mining branches, decreasing ore reserves do not limit the growth of potash fertilizer production. The economy of potash ore extraction and processing in the classical mining districts is strongly influenced by operational and transportation costs, more than by geological parameters. Important influencing factors result from the environmental impact of the underground extraction and the ore processing. In the case of underground mining extraction, this impact is mainly determined by solid and liquid processing residues at the surface and the time-dependent subsidence of the surface above the mines (for example see Bodenstein et al., this volume).

During the last decade, some potash producers have reduced these effects step-wise by several activities. For example, in the Werra Production Centre of the unified German Potash Industry, the salt content of the river Werra was reduced by introduction of modern processing technology after Germany´s political re-unification.

In general, a further reduction of the environmental impact is possible by a more efficient waste management coupled with consistent backfill of the extraction rooms. This technology reduces both the tailings piles at the surface and the surface subsidence above the mines. Firstly, this contribution will explain the processing technologies on two common examples and their control by the ore quality, respective the chemical composition of the ore. Secondly, we will illustrate the mass and volume balance between mine openings and residues. Finally, we will discuss the positive effect of the slurry backfill technology on both the environmental impact of potash processing as well as the economics of potash mining.

2 COMPOSITION AND PROCESSING OF POTASH ORE

Modern potash production is not simply an extraction and processing business alone. The environmental impact and the economics of modern potash production as well, are strongly influenced from the tailings and disposal brine management. The reason behind this comes from the chemical, respective mineralogical composition of the potash ore.

This fact should be explained by the following figures: the most important potash mineral is sylvite (KCl) with an mineral K_2O content of 63.2 mass percent, followed by carnallite (KCl x $MgCl_2$ x 6 H_2O) with an mineral K_2O content of 16.9 mass percent. Potash deposits around the world contain sylvinitic and/or carnallitic crude salt with K_2O contents in the range from 8 up to 30 mass percent. These figures illustrate the amount of waste material, which is separated from the potash components by mineral salt processing, e.g. hot by leaching and/or flotation of the extracted crude salt. These processing residues are liquid (disposal brine) and solid (tailings). Their specific amount per each ton of product depends on the mineralogical composition of the crude salt and/or the processing technology, and will illustrate by two examples:

Example A (Fig. 1a) is a low-grade carnallitic ore with an average K_2O content of 9.0 mass percent from a sulphate-type deposit (e.g. Zechstein 2, Germany). The main component and the ore mineral is carnallite. The sulphate minerals are kieserite ($MgSO_4$) and anhydrite ($CaSO_4$). Additionally tachyhydrite ($CaCl_2$ x $MgCl_2$ x 12 H_2O) occur also. The normative sylvite content is zero.

Example B (Fig. 1b) is a high-grade sylvinitic ore with an average K_2O content > 22.0 mass percent from a chloride-type deposit (e.g. Russia, western Canada, north central Thailand). The main component is halite. The ore minerals are sylvite and carnallite. Small amounts of unsolvable clay material and a small brine content in the pore volume occurs in both ores.

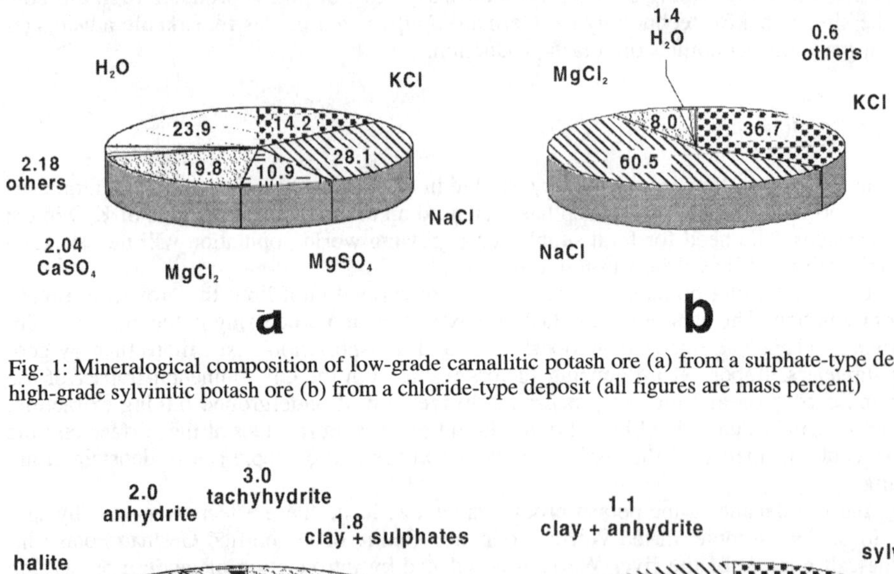

Fig. 1: Mineralogical composition of low-grade carnallitic potash ore (a) from a sulphate-type deposit and high-grade sylvinitic potash ore (b) from a chloride-type deposit (all figures are mass percent)

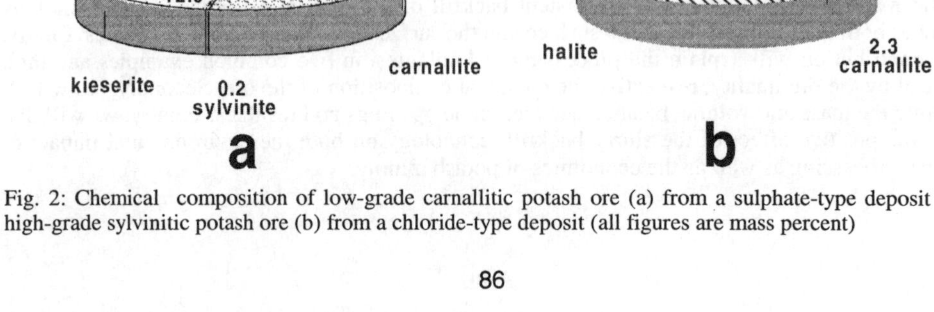

Fig. 2: Chemical composition of low-grade carnallitic potash ore (a) from a sulphate-type deposit and high-grade sylvinitic potash ore (b) from a chloride-type deposit (all figures are mass percent)

Table 1. Composition of solid and liquid residues from the processing of low-grade carnallitic ore.

	KCl	NaCl	$MgSO_4$	$MgCl_2$	$CaSO_4$	Other	H_2O	*Total*
Disposal brine (g/l)	17	16	31	362.6	-	-	879.9	1,306.5
Wet solid residues (kg/t)								
Fine-grained	7.5	63.1	21.3	11.9	5.0	4.7	38.5	151.8
Coarse-grained	3.9	185.6	62.5	8.5	14.9	14.1	45.4	334.8

The main differences of these ores are given by the percentage of valuable material (K and/or Mg) and waste material, in general shown by the K_2O content, and by the mineral bound water content. These differences are shown by the chemical compositions (Fig. 2). According these differences the processing for both ores needs different technological steps. The main technological step to extract the valuable potassium content are described briefly in the following subchapters.

2.1 *Hot Leaching for the Processing of Low-grade Carnallitic ore*

This process requires an complete dissolution of the crude salt by aqueous solution. The result of the first processing step, the dissolution process himself, and the following cooling and crystallisation is KCl and a mother liquor, for this example here it contain 52 g/l of KCl and 292 g/l $MgCl_2$ as main chemical components. Evaporation of the surplus of the mother liquor by an further processing steps leads to the crystallisation of the valuable potassium as artificial carnallite. The final product results from decomposition of this carnallite. Residues of this processing are 288 m³ wet solids with an high pore volume and 399 m³ disposal brine per 1000 tons of processed carnallitic ore.

2.2 *Flotation for the Processing of High-grade Sylvinitic ore*

The backbone of this processing technology is flotation. The separation starts after repulping of the crushed crude salt with process brine and leads to a KCl enriched concentrate and flotation residues. Leaching by water effect an enrichment of KCl up to the final product. This leaching water and further water sources leads to a surplus of processing brine which should be disposed. In our example, the amount of disposal brine is 268 m³ per each ton of processed sylvinitic ore and the amount of wet solid flotation residues with an high pore volume is 494 m³ per 1000 tons of processed sylvnitic ore.

3 VOLUME BALANCE OF EXTRACTED ORE AND PROCESSING RESIDUES

The mass and volume balance between extracted ore, products and residues is shown in Fig. 3. At first, the figures show that the amount of product is relative small compared with the amount of wet and liquid waste material. These relations are strongly controlled by the quality of the ore. The comparison of both processing technologies shows that the amount of disposal brine is higher for the carnallitic ore, as expected in respect to the chemical composition. The amount of wet solid residues is higher for the sylvinite processing by flotation. In both cases the solid residues are wet and contain absorbed brine from the processing. The pore volume of these processing residues is in the range from 40 up to 55 percent and depends from the grain size and the grain size distribution.

These figures illustrate too that the volume of the extracted ore in the mine is much more smaller than the sum of the final potash product and the different waste materials. Because processing needs additional water and leads to a decomposition of hydrate minerals the volume of the solid residues and the disposal brine is also larger than the volume of the extracted ore. The high and mainly air-filled pore space of the wet residues cause the large volume of the initial tailings. In case of the high-grade sylvite processing this initial residue volume is larger than the volume of the extraction rooms in the mine. For these reasons, present potash production needs tailings piles for the wet solid residues and underground injection wells and/or other solutions for the disposal brine problem.

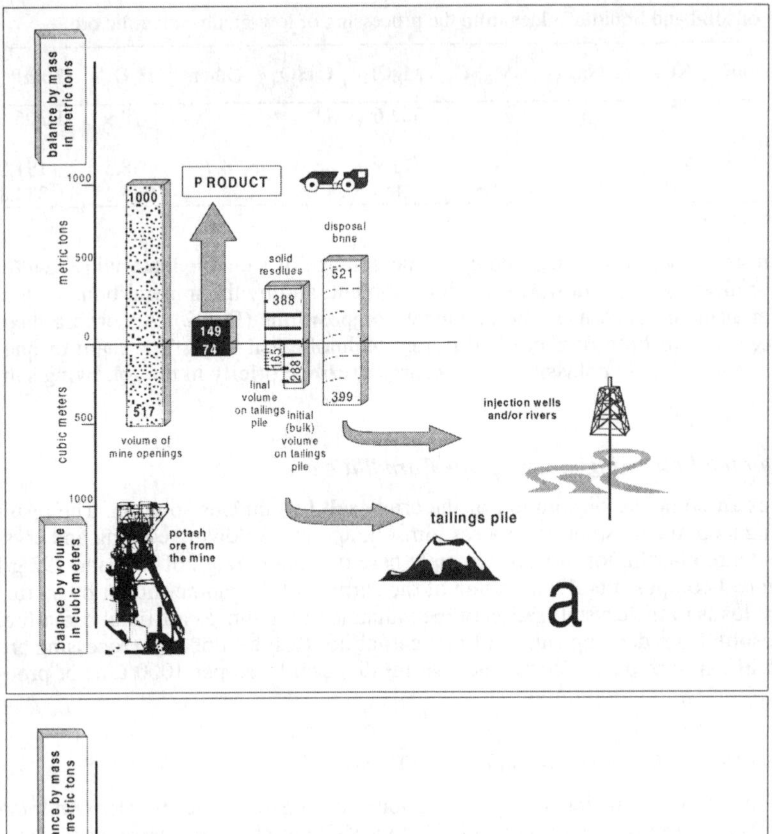

Fig. 3: Mass and volume balance between extracted ore, final potash product, solid and liquid waste material for hot leaching of low-grade carnallitic ore (a) and flotation of high-grade sylvinitic ore (b)

4 RE-USE OF PROCESSING RESIDUES AS BACKFILL MATERIAL

Re-use of tailings for backfilling the extraction rooms started in the early years of potash industry by use of dry and wet residues. Different technologies are developed during the last 80 years. One of these, the slurry backfill technology (in German: *Spülversatz*) is a very efficient backfill method, because of its combination of tailings and brine disposal.

Since 1908, this special hydraulic backfill technology was adapted from the Upper Silesian coal mines into the young German potash industry. Since this time, the slurry backfill technology has been successful practised in the German potash industry, especially in flat-lying deposits of the South Harz potash district.

In slurry backfill technique the waste product of potash processing (e.g. halite and anhydrite) is mixed with a flushing brine and reintroduced as backfill in a liquid form into previous mine rooms. The flushing fluid comprises residual brine that is also a by-product of the processing of the crude salt and acts as the transport medium for the solid crystalline tailings. The hydraulic transport of this slurry backfill is accomplished via a pipeline system from the surface to the backfill panel where the gravitational potential energy of the introduced medium ensures its transport.

Experience from working simple slurry backfill facilities shows that in a routine mine environment with a vertical height difference of 600 m, transport over 5 km is possible. For transport to the further reaches of the mine system then a pump system needs to be installed to overcome pipe friction. In general, the horizontal transport distance without additional pumps is 8 ... 10 times more than the geodetic difference.

A turbulent flow within the pipeline system avoids the settling of the solids out of the slurry mixture and any gravitational sorting of the solid material and ensures that the quality of the backfill slurry remains constant during transport (Fulda 1966). The introduction or stowage of the backfill is achieved by an inclined build-up in the dip direction. The further lithification of the backfill is achieved through the gravitational separation of the solid from the brine.

The density of the backfill body is dependent upon the waste solid material used and the on-site condition of the mine rooms. For the waste tailings of the high-grade sylvinitic ore a density higher than 1.80 t/m^3 is expected. This density is usually achieved after several hours to days after introduction of the slurry backfill, dependent upon local working and/or geological conditions.

The lithification of the backfill body is also dependent upon the amount pore space reduction, and more importantly the crystallisation of surplus brine within the pore space. As this is time dependent, the backfill body reaches its maximum lithification only after a significant time period. In situ measurements indicate that in a routine working environment the maximum lithification of the introduced backfill is reached after few weeks to months, and is 50 to 75 % the original density of the originally mined potash and rock salt (Eulenberger 1999).

The backfill body is further compressed by convergence, leading to increased lithification and strength.

Early slurry backfill installations, created to allow the re-use of residue from German *Hartsalz* (sylvite + halite + anhydrite) processing, produced backfill densities over 1.90 t/m^3. High-grade sylvites contain lesser anhydrite than *Hartsalz*. Due to this lower anhydrite content, the density of the backfill in a high-grade sylvite mine will be less, probably over 1.85 t/m^3. The stowage capacity (degree of filling) is dependent upon from the morphology of the excavated rooms and varied in German mines between 85 % and 98 % as average value for the whole backfilled panels.

5 ADVANTAGES OF SLURRY BACKFILL TECHNOLOGY

The comparison of the material parameters of dry or paste backfill and slurry backfill shows that the first technology is not so efficient in sense of tailings and disposal brine reduction. The dry backfill method involves the stowing of artificially dampened residues into previous mine rooms. The transport of such backfill material requires a system of conveyor belts stretching from the mine head to the area being backfilled. The stowing of the dry backfill to within 15 cm of the mine rooms roof is performed using flingers. The degree of filling of the extraction rooms

Table 2: Main differences between slurry and paste backfill technology, based on the calculations for residues from flotation of high-grade sylvinitic ore.

	Paste backfill material	Slurry backfill material
In-place density	1.36 t/ m^3	> 1.80 t/ m^3
Moisture content	6 wt %	5 wt %
Brine handling	open system	closed system (circuit)
Transport technology	open conveyer belts	closed pipes
Stowing rate	up to 15 cm under roof < 90 %	80 to 95 %
Stowing technology	flingers	pipes and dams

is up to 90 % under good geometrical conditions. The in place the density of this backfill is about 1.35 t/m^3, for example of the flotation residues from a high-grade sylvinite.

The figures shown in Table 2 express the better efficiency of slurry backfill technology from the volume-mass balance point of view, which is mostly important for the environmental impact. That means, a larger amount of residues can be deposited in the mine openings by slurry backfill technology than by dry and/or paste backfilling.

Advantages on the rock mechanical effect of backfilling comes from the higher mechanical strength of slurry backfill as discussed by Bodenstein et al. (this volume) in detail. This property is induced by the lithification of the slurry material which is realised by sedimentation of residual fragments from the suspension and the crystallisation of new solids in pore space from the brine. Both processes produce an artificial salt rock, which shows mechanical and hydraulic properties like or better the natural salt deposit (Rauche & Mierdel 1999, Eulenberger 1999).

6 SUMMARY AND CONCLUSIONS

The slurry backfill technology can contribute to an reduction of the environmental impact of potash fertilizer production by two different ways: (i) The higher degree of filling, the higher density, compared with dry and/or paste backfill, and the brine-filled pore space leads to a better utilization of the mine openings by backfilling and to a reduction of tailings and/or disposal brine at the surface. (ii) The higher degree of filling paired with the higher mechanical strength of the slurry backfill body in the mine openings reduce the surface subsidence above the mines.

Furthermore, the positive rock mechanical influence of the slurry backfill material allows a reduction of the extraction losses by

- smaller pillar sizes during the primary mining phase

and/or

- pillar re-extraction during a secondary mining phase

without any risk of surface subsidence and/or instability of the mine. Both, the reduction of the tailings and brine disposal at the surface and the higher extraction ratio influence the economics of the production by lower overall operating costs and/or longer lifetime of the resources.

REFERENCES

Bodenstein, J., K.-H. Eulenberger, H.A.M. Rauche, & W. Schreiner. 2000. Reduction of Surface Subsidence and Brine Inflow Prevention in Potash Mines by Subsequent Backfilling. *This volume*. Rotterdam: Balkema.

Bodenstein, J. & Schreiner, W. 2000. Reduzierung der Bergsenkungen durch Einbringen von Spülversatz – Gebirgsmechanisch-bergschadenkundliche Aspekte der Anwendung des Spülversatzverfahrens im Kalibergbau.- Kolloquium Bergakademie Freiberg, *Freiberger Forschungshefte*, Reihe A, 891: 101-111. Freiberg.

Eulenberger, K.-H. 2000. Untersuchungsergebnisse über die Eignung von Spülversatzmaterial aus Rückständen der Kalirohsalzverarbeitung als Verfüll- und Verschlussmaterial in UTD im Salinar. Kolloquium Bergakademie Freiberg, *Freiberger Forschungshefte*, Reihe A, 891: 45-59. Freiberg.

Fulda, D. 1966. Beitrag zur Berechnung von Spülversatzanlagen im Kalibergbau. *Freiberger Forschungshefte*, A 374. Freiberg: Mining Academy of Freiberg.

Mierdel, K. 1999. Mikrostrukturen in künstlichen Salzgesteinen. *Diploma Thesis*, Jena: Friedrich-Schiller-University.

Rauche, H. & K. Mierdel. 2000. Von der Festkörpersuspension zum tragfähigen Spülversatzkörper – Einfluß der strukturellen Entwicklung auf die mechanischen und hydraulischen Eigenschaften von Spülversatzmaterial.. Kolloquium Bergakademie Freiberg. *Freiberger Forschungshefte*, Reihe A, 891: 32-45. Freiberg.

Sitz, P. 2000. Ausblick auf die weitere Nutzung und anwendungsorientierte Forschung zur Spülversatztechnologie. Kolloquium Bergakademie Freiberg. *Freiberger Forschungshefte*, Reihe A, 891: 135-147. Freiberg.

Geotechnical considerations

Tailings and Mine Waste '01 © 2001 Balkema, Rotterdam, ISBN 90 5809 182 1

Geotechnical requirements for tailings dams in Western Australia

J. Ranasooriya
Department of Minerals and Energy, Perth, W.A., Australia

ABSTRACT: Due mainly to the increased community concerns on the potential safety and environmental hazards associated with mine tailings, the tailings management practices in Western Australia (WA) were reviewed in 1989. The review identified significant deficiencies in tailings management and the need for improvements. In response to the findings of the review the regulatory approach to the tailings management in WA has been changed during the 1990s. The changes included formulation of two sets of guidelines and introduction of tenement conditions specific to tailings management. The objectives of the guidelines are to encourage a well-engineered uniform approach to tailings management based on the application of geotechnical engineering. The guidelines have been readily accepted by the mining industry and the consultants engaged in tailings management. Since the introduction of the guidelines there have been significant improvements in tailings management in WA.

1 INTRODUCTION

Mine tailings management in Western Australia (WA) can be subject to various pieces of legislation covering land rights, heritage issues, safety and the environment, depending on the location of the tailings storage site and the materials involved. In general, all mine tailings storage systems in WA are subject to the relevant sections of the following Acts of Parliament and associated regulations:

- Mining Act, 1978 (MA) and Mining Regulations, 1981.
- Mine Safety and Inspection Act, 1994 (MSIA) and Mine Safety and Inspection Regulations, 1995 (MSIR).
- Environmental Protection Act,1986, Parts IV and V (EPA) and Environmental Protection Regulations, 1987.

The MSIA and the MA are administered by the Department of Minerals and Energy of WA (DME) which is responsible for regulating the mining industry in WA. The EPA is administered by the Department of Environment Protection of WA (DEP). Under the relevant provisions of the MSIA, MA and EPA and associated regulations adequate consideration of geotechnical issues is a direct or indirect requirement to assure the structural integrity of a tailings dam. In general, DME ensures that geotechnical issues of tailings management systems are adequately addressed to meet the requirements of all relevant pieces of legislation.

A review of the tailings management practices in the 1980s (DOMWA 1989) identified significant deficiencies associated with tailings management at that time and the need for improvements. To address the findings of the review and to assist the industry to comply with the relevant pieces of legislation by better planning and managing mine tailings, DME formulated a set of guidelines on safe design and operating standards for tailings storage. Initially an interim version of the guidelines was issued in 1992. Following the review and valuable contributions by the industry and relevant specialist consultants, the guidelines were formally published in March 1996. Three years later a revised and updated version (DME 1999) was issued. In addition, a generic guideline on the preparation of site specific tailings storage operating manuals (DME 1998) was also produced. Since 1993, specific tenement conditions have been imposed under the MA to ensure appropriate construction, operation and decommissioning of all new tailings management systems with emphasis on the application of geotechnical engineering. Following the introduction of the guidelines and the specific tenement conditions, there has been significant improvement in tailings management in WA.

This paper outlines the key geotechnical requirements for tailings storage design, construction and operation and the associated administrative procedures. The paper does not attempt to discuss other requirements for tailings storage that should be complied with as per the relevant pieces of legislation administered by DME, DEP or other agencies.

2 TAILINGS STORAGE FACILITIES

In WA, the term "tailings storage facility" (TSF) includes all types of mine tailings retention and/or storage systems. There are some 360 registered TSFs in WA, of which about 75% are ring-dyke dams (locally known as paddock dams). Side-hill and cross-valley dams represents approximately 7% and 8% respectively. The remaining 10% include in-pit storage and facilities that are unclassified due to complex and short-term construction primarily associated with mineral sand mining. In general, 90% of the TSFs in WA are embankment dams.

3 GENERAL REQUIREMENTS

Before building a TSF in WA it is a regulatory requirement for TSF operators to provide a Notice of Intent (NOI) document. The NOI should address all environmental issues associated with the proposed TSF. Full NOI requirements are given in DME (1998a). When an embankment dam is constructed for tailings storage a design report must be provided as an essential part of the NOI to geotechnically justify the design and the proposed methods of construction. To determine the design philosophy, the technical input and the type of construction appropriate for a particular tailings dam, the dam is assigned a hazard rating and classified into one of tree categories (DME 1999). The hazard rating and classification are based on generally accepted standards for potential hazards assessment of dams.

3.1 *Hazard rating*

Each tailings dam is given a hazard rating of *High, Significant or Low* (DME 1999) based on the assessment of the potential impact on human life, property and the environment which may result from uncontrolled release of the stored material (eg: by failure of the dam). The hazard rating is not an assessment of the risk of failure. It is only an assessment of the potential impacts. Therefore, a well designed and well constructed tailings dam may still receive a high or significant hazard rating.

3.2 *Derivation of categories*

The hazard rating and the height of the tailings dam are used to define three categories of dams. The matrix for defining the three categories of dams is shown in Table 1. This classification system recognises that different levels of investigation, design and construction as well as operating procedures are required to manage varying safety and environmental hazards posed by tailings dams of different categories. A summary of the design, construction and operating requirements for the three categories is given in Table 2.

Table 1 Hazard rating versus height matrix to derive tailings dam categories.

Factors affecting categories		Tailings dam categories		
Hazard rating[1]		High	Significant	Low
	> 15 m	1	1	1
Maximum embankment height[2]	5 - 15 m	1	2	2
	< 5 m	1	2	3

1. Hazard ratings are assigned on the basis of potential impact on human life, property and the environment as per DME (1999).
2. All cross-valley dams, which block or significantly impede the flow in natural drainage lines, are to be considered as Category 1 dams regardless of the height.

Table 2 Design, construction and operating requirements for tailings storage dams in WA.

Requirement	Category 1	Category 2	Category 3
Design report	Prepared in detail by a geotechnical specialist	Prepared by a geotechnical specialist	Brief design report
Construction supervision by	Geotechnical specialist	Experienced person with input from geotechnical specialist	Experienced contractor
Construction report	Detailed report with as-built drawings	Brief report with as-built drawings	Optional
Operating manual	Yes	Yes	Yes
Emergency action plan	Yes	Yes	Yes
Daily inspection by site personnel	Yes	Yes	Yes
Operational audit and review by geotechnical specialists	Annually	Biennially	Triennially
Pre-decommissioning status report by specialists	Yes	Yes	Yes

4 DESIGN CONSIDERATIONS

The above-mentioned three categories of tailings dams weigh-up the potential safety and environmental impacts and provide a basis for design considerations. For Category 3 dams, generally less design and construction details are required than that of Category 1 and 2 dams.

For both Category 1 and 2 dams DME requires design documentation from suitably qualified geotechnical specialists. It is the responsibility of the designer to determine the level of technical input appropriate to the specified category of a tailings dam. In general, considerably more technical details (ie: site investigation, material testing and proposed construction procedures) are required for Category 1 dams than that for Category 2 dams. Detailed information on the technical issues that should be addressed at the design stage may be found in DME (1999) and EPA (1995). The minimum technical issues that should be covered in the design reports are outlined in the following sub-sections.

4.1 *Geotechnical investigations*

The extent of geotechnical investigations required to design a tailings dam is dependant on the category of the dam, height and length of the dam and the complexity of the local geotechnical conditions. In general, the level of investigations conducted should be sufficient to determine the local hydrogeological conditions and geotechnical properties of the foundation and construction materials. The investigations should be conducted in accordance with the relevant Australian Standards. Details of the field and laboratory techniques used and the results obtained should be provided for Category 1 and 2 dams. For tailings dams of all three categories the following information should be provided:

- geotechnical properties of the foundation materials;
- depth to the natural groundwater tables(s);
- estimated in situ permeability and details of potential seepage paths, if present;
- locations of proposed borrow areas and estimated volumes of reserves;
- geotechnical properties of the construction materials.

4.2 *Hydrological analysis*

The designs must take account of the impact of flooding in terms of external toe erosion as well as overtopping. Sufficient freeboard and, where applicable, spillways and stream or runoff diversion measures should be provided. The local and regional hydrological conditions must be considered to assess embankment stability, surface runoff erosion, reclaimed water management and decant outlets and spillway design. At least 1 in 50-year rainfall event should be considered during the operation of Category 3 dams. For assessment of flood capacity after decommissioning of Category 3 dams, a longer-term approach with a 100-year return period wet season should be used. For Category 2 dams a minimum 1 in 100-year wet season is recommended for all situations. A minimum 100-year return period wet season is recommended for the operating life of Category 1 dams and a probable maximum precipitation event for designing decommissioning measures.

If natural drainage diversion becomes necessary, all safety issues arising from water diversion measures must be suitably addressed at the design stage. Special attention must be given to minimise the potential for flooding nearby operating mines, residential areas and access roads due to a combined effect of extreme rainfall events and natural drainage diversion measures associated with the TSF. This is a general mine safety requirement, regardless of whether a tailings dam is constructed or not, as per the Regulation 4.11 of the MSIR.

4.3 *Design details*

The design of tailings dams should be based on standard methods of stability analysis used for embankment dam design. The stability analysis of Category 1 and 2 dams must consider both static and dynamic loading conditions. The tailings dam design report should include the following information:

- geotechnical parameters of embankment and foundation materials used for the analysis;
- phreatic surface(s) and/or pore water pressures considered in the analysis;
- the methods of stability analysis used and the methods adopted to incorporate the effects of dynamic loading on the stability of the embankment; and
- the minimum factor of safety obtained, the locations of the critical failure surfaces, and results of any other methods of assessment of the stability of the embankment.

Selection of suitable methods of analysis and appropriate factors of safety for both static and dynamic loading conditions are the responsibilities of the designer. As a general rule for static and dynamic loading conditions, respective minimum factors of safety of 1.5 and 1.2 are recommended for the downstream slope of a tailings dam.

In addition to the above, the design details of the following are required:

- full embankment geometry with runoff control measures;
- tailings delivery and discharge systems;
- water reclaim systems and return water holding ponds;
- seepage control measures and under-drainage systems, where applicable;
- methods of managing pipelines and procedures to contain and/or provide early warning or emergency shut-off in the event of pipeline leakage or failure; and
- instrumentation for monitoring embankment performance and environmental impact.

4.4 *Dam-break studies*

Where important installations or infrastructures such as underground mines, processing plants and public facilities are located close to a tailings dam, a dam-break study should be carried out to assess the potential impact of a failure of the dam considering the possible worst case scenarios during the active life of the tailings storage. The scope of a dam-break study should include:

- potential failure modes and probability and significance of each;
- potential quantity of material that could flow;
- how far the material would travel;
- consequences of failure; and
- strategies that could be taken to minimise the risk of failure and the potential impact.

4.5 *Decommissioning proposals*

Proposals for decommissioning of a TSF should be provided at the design stage, and should include the following:

- a staged plan to rehabilitate the embankment faces and storage surface area;
- measures to control dust generation and water erosion;
- measures to minimise surface and sub-surface water contamination in the long-term; and
- decommissioning measures for water recovery structures and under-drainage systems.

As each TSF differs from the next in terms of its mineralogy, process, management, design, climatic conditions and location, it is expected that rehabilitation solutions may also differ.

5 CONSTRUCTION QUALITY CONTROL

It is a generally accepted practice for any construction project that suitably qualified and experienced personnel should (a) ensure that the construction is performed in accordance with the design, and (b) record as-built details and the actual ground conditions found during construction. Similarly, to ensure that construction of Category 1 and 2 dams meet the design submitted for approval, DME requires suitably qualified specialists to supervise the construction and to prepare as-constructed reports.

The extent of construction supervision required depends on the complexity of the design, and the size and the hazard rating of the dam. It is the responsibility of the designer to specify the level of supervision required assuring adequate construction quality control and reporting. The following information should be included in a construction report:

- foundation conditions as exposed, and their impact on the structure if the conditions were different to what was expected;
- details of backfilling and sealing of investigation boreholes and test pits etc;
- methods of construction quality control and results of testing;
- variations from the proposed design, reasons and impacts;
- locations and construction details of all monitoring instrumentation; and
- as-constructed survey drawings showing the locations of buried pipes, water recovery systems and the embankment profile.

6 OPERATION

6.1 *Operating manuals*

The operation of a TSF should be of an appropriate standard to ensure that the tailings material is deposited in a manner to achieve the objective of safe, long-term storage with minimal environmental impact. Preparation of site-specific Tailings Storage Operating Manuals for TSFs of all categories is an important part of achieving this objective. A detailed list of items that should be addressed in an operating manual is given in DME (1998b). In general, an operating manual should cover at least the following:

- deposition methodology;
- pond control and water management;
- seepage control and management;
- pipeline management;
- inspection checklists, and reporting requirements; and
- remedial action plans.

The operating manual should be used as a basis for daily operation of a TSF and training the operating personnel. Preparation of an operating manual is not mandatory, but it may be used as evidence of good practices employed by the operator when required.

6.2 *Emergency action plans*

An emergency action plan is required for tailings dams of all categories. This may be produced as part of the mine-wide emergency action plan required by Regulation 4.30 of the MSIR. The emergency action plan should provide an assessment of the possible danger to life, property, and features of environmental significance in areas situated adjacent to, and downstream of the dam. Information relating to any warning or emergency alarm systems used, with a description of proposed emergency procedures should be clearly outlined. The emergency action plan should consider a range of emergencies including the following possibilities:

- blocked decant structure with the danger of overtopping the embankment;
- development of significant erosion gullies;
- seepage from the embankment faces, with or without erosion of materials;
- tension cracks observed on the outer face or crest of the embankment;
- breaks or bursts in the tailings delivery and return water pipe lines;
- overtopping the return water storage pond;
- flooding with potential for overtopping and/or embankment instability;
- shear failure or "breakaway" of part of the embankment face; and
- embankment failure or overtopping with massive loss of contained material.

6.3 *Operational audits*

An operational audit conducted by geotechnical specialists provides a measure of performance against the design expectations. The results of operational audits can be used for further improvement of tailings management and future planning, if required. Table 2 shows the recommended frequency of operational audits for TSFs of different categories. The information expected from an operational audit report includes:

- reconciliation of stored volume and an assessment of the remaining capacity;
- assessment of in situ properties and phreatic surfaces within the facility;
- validation of the design, using updated input parameters derived from site measurements and testing, and recommendations for any necessary modifications;
- water balance studies - this may indicate seepage losses;
- groundwater monitoring results with interpretation; and
- a discussion on the tailings management and recommendations for improvements where required.

7 DECOMMISSIONING

At the completion of operation, a status review should be conducted by geotechnical and other relevant personnel to assist with rational planning for safe and environmentally acceptable decommissioning. The information expected from a decommissioning status review is similar to that expected from an operational audit with more emphasis on long-term issues. The status report should review the decommissioning proposals provided at the design stage and any ongoing rehabilitation measures to validate their suitability to the conditions of the TSF at the completion of operations.

8 ADMINISTRATIVE PROCEDURE

Before granting approval for TSF construction, the design reports are generally scrutinised by DME to verify the technical basis of the design and proposed construction procedures. To expedite the approval process the proponent of a TSF may include a *Certificate of Compliance for TSF Design*, duly signed by the designer, to certify that the design meets the relevant guideline recommendations. If a Certificate of Compliance is provided with the design report, the submission may be considered suitable for direct processing and approval. However, the design may still be audited before or subsequent to the granting of approval, and additional work/information may be requested by DME if required.

At the time of granting approval for construction of a TSF, appropriate tenement conditions are imposed under the MA to ensure adequate construction quality control and reporting, periodic audits during operation and status review at decommissioning. In some circumstances, where significantly large long-term mining projects involving extensive public infrastructure development, the mining company may seek and enter into a State Agreement with the State Government. In this situation the project is primarily developed and regulated under the State Agreement and not the MA. Nevertheless, the MSIA will still be applicable. To ensure environmentally acceptable management of mine tailings, appropriate conditions may be included in the State Agreement or imposed under the EPA.

As per Section 43(2)b of the MSIA the manager of a mine must ensure that every person who is appointed to perform any duty understands the nature and the scope of that duty. This section of the MSIA may also be applicable to the construction of a tailings dam. In addition, attention is drawn to Regulation 4.18(f), 4.20 and 4.21 of the MSIR. These regulations deal with construction work at mine sites in general.

The as-constructed reports for Category 1 and 2 dams also receive DME assessment. In some circumstances tailings may not be deposited in the new facility unless the as-constructed report confirms that the construction meet the design intent. With the as-constructed report, the operating company may provide a *Certificate of Compliance for TSF Construction* duly signed by a registered engineer and the site manager of the TSF operating company to certify that the construction met the design intent. The Certificate of Compliance will expedite additional approvals that may be required before commissioning the new TSF. The as-constructed report may still be audited as part of DME's commitment to quality control.

Operational audits are to be conducted at regular intervals (eg: for Category 1 dams annually). A copy of the operational audit report, together with a proposed action plan to remedy any identified deficiencies of the TSF, should be provided to DME. Any deficiencies identified in the audit report should be suitably addressed and improved by the operating company within an agreed time frame. Status review reports are to be provided before decommissioning and any deficiencies identified in the report should be improved by the tenement holder.

Since 1993, the tenement conditions mentioned above have been imposed under the MA to ensure adequate construction, operation and decommissioning of all new TSFs with emphasis on the application of geotechnical engineering. Following the introduction of the guidelines and the tenement conditions specific to tailings management the industry has been adopting the guideline recommendations for construction and operation of all new TSFs in WA. As a result there has been a significant reduction in the incidents of tailings spillage, seepage, runoff erosion, piping failures and dust generation. Further, the conditions of the TSFs at decommissioning have significantly improved.

9 CONCLUSIONS

Two sets of guidelines have been prepared to establish a well-engineered uniform approach to design, construction and operation of TSFs in WA. The first addresses the geotechnical issues relevant to design and construction and the second addresses the routine operational issues. The guidelines have been readily accepted by the mining industry and the consultants engaged in tailings management. Following the introduction of the guidelines and the tenement conditions specific to TSF construction and operation, there has been significant improvement in tailings management in WA.

10 ACKNOWLEDGEMENTS

This paper is published with the permission of the Director General, Department of Minerals and Energy, Western Australia.

11 REFERENCES

DME, 1998a. Guidelines to help you get environmental approval for mining projects in Western Australia. Dept of Minerals and Energy Western Australia, 73p.

DME, 1998b. Guidelines on the development of an operating manual for tailings storages. Dept of Minerals and Energy Western Australia, 33p.

DME, 1999. Guidelines on the safe design and operating standards for tailings storages. Dept of Minerals and Energy Western Australia, 54p.

DOMWA, 1989. Review of tailings management practices, Report by A G Consulting Group for the Dept of Mines Western Australia (Now Dept of Minerals and Energy).

EPA, 1995. Tailings Containment, Best practice environmental management in mining. Environmental Protection Agency, Australia, 36p.

Tailings and Mine Waste '01 © *2001 Balkema, Rotterdam, ISBN 90 5809 182 1*

Evaluation of historic and future performance of a partially reclaimed tailings impoundment, Uravan, Colorado

D.D.Overton & K.C.Chao
Shepherd Miller Incorporated, Fort Collins, Colo., USA
D.B.Durkee
Golder Associates, Phoenix, Ariz., USA
T.E.Gieck
Umetco Minerals Corporation, Grand Junction, Colo., USA

ABSTRACT: This paper presents the results of analyses performed for reclamation of Tailings Pile No. 3 at the Uravan, uranium mine and millsite in Western Colorado. The analyses were performed to evaluate stability concerns related to anomalous pore pressure data, restriction of piezometers, separation of jointing in the sandstone abutment, and to evaluate differential settlement due to proposed placement of riprap. Tailings settlement was analyzed using coupled two-dimensional finite element seepage and stress-strain models (SEEP/W-SIGMA/W). Lateral inflow from the up-gradient abutment of the tailings pile was considered in the analyses. The results were also used to determine cracking potential of the existing clay radon barrier and the development of flow concentration zones in the proposed riprap erosion protection. These analyses were presented to the State of Colorado as part of a performance evaluation of the tailings pile.

1 INTRODUCTION

The Uravan site began operations in 1912 as a radium recovery plant called the Joe Jr. Mill. The operation was converted to vanadium extraction in 1928 and a uranium-vanadium processing facility from 1938 to 1984. The town, named Uravan in 1935, housed workers and their families, and included residences for approximately 800 people, medical facilities, schools, a boarding house, a town dump, community recreational facilities, and other amenities. Union Carbide operated the mill until 1984. Tailings piles were constructed on the mesa in the 1950s to contain the residues from ore processing. The original operating portion of the mill, known as the A-Plant, was located along the west bank of the San Miguel River. The B Plant was constructed on the mesa above the west bank of the river. Starting in the mid-1970s, raffinate was also disposed of by spray evaporation on the Club Mesa fields west of the tailings disposal areas and evaporated in the Club Ranch Ponds area.

In 1984 Union Carbide created a subsidiary, Umetco Minerals Corporation (Umetco), and one of their missions was to reclaim the Uravan site. The town was demolished and closed after termination of ore processing at the mill. The Uravan site was placed on the National Priorities List (NPL) of Superfund sites in 1986. Umetco and the State of Colorado agreed to a Remedial Action Plan (RAP) for the site in 1987 in a Final Consent Decree (Civil Action No. 83-C-2384). The plan was designed to control radon emissions and the migration of heavy metals, residual salts, and radionuclides from the site. Reclamation activities were conducted during mill operation and after the mill was shut down. As facilities were no longer needed, steps were taken to reduce potential emissions. By the end of 1999 nearly all of the plant and tailings disposal areas, the town site, town dumps, and Club Mesa had been reclaimed to meet regulatory requirements.

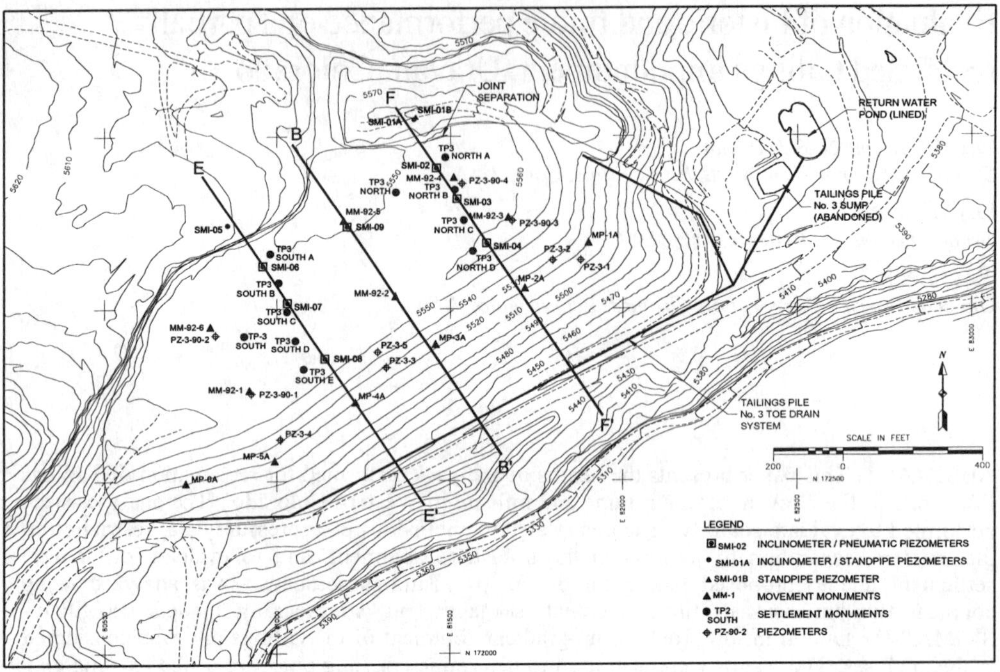

Figure 1 Cross Section and Instrumentation Locations, Tailings Pile #3

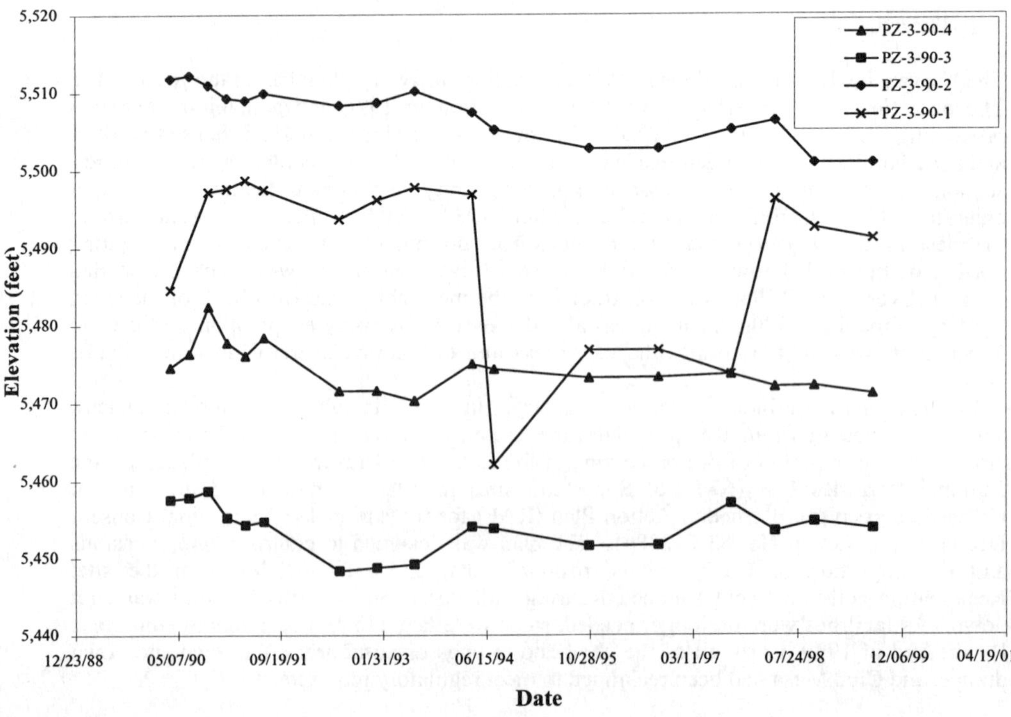

Figure 2 Historic Stand-Pipe Piezometer Data

Tailings Pile No. 3 was constructed in the mid-1950's using upstream construction methods to store tailings generated by the Uravan Mill (Umetco, 1997). Tailings slurry was discharged from the crest of the embankment. The coarse fraction of the tailings sands settled out first and formed beaches at the upstream slope of the embankment that sloped toward the center of the impoundment. The fines remained suspended in the tailings slurry and settled out in the pond that formed in the center of the impoundment. Embankment raises were constructed by regrading and compacting the beach sands.

In early 1980, toe berms were constructed in areas identified as unstable and a series of horizontal drains were installed beneath the toe of the slope (American Acres, 1980). Discharge of tailings into the impoundment ceased on November 15, 1984, the pile was regraded, and an interim soil cover was placed. In 1989, cover soils were placed on the sideslopes of the pile and riprap was completed to the top of the rockfill buttress. A design interim cover was placed over the impoundment in 1991. Final grading and riprap placement will be completed on the top of the pile after it has been determined that the rate of settlement is low enough to not adversely affect performance of the final cover.

2 FIELD INVESTIGATION

Umetco installed 12 movement monuments, 11 settlement monuments, and 8 piezometers during reclamation. Figure 1 shows a plan view of the regraded pile and the locations of cross section F-F' and instrumentation. Piezometer PZ-3-90-1 showed anomalous readings between July 22, 1994 and April 8, 1998. However, as seen in Figure 2, readings since April 1998 are consistent with readings before July 1994. A piezometer probe could not be advanced past a depth of 33 feet in piezometer PZ-3-3 from July 1992 through October 1995. In addition, joint separation in the sandstone abutment occurred prior to August

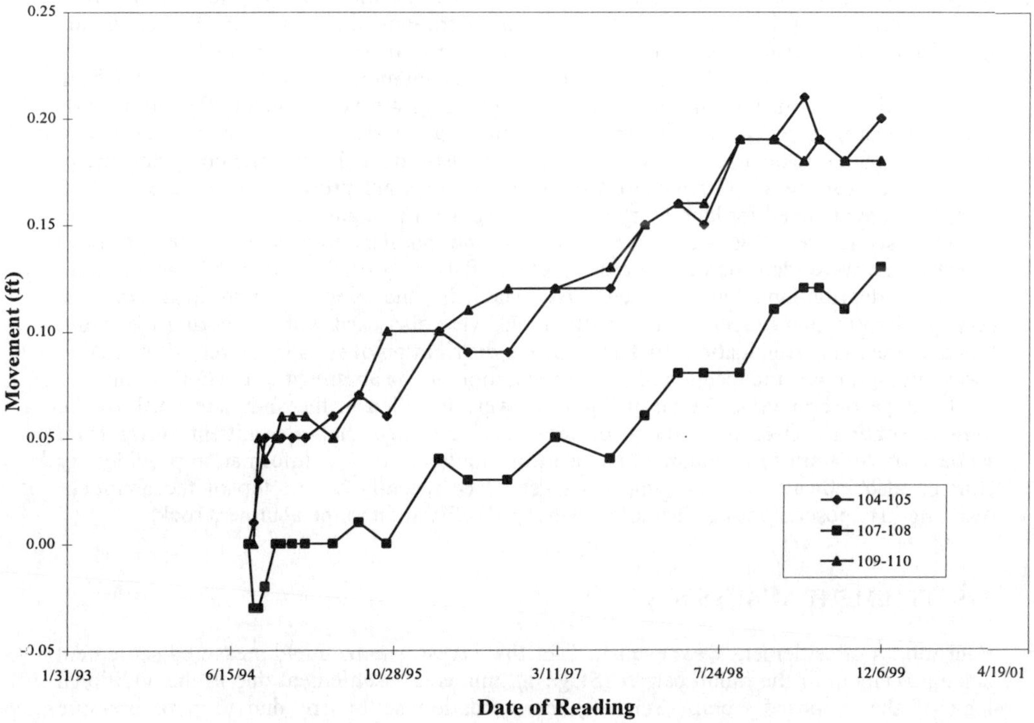

Figure 3 Movement of Joint Separation in Sandstone Abutment

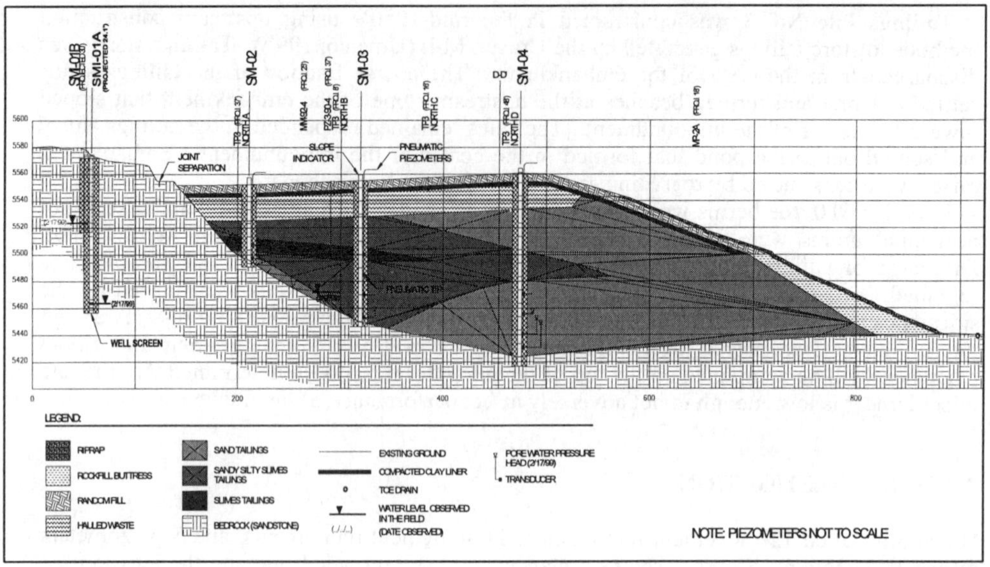

Figure 4 Cross Section F-F'

1994. Umetco monitored the joint separation and the measured movement is shown in Figure 3.

The occurrence of the joint separation, the anomalous piezometer data, and the restriction in piezometers initiated a performance investigation of the tailings pile. Since final reclamation of the pile is pending settlement considerations, the geotechnical investigation included the long-term settlement analysis that is also presented in this paper.

Shepherd Miller, Inc. (SMI) installed field instrumentation and collected samples from Tailings Pile No. 3 in June and July 1998. Seven borings were excavated in the tailings pile and were instrumented with inclinometers and pneumatic piezometers. Three borings were excavated in the Club Mesa abutment and were instrumented with inclinometers and/or standpipe piezometers. The locations of SMI's borings are presented in Figure 1. Soil samples were obtained for laboratory testing during the investigation.

Cross section F-F' is shown in Figure 4 and similar pore pressure and tailings distributions were identified along cross sections B-B' and E-E'. These tailings and pore pressure distributions were used to establish the initial conditions in the seepage/deformation model. Aerial photographs were also used with the boring logs from this and previous investigations to determine the distribution of soils in the repository. Also shown in Figure 4 is the location of a joint separation on the abutment of the tailings pile.

Standpipe piezometers, shown in Figure 4, were installed in the abutment north of the joint separation. Data from the two standpipes indicated that intermittent flows from seepage on the top of the abutment occur in the abutment rock . Information provided by Umetco also indicates that grading had taken place recently on the top of the abutment resulting in exposed fractures that allow increased infiltration to the abutment rock.

3 SETTLEMENT ANALYSIS

Four modes of settlement were evaluated for this investigation. Field measured settlement since placement of the radon barrier (Stage 1), immediate settlement due to the additional load of the proposed riprap (Stage 2), consolidation settlement due to pore pressure dissipation (Stage 3), and long-term settlement or continued deformation of the soils at a constant state of stress (Stage 4). The analyses were conducted on the cross sections shown

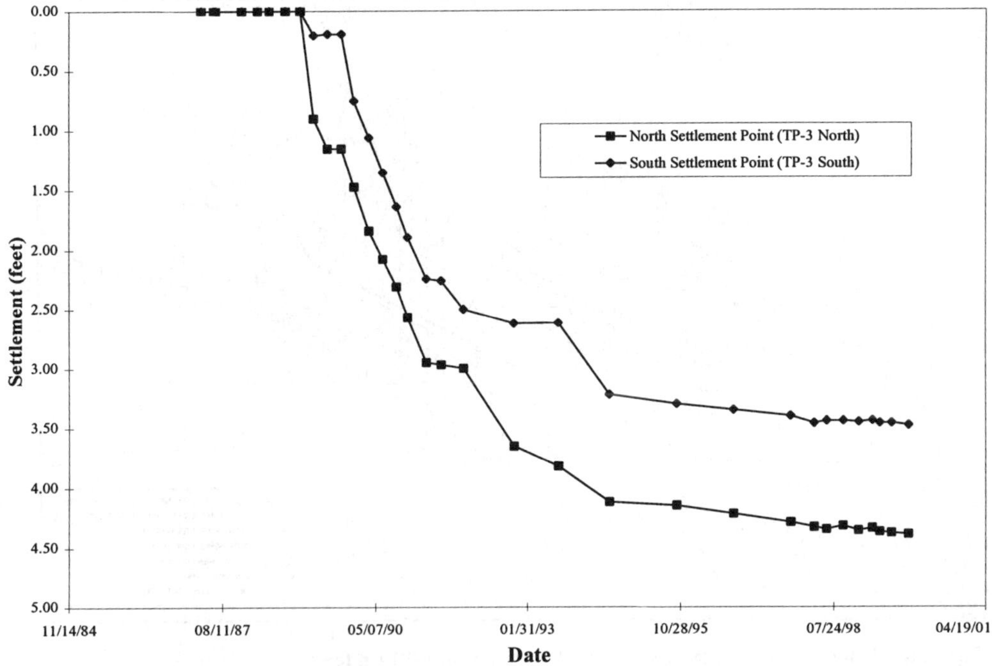

Figure 5 Settlement Data from TP-3 South

in Figure 1. The differential settlement, which is used to evaluate cover cracking potential and potential for development of surface water flow concentration zones, was also evaluated along each cross section.

3.1 *Settlement since placement of the radon barrier*

Settlement data has been collected at 11 settlement monuments and 13 movement monuments since reclamation was initiated. Settlement data for monuments TP-3 South and North are presented in Figure 5. The radon barrier and frost protection materials were completed in 1991. Settlement data collected since 1991 were used to develop field measured settlement contours since placement of the interim cover. The measured settlement contours, shown in Figure 6, indicate that two primary slimes deposits are located on the north-northwest side of the pile near the Club Mesa abutment and that a maximum of 1.2 feet of settlement has occurred since 1991.

3.2 *Immediate settlement*

Immediate settlement due to placement of riprap was predicted using the SIGMA/W computer program. The soil strains contributing to immediate settlement were considered elastic. Overburden stresses from placement of riprap were determined by the model based on the unit weights of the material. The initial pore pressure distribution was obtained from measured piezometer data as shown in Figure 4. Placement of the riprap was simulated with a uniform loading. The immediate settlement due to the riprap loading for cross section F-F' is shown in Figure 7. Similar plots were generated for immediate settlement along the other cross sections. The maximum immediate settlement predicted, due to the riprap loading was less than 0.7 inches.

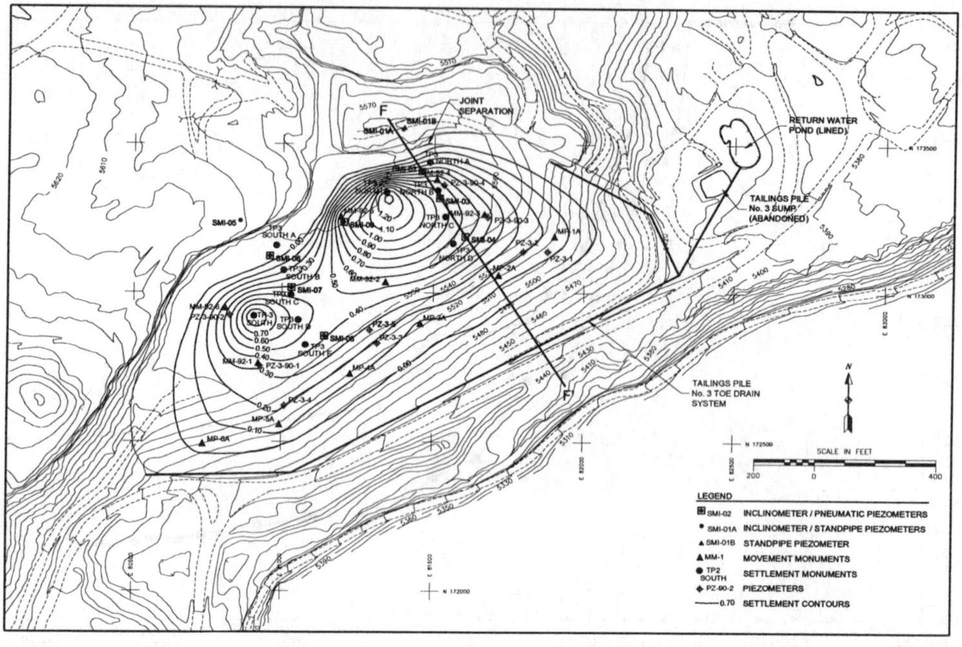

Figure 6 Settlement since Placement of Radon Barrier in 1991 (in feet)

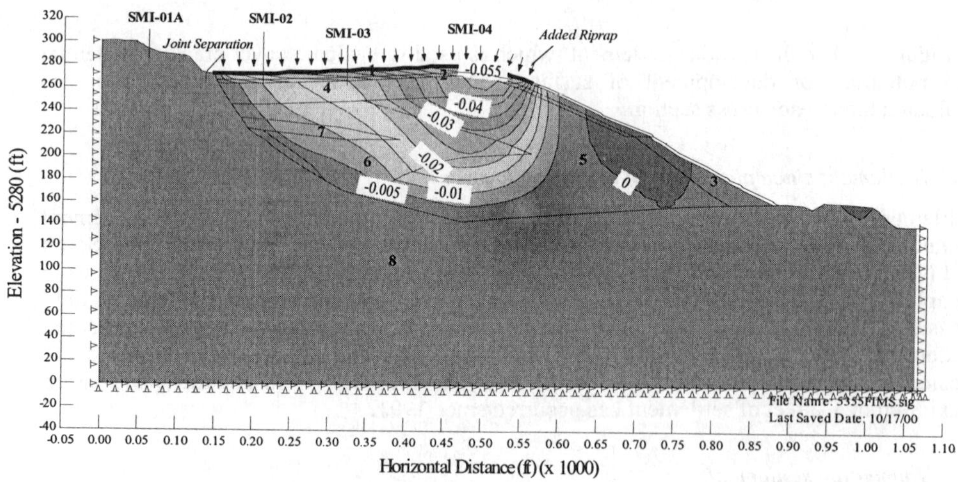

Figure 7 Immediate Settlement due to Riprap Loading (in feet)

3.3 *Consolidation settlement*

The consolidation settlement was predicted using a coupled analysis of seepage and deformation with the computer programs SEEP/W and SIGMA/W. SIGMA/W computes displacements and stresses while SEEP/W computes the changes in pore-water pressure with time to form a coupled consolidation analysis. Considering the soil properties as well as the stress range applied in the field, a nonlinear elastic-plastic soil model was used to predict the consolidation settlement due to dissipation of pore pressure in the saturated tailings. A linear elastic soil model was used to model the other materials.

Initial pore pressure conditions in the repository were determined from existing pneumatic piezometers installed by Umetco and strings of pneumatic piezometers installed by Shepherd Miller, and checked visually by correlation with areas of ponded water observed on historic aerial photographs. Based on the soil profiles and the pore pressure distributions in the repository and historic records of operation, the critical area in terms of future settlement was determined to be in the center of the impoundment. Stress strain and hydraulic parameters were obtained from laboratory triaxial tests, flexible-walled permeameter tests, and water retention tests conducted on samples collected during the installation of pneumatic piezometers. The deformation model was calibrated using the results from the laboratory triaxial tests.

The intermittent flows from the abutment have a large influence on the pore pressure distribution and time rate of seepage in the pile. Therefore it was necessary to calibrate the inflow boundary flux from the abutment and the K_h:K_v ratio of the abutment rock and tailings, to measured pore pressure data. The occurrence of intermittent flows was simulated in the model by applying a constant boundary flux along the upstream edge of the model mesh. The seepage model was calibrated using pore pressure data from 1991 as an initial condition in the pile. Transient seepage analyses were then conducted through 1998. The boundary flux was varied during the calibration phase until the simulated 1998 pore pressures in the tailings matched the measured 1998 pore pressures.

The actual preferred flow paths and transient flow conditions along fractures and discontinuities are not known. Thus, to model the foundation soils as one unit, the K_h:K_v ratio was assumed to be 50. A fitted boundary flux of 5 gallons per minute resulted in a reasonable simulation of the measured pore pressure distribution.

The consolidation settlement along cross section F-F' predicted by the model is shown in Figure 8. Similar plots were developed for the other cross sections. The maximum predicted consolidation settlement is approximately 6 inches.

3.4 *Long-term settlement*

In order to extrapolate the strain data to predict the long-term strain it was assumed that the strain data is represented by a hyperbola of the form:

$$\varepsilon_c = \frac{t}{a + bt}$$

where:

t = time since the end of consolidation settlement, and

a and b = hyperbolic parameters.

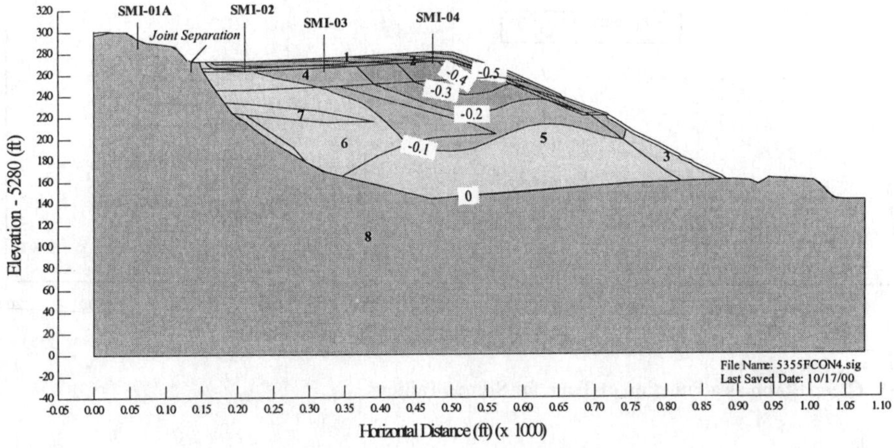

Figure 8 Consolidation Settlement, Cross Section F-F' (in feet)

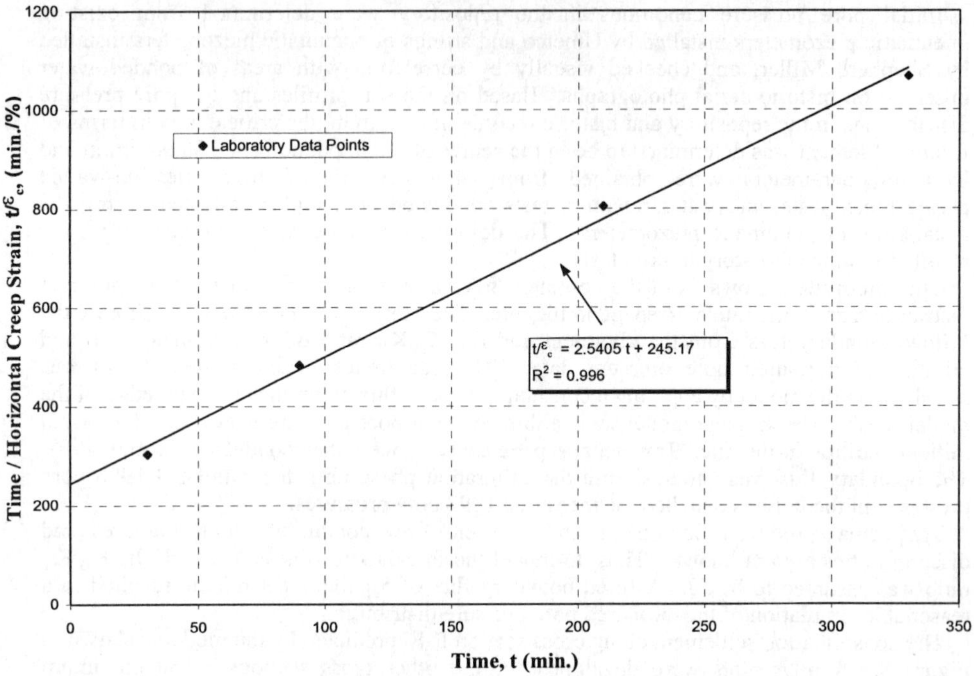

Figure 9 Hyperbolic Fit of Secondary Compression Data of Slimed Tailings

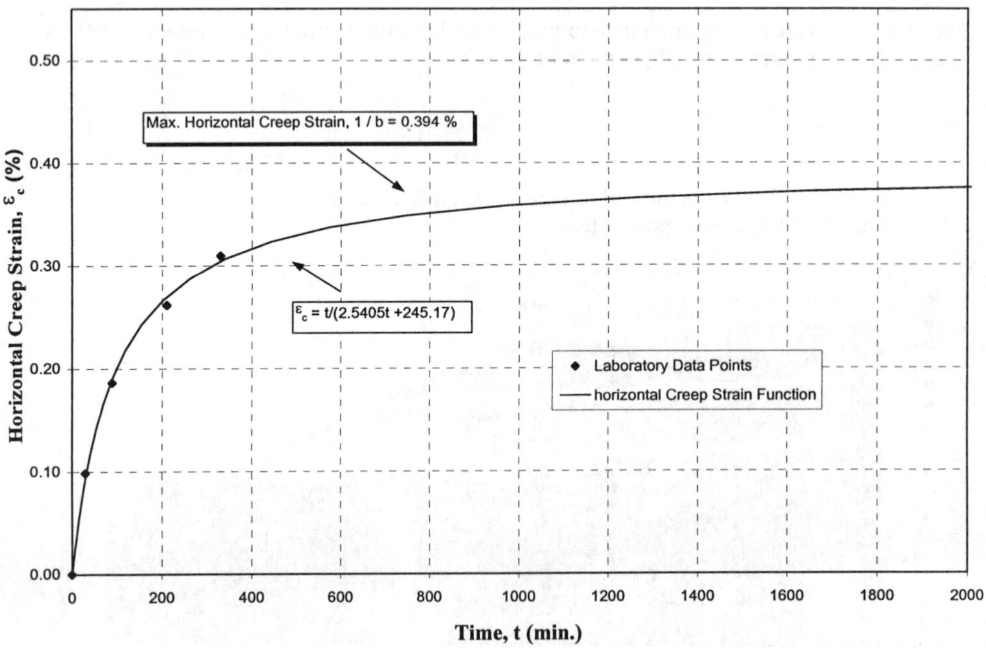

Figure 10 Creep Strain as a Function of Time for Slimes Tailings

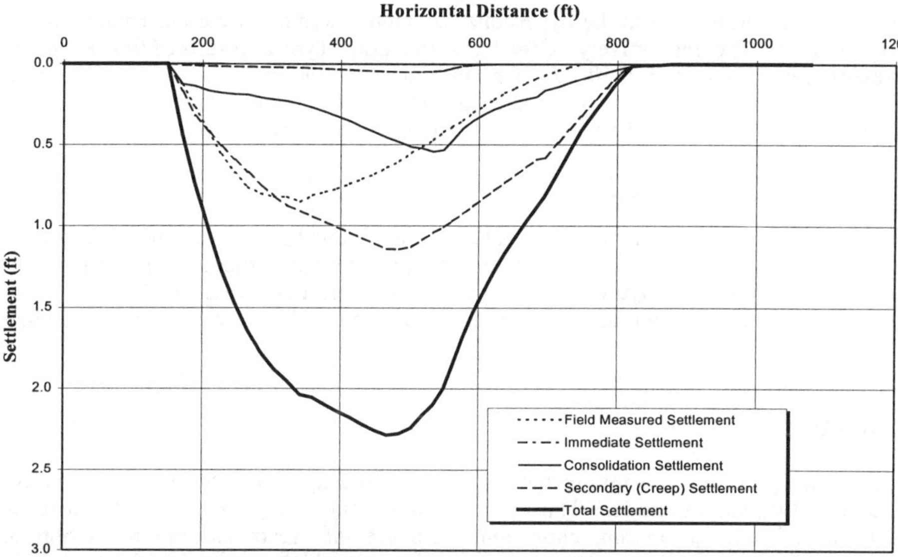

Figure 11 Settlement Along Cross Section F-F' (in feet)

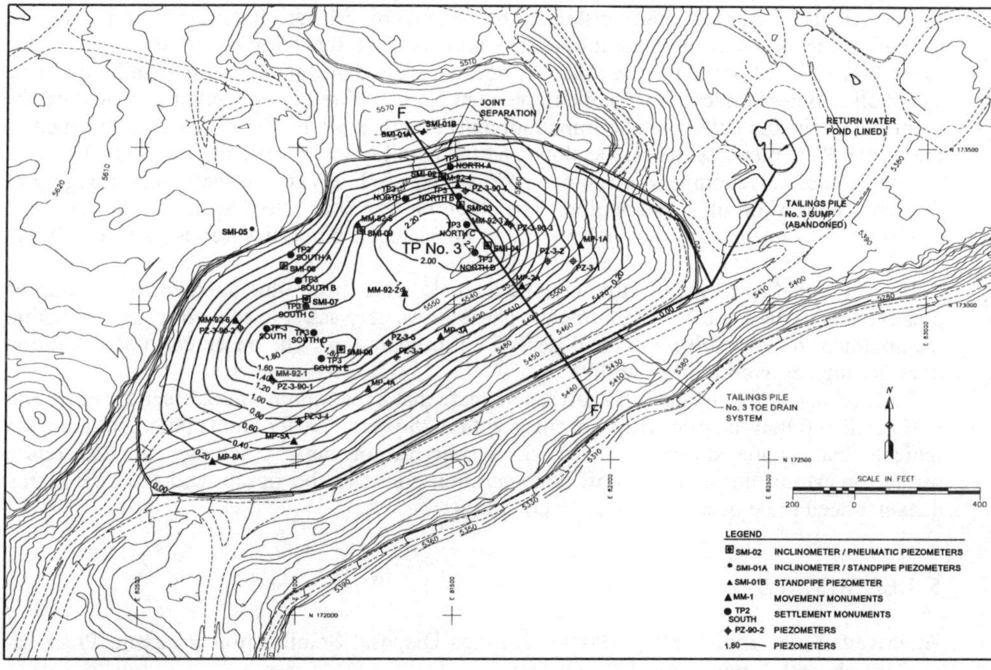

Figure 12 Predicted Total Settlement

From the hyperbolic strain equation a plot of t/ε_c as a function of t would be of the form:

$$\frac{t}{\varepsilon_c} = a + bt$$

This equation shows that t/ε_c is a linear function of t if the data fit a hyperbolic relativity. From consolidation tests the strain data after 90 percent of consolidation occurred were plotted in the above form, and the hyperbolic parameters a and b were determined as the intercept and the slope, respectively, of the linear function. Typical results of this technique are presented in Figures 9 and 10. As seen in Figure 10, the maximum long-term strain approaches a value of 1/b as time approaches infinity. The maximum long-term settlement predicted from the hyperbolic fit method for Tailings Pile No. 3 was 1.15 feet.

3.5 *Summary of settlement analyses*

The total settlement is the sum of the field measured, immediate, consolidation, and long-term settlement. The individual settlement profiles and the total settlement profile for cross section F-F' are shown in Figure 11. Similar profiles were developed for the other cross sections. Total settlement contours for the impoundment are shown in plan view in Figure 12.

4 SUMMARY

The investigation presented herein includes analyses of seepage and deformation of Tailings Pile No.3 at Uravan, Colorado. The analyses were conducted to evaluate the possible mechanisms of joint separation, anomalous pore pressure readings, and restriction of piezometers that were previously observed. In addition, cracking potential of the clay radon barrier and development of concentration zones due to placement of the riprap on the existing interim cover system were evaluated from the results.

The field instrumentation installed by Shepherd Miller (pneumatic piezometers, standpipe piezometers, and inclinometers) indicate that the tailings are saturated in the lower 20 to 30 feet of the pile and that the pore pressure distribution is less than hydrostatic. The inclinometers indicate that slope movement is in the direction of settlement (toward the slimes areas in the center of the impoundment). This explains the restriction observed in the 1-inch PVC piezometers on the slope and the inability to measure the water level.

The anomalous readings in piezometer PZ-3-90-1 are likely due to temporary clogging of the screened portion of the standpipe. The clogging was later (after April 8, 1998) relieved as observed from the more recent data that is consistent with earlier trends (before July 22, 1994).

Intermittent flows within the abutment rock have been documented by standpipe piezometers in the Club Mesa abutment. These flows may assist in relieving stresses within the abutment rocks resulting in the observed joint separation. The observed joint separation does not impact overall pile stability.

The seepage and deformation analyses indicate that a maximum of 2.25 feet of settlement will occur and that the pile will be drained within 25 years. Differential settlement analyses indicate that tensile strains in the radon barrier do not cause cracking, and that future settlement would not create runoff concentration zones in the riprap, and thus the riprap does not need to be oversized for such conditions.

5 REFERENCES

American Acres, Inc. 1980. "Uravan Tailings Disposal Stabilization Program, Phase II-Geotechnical Report," April.

GEO-SLOPE International, Ltd. 1998a. SEEP/W Software Package for Seepage Analysis, Version 4.20. Calgary, Alberta, Canada.

GEO-SLOPE International, Ltd. 1998b. SIGMA/W Software Package for Deformation Analysis, Version 4.20. Calgary, Alberta, Canada.

Shepherd Miller, Inc. (SMI), 1999. "Geotechnical Engineering Investigation, Tailings Pile No. 3, Uravan, Colorado," Fort Collins, Colorado, March.

Umetco Minerals Corporation, 1997. "Joint Separation, Tailings Pile Nos. 3, Uravan, Remedial Action Project, Uravan, Colorado," December.

Tailings and Mine Waste '01 © 2001 Balkema, Rotterdam, ISBN 90 5809 182 1

Geotechnical characterisation of in-pit tailings at Ranger Uranium Mine, Northern Australia

H. Li
Earth-Water-Life Sciences Pty Limited, Darwin, N.T., Australia (Currently: Hatch Associates, Brisbane, Qld, Australia)

G. Cramb & A.R. Milnes
Earth-Water-Life Sciences Pty Limited, Darwin, N.T., Australia

ABSTRACT: A number of field investigations have been carried out to determine the geotechnical behaviour of in-pit tailings at Ranger uranium mine in Northern Australia. The investigation results show that the bulk dry density of tailings in the central parts of the pit are low, varying between 0.4 and 1.0t/m^3 in the top 10 metres. The material is fine clayey silt/silty clay. The self-weight consolidation is not completed. By contrast, tailings in the beach areas are much coarser, and have a much higher dry density with the average of 1.48 t/m^3 over a 30m profile. It is considered that the self-weight consolidation in those areas is virtually complete. This information forms the basis for developing strategies and techniques to expedite consolidation and increase the density of the tailings in the pit for effective and timely rehabilitation of the pit after filling.

1 INTRODUCTION

1.1 Tailings disposal history

Ranger Mine is operated by Energy Resources of Australia (ERA Limited). The mine lease lies within the boundaries of Kakadu National Park, a designated world Heritage Area, in Northern Territory, Australia. Long-term security and effective rehabilitation of the tailings storage facilities have been a vital part of mining operations and rehabilitation planning for the mine. The tailings represent the residues of uranium-bearing ore that has been mined, milled, treated with sulfuric acid to dissolve uraninite, separated from the pregnant solution by countercurrent decantation, neutralised with lime to minimise the potential for mobilisation of metals and radionuclides remaining in the pore fluids (and to provide neutral decant water for grinding ore), and pumped as a slurry to the tailings storage facility.

Initially, tailings were deposited in an aboveground tailings storage facility which is a water-retaining structure (dam) covering an area of approximately 1 km^2. Tailings were deposited beneath 2m of water in the tailings dam to ensure that radon emanation was suppressed. The measured dry density of these tailings deposits was less than 1.0 t/m^3. In the mid 1980's a series of investigations and trials demonstrated that radon emanation from beached tailings was not a significant hazard, and that subaerial deposition could considerably improve the settled density of the tailings. Subaerial deposition was established in the 1987 dry season following approval obtained in January 1987. However, the results did not meet expectations and the measured dry density, even in the beaches near the tailings dam walls, did not exceed ~1.04t/m^3 on average. The measured values in the dam were generally between 0.85 and 0.95t/m^3 with the lowest value being 0.68t/m^3 (Sheng *et al*, 1997).

Since August 1996, tailings at Ranger have been deposited directly from the mill into the mined-out Pit #1 by cascading the tailings slurry (~45% solids) down the pit walls from various takeoff points along a ring pipeline located on the uppermost benches. In-pit disposal has the advantages of long-term secure containment, a very low possibility of structural failure of the impoundment, and low visual and air pollution effects. The primary concerns for disposal in the pit are the large post-deposition settlement, and the potential for tailings pore fluids to escape the pit and impact on groundwater quality.

Following a comprehensive review and assessment of best practicable technology, a decision was made late in 1997 to return all tailings currently in the tailings dam to the mined out pits to maximise the potential for secure long term storage (Milnes 1998). During 1998 this process was started when the elevated beach areas in the dam were removed by dredging and transferred to Pit #1 in order to maximise the surface area of process water and so enhance evaporative losses. Process water must be retained on site and cannot be released: evaporation is the only mechanism available for its disposal at this time.

1.2 *In-pit disposal*

The pit is the mined-out void of orebody #1. It has the shape of an inverted cone with a final depth of 170m below the surrounding land surface at RL 22m. It is generally circular in plan view with a diameter of approximately 750m and a surface area of 42 ha. The maximum depth of the tailings deposits has been approved to be 142m to RL 0.0 m. Investigations are being carried out to explore the possibility of disposing of tailings above RL 0.0 m. The storage capacity is 15.2 Mm^3 to RL 0.0m and 19.4 Mm^3 to RL 14.0m.

There is an underdrainage system installed to promote consolidation of tailings. The underdrainage layer occupies a small area at the bottom compared to the area of the pit at the original landsurface level and was extended to RL –108m with a slope of 1:3.5 on the northern side to increase its effective area. The underdrainage is connected to a rock-filled horizontal adit in the southwestern corner in the pit and thence to the landsurface via a vertical bore. Collected water is removed using a pump installed in the bore. Pumping water from the underdrainage system provides a zone of low pore water pressure at the base of the tailings. Under ideal conditions the pressure in the underdrainage system should approach atmospheric pressure. The hydraulic gradient induced by this low pressure zone causes flow towards the base of the tailings, reducing the drainage path length and the time required for the dissipation of excess pore water pressure.

Consolidation analysis prior to commencement of deposition in the pit suggests that the final settlement could be as great as 40m (Knight Piesold, 1992) and that settlement is likely to be non uniform across the tailings surface (Craig, 1998). Thus, it could take many years for the tailings to achieve 95% of final consolidation. Such a large differential residual settlement and consolidation time would have significant impacts on the long-term effectiveness of the required capping over the tailings and on the stability of the rehabilitated pit landform. To avoid this, intervention to enhance consolidation may be required during the operational stages of tailings deposition. Geotechnical characterisation and monitoring of tailings are critical in the context of choosing and assessing consolidation-enhancing measures. A research program has been carried out to investigate the geotechnical behaviour of the in-pit tailings. The program includes *in situ* testing, sampling and instrumentation, and laboratory testing. This paper describes the program, presents the results and discusses some strategies and measures to enhance tailings consolidation.

2 GEOTECHNICAL INVESTIGATION TECHNIQUES

2.1 *Undisturbed sampling and sludge sampler*

Measurements of *in situ* density were made on undisturbed cores sampled during early stages of tailings deposition in 1996 and on beached areas in 1999. A lightweight portable soft sediment, piston sampler was used to obtain undisturbed samples of tailings. The sampler consists of a double rod system connected to a sampling tube. The double rod system is made up of outer hollow rods and inner solid rods (stainless steel). The sampling tube is made of thin-walled PVC (internal diameter 42.5mm, length 500mm) and has a machined cutting edge.

During operation, the inner and outer rods are locked together, and jacked into the ground with the piston protruding from the front end of the sampling tube. The front face of the piston is conical, to aid penetration into the tailings and to also prevent tailings entering the sampling tube. At the desired depth, the inner rods are held stationary while the outer rods are jacked further down. The inner and outer rods are locked together upon retrieval of the sample. With this system, undisturbed samples can be obtained at exact target depths from the tailings mass. The piston system ensures that the sample is held in the container by suction.

When tailings are in slurry form and/or below a deep cover of supernatant water, tailings samples cannot be obtained using the above technique. A sludge sampler produced by CSIRO was made to adapt to the existing coring rig to sample slurry from the tailings mass, metres below the water surface. The sludge sampler is attached to the first inner rod. Unlike the setup for higher density tailings sampling, the sludge sampler is positioned below the sampling tube. Once at the desired depth in the tailings mass, the sampling tube (connected to the outer rods) is then pushed over the sludge sampler, trapping the tailings inside the tube. When retrieving the sample from the tailings surface, care was required so that the sludge sampler did not slide down proud of the sampling tube and allow the tailings sample to drop out.

2.2 *Piezocone testing*

Piezocone tests were conducted in the beached area and in the central part of Pit #1. The piezocone test is formally referred to as a static cone penetration test with direct measurement of pore pressures (CPTU). A lightweight penetrometer is used, consisting of three main parts: piezocone, data logging system and pushing system. The cone has an apex angle of 60° and base area of 1000 mm^2. The friction sleeve has an area of 15,000 mm^2. The filter for pore water pressure measurement is located behind the cone shoulder.

During CPTU, the cone is driven into the ground by a hydraulic system via a series of hollow rods, at a constant rate of approximately 20mm per second. A cable located inside the rods, is connected to a computer-controlled data logging system. Continuous or intermittent measurements are made of: resistance to cone penetration (cone resistance, q_c), resistance of surface sleeve (sleeve friction, f_s), and pore water pressure behind the cone (u_2). The maximum measuring capacity for cone resistance is 50MPa and 1MPa for both sleeve friction and pore pressure. The accuracy of measurements is 0.5% (or 25kPa) for cone resistance and 1% (or 10 kPa) for both sleeve friction and pore pressure.

Dissipation tests were also conducted during CPTU. Here the penetration of the cone is halted at the desired depth, and the decay of pore water pressure (u_2) is monitored. If sufficient time is allowed, pore water pressure in the cone will equilibrate with the in situ pore water pressure of the tailings.

In the central part of Pit #1, the piezocone apparatus was deployed from a purpose-built pontoon with a specially devised anchoring system. A special rod clamp system was installed to hold the rods over a 20m deep cover of supernatant water.

2.3 *Plummet penetration system*

For investigations of the properties of very soft and slurry-like tailings under a deep supernatant water cover, the sludge sampler technique is tedious and somewhat effective, and the piezocone is not accurate enough. A new instrument, called plummet penetration system, was developed to address the problem. In the plummet penetration system, a plummet, a conical lead weight, is attached to a stainless steel tape which applies forces via two pulleys to a force gauge. When the plummet is lowered into tailings slimes, the resistance it encounters is registered in the force gauge. The relationship between the plummet resistance and tailings dry density has been calibrated in the laboratory and in the field. Tailings dry densities can be obtained from plummet penetration tests by using this relationship. The error of such determined density is estimated to be not greater than ±10%. The maximum penetration depth of 10m has been achieved using the system. The maximum dry density was approximately 1.0 t/m^3.

The plummet is made of lead and has a cylinder in the middle with two cones at top and bottom. The apex angles of the bottom and top cones are both 60° in line with the standard apex angle of a conventional cone penetrometer. The top cone is designed such that the plummet can be pulled out easily. Two such plummets were made. The first has a diameter of 35 mm with a volume of approximate 100 cm^3. The second has a diameter of about 70 mm with a volume of approximate 500 cm^3. The former has weight of 1.1 kg and the latter 5.5 kg.

2.4 *Instrumentation*

Before and during the early stages of tailings deposition in Pit #1, a program was carried out to monitor the tailings behaviour (Sheng *et al*, 2000). The instruments installed include vibrating wire piezometers and settlement stations using magnetic rings. These instruments provided valuable information on the tailings behaviour during the early stages of deposition. Unfortunately, they did not last long due to large lateral movements of tailings and the deep supernatant water cover.

A further geotechnical instrumentation program was carried out in 2000 to characterise the geotechnical properties of the tailings and to provide data to assess the effectiveness of a field trial of a floating prefabricated vertical drain (PVD) system to enhance consolidation. Two nests of instruments were installed in Pit #1. The first was centrally located within the grid of twenty-four floating PVD systems. The second was installed outside of the influence of PVDs.

The three instruments used at each nest were: vibrating wire settlement cells – measuring settlement occurring in the tailings; vibrating wire total pressure cells – measuring the total pressure of tailings and water above the cells; and vibrating wire piezometers – measuring pore-water pressure in the tailings. The measuring ranges are 66m between the settlement cell and the reservoir for the settlement cell, 1750kPa for the total pressure cell and 2000 kPa for the piezometers.

The instruments were mounted on an instrument plate, which was balanced with a weight. The plate had a hole centrally drilled (60 mm i.d.), so it could slide over a pre-installed access tube, which was used as a guide to ensure that the instruments were installed at desired locations. The instruments were installed in the middle of the pit and the instrument cabling was placed in a protection box on a floating pontoon which provides a working platform for taking readings and storing cables. The instrument cables were not designed to run into the pit walls but stay near the instruments. This was to avoid potential damage to the cables caused by the large movements of tailings in both vertical and horizontal directions over the long distance between the instruments and the pit walls.

3 TAILINGS BEHAVIOUR IN THE RANGER PIT

3.1 *Particle size distributions*

Samples were taken at four different locations in the pit. Three of these were in the beached area (at the top of the beach, i.e., near the discharge point; in the upper part of the beach; and at the bottom of the beach), and the fourth was in the centre of the pit. Particles larger and finer than 45μm were separated using the wet sieving method. Dry sieving was used for particles larger than 45μm. The results are shown in Figure 1. The tailings from the central part of the pit are much finer than the tailings near the discharge point. This clearly shows that particle segregation exists. This is not surprising because the tailings are discharged from pipes located at the topmost edges of the pit and cascade down slopes across the former benches. This causes beach-like deposits of coarse particles directly adjacent to the discharge points and the accumulation of fines in the central parts of the pit, similar to deltaic systems, as shown from geophysical surveys (Ringis, 1997).

3.2 *Densities of tailings*

The densities of tailings were determined using four methods: direct, indirect, plummet, and piezocone method. In the direct method, tailings density was obtained from the direct measurements of mass and volume of undisturbed samples. In the indirect method, tailings density

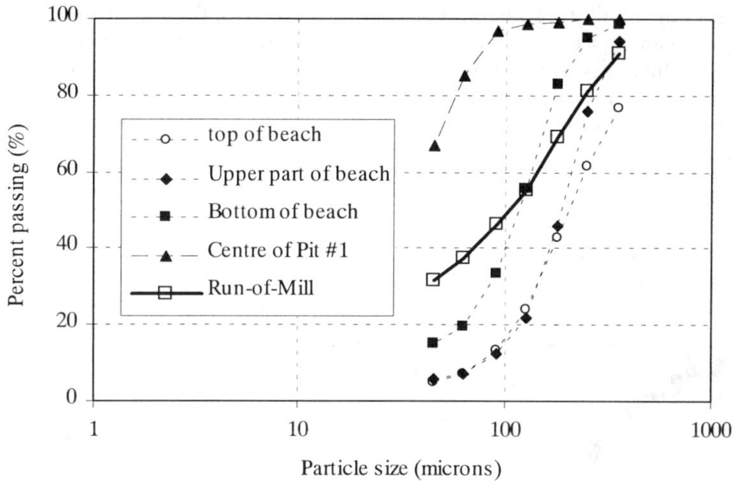

Figure 1 Particle size distribution of tailings at different locations in Pit #1

was estimated from measured moisture content and assumed/measured specific gravity of disturbed samples assuming full saturation. In both the plummet and piezocone methods, tailings density was estimated from the measured cone resistance of tailings and calibrated empirical relationships between tailings density and resistance. In the piezocone method, iterations were required since density is required to determine overburden pressure to correct cone resistance.

Initial measurements of *in situ* density during the earliest stages of in-pit deposition (Cramb, 1997) were made on undisturbed cores sampled to 15m depth from the surface of the tailings deposits. Results indicated that dry densities up to ~1.6t/m^3 were achieved. This might have resulted from an effective underdrainage and the movement of the whole tailings mass due to the kinetic energy of the slurry cascading down the pit walls.

From the indirect method, the tailings density was 1.37 t/m^3 at the surface near the discharge point, 0.92 t/m^3 at a depth of 0.2 m in the upper part of the beach, and 0.65t/m^3 at a depth of 1.0 m in the central part of the pit. This clearly shows that tailings density decreases as the distance from the discharge point increases. This may result from particle segregation.

Results from plummet and piezocone methods show that the density of tailings is much higher in the beached area than in the middle of the pit (Figure 2). Piezocone measurements to ~30m depth have demonstrated that the tailings deposits in the beached 'deltas' have an average dry density of 1.48 t/m^3 over the whole profile. By contrast, the slimes in the middle of the pit, as measured using the plummet technique, have a low density ranging from 0.4 to 1.0 t/m^3 in the top 10 metres of the deposits. A sounding technique was also adopted to estimate the overall average density of tailings in Pit #1. This involved locating the surface of the tailings deposits beneath the supernatant water cover from a boat, the position of which was mapped using survey techniques. The average density of the tailings deposits was calculated from measured tonnages of tailings in the slurry input to the pit and the volume of tailings measured in the pit at given times. The measured average tailings densities in Pit #1 has reached or exceeded the target of 1.2 t/m^3 since deposition commenced.

3.3 *Consolidation behaviour*

Tailings self-weight consolidation is quite different in the beached area, compared to the central parts of the pit (Figure 3). For the beached tailings the *in situ* pore water pressure, estimated from piezocone dissipation test results, and the pore water pressure behind the cone, are very close to the hydrostatic pressures. This means that materials in the beached area are permeable and that self-weight consolidation is virtually complete. In contrast, the *in situ* pore water pressure in the centre of the pit is much higher than the hydrostatic pressure. This shows that the tailings are underconsolidated.

Figure 4 shows the results of monitoring instrumentation installed near the centre of Pit #1 (outside of the PVD trial). It can be seen that the instruments (and therefore tailings) have settled with time. However, the thickness of tailings which accumulated above the instruments is now known. The effective stress and excess pore water pressure are shown in Figure 4 to increase with time and therefore general depth. This again shows that the tailings in the centre of the pit are underconsolidated.

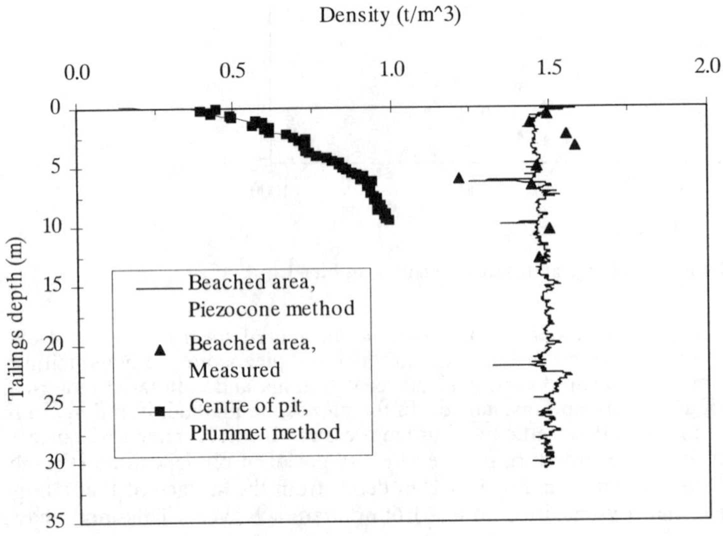

Figure 2 Tailings dry density profiles in beached area and in the middle of Pit #1

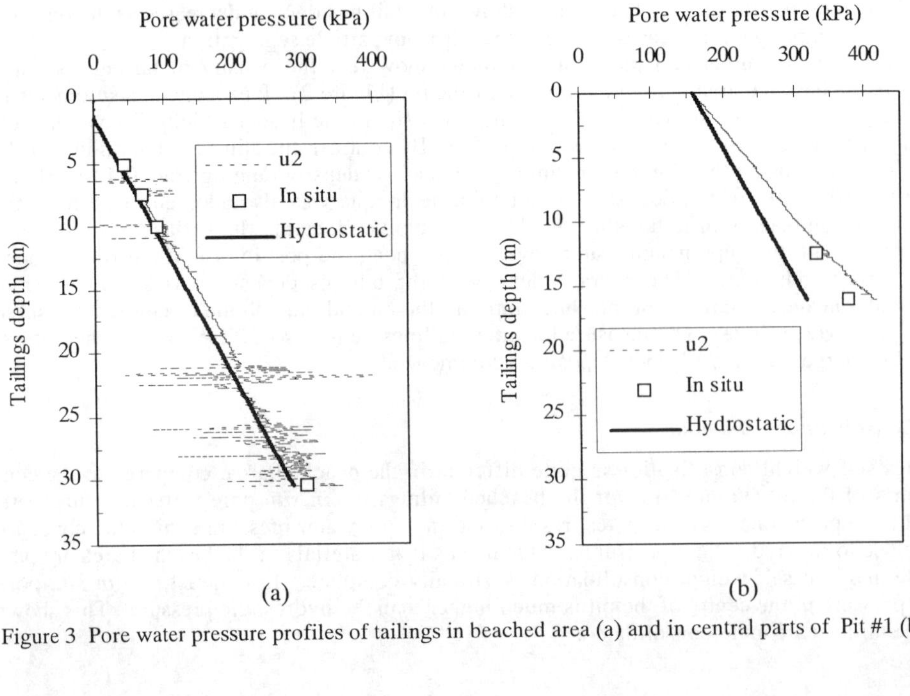

Figure 3 Pore water pressure profiles of tailings in beached area (a) and in central parts of Pit #1 (b).

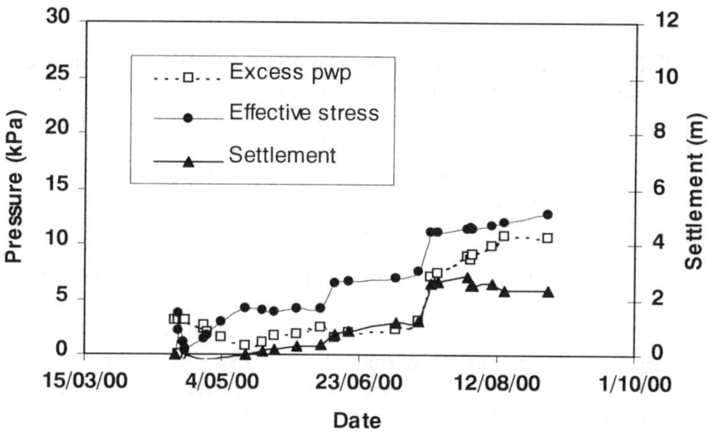

Figure 4 Measured settlement, effective stress and excess pore water pressure of tailings in the centre of Pit #1

4 DISCUSSIONS ON DISPOSAL METHODS AND INTERVENTION STRATEGIES

A number of disposal methods and intervention strategies have been extensively discussed (Li & McGill, 1999) to help plan for cost-effective and sustainable rehabilitation of the tailings-filled pit. In conjunction with monitoring and modelling investigations, the effectiveness of these strategies will be assessed as they are implemented.

4.1 *Floating central discharge system*

As discussed in the previous section, there exists significant particle segregation, resulting in coarse particles deposited near discharge points and fine particles accumulated in the central part of the pit. Tailings have low density and are unconsolidated in the central part of the pit while tailings near discharge points have higher densities and are normally consolidated. Clearly, if the peripheral disposal is continued, a very thick mass of slimes would form in the central parts of the pit. This would have significant impact on decommissioning and rehabilitation of the pit after filling due to substantial settlement and longer term consolidation. A floating central discharge system has been implemented. In this system, tailings are discharged into the central parts of the pit from a floating pipeline and particle segregation is encouraged through the deep supernatant water cover. It is anticipated that once a cone of material has accumulated at the centre of the pit, deposits of coarse material will be generated immediately below the discharge point and that slimes will consequently be transported out towards the pit margins. The increase effective stress at the centre of the pit will also assist in consolidation of the tailings.

4.2 *Paste tailings disposal*

Investigations have also been undertaken of paste tailings disposal wherein tailings are deposited as a paste rather than in a slurry form. Pastes are defined as dense, viscous mixtures of tailings and water which, unlike slurries, do not segregate when allowed to rest (Cincilla *et al*, 1997). Pastes can have solids contents as high as 85% by weight. Due to the significant improvements in paste production achieved in the early 1990s in Canada and other countries, this technology is being extensively used in many North American underground mines and has been adopted in several Australian underground mines. Cost-benefit analyses of this technology are likely to be undertaken at Ranger.

4.3 *Floating Wick System*

Prefabricated vertical drains (PVD), or wick drains, are a proven effective and economic technology for improving the geotechnical properties of soft soils and similar materials by dramatically accelerating consolidation. Wicks are made of plastics wrapped in geofabric. Wick drains expedite tailings consolidation by providing high permeability channels and reducing the drainage path length for removal of excess pore water from unconsolidated materials.

The installation of wicks is straightforward if the surface is trafficable. Wicks are installed with specialised equipment called stitchers that are mounted on standard cranes or backhoes. Because the tailings deposits in the Ranger pit are covered by a considerable depth of water, access is difficult. However, experiments have commenced with a floating system by which wicks are dragged down into soft and loose tailings by means of a heavy cone-shaped weight attached to the head of the wick. The rest of the wick is attached to a small float on the surface of the water. As the level of deposited tailings and supernatant water rises, the top part of the wick will be extended from the floating pontoon as the lower part progressively buried. In this way, wicks will ideally be extended upwards to the final surface of the tailings deposits and will remain within the tailings deposits as channels for pore water to escape to the surface.

5 CONCLUSIONS

Field geotechnical investigations show that the bulk dry density of tailings in the central part of the pit are low, varying between 0.4 and 1.0 t/m^3 in the top 10 metres of the deposit. The material is a fine clayey silt/silty clay. Tailings are unconsolidated. By contrast, tailings in the beaches are much coarser, being silty sand, and have much higher dry density with the average of 1.48 t/ m^3 over a 30m profile in depth. The self-weight consolidation in this area is virtually finished. This may result from the peripheral tailings disposal method where by the tailings are discharged from pipes located at the topmost edges of the pits and cascade down the slopes across the former benches. This causes beach-like deposits of coarse particles directly adjacent to the discharge points and the accumulation of fines in the central parts of the pit. This would have significant impact on decommissioning and rehabilitation of the pit due to greater settlement and a longer time for consolidation. In order to create sufficient bulk density and strength to expedite capping of the deposits using waste rock for timely rehabilitation, some intervention measures such as central deposition, prefabricated vertical drains and paste technology are required. The former of these two measures are currently being trialed.

6 ACKNOWLEDGEMENTS

This paper is based on results of research projects carried out at ERA Ranger Mine by EWL Sciences. ERA has kindly consented to publication of the paper. The authors would like to thank managers of the Mill and Environment Departments at Ranger Mine for their assistance and for valuable discussions in the course of research. The paper was reviewed by Mr I Gordon from Hatch Associates.

REFERENCES

Cincilla, W A, Landriault, D A & Verburg, R. 1997. Application of paste technology to surface disposal of mineral wastes. *Proceedings of the 4th International Conference 'Tailings and Mine Waste '97'*: 343-355. Rotterdam: A.A.Balkema.

Craig, M. S. 1998. *A conceptual model for the prediction of consolidation of the Ranger tailings*. CSIRO Minesite Rehabilitation Research Program, June 1998. Unpublished Report No 502C. 40pp.

Cramb, G. 1997. *Pit #1 tailings density determination*. Unpublished Report to ERA Ranger Mine, July 1997. 17pp.

ERA Ranger Mine, 1995. *Application for Approval, Section 13, Uranium Mining (Environmental Control) Act. Deposition of Neutralised Tailings in Ranger #1 Pit.* Unpublished Report, August 1995. 47pp.
Knight-Piesold, 1992. *Preliminary design for in-pit tailings disposal.* Unpublished Report to ERA Ranger Mine, December 1992. Ref. 518/2. 20pp.
Li, H. & Cramb, G. 1999. *Geotechnical characterisation of in-pit tailings.* Unpublished Report to ERA Ranger Mine, July 1999. 27pp.
Li, H & McGill, J (1999). *Managing tailings at Ranger: strategies to tackle low settled density of tailings in Pit #1.* Report to ERA Ranger Mine. May 1999. 15pp.
Milnes, A. R. 1998. Ranger Mine closure strategies for tailings. *Australian Centre for Geomechanics Seminar on Tailings Management for Decision Makers, 30 April - 1 May, 1998*, Perth. 26pp.
Ringis, J. 1997. *Echo sounder, Pinger and Boomer Surveys in Ranger No 1 Pit and Tailings Dam.* Fugro Survey Pty Ltd Report HY13261, October 1997 (unpublished). 33pp.
Sheng, Y, Peter, P, Wei, Y, Nefiodovas, A & Wright, M. 1997. A rational approach to the prediction, modelling and monitoring of the stability of extremely soft tailings. *Proceedings of 9th International Conference on Computer Methods and Advances in Geomechanics*, Wuhan, China. 7pp.
Sheng, Y., Peter, & Milnes, A. R. 2000. Instrumentation and monitoring behaviour of uranium tailings deposited in a deep pit. *Tailings and Mine Waste'00*: 91-100. Rotterdam: A.A.Balkema.

Tailings and Mine Waste '01 © 2001 Balkema, Rotterdam, ISBN 90 5809 182 1

Closure of the Atlas Uranium Tailings Impoundment

Michael E. Henderson
SRK Consulting, Lakewood, Colo., USA

ABSTRACT: The Atlas Uranium Tailings Facility is a large uranium mill tailings impoundment located on the banks of the Colorado River near Moab, Utah. The tailings facility was operated from the 1950's to the mid-1980's, during which approximately 10.5 million tons of tailings were deposited.

The facility is currently being closed under a NRC-approved plan. Closure of the site is being conducted under the supervision of PricewaterhouseCoopers (PWC), who is the trustee. SRK Consulting (SRK) was selected by PWC to provide geotechnical, hydrogeologic and geochemical consulting services related to characterization and closure planning for the tailings facility.

This paper describes the recent work at the site, including a detailed characterization of the tailings, which was performed using a combination of piezocone and geotechnical drilling techniques. Data from this characterization were used to evaluate various methods of dewatering and speeding consolidation of the tailings. The methods considered included vertical wicks, vacuum-assisted wicks, vibro-consolidation, and a combination of air sparging and vacuum wick drainage. Data will be provided from the various field test programs, comparing modeling and pilot-scale tests.

1 BACKGROUND

1.1 *Site history*

The Atlas Moab Uranium Mill Site is located approximately 3 miles northwest of Moab, Utah along Highway 191 (Figure 1). The Uranium Reduction Company (URC) began operations at the Moab mill in October 1956. Atlas Minerals Company acquired URC in 1962 and operated the Moab mill until 1984 when it was placed on stand-by. Tailings were disposed in the unlined impoundment from October 1956 almost continuously until 1984 using both acid and alkaline circuits. The impoundment covers approximately 130 acres and contains approximately 13.1 million tons of tailings and impacted soils (e.g. interim cover).

Decommissioning of the mill began in 1988, and an interim cover was placed on the impoundment between 1989 and 1995. In 1996, Atlas submitted a revised Reclamation Plan and Technical Specifications.

Atlas submitted an application to the Nuclear Regulatory Commission (NRC) for an amendment to its existing NRC License No. SUA-917 to allow for reclamation of the site. The proposed amendment would allow Atlas to reclaim the tailings pile for permanent disposal and long-term custodial care by a government agency in its current location, and prepare the entire 400-acre Moab site for closure. The amendment to the NRC license required preparation of an Environmental Impact Statement (EIS) to assess potential impacts from the proposed reclamation plan.

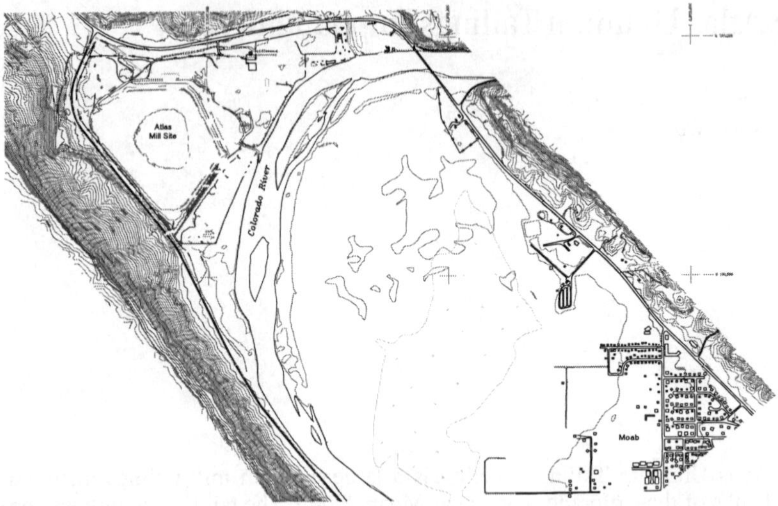

Figure 1.

In September 1998, Atlas filed for bankruptcy and in March 1999 a trust was created to fund future reclamation and site closure. Atlas was released, in principle, from all future liability with respect to the uranium mill facilities and tailings impoundment at the Moab site. PricewaterhouseCoopers (PWC) became custodian of the Moab Mill Reclamation Trust and site closure in 1999.

1.2 *License and regulatory requirements for closure*

Source Material License SUA-917 applies to the Atlas Moab mill site. Under the license, the licensee is required to reclaim the tailings impoundment in accordance with the October 1996 submittal to the NRC entitled "Final Reclamation Plan, Atlas Corporation Uranium Mill and Tailings Disposal Area" and the following modifications outlined in the license:

- Actively dewater the tailings pile by July 1, 2002;
- Cleanup contaminated groundwater to the extent necessary to meet relevant standards within 7 years of Atlas's receipt of the NRC approval of the revised groundwater Corrective Action Plan;
- Before construction of the final radon barrier, provide analyses, appropriately supported by necessary data, showing that the ammonia standards identified in item 2 of the Reasonable and Prudent Alternative of the Final Biological Opinion, (FBO) issued by the US Fish and Wildlife Service will be met over the design life of the reclamation;
- Before commencing with reconfiguration of Moab Wash, the licensee shall provide a design that NRC, after consultation with FWS, finds acceptable to compensate for the loss of 0.5 acre of critical habitat, in conformance with item 5 of Terms and Conditions of the FBO;
- Comply with item 6 of Terms and Conditions in the FBO by:
  1. Monitoring for the southwest willow flycatcher;
  2. Implement construction activities in such a way as to minimize loss of southwestern willow flycatcher habitat and revegetate disturbed area with willow plantings.
- Before commencing construction activities, the licensee shall obtain NRC approval of a plan to implement the mitigative measures identified in the Final Environmental Impact Statement (FEIS).

1.3 *Closure and reclamation requirements (from previously accepted design)*

A tailings closure design proposed by Atlas underwent a detailed review as part of the final EIS and was accepted by the NRC. The soil cover design consists of a layer of coarse tailings and affected soils from the mill area overlain by compacted clay, a layer of sandy soil, and a layer of rip-rap (Figure 2).

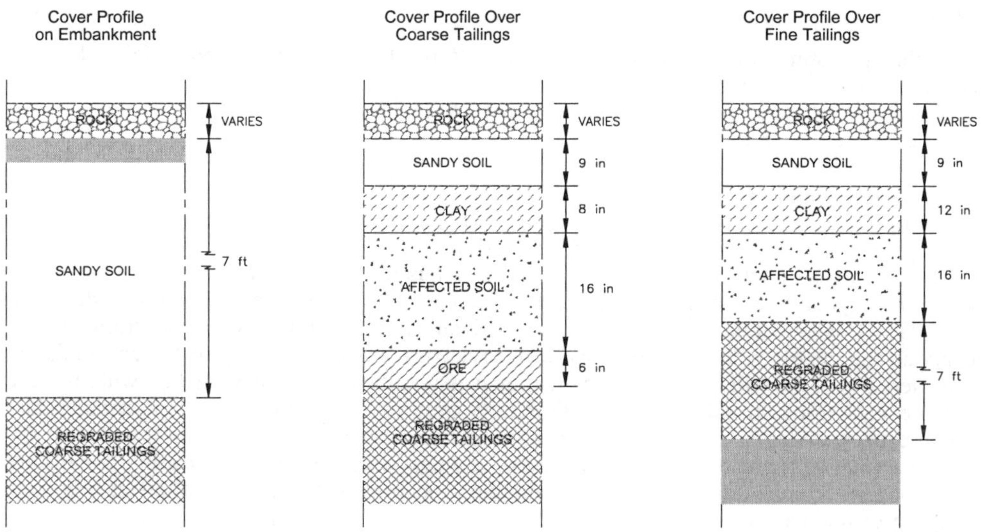

Figure 2.

2 GEOTECHNICAL INVESTIGATION

2.1 *Goals*

The objectives of the geotechnical investigation were to determine the spacial variability and physical characteristics of the tailings, such as grain size and plasticity, as well as global characteristics of the tailings such as permeability and consolidation. This information was then used to assess the potential effectiveness of the different dewatering options.

2.2 *Current field program*

The field program consisted of direct-push, Cone Penetrometer Testing (CPT) to collect in-situ geotechnical data on the tailings pile, and hollow-stem auger drill continuous sampling of tailings for geotechnical and geochemical testing. The CPT was carried out by ConeTec Inc. using a low ground-pressure, track-mounted, cone penetrometer rig. Auger drilling was performed by Agapito Drilling Inc. using a truck-mounted rig. To obtain data evenly across the pile, holes were drilled along an orthogonal grid with a 200-foot spacing surveyed and staked across the pile surface.

The CPT drilling was conducted first to provide the basis for identifying representative conditions within the pile and any zones of specific interest. A total of 96 CPT holes were driven. The in-situ data from the CPT was then used to guide auger drilling locations and sample collection. Ten borehole locations were drilled using the hollow-stem auger rig.

Tailings samples were collected using the auger rig by driving a sampling tube into the tailings and then augering down around it to enable the tube to be extracted. The drive tube was five feet in length. The cores were then described on geological log forms, including such information as: color, grain size, hardness, friability, and relative water content. Notable odors, colors, and the presence of precipitates were also noted. Samples were then placed in plastic bags and either sent to the laboratory for testing or stored on site.

The in-situ data collected using CPT included dynamic porewater pressures, cone-tip and sleeve resistance, and electrical resistivity. In addition, approximately 300 pore pressure dissipation tests were also conducted using the CPT rig. From these data, estimates of material strength, texture, consolidation, and permeability were obtained.

2.3 *Laboratory program*

Geotechnical testing was conducted at Advanced Terra Testing in Lakewood, Colorado. Selected samples were subjected to the following tests:

- Grain size analysis, including hydrometer (ASTM D-422);
- Specific gravity test (ASTM D-854);
- Atterburg limits; (ASTM D-4318);
- Consolidation test (ASTM D-2435);
- Permeability test (flexible wall apparatus) (ASTM D-5084);
- Capillary moisture retention test (ASTM D-3152).

The grain size analysis (including hydrometer) was used to determine material variability within the impoundment and was correlated with the empirical soil classification provided from the CPT. Consolidation testing was conducted on tailings samples obtained from various locations and depths within the impoundment to determine the variability of the consolidation characteristics of the tailings. The consolidation data was then used in conjunction with the pore pressure data from the CPT and the laboratory grain size data to estimate potential future settlement/consolidation of the tailings due to the dewatering and construction of the final cover.

Representative samples of fine and coarse tailings were subjected to capillary moisture retention tests to aid in determining the hydraulic properties of the tailings. The test was done using a pressure plate apparatus.

2.4 *Laboratory data*

Grain Size Analysis

A total of 15 grain size analyses were completed following ASTM D-422 standard. The sieve analyses concluded that the tailings materials could be categorized as either fine-grained silty sands or clayey silts.

Specific gravity

A total of nine specific gravity analyses were performed following ASTM D-854 standard. The results of these tests are presented in Table 1.

Table 1. Results of specific gravity analyses

Property	Lithology	No. of Tests	Value	Mean	Geometric Mean
Specific Gravity	Unconsolidated Silts	1	2.73	-	-
	Consolidated Silts	6	3.01 2.54 2.56 2.72 2.87 3.03	2.79	2.78
	Wet Sand	1	2.7	-	-
	Native Material	1	2.72	-	-*

*Note: Does not include data from piezocone tests.

Consolidation Tests

Consolidation tests were performed following ASTM D-2435 standard. A total of five tests were conducted, resulting in the determination of initial void ratio, dry density, wet density, coefficient of consolidation, and permeability. A summary of these results is presented in Table 2 and Table 3.

Table 2. Results of testing

Property	Lithology	No. of Tests	Value	Mean	Geometric Mean
Initial Void Ratio	Unconsolidated Silts	1	1.9702	-	-
	Consolidated Silts (shallow)	1	2.4374	-	-
	Consolidated Silts (deep)	1	0.9165	-	-
	Native Material	2	1.1047 1.148	1.13	1.13
Dry Density	Unconsolidated Silts	1	57.2	-	-
	Consolidated Silts (shallow)	1	49.4	-	-
	Consolidated Silts (deep)	1	88.6	-	-
	Native Material	2	80.7 79.1	79.9	79.9
Wet Density	Unconsolidated Silts	1	99.5	-	-
	Consolidated Silts (shallow)	1	98.4	-	-
	Consolidated Silts (deep)	1	111.2	-	-
	Native Material	2	85.4 86	85.7	85.7

Table 3. Results consolidation testing

Lithology	Final Water Content (%)	c_v (m^2/day)	K (cm/s)
Unconsolidated Silts	37.5	0.0042	5.68E-6
Consolidated Silts (shallow)	41.3	0.0094	1.57E-5
Consolidated Silts (deep)	28.9	0.7571	1.98E-4
Native Material	19.1	0.1893	2.34E-4

3 HYDROLOGIC EVALUATION

3.1 *Goals*

The goal of the hydrologic evaluation was to understand the existing seepage from the impoundment and to predict the impact that the various closure options will have on this seepage quantity.

3.2 *Phreatic surface in tailings*

Data from the nine temporary monitoring wells installed in the tailings pile as well as piezometric pressure data from the CPT was used to delineate the phreatic surface. The data show a zone of saturation in the slimes area of the tailings up to 40 feet thick, as is illustrated in Figure 3. This zone of saturation is expected since the slimes have a significantly lower permeability than the surrounding coarse tailings and underlying alluvium.

The groundwater table is located 40 feet below the tailings impoundment, and slopes south-southwest toward the Colorado River at a gradient of 0.04%.

3.3 *Modeling methodology*

For determining consolidation, coupled SEEP/W and SIGMA/W codes were utilized. SIGMA/W computed displacements and stresses while SEEP/W computed the changes in pore pressure with time. The following conditions were evaluated:

1. Existing seepage conditions were assessed by using SEEP/W;

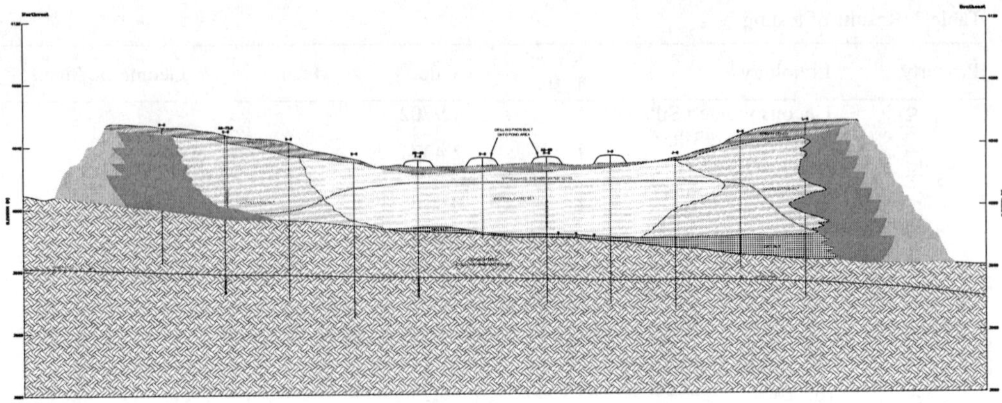

Figure 3.

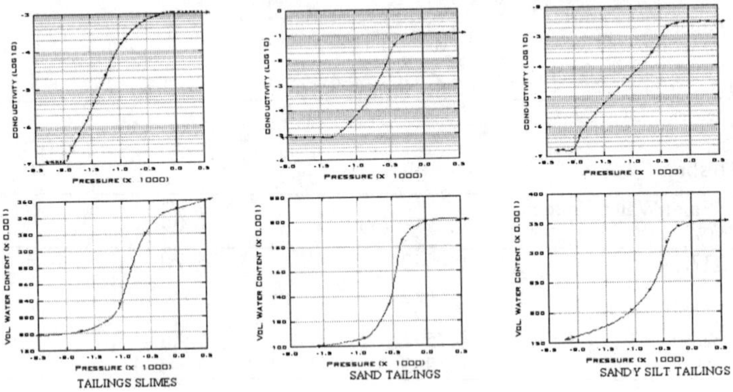

NOTES:

1) The pressure is in pounds per square feet (psf);
2) The hydraulic conductivity is in ft/day.

Figure 4.

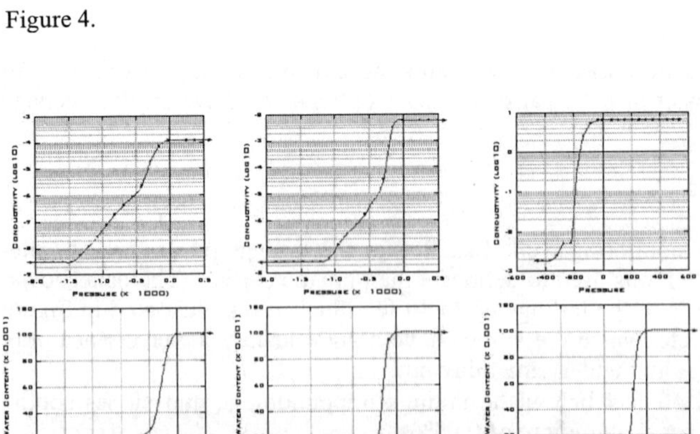

NOTES:

1) The pressure is in pounds per square feet (psf);
2) The hydraulic conductivity is in ft/day.

Figure 5.

2. Total settlement due to soil cover without wick drains was evaluated by SIGMA/W. Seepage into the groundwater under the settlement conditions was assessed using SEEP/W.

The details and results of these analyses are presented below.

3.4 *Existing seepage*

SEEP/W was utilized to estimate the existing seepage into the groundwater at the Atlas Uranium Tailings Facility. SEEP/W is a two-dimensional finite element code that can be used to model the movement of water and the distribution of pore-water pressures within porous materials. It can simulate water flow under both saturated and unsaturated conditions. As water flows through soil, certain amounts of water are stored or retained within the soil structure. The amount of water stored or retained is a function of pore-water pressure and the characteristics of the soil structure. SEEP/W considers the relationships between pore pressure–water content and pore pressure-hydraulic conductivity as continuous functions. Specifying these functional relationships is part of the input requirement for the program.

A finite element mesh was constructed for a typical cross section through the tailings. The material properties and boundary conditions were specified using a total of eight soil types. These included:

1. Tailings slimes;
2. Sandy silty tailings;
3. Sand tailings;
4. Consolidated dry clay tailings at the interface between tailings and the alluvium;
5. Natural fill;
6. Gypsum precipitates in the pond area;
7. Consolidated alluvium; and
8. Alluvium.

For each soil type, hydraulic parameters are specified. These parameters include:

- Hydraulic conductivity function (k_x);
- A volumetric water content function;
- The ratio of the k_y to k_x hydraulic conductivities; and
- The direction of the k_x hydraulic conductivity.

The functional relationships defining these properties are presented in Figures 4 and 5.

The values are obtained from CPT tests, laboratory tests, or field tests. For materials without any site-specific data, typical literature values were used.

In SEEP/W, the boundary conditions may be specified as total head (H), total nodal flow or flux (Q), or flow per unit length along the side of an element (q). The left model boundary for the alluvium was specified as constant head at 3,958 ft amsl (above mean sea level) based on the water levels observed upgradient of the tailings in well ATP-3. The downgradient (right) boundary was also constant head at 3,950 ft amsl based on the river levels.

The area receives about 8-inches of rain annually. It was assumed that approximately 30 percent of the precipitation is currently recharged into the tailings in the vicinity of the tailings pond where gypsum usually precipitates. Outside the area of gypsum precipitation, the recharge is assumed as about 10% of the precipitation. Recharge values were input as a constant flux at the top of the tailings pile. The recharge values of 6.0e-04 ft/d for the center area and 1.8 e-04 ft/d for areas outside the pond area were input to the model.

Results

The model was run as steady-state to estimate the seepage volumes.

The water levels obtained by the model simulations are compared with the field observed water levels (Table 4). The simulated water levels were in close agreement with the field observations, validating the assumption regarding infiltration of precipitation.

Table 4. Comparison of model simulated and field observed water levels

Boring ID	Field Observe Water Levels (amsl) (ft-amsl)	Model Simulated Water Levels (amsl) (ft-amsl)	Difference (ft)
<u>Slimes</u>			
AR7	3997.74	3993	4.74
AR8	4018.1	4018.7	-0.6
<u>Alluvium</u>			
AR4	3956.77	3956	0.77
AR7	3957.27	3956.8	0.47

Seepage from the tailings into the groundwater was estimated by a two-dimensional flux section specified through the tailings in the model. The flux section provides seepage fluxes through that section as ft^3/day/ft of cross-section thickness. This value was 0.95 ft^3/day/ft. Assuming that the tailings is almost circular, the seepage flux is converted to the actual flux by first dividing this value by the length of the pile (~2,800 ft), and then multiplying the resulting value with the bottom surface area of the tailings (6,155,000 ft^2). The estimated seepage flux into the groundwater was about 2,090 ft^3/d (~11 gpm).

3.5 *Settlement analysis following cover placement*

The settlement analysis was conducted by coupling SIGMA/W and SEEP/W. Both programs utilized the same finite-element mesh. SIGMA/W computed the displacements and stresses while SEEP/w estimated the pore pressure at each time interval. Soil cover materials were added to the existing finite element mesh that was described above.

After the cover placement, the recharge through the cover was specified as 3.7 gpm for the cover permeability of 10^{-7} cm/sec (ref: Final Environmental Impact Statement Related to Reclamation of the Uranium Mill Tailings at the Atlas Site, Moab, Utah, NUREG-1531, Vol. 1, March 1999). As a result, the recharge boundary (top surface) was adjusted to this value in SEEP/W. The remaining parameters in SEEP/W were the same as in the existing case.

The SEEP/W mesh was exported into SIGMA/W for settlement analyses. In SIGMA/W, the input parameters for each soil type included the following:

1. Soil model type. For each soil type a linear-elastic model was selected with the exception of slimes. For slimes, a nonlinear-elastic model was specified. For each model type the geotechnical parameters were input. These parameters include a modulus of elasticity (E) and Poisson's ratio (ν). The slope of the soil-water characteristic curve, m_v , in the SEEP/W must reflect the Young's modulus and Poisson's ratio values in SIGMA/W. The relationship defined for these parameters was satisfied in defining the values for E and ν. The following equations was used:

   $$E= \{ (1-\nu)\}/\{m_v * (1+\nu)* (1-2\nu)\}$$

2. The boundary displacement and stress conditions were specified. It is assumed that there will not be any horizontal and vertical displacement in alluvium outside the tailings area. Secondly, no horizontal displacement was allowed on top of the tailings pile.
3. The initial conditions for soil-cover loading were assumed to be represented by the existing conditions in the pile, which are hydrostatic conditions with no excess pore pressure. In actuality, the data shows that there is some excess pore pressure in the existing pile.
4. In order to simulate self-weight consolidation, body forces were specified for the soil cover and the slimes. The assumed unit weights were 105 psf for the soil cover and 99 psf for the slimes.

Results

Settlement in the tailings pile can be divided into three segments. These are initial settlement, primary settlement and long-term settlement. SIGMA/W analyzed all of these settlement cases. The immediate settlement was observed starting from the first time step in the model. The immediate settlement values were about 0.2 ft for the slimes. Time steps in the model were incremented up to about 70 years. The settlement values observed at various depths at F-5 located in the center of the impoundment, are shown in Figure 6. The primary consolidation settlement at 26 feet below ground surface (at top of the slimes) was about 3.7 ft approximately 6 years after cover placement. An additional 0.8 ft of settlement was predicted to occur between year 6 and year 70. The total settlement at F-5 was about 4.7 ft in 70 years.

The seepage flux values obtained by SEEP/W were converted to gpm and plotted against time. The plot is shown in Figure 7. Following the placement of the load, the seepage values increased to about 40 gpm for about two years. Then the seepage values decreased. However, even after 70 years, the seepage values are above the hydrostatic seepage value of 11 gpm.

After the cover placement, pore pressures in the slimes are seen to increase. Then, as the slimes consolidate, the pore pressures gradually dissipate. The dissipation rate decreases after about 6 years. At 70 years after cover placement, pore pressure values are slightly above the initial values.

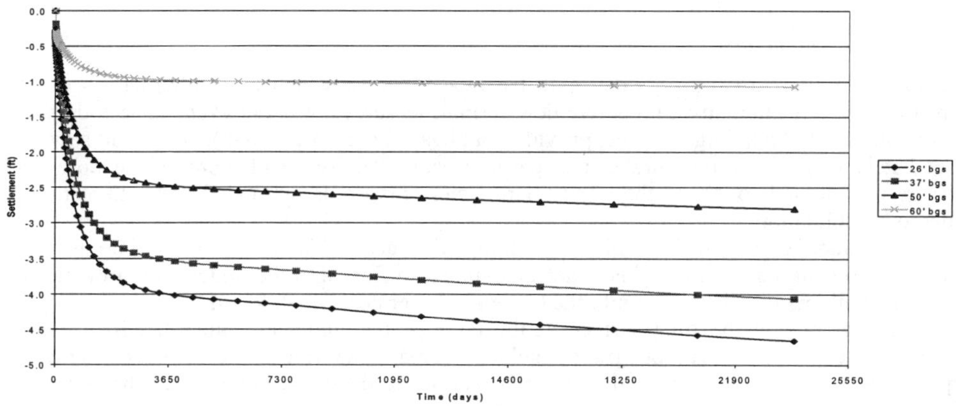

Figure 6.

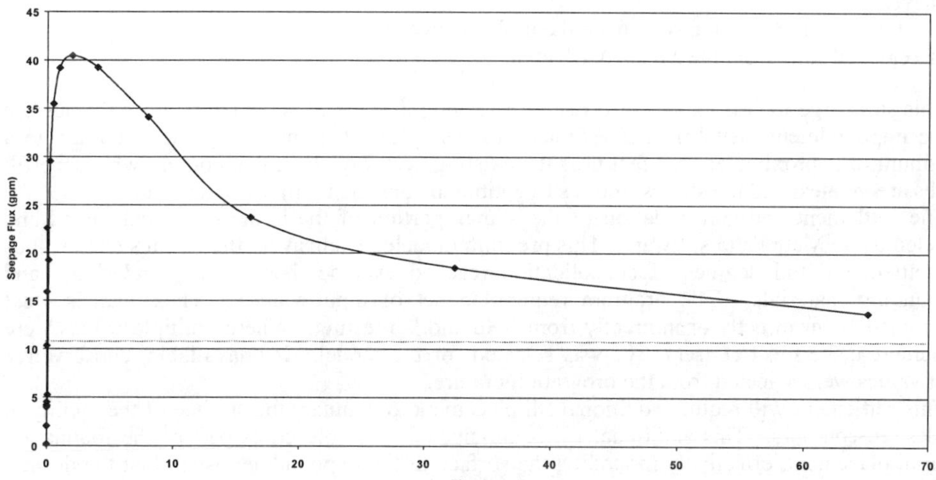

Figure 7.

4 DEWATERING OPTIONS

The various options for dewatering the tailings are primarily aimed at dealing with the volume of water that is expected to be displaced (e.g. squeezed out) as the foundation layer is placed. This foundation layer is the layer of material that is placed on the top and sides of the tailings impoundment to achieve the final grade prior to placement of the final cap.

A condition related to placement of the final cap on the foundation layer is that 90 percent of the anticipated settlement must have occurred prior to cap placement. The rate at which settlement occurs is directly related to the rate at which water is squeezed from the saturated soils. Conversely, partially saturated soils are expected to consolidate relatively quickly, as primary consolidation occurs.

Each of the dewatering alternatives will result in a similar final surface, with similar total settlement. However, the time for 90 percent of the settlement to occur is approximately 6 years under option 1 versus 18 months for the remainder of the options.

The effectiveness of these options to limit solute release from the tailings is also discussed in the following sections. Solute release was assessed by comparing solute loading with the 'no-action' (i.e. no wicks) option for the cover construction and initial consolidation period.

4.1 *Option 1 – no action*

The no-action alternative is provided as a base case against which other alternatives can be compared. In this alternative, no active dewatering measures are undertaken. The foundation layer is placed on the tailings surface, providing a physical surcharge (load). This load serves to immediately consolidate unsaturated and partially saturated soils and increase the porewater pressure in saturated soils. As the excess pore water pressure is dissipated, the saturated soils undergo consolidation.

The consolidation of saturated soils (e.g. slimes) displaces a corresponding volume of water. Under the no-action alternative, this water is assumed to flow into the underlying groundwater table, which ultimately discharges into the Colorado River near Moab Wash.

The topography of the final surface of the tailings impoundment, under the final cap, is shown in Figure 3. This final topography includes a small area near the northwest corner of the impoundment that will be excavated to bring the surface to the desired contours. The remainder of the surface is filled with a mixture of soils, including:

- Excavated tailings and interim cover from the northwest corner of the facility;
- Barium-impacted soils from the river bank between the tailings facility and the Colorado River;
- Radiologically-impacted soils from the mill site area; and,
- Excavated soils from the Moab Wash area.

This surcharge load is expected to result in varying degrees of settlement across the surface of the impoundment. Settlement is estimated to range from 0 near the northwest corner, to a maximum of approximately 5.5 feet near the existing center of the impoundment where the fill depths are greatest and the slimes (more susceptible to consolidation) are extensive.

The settlement and consolidation of the slimes portion of the tailings was calculated and modeled using MebraWin software. This program modeled exclusively the tailings slimes, thus the settlements and degrees of consolidation reached exclude those of the underlying and surrounding materials. The program required a set of input values. These values were determined either directly or indirectly from field and lab testing. Where multiple values were encountered, the most conservative was selected for the model. If unavailable, conservative input values were selected from the program literature.

This settlement will require additional fill placement to maintain the surface of the facility at the desired contours. This additional fill is anticipated to be placed as part of the foundation material placement, effectively mounding the surface of the impoundment such that the desired levels are achieved when the consolidation has stopped.

The time required for 90 percent of the final consolidation to occur is 6 years, as was described in Section 4.

This settlement will effectively squeeze or displace water from the voids in the saturated material. Under alternative 1, this water will flow into the underlying water table.

The total volume of water that will be displaced into the groundwater table under this alternative is 6.5 million cubic feet (cf). On average, this means that the rate of discharge to the underlying groundwater table will increase to 40 gpm over the 18 to 24 months under this option.

The "no action" option will not impose a significant geochemical change in the saturated tailings. Furthermore, assessment of the tailings pore water showed that process water dominates current water quality. This indicates that the rate of displacement of the tailings water is relatively slow. Consequently, because of the slow rate of evolution of the pore water quality, the solute concentrations in water that will be displaced over the construction period (approximately 18 months) is not expected to change significantly from the current average pore water quality.

The pore water discharge to the groundwater regime will occur by one of two pathways. The first is direct discharge at the base of the tailings. The second pathway will involve the lateral displacement of porewater through the unsaturated sandy tailings. These tailings, as indicated by paste pH and conductivity measurements, likely contain accumulated secondary mineral phases that may readily be dissolved as the flow passes through the tailings. Solute concentrations for some parameters may increase in this discharge. However, overall, the flow that will follow this pathway is expected to be small. Therefore, the solute concentrations in the discharge can reasonably be estimated from the average current pore water quality.

4.2 *Option 2 – wick drainage*

Option 2 consists of using conventional vertical wicks to relieve the pore-water pressure that is created from placing the foundation layer, as described above. In concept, in a low permeability soil such as the tailings slimes, the pore water pressure immediately following placement of a surcharge increases an amount equal to the surcharge pressure. Wick drains function by providing a drainage pathway in the tailings, significantly decreasing the time required for pressure (and flow) to be relieved.

The wicks are installed from the surface of the tailings impoundment, with a backhoe-mounted device. In this case, the wicks will be installed through the foundation layer, following its placement. The flow exiting from the wicks will be collected in a surface sump, and pumped to an evaporation pond. This evaporation pond will be located on top of the final cap, in the northwest corner of the impoundment, where additional settlement is not expected.

Based upon the slimes properties obtained during the field program, the wicks will be installed on 3-meter intervals. This spacing allows for 90 percent of the expected consolidation to occur within about 18 months.

In principle, the downward flow of pore water into the underlying aquifer will not increase significantly. The current estimate of seepage utilizing the wick drains is 12 gpm, as opposed to 11 gpm under existing conditions. Most, if not all of the excess pore-water pressure will be collected in the wick system and discharged on the surface of the impoundment, where it can be evaporated.

Based upon a conservative estimate of wick efficiency, it is estimated that 5.85 million cubic feet will ultimately report to the surface of the tailings and 650,000 cubic feet of water will exit into the underlying aquifer.

The wick drains are expected to remain saturated during the dewatering period and, as such, will not result in the influx of air to the tailings. Consequently, the geochemical conditions in the pore water will not change, and similar to the "no action" option, the water quality that will be discharged to the groundwater regime will be similar to background.

4.3 *Option 3 – vacuum assisted wick drainage*

Under this scenario, flows from the wicks described under Option 2 are collected in a surface header system and discharged to the evaporation pond as described above. The collection piping system consists of a network of solid PVC (polyvinyl chloride) pipes that are connected to a vacuum pump. This system serves to decrease the hydraulic head at the top of the wick to below atmospheric pressure, increasing the flow and capture efficiency of the wick system.

The vacuum system envisioned is capable of providing a decrease in atmospheric pressure at the top of the wick of 5-10 psi.

The effect of the vacuum system is that the flow entering the underlying aquifer is expected to significantly decrease, to about 6-8 gpm. The net increase in volume of water discharged to the underlying aquifer is estimated to be 0 in this case.

Depending on how the wick drains are operated, vacuum assistance potentially may cause air influx to the upper regions of the tailings deposit. This may result in oxidation and additional solute release to the pore water. However, the pore water that may be affected will likely be captured by the wick drains. The pore water that will be released to the groundwater regime will originate at depth within the saturated tailings, and is not likely to be affected by any air ingress. Therefore, the solute concentrations in the discharge to the groundwater will be reflected by the average pore-water quality as summarized in Table 5.

4.4 *Option 4 – air sparging*

Under this option, the surface manifold system is selectively connected to either a vacuum pump or to a compressor. The function of this system is to selectively draw water from the tailings and introduce air, removing the ammonia via an air sparging system.

This system would be operated similarly to Option 3, except for the air introduction. However, the slimes would be subjected to cycling of vacuum extraction and air pressurization, which would likely produce a flow to the underlying aquifer similar to Option 2 (12 gpm).

The primary purpose for considering air sparging is the potential for removing excess ammonia from the pore water. Ammonium (NH_4^+) converts to ammonia (NH_3) at alkaline pH values. At high pH, NH_3 can readily be volatilized and vented to the atmosphere through sparging with air. However, this reaction is pH dependent. At the estimated average pore-water pH, only approximately 10 percent of the ammonium-nitrogen will be present as ammonia. At the highest observed pH (9 s.u.), the proportion increases to approximately 15 percent. Irrespective, stripping of ammonia is likely to proceed, but as more ammonium is converted to ammonia, the pH will decrease and the rate of volatilization will slow. Thus, not all of the ammonium-nitrogen will be stripped from the residual pore water. The stripping will further be hampered by the fact that sulfide oxidation will further decrease the pH. The effectiveness of air sparging will, therefore, be limited at best. Air sparging also has significant implications on the pore water quality. Apart from increased solute release from sulfide mineral oxidation, aeration will cause an increase in the redox conditions, and decrease in pH conditions. Solute concentrations were estimated from the pore water quality in samples which indicated higher redox conditions from the MINTEQA2 modeling.

4.5 *Performance evaluation of dewatering options*

For the purpose of this evaluation, in the case of unassisted wick drains, it was assumed that the pore pressures in the tailings would increase slightly in the short term. This would result in approximately a 50% increase over current estimates in the discharge rate through the bottom of the tailings. Furthermore, it was assumed that by implementing vacuum assistance the net increase of seepage out the bottom of the pile would be zero since the vacuum extraction would remove all excess (consolidation) water from the pile.

Selective air-sparging and vacuum assistance would result in a cyclical increase and decrease of pore pressures. The net effect is that the overall discharge rate would be similar to Option 2 with a net increase rate of 6.3 gpm.

Estimated net solute loading (additional to the current seepage) for selected parameters are summarized in Table 5.

The results in Table 5 suggest that the installation of wick drains could decrease solute loading in the short term to about 10% of that which would result from consolidation and construction without any drains. It should, however, be noted that even after 70 years, the seepage values for the 'no action' option would be above the hydrostatic seepage value of 11 gpm, and that the calculations presented in the table only considered short term loading.

Table 5. Estimated cumulative net short term solute loading

Parameter	Estimated Loading Over Cover Construction Period (kg) Basecase – Current Conditions	Case 1 – No Dewatering - Place Cap	Case 2 – Wick Drainage	Case 3 – Vacuum Assisted Wicks	Case 4 - Air Sparge / Vacuum Wicks
Period (months)	18	24	18	18	18
Aluminum	222	1697	127	0	127
Ammonia-N	45352	346876	26016	0	26016
Arsenic	19	147	11	0	11
Boron	11	81	6	0	6
Cadmium	10	77	6	0	6
Calcium	7318	55973	4198	0	4198
Cobalt	12	91	7	0	7
Copper	24	181	14	0	14
Chloride	42726	326787	24509	0	24509
Sulfate	439853	3364208	252316	0	252316
Iron	1465	11202	840	0	840
Lithium	45	348	26	0	26
Magnesium	14179	108445	8133	0	8133
Manganese	603	4610	346	0	346
Molybdenum	194	1485	111	0	111
Nickel	8	65	5	0	5
Nitrate + Nitrite as N	1095	8372	628	0	628
Uranium Natural	706	5399	405	0	405
Vanadium	49	375	28	0	28
Zinc	30	232	17	0	17

Overall, loading for the 'no action' case would be significantly higher than indicated.

Vacuum assistance would result in a decrease in the loading to about 0. The selective air-sparging and vacuum assistance option would have varying impacts compared to the other two drainage options. Overall, solute release for most trace elements that normally would be present as primary sulfide minerals would be expected to increase, whereas only iron (due to solubility control) and ammonium-nitrogen (venting) loading would be expected to decrease. Therefore, the loading will be similar to those expected using the wick drains alone.

The benefits from the wick drains and from vacuum assistance are clearly demonstrated by the table, as are the disadvantages for aeration.

5 RECOMMENDATIONS

Based upon the discussions provided in this paper, the vacuum-assisted wick drainage system was recommended for enhanced tailings dewatering. This recommendation was based upon:

- The discharge flowrate into the underlying aquifer is minimized;
- The mass loading of undesirable chemical constituents into the underlying aquifer is minimized;
- The vacuum system serves to collect and control surface discharge from the wicks; and,
- The system is capable of obtaining the desired 90 percent consolidation level within about 18 months.

The proposed system would consist of approximately 869,000 feet (265,000 meters) of wicks, a surface collection piping system, a vacuum extraction system, and a surface evaporation pond.

This system can be placed immediately following placement of the foundation layer. No significant impacts to the project schedule should result from installation of this system.

6 REFERENCES

ONRL, 1998. *Limited Groundwater Investigation of the Atlas Corporation Moab Mill, Moab, Utah.* Prepared for the U.S. Fish and Wildlife Service, January 9, 1998.

NRC, 1979. *Operation of the Moab Uranium Mill, Atlas Minerals Division.*

NRC, 1999. *Final Environmental Impact Statement Related to Reclamation of the Uranium Mill tailings at the Atlas Site, Moab, Utah.*

Tailings and Mine Waste '01 © 2001 Balkema, Rotterdam, ISBN 90 5809 182 1

Slope stability performance of high heap fill structures for closure

A.J. Breitenbach
AB Engineering Incorporated, Littleton, Colo., USA

ABSTRACT: The exterior slopes of high heap fill structures are typically constructed as steep as practical during mining operations to maximize the amount of ore tonnage on the geomembrane-lined leach pad foundations. The maximum allowable heap slope angle is determined in design stability analyses for maintaining stable slope conditions to the planned ultimate heap height.

A major environmental concern for closure is the long-term slope stability performance of the relatively high granular fill structures at the end of operations. This paper presents a post-mining slope stability study of heap fill slopes after completion of ore lift loading and leaching operations. The study considers the operational change in heap fill strengths and geomembrane liner interface strengths with respect to high fill loads. This study shows that the post-mining heap slope stability for closure should increase with time, which is in agreement with the historical performance of geomembrane-lined fill structures.

1 INTRODUCTION

Gold, silver and copper ore heap fill structures generally have existing or planned final fill heights in the range of 100 to 200 feet (30 to 61 meters). Some of the highest heap fills exceed 300 feet (91 meters) in vertical height, especially at the copper mine sites. The ore heap fills consist of granular crushed or run-of-mine rock materials placed in controlled loose lifts and wetted during leaching operations for removal of the target ore metals.

The gold and silver heap leach pads have been constructed and operated within the last 20 years with geomembrane-lined foundations. The copper heap leach dump operations have generally changed to geomembrane-lined foundations and interlift liners within the last 5 to 8 years. Only a few of the lined heap fill operations are in the final stages of leaching and/or rinsing for reclamation and closure.

This paper focuses primarily on the long-term slope stability of heap fill structures on geomembrane-lined leach pads at the end of operations for post-mining closure. The underlying heap foundation conditions are assumed to be high strength and stable for this study. Some aspects of the heap slope stability analyses may be applicable to the post-mining conditions of leach dump and waste dump fill slopes, as well as granular landfill slopes.

2 BACKGROUND

2.1 *General*

Limited published information is available concerning the estimated long-term slope stability performance of high fill structures on geomembrane-lined foundations. This section provides an overview of heap construction with references to the historic performance and past field and laboratory test studies of granular fill materials and geomembrane liner systems under high fill load conditions.

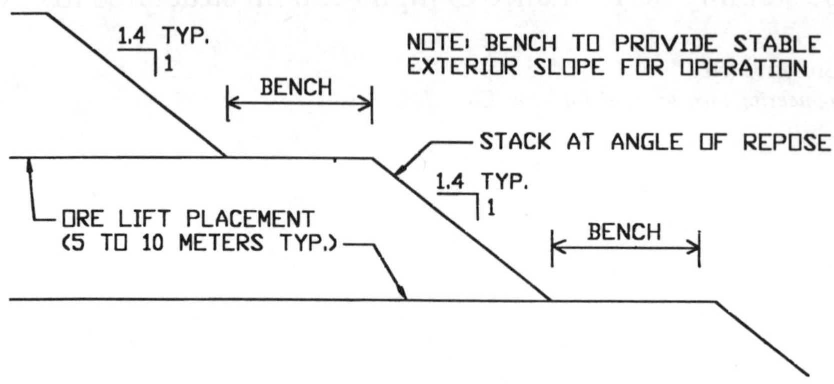

FIGURE 1 - HEAP SLOPE SECTION

2.2 *Heap Construction*

The construction of heap fills involves placement of precious or base metal ore materials in controlled individual loose and relatively dry fill lifts stacked at the natural angle-of-repose. Each ore lift surface is wetted uniformly by using irrigation drip emitters or sprinklers and leaching with alkaline (gold and silver) or acidic (copper) process solutions. The maximum rock particle size of the granular ore materials range from run-of-mine cobble and boulder rock sizes to crushed sand and gravel sizes. The crusher operations may include agglomeration as needed to provide a more efficient distribution of fines (minus No. 200 sieve size) for improved permeability and recovery of the target metals. The individual ore lifts are offset with benches along the exterior slope as required for establishing the overall stable design slopes for operations. A schematic section of the exterior ore heap slope is shown in Figure 1.
Each individual ore lift is typically placed at 15 to 30 feet (5 to 10 meters) in height and thoroughly wetted during prolonged leach cycles of generally 3 months or longer. The controlled leach application flow rates are in the range of 0.002 to 0.005 gpm/sf (less than 0.2 lit/min/sm), which keeps the granular ore materials moist and unsaturated. The multiple ore lift fill placement and controlled wetting from leaching causes the fully drained granular fills to consolidate and gain strength with each lift from self-weight settlement and densification. The ore fill settlements from wetting and densification generally range from 7 to 10 percent for silver and gold ore heaps and 10 to 15 percent for copper ore heaps. The relatively long operational leach cycles allow any excess fill or foundation pore pressures to dissipate with time from each successive heap ore lift load.

2.3 *Historic Performance*

The past slope failures on geomembrane-lined fill structures, such as solid waste landfills, heap leach pads, and reclamation cover fill caps, have shown that liner induced slides generally occur at the planar geomembrane liner interface contact with weaker underliner or overliner materials. One of the earliest and most known geomembrane induced slope failure was the Kettleman Hills landfill slope failure in Northern California in 1988 (Mitchell et al. 1990). Other lesser known leach pad slope failures, occurring between 1985 and 1993 at several mine sites located in North America, South America and Australia, have also shown the importance of stabilizing the downhill toe of geomembrane-lined leach pad surfaces (Breitenbach 1997). The Northridge earthquake in Southern California in 1994 (Matasovic et al. 1995) and subsequent earthquakes in Chile and Peru in 1995 and 1996 gave some insight into the seismic behavior of high fills on geomembrane liner systems.

The end result of the historic performance of fill structures on geomembrane liner systems is that wedge slip failures along the planar liner interface contact with weak soils or synthetic geotextiles generally occur during the initial fill placement operations, rather than at the higher fill lift heights. The only exceptions were fill structures with weak foundation conditions beneath the liner system. These foundation failures beneath the liner system included fill loads that reactivated existing landslides, thawing of frozen subgrade soils, subgrade subsidence of compressible soils, or excavations at critical downgradient fill toe areas that caused unloading of support fill materials.

2.4 *Test Studies*

2.4.1 *Granular Fill Strengths*

The estimated strengths for coarse grained granular soil fills or rock fills can be determined by surveyed field measurements of the natural angle-of-repose of the actual fill slope surfaces or a review of past large

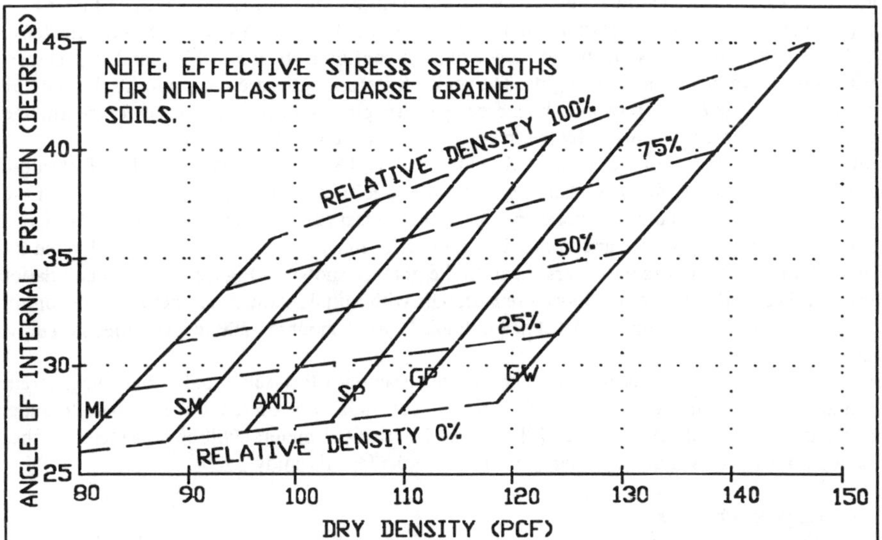

FIGURE 2 - FRICTION STRENGTH VS. DENSITY

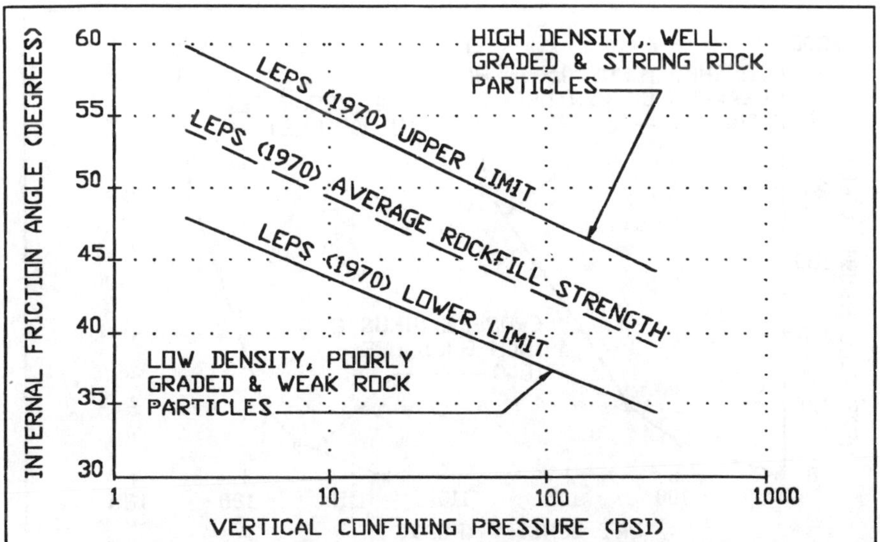

FIGURE 3 - ROCKFILL STRENGTH VS. PRESSURE

scale laboratory direct shear box and triaxial cell strength testing. The difficulty in accurately sampling and testing the larger rock sizes in the heap fills for strength, density and gradation generally limit the engineer to past experience and published test results for comparison to site specific conditions.

The rock size (small versus large), rock quality (weak versus strong), rock shape (round versus angular), particle size distribution (poorly graded versus well graded), fines content (clayey and silty versus clean), plasticity (plastic versus non-plastic), moisture content (wet versus dry), confining stress (low versus high), and relative density (loose versus dense) influence the fill material strength properties. The relative change in strength with respect to rock particle size, distribution and relative density for coarse grained soils is illustrated graphically (NAVFAC 1982), as shown in Figure 2. The graph shows that the densification of coarse-grained fills and an increase in the rock particle size from sands to gravels significantly increases the shear strength of the fill materials.

Large scale rock fill and granular soil fill test strength studies by Leps (Leps 1970) and later by Marachi (Marachi et al. 1972) and Barton (Barton and Kjaernsli 1981) are summarized graphically in Figure 3. These test summaries show the influence of high confining stresses, which tend to decrease the rock fill strengths with increasing fill heights. The rock fill test strengths are more applicable to the run-of-mine ore fills, but generally show that the increase in strength from densification of coarse grained soils and rock fills eventually is offset by the increase in vertical confining stress loads.

In addition to the strength summaries presented in Figures 2 and 3, a determination of the fill strength with respect to depth requires an understanding of the site specific consolidation characteristics of the wetted fill materials. Large diameter laboratory consolidation test results of crushed angular minus 0.5 inch (12.7 mm) leached copper ore and minus 2.25 inch (57.2 mm) gold ore rock particles, loaded to an equivalent 200 to 300 feet (61 to 91 meters) heap fill height, are shown in Figure 4. The consolidation tests show a rapid densification of the loose ore materials with initial loading, and generally no significant additional increase in the fill density for fill height loads beyond about the 200 feet (61 meters) equivalent fill height.

The test studies indicate that a loosely placed and wetted granular fill should rapidly increase in strength during initial fill lift placement to an optimum depth, and eventually begin to reduce in strength under higher confining stress loads. An assumed fill depth of 200 feet (61 meters) will be considered as the optimum depth for the maximum granular heap fill strengths for this study.

2.4.2 *Liner Interface Strengths*

Large scale laboratory direct shear box test results for long-term liner performance under high fill loads indicate the geomembrane liner interface contact with underlying and overlying fill materials gains

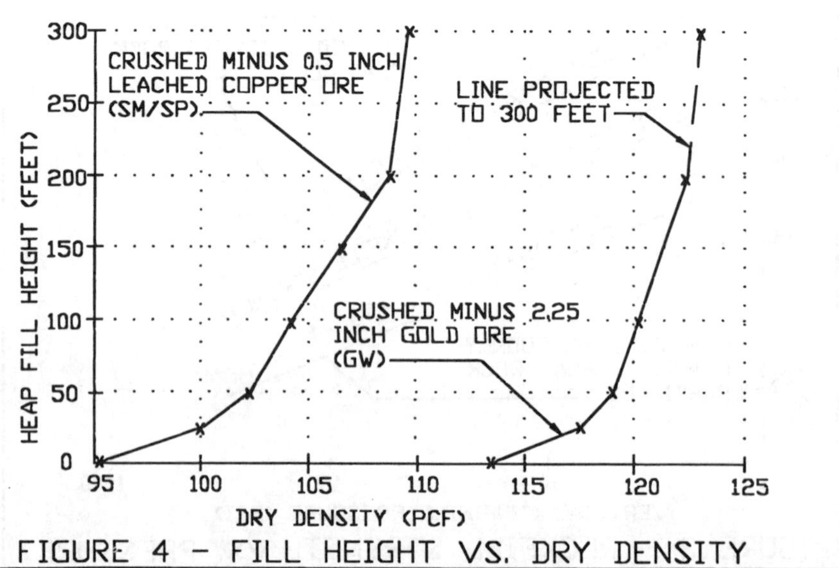

FIGURE 4 – FILL HEIGHT VS. DRY DENSITY

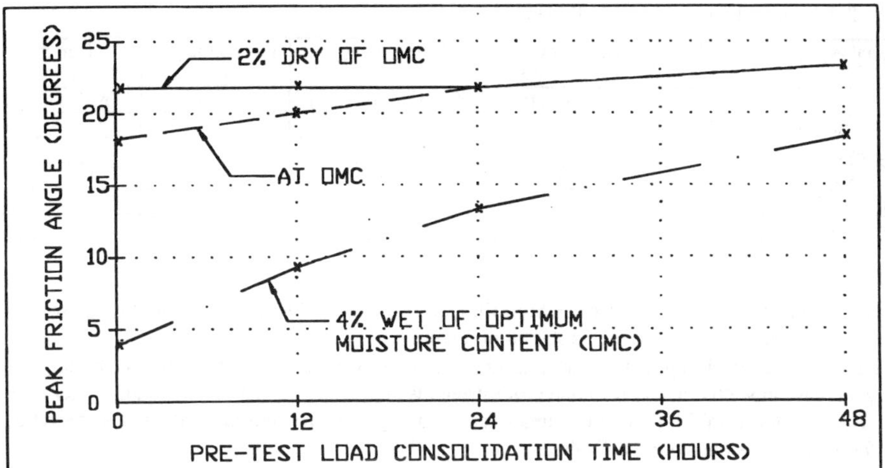

FIGURE 5 - INTERFACE LINER STRENGTH VS. TIME

strength with time (Breitenbach and Swan 1999). The long term increase in liner interface strength is mainly due to the apparent affect of load deformations or dimpling on a micro scale in the planar geomembrane liner surface and a reduction in underliner excess pore pressure conditions with time. The high load deformations cause the interface sliding failure surface to shear through a portion of both the underlying and overlying soil materials. This dimpling affect results in an overall measured increase of typically about 5 degrees of interface friction strength in a 24 to 48 hour period following placement of the high test loads on a composite liner system at optimum underliner soil moisture conditions, as shown in Figure 5.

The measured dynamic laboratory geomembrane liner interface strengths from simulated cyclic earthquake conditions on several composite geomembrane liner systems indicates the interface strength remains constant or increases from dynamic earthquake loads (De and Zimmie 1997, Nicola and Filippo 1997). Therefore dynamic interface strengths were not considered for the long term stability analyses.

3 SELECTED STRENGTHS

The selected strengths for long-term heap fill slope stability analyses are based in part on the background information obtained from the historic performance and test studies, and in part on site specific material characteristics and engineering judgement in the design and construction of heap and mine waste dump fills. For example, copper ore heap fills are generally weaker than silver and gold ore heap fills due to copper acid leaching and accelerated chemical weathering, which affect the rock quality of the granular fill particles. Typical ore fill strengths were selected for this analyses.

The selected ore strengths for design analyses are generally conservative and do not account for the densification and gain in heap fill strengths with depth or the added interface liner strength from the non-planar liner dimpling affect. For simplification purposes, a moderate and uniform heap fill and liner strengths were assumed for the design slope stability analyses with both smooth (design) and dimpled (construction) interface liner strengths, as presented in Table 1. Note that the interface dimpling strength

TABLE 1 – Assumed design and construction strength parameters

Material type	Heap height		Moist density		Friction angle (degrees)	Cohesion	
	(feet)	(meters)	(pcf)	(g/cm3)		(psf)	(kPa)
Liner surface (smooth)	0 to 300	0 to 91	62.4	1.00	18	0	0
Liner surface (dimpled)	0 to 300	0 to 91	62.4	1.00	23	0	0
Heap layer	0 to 300	0 to 91	115	1.84	36	0	0

TABLE 2 – Assumed operation strength parameters

Material type	Heap height		Moist density		Friction angle (degrees)	Cohesion	
	(feet)	(meters)	(pcf)	(g/cm3)		(psf)	(kPa)
Liner surface (dimpled)	0 to 300	0 to 91	62.4	1.00	23	0	0
Heap layer 1	0 to 25	0 to 8	115	1.84	36	0	0
Heap layer 2	25 to 50	8 to 15	119	1.91	38	0	0
Heap layer 3	50 to 100	15 to 30	121	1.94	39	0	0
Heap layer 4	100 to 300	30 to 91	122	1.96	40	0	0

assumed for construction is dependent on high fill loads, as well as the moisture condition of soils in contact with the geomembrane liner, as shown on Figure 5.

The selected operation strengths for post-mining slope stability at closure reflect both the anticipated relative increase in wetted and consolidated ore strengths and densities with depth and the liner dimpling affect. The long-term stability section assumes four heap layers increasing in fill density and strength with depth for analyses, as presented in Table 2.

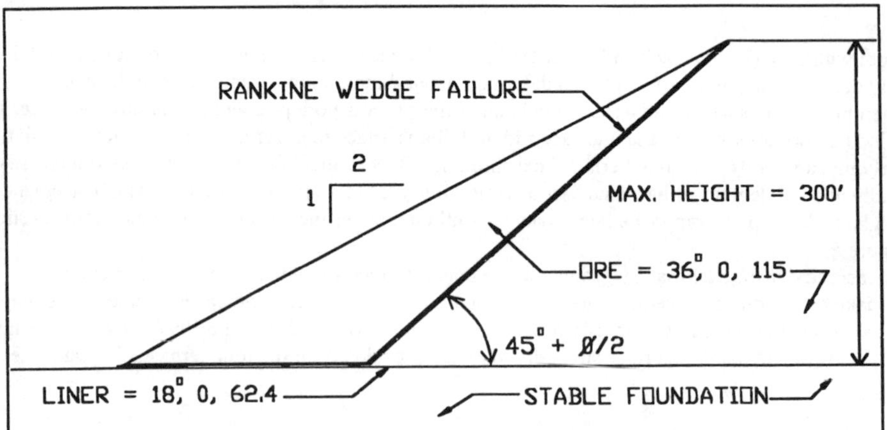

FIGURE 6 – DESIGN STUDY SECTION

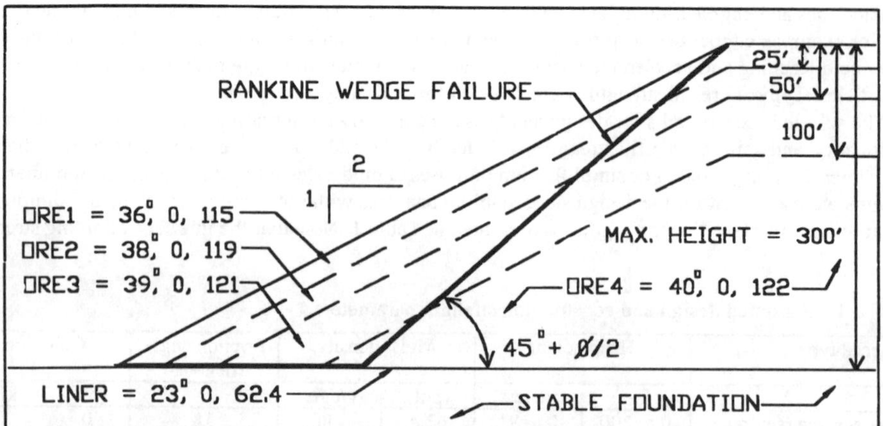

FIGURE 7 – OPERATION STUDY SECTION

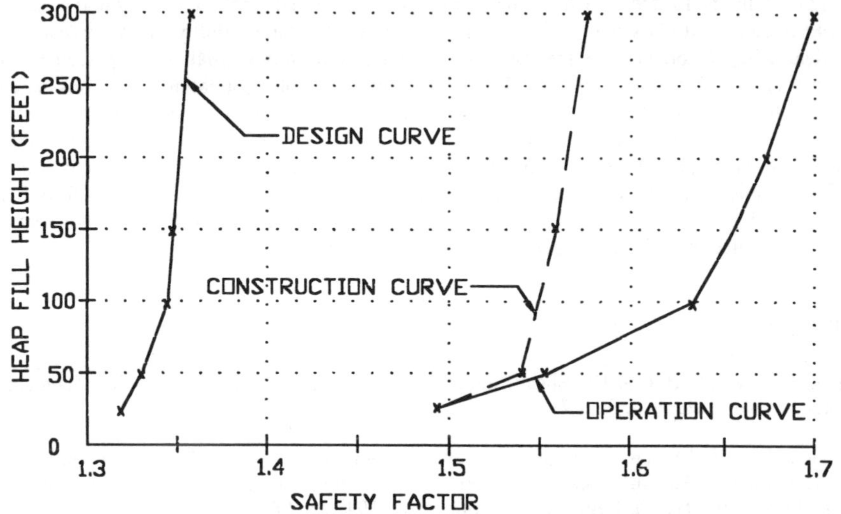

FIGURE 8 - FILL HEIGHT VS. SAFETY FACTOR

4 IDEALIZED STUDY SECTION

The idealized design and operation study sections for slope stability focus on the relative change in ore heap fill densities and related fill and liner strengths under static wedge failure conditions. The idealized study sections assume a maximum heap height of 300 feet (91 meters), 2 horizontal to 1 vertical overall exterior slopes, a flat geomembrane liner foundation surface grade, and fully drained conditions with no phreatic water level. The foundation was assumed competent and stable beneath the heap fill and liner system.

Static wedge planar failure conditions were analyzed at vertical heap height intervals ranging from 25 to 300 feet (8 to 91 meters) in vertical height for both the design and operation study sections. The assumed failure wedge extends from the heap crest height to the interior liner surface and then to the heap toe area, as determined by the Rankine Method. The maximum design and operation study sections for analyses are shown in Figures 6 and 7, respectively.

5 STABILITY ANALYSES SUMMARY

A design, construction and operation stability analyses summary of safety factors (SF) versus fill heights is shown in Figure 8. The stability analyses summary reflects static wedge failure planes from the crest of each heap height to the geomembrane liner surface. Underlying foundation conditions were assumed to be stable for this analyses.

The design stability analyses show a slight increase from SF = 1.32 to 1.36 for a uniform ore fill strength over an incremental fill height of 25 to 300 feet (8 to 91 meters), respectively. Modifying the interface liner strength by 5 degrees friction to reflect micro scale liner dimpling under a high heap fill load condition during construction significantly increases the heap stability to SF = 1.57 at the 300 feet (91 meter) fill height. Additionally modifying the ore fill strength to reflect operational wetting and densification of the ore heap fill increases the end of operation stability analyses to SF = 1.70 at the 300 feet (91 meter) fill height.

6 CONCLUSIONS

The design stability analyses for construction of loosely placed ore heap fills is typically conservative for safe operations and do not consider the increase in fill and liner strength with time during operations. The operational conditions for the fully drained granular ore fill materials include wetting and consolidation under controlled ore lift placement and leaching operations. The slope stability summary results show the most critical safety factors occur at the startup of operations during the initial ore lift placement. The heap stability increases with continued ore lift placement, assuming stable foundation conditions beneath the heap fill and liner system. The overall increase in the ore heap slope stability for the idealized study section from startup construction to final operations shows a 25 percent increase in the safety factor.

The long-term slope stability of high fills on geomembrane liner systems for closure is anticipated to increase with time, assuming the slopes are protected from erosion and the heap remains fully drained on a stable underlying foundation. The stability analyses summary results are in general agreement with the historic performance of high fills on geomembrane liner systems.

7 REFERENCES

Barton, N. & Kjaernsli, B. (1981); "Shear Strength of Rockfill", Proceedings of ASCE, Geotechnical Testing Journal, Volume 7, pp. 873 to 891.

Breitenbach, A.J. (1997); "Overview Study of Several Geomembrane Liner Failures Under High Fill Load Conditions", Geosynthetics '97 Conference Proceedings, Industrial Fabrics Association International, Long Beach, California, Volume 2, pp. 1045 to 1062.

Breitenbach, A.J., & Swan Jr., R.H. (1999); "Influence of High Load Deformations on Geomembrane Liner Interface Strengths", Geosynthetics '99 Conference Proceedings, Industrial Fabrics Association International, Boston, Massachusetts, Volume 1, pp. 517 to 529.

De, A. & Zimmie, T.F. (1997); "Factors Influencing Dynamic Frictional Behavior of Geosynthetic Interfaces", Geosynthetics '97 Conference Proceedings, Industrial Fabrics Association International, Long Beach, California, Volume 2, pp.837 to 849.

Leps, T.M. (1970); "Review of Shearing Strength of Rockfill", Proceedings of ASCE, Soil Mechanics Journal, Volume 4, pp. 1159 to 1170.

Marachi, N.D., Chan, C.K., & Seed, H.B. (1972); "Evaluation of the Properties of Rockfill Materials", Proceedings of ASCE, Soil Mechanics Journal, Volume 1, pp. 95 to 114.

Matasovic, N., Kavazanjian Jr., E., Augello, A.J., Bray, J.D., & Seed, R.B. (1995); "Soil Waste Landfill Damage Caused by 17 January 1994 Northridge Earthquake", Woods & Seiple, (eds), California Department of Conservation, Division of Mines and Geology.

Mitchell, J.K., Seed, R.B., & Seed, H.B. (1990); "Kettleman Hills Waste Landfill Slope Failure", Volume I: Linear Systems Properties, Geotechnical Engineering Journal, ASCE, 116 (4), pp. 647 to 668.

NAVFAC (1982); "Correlation of Strength Characteristics for Granular Soils", Department of the Navy, Naval Facilities Engineering Command, Alexandria, Virginia, Soil Mechanics Design Manual 7.1, Figure 7, p. 149.

Nicola, M. & Filippo, M. (1997); "Behavior of Geogrids Under Cyclic Loads", Geosynthetics '97 Conference Proceedings, Industrial Fabrics Association International, Long Beach, California, Volume 2, pp. 961 to 983.

Tailings and Mine Waste '01 © 2001 Balkema, Rotterdam, ISBN 90 5809 182 1

Rockfall analysis and prediction at the Summitville Mine Superfund Site

Steve Laudeman
Colorado Department of Public Health and Environment, Denver, Colo., USA

ABSTRACT: Past open pit mining at the Summitville Mine Superfund Site has left an 800 foot high pit highwall that has been and continues to be a source of rockfall. Much of the source area for rockfall includes rock with a high sulfur content, which, when exposed to water, produces acid rock drainage. Future reclamation of the area located immediately adjacent to the toe of the highwall will include a rock catchment area, and regrading and revegetation of the area beyond the catchment. A means was required to determine if revegetation in the area beyond the catchment would be adversely impacted by acid generation from rockfall debris. The Colorado Rockfall Simulation Program (CRSP) was used to estimate current rockfall parameters at the highwall and predict future rockfall behavior. Rockfall parameters were initially estimated from literature guidance and field observations. These parameters were further refined by calibrating the model to observed locations and sizes of rockfall debris currently in place at the site. The results of the calibration were used to predict future rockfall stopping locations for different diameter rocks. The predicted stopping locations suggest that the size of the design catchment area is adequate and that reclamation of the area beyond the catchment should not be adversely impacted by future rockfall.

1 INTRODUCTION

Mining for precious metals has occurred at the Summitville area since the 1870's. The most recent mining activity occurred from 1984 to 1992, during which time the Summitville Consolidated Mining Company, Inc. operated an open pit mine and heap leaching operation (Gormley, et al., 1994). Excavation of the mine pits included removal of the northeast flank of South Mountain, which resulted in a pit highwall which ranges up to approximately 800 feet in height and extends laterally around 3000 feet. The overall angle of the highwall is approximately 45 degrees, with locally steeper sections separated by benches. The face of the highwall consists of several rock types, which are the result of the hydrothermal alteration associated with the Summitville mineralized zones. Generally, resistant silicified rocks are the more prominent features on the highwall, and are surrounded by argillically altered material which weathers easily (Rocky Mountain Consultants, 2000). Observations of rockfall at the base of the highwall suggest that the silicified outcrops are the source of most rockfall debris.

2 CURRENT RECLAMATION PLAN

The reclamation plan currently being implemented at Summitville was developed by the Bureau of Reclamation under contract to the Colorado Department of Public Health and Environment ("the Department"). The reclamation is taking place under Operable Unit IV of the Interim Record of Decision, signed in 1994. Previous work by contractors to the

Environmental Protection Agency included backfilling of the mine pits with waste rock. This operation was completed in 1995. The current reclamation of the mine pit area includes placing additional fill over the mine pits and grading and ditch construction to promote surface water drainage off of the pit area. After final grading, the subgrade will be amended with compost and granular limestone, amended topsoil will be placed at a thickness of six inches, and a site-specific seed mix will be drill-seeded into the topsoil. Figure 1 shows the grading plan for the South Pit area. Note the proximity of the westernmost edge of the fill to the toe of the highwall.

During reclamation design, there was some concern regarding the occurrence of rockfall from the highwall and its potential impact on the reclaimed areas at the base of the highwall. To address these concerns, the fill in the South Pit area was designed to include a reverse slope near the base of the highwall, creating a catchment area between the fill area and the highwall. The fill slope on the downslope side of the catchment would be constructed of waste rock removed from other parts of the site, and would be sloped at 3H:1V, with a maximum height of about 15 feet. The bottom width of the catchment area formed would vary in width from 25 feet up to 100 feet.

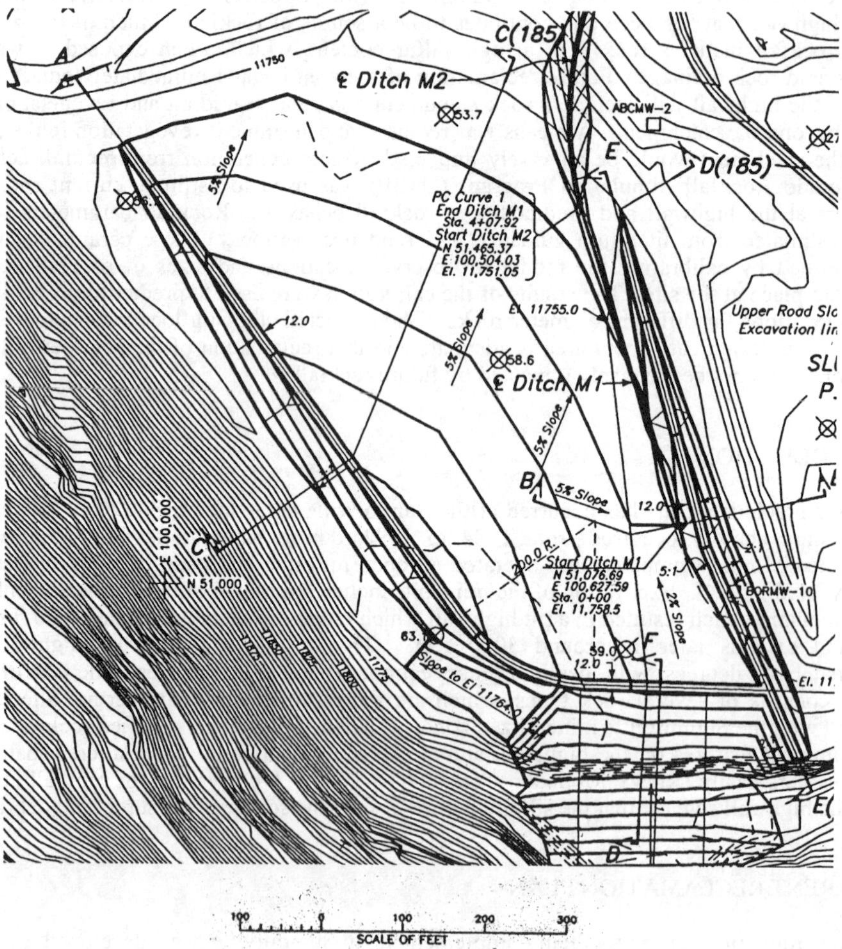

Figure 1. Grading Plan for South Pit Area

3 ROCKFALL RECONAISSANCE AND PREDICTION

Subsequent to the completion of the reclamation design, further discussion regarding the potential from rockfall in the former mine pit area prompted further investigation. Some members of the public advisory group and technical consultants to the Department expressed concern regarding future rockfall impacting reclaimed areas. Also, the significant fill being placed near the toe of the highwall raised some concerns regarding construction safety during fill placement. To address these concerns, the Department undertook to record observations of current rockfall debris and catchment areas, and to attempt to predict future rockfall stopping locations.

3.1 *Reconaissance*

The author completed an initial field reconaissance of the highwall and rockfall catchment areas in July, 1999. This included mapping approximate catchment areas at two locations: near the toe of the highwall at the South Pit, and on a large bench above the North Pit. Specific data collected during this phase included the approximate limits of each catchment area and the size of rocks within each catchment area. Information on the approximate size, shape, and location of the largest rocks in each catchment area was also recorded. Due to limited time and resources, no quantitative ground surveys were performed. However, the data collected appears to be adequate for the analyses performed.

Results of the July, 1999 mapping are shown in Figures 2 and 3. The catchment area at the South Pit ran along the toe of the highwall for approximately 900 feet. The furthest rockfall debris was located approximately 110 feet from the toe. The majority of the rocks were in the 8 inch to 10 inch diameter range. The largest rocks were approximately 2.5 feet in diameter. The location of these rocks was noted on the field maps. In the South Pit catchment area, eight rocks of this size were noted. The North Pit bench catchment area catches rockfall from a slope of the highwall which trends almost due north, as opposed to the larger slope above the South Pit which slopes toward the northeast. The North Pit catchment area runs along the toe of this slope for approximately 500 feet. The furthest rocks are located approximately 80 feet from from the toe of the slope. As at the South Pit, most rocks in this catchment area were in the 8 inch to 10 inch diameter range. The largest rocks were 2.5 feet in diameter. Two of these rocks were present, at the locations shown on the map. At both catchment areas, the rockfall present was characterisitic of the silicified zones of the highwall mentioned above: angular, iron stained, and relatively durable compared to the highly altered material.

The initial reconaissance of the highwall showed that rockfall was occurring as expected, and the results indicated that some future rockfall may impact reclaimed areas. The quantity of rockfall debris accumulated in the South Pit catchment area over a four year period (1995 to 1999) suggested that a significant quantity of rockfall may be expected in the future. Similarly, the North Pit catchment area, which has been relatively undisturbed since active mining occurred - approximately 12 years - showed a notable quantity of rockfall. Due to the likelihood of ongoing rockfall, and to address concerns raised as noted above, the initial data collected during the July, 1999 reconaissance was used to attempt to predict future rockfall occurrence.

3.2 *Colorado Rockfall Simulation Program*

The method selected for rockfall simulation was the Colorado Rockfall Simulation Program (CRSP) (Colorado Department of Transportation, 2000). CRSP was selected as the simulation program because it was readily available, it has been verified by a number of researchers (ibid, Chapter 2), and the author had some limited previous experience with it.

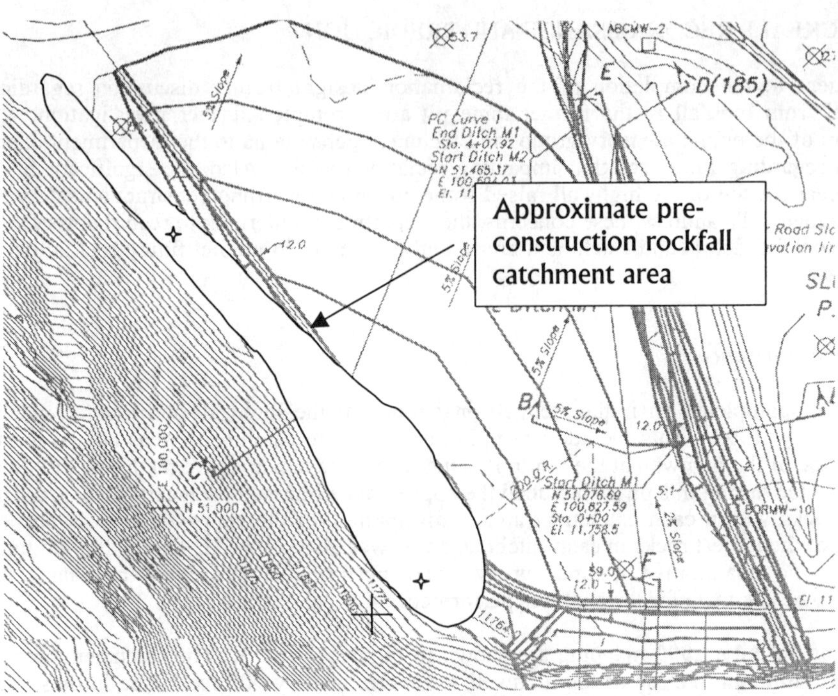

Figure 2. Rockfall Catchment Area at South Pit. (Note locations of 2.5 ft diameter rocks at stars.)

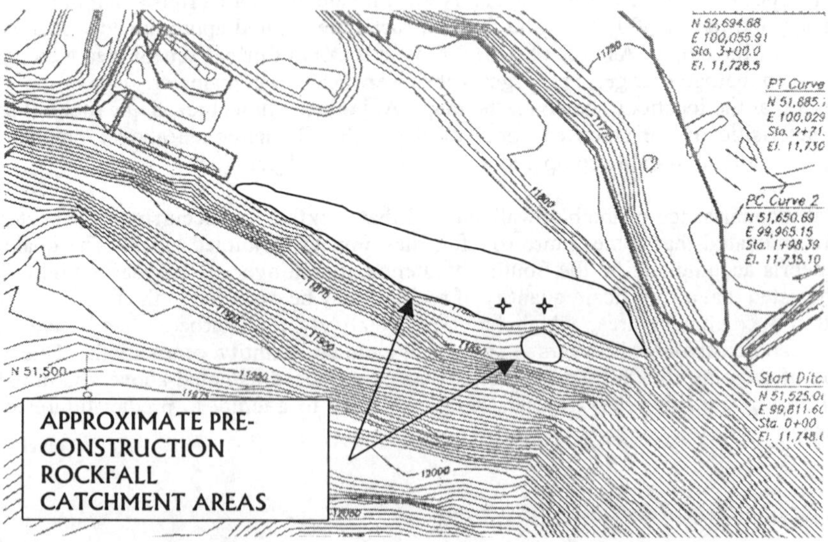

Figure 3. Rockfall Catchment Area at North Pit. (Note locations of 2.5 ft diameter rocks at stars.)

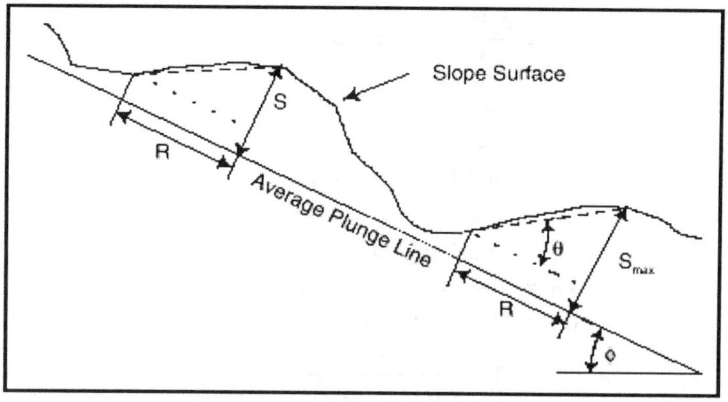

Figure 4. Surface roughness (S) established as the perpendicular variation from an average plunge line (defined by slope angle ϕ) over a distance equal to the radius of the rock R. Maximum slope variation (θ_{max}) is defined by S and R. (from: Colorado Department of Transportation, March, 2000)

Rockfall from manmade and naturally occurring slopes is controlled by a number of slope and rock factors, namely, slope geometry, slope material properties, rock geometry and rock material properties (Ritchie, 1963). Under each of these broad headings are several specific slope and rock parameters which are of greater importance to the CRSP algorithm. Among these are slope surface roughness, slope material coefficients, and the size and shape of the rock. Greater emphasis is placed on these parameters because selection of representative values can be somewhat subjective, and simulation results are often very sensitive to variation in these parameters. Other parameters, specifically slope inclination and length, are very important to the simulation, but are often predetermined for a selected slope profile through survey data or from topographic maps. In CRSP, slope surface roughness is defined as an angular variation from the overall slope angle. A maximum variation to be used is specified in the program input. The method to determine the maximum slope angle variation is shown in Figure 4. Slope material coefficients control the change in translational and rotational velocity as a falling rock impacts the slope surface. The two coefficients are thus designated as R_t, which is the measure of frictional resistance parallel to the slope, and R_n, which is the measure of the degree of elasticity of a collision normal to the slope. The properties of the rock, as opposed to that of the slope, are also critical in a rockfall simulation. The most critical rock properties for CRSP are the size and shape of the rock. Rock material properties also effect a rock's behavior, but CRSP accounts for such effects in the vaules of the slope coefficients R_t and R_n (Colorado Department of Transportation, 2000).

The CRSP algorithm begins rockfall with a rock departing the slope within a specified source area and at a nominal horizontal and vertical velocity. Gravitational acceleration acts upon the rocks trajectory until it intercepts the slope. Upon impact, the pre-impact values for rotational and translational velocity are used, in conjuction with slope coefficients, to determine the post-impact rotational and translational velocities, via the following relationship:

$$\left(\frac{1}{2}I\omega_1^2 + \frac{1}{2}MV_{t_1}^2\right)f(F)SF = \frac{1}{2}I\omega_2^2 + \frac{1}{2}MV_{t_2}^2 \quad (2)$$

where:

M = rock mass

I = rock moment of inertia

$$I = \frac{2MR^2}{5} \text{ (for a sphere)}$$

$$I = \frac{MR^2}{2} \quad \text{(for a disk)}$$

$$I = \frac{MR^2}{4} + \frac{ML^2}{12} \quad \text{(for a cylinder, L = length)}$$

ω_1 = initial rotational velocity

ω_2 = final rotational velocity

V_{t1} = initial tangential velocity

V_{t2} = final tangential velocity

$f(F)$ = friction function

$$= R_t + \frac{1 - R_t}{\left(\frac{V_{t_1} - \omega_1 R}{20}\right)^2 + 1.2}$$

SF = scaling factor

$$= \frac{R_t}{\left(\frac{V_{n_1}}{250 R_n}\right)^2 + 1}$$

In this expression, the effect of the slope material coefficients in seen in the friction function and in the scaling factor. As can be seen, the friction function is a function of the difference between the translational and rotational velocities of the rock prior to impact. The scaling factor is a function of the velocity of the rock normal to the slope prior to impact. The numerical expressions for both of these parameters are based in part on experimental results.

After an impact, the new tangential and normal velocities are determined by the following expressions:

$$V_{t_2} = \sqrt{\frac{R^2\left(I\omega_1^2 + MV_{t_1}^2\right)f(F)SF}{\left(I + MR^2\right)}} \qquad (4)$$

and

$$V_{n_2} = \frac{V_{n_1} R_n}{1 + \left(\frac{V_{n_1}}{30}\right)^2} \qquad (5)$$

Expression (4) is arrived at by solving equation (2) for V_{t2}. Expression (5) is a function of the coefficient of normal restitution, the pre-impact velocity, and a velocity-dependent scaling factor.

For each simulation run, CRSP models 100 rockfall events with the given input data. To represent the randomness of natural rockfall occurrences, CRSP randomly varies the slope angle at each impact within the range allowed by the surface deflection S specified by the user.

3.3 *Data Collection*

Data required for input into CRSP consists of the slope section topography, slope roughness, tangential and normal coefficients, and designation of the rockfall source zone. The topographic information was obtained from the 1997 topographic map for the site. The two sections selected were those above the North Pit bench, and above the South Pit at the point where the fill reaches the closest point to the toe of the highwall. Surface roughness

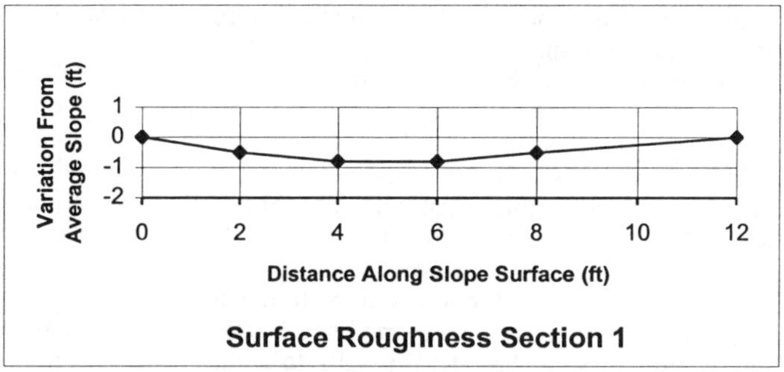

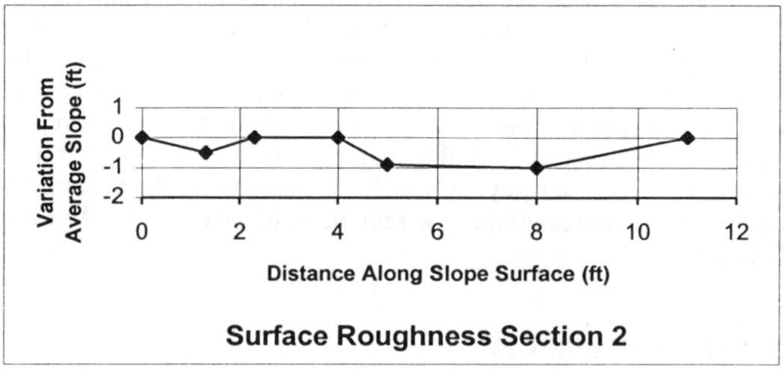

Figure 5. Surface roughness sections near base of highwall.

information was obtained at two locations on near the base of the highwall, using the method described in the CRSP manual, which involves laying a survey tape along the slope at an angle parellel to the overall slope angle, and measuring offset from the tape to the slope surface. Only two locations were measured due to difficult access and dangerous conditions at the highwall.

The data thus collected are shown graphically in Figure 5. The data for the surface roughness was also plotted as rock radius (R) versus slope offset (S). This allowed selection of the most representative S value for each rock size analyzed, as described below, for input into CRSP.

In addition to surface roughness measurements, visual observations of the slope material were made in order to estimate values for the slope coefficients. The slope conditions were compared to the descriptions for typical slopes provided in the CRSP manual, which allowed selection of initial values for these parameters.

3.4 *Simulation and Results*

The data collected were input into CRSP, and several trial runs were performed for the North Pit section, initially, and for the South Pit section after some refinement of input parameters. At the North Pit section, field observations showed two 2.5 ft diameter rocks located at approximately 25 feet from the toe of the highwall. In the CRSP input file, this location equated to 435 feet along the rockfall slope cross section, i.e., the rocks were location 435 feet from the crest of the highwall slope. Linear regression was used to select a the best fit line for the R vs. S data at the two surface roughness sections, thus allowing selection of an S value to input into the program for a rock size of 2.5 feet. The source area for the North Pit section was selected based on field observations and review of aerial photographs. After the input of the initial data, several minor refinements of the surface coefficients were required to produce rock stopping locations from CRSP that matched the stopping locations observed in the field. After these refinements, the CRSP results for rock stopping location matched well with the field observations, i.e., a majority of the rocks in the simulation stopped at locations between 430 and 440 feet along the rockfall section.

The refined input values from the North Pit calibration were used as input for the South Pit simulations. The section for the South Pit simulations was selected as the most critical for rockfall impact to reclaimed areas, as the South Pit fill is closest to the toe of the highwall at this location. If rockfall were to overtop the slope on this side of the South Pit fill, it was assumed that this would happen at this location. The South Pit simulation started with modeling rockfall for 2.5 ft diameter rocks, as this was the size of rock observed on the ground at this location during the initial field reconaissance. The source zone for the South Pit section was selected as reaching from the top of the lower slope to the crest of the highwall, based on field observations and review of aerial photographs. Some minor adjustment of slope coefficients was required to calibrate the CRSP results to actual observations. These same input values were used to model rockfall for 5.0 ft diameter rocks at the same location. The 5.0 ft diameter size was selected as twice the size of any observed rockfall during the preceding 4 years, and was assumed to be a representative worst case for future rockfall events.

The simulation results for 2.5 ft diameter rocks for both the North Pit and South Pit sections matched well with observations from the field. These results provided a degree of confidence that the CRSP algorithm and selected input values were able to realistically represent rockfall conditions at the highwall. The results for 5.0 ft diameter rocks at the South Pit location show that rocks of this size did not pass far beyond the 2.5 ft diameter rocks in the earlier analysis. For both rock sizes, the results indicated that the reclaimed areas of the South Pit would not be impacted by rockfall events.

4 SUMMARY AND CONCLUSIONS

Due to conerns that future rockfall might impact reclaimed areas adjacent to the mine highwall at the Summitville Mine site, the Colorado Rockfall Simulation Program was used to model rockfall events. Input parameters for CRSP were determined from literature guidance and field observations. Also, field reconnaissance was performed to determine current rockfall behavior form the highwall.

Selected CRSP input parameters were modified slightly to calibrate simulation results with observed rock size and rock stopping locations. After this initial calibration, future worst case rockfall was modeled to predict if reclaimed areas could be impacted from future rockfall. The results from all simulations show that reclaimed areas adjacent to the mine highwall will not be impacted by future rockfall events.

REFERENCES

Gormley, John T., and Barbara A. Filas, *The Summitville Mine - What Went Wrong?*, in *New Directions in Geotechnical Engineering*, ASCE Conference, April 1994

Rocky Mountain Consultants, *Draft Remedial Investigation Summitville Mine Superfund Site*, February 2000

Colorado Department of Transportation, *Colorado Rockfall Simulation Program, Version 4.0*, March 2000

Ritchie, A. M., *The Evaluation of Rockfall and Its Control*, Highway Research Record, National Academy of Sciences - National Research Council, Washington, DC, Number 17, pp 13 - 28, cited in Colorado Department of Transportation, March 2000

Tailings and Mine Waste '01 © 2001 Balkema, Rotterdam, ISBN 90 5809 182 1

Assessing the engineering performance of tailings using a centrifuge

K.J. Stone
Division of Civil Engineering, School of the Environment, University of Brighton, UK
T.A. Newson
Department of Civil Engineering, University of Dundee, UK

ABSTRACT: This paper presents the results of a programme of centrifuge experiments performed to provide well-controlled, highly instrumented test data on the engineering performance of crusted tailings. This type of information is vital for the development of analytical and numerical methods for the prediction of bearing capacity and settlement of crusted tailings storage surfaces. Ultimately this will provide accurate analytical tools to assist in the assessment of tailings disposal and rehabilitation strategies for storages. Different mixtures of sand and kaolin clay were used to represent a range of typical tailings materials. The performance of shallow footings and cone penetrometer tests placed on the consolidated and crusted tailings mixtures were evaluated. The effects of changes in groundwater level on the footing behaviour and tailings strength were also assessed during the testing programme.

1 INTRODUCTION

The decommissioning and rehabilitation of tailings storage facilities is an issue of major concern to mine operators. Once the facility has been decommissioned, it is common for part of the rehabilitation process to include a self-generating top cover for the storage compatible with the surrounding area. This will typically comprise of some form of capping layer, e.g. a thickness of 1-2m of suitable rock dump, covered with a layer of seeded topsoil (or growth medium). Implicit with these requirements are the conditions that the tailings surface has sufficient strength to support vehicular traffic during the placement of the capping layer (see Figure 1) and that primary consolidation has essentially finished, so that further large surface deformations do not occur under the additional loads imposed. The economic penalties faced by mine operators who cannot complete this task during the life of the mine can be quite severe.

Many tailings disposal management methods rely on the beneficial effects of self-weight consolidation (and in arid climates the additional surface evaporation) to generate sufficient strength in the tailings deposit such that rehabilitation can proceed. The consolidation behaviour of tailings has been previously investigated using both physical and numerical models, e.g. Fahey and Toh (1991, 1992) and Stone et. al (1999). Similarly, the contribution of surface crusting due to evaporation has also received attention, e.g. Fahey and Fujiyasu (1999), Rassam and Williams (1997) and Newson et at (1997). More recently Rassam and Williams (1999) have presented work on the bearing capacity of desiccated tailings whereby the matric suctions generated in the crustal zone are represented as a contribution to shear strength of the material. This data was interpreted using numerical analysis and traditional bearing capacity formulae to assess the bearing capacity of desiccated tailings.

The work reported in this paper was performed to provide well-controlled, highly instrumented test data on the engineering performance of crusted tailings. This will allow development and validation of analytical and numerical methods for the prediction of bearing capacity and settlement of tailings storages. Ultimately this will provide suitable analytical tools to assist in the assessment of tailings disposal and rehabilitation strategies for storages. The engineering performance of the crusted tailings has been assessed using centrifuge modelling of the consolidation behaviour, strength gain and bearing capacity of different mixtures of sand and fine grained (kaolin clay) tailings material subject to various boundary conditions.

2 METHODOLOGY

2.1 *Material properties and characterisation*

The fundamental properties of the sand and kaolin clay fractions were determined by laboratory testing and are summarised below in Table 1.

Table 1: Material properties of sand and clay fractions

Sand		*Kaolin Clay*	
Specific Gravity (Gs)	2.65	Specific Gravity (Gs)	2.7
Median Diameter (D_{50})	0.37 mm	Liquid Limit (LL)	99.0 %
Effective Diameter (D_{10})	0.17 mm	Plastic Limit (PL)	25.0 %
Coefficient of Uniformity	2.5	Linear shrinkage (LS)	15.5 %
Maximum void ratio (e_{max})	0.83		
Minimum void ratio (e_{min})	0.56		

The basic consolidation characteristics of the tailings mixtures were determined in the laboratory through a series of Rowe cell tests. The results of these tests are summarised below in Table 2. The different mix ratios were chosen to represent a range of typical tailings materials and the proportions shown refer to the dry weights of each material.

Table 2: Summary of laboratory consolidation tests

Ratio of clay to sand (% clay)	Compression Index, C_c	Swelling Index, C_s	Consolidation coefficient, c_v (m^2/yr) [at 37.5 kPa]	Consolidation coefficient, c_v (m^2/yr) [at 150 kPa]	Compressibility m_v (m^2/MN)	Permeability k (m/yr)
1:2.5 (29%)	0.333	0.0183	1.4	3	3.85	0.054
1:3.1 (24%)	0.185	-	6	11	0.83	0.050
1:4.12 (20%)	0.330	0.026	15	55	2.84	0.42
1:5.16 (16%)	0.249	0.024	45	225	2.13	0.94

2.2 *Centrifuge modelling*

A potentially highly effective experimental technique is available for consolidation studies using scaled modelling under enhanced gravitational fields in a large centrifuge. Centrifuge modelling has increasingly been recognised as a powerful tool in geotechnical engineering (Schofield, 1980; Taylor, 1995), which allows rapid, repeatable and relatively inexpensive modelling with well-defined boundary conditions. Since the stress-strain behaviour of soils is highly non-linear and therefore stress-level dependent, accurate scaled modelling requires that the *in situ* stresses at prototype (full) scale can be accurately reproduced in the model. This can only be achieved if models are tested in an elevated gravity field, such as that produced by the rotation of a centrifuge. Thus the stress distribution within a 1/N scale model tested at N gravities will be identical to that in the prototype, given similitude in soil properties. For centrifuge modelling, time for seepage force similarity and conductive heat transfer both scale as $1/N^2$, where N is the number of gravities. The centrifuge technique therefore offers an ideal approach to the study of tailings settlement processes. These advantages have been exploited previously in the analysis of consolidation and transport processes, Arulanandan et al (1988), Hensley (1989) and in particular for the consolidation behaviour of tailings, Fahey and Toh

(1991, 1992). A summary of the scaling relationships relevant to the current study are presented below in Table 3.

Table 3: Centrifuge scaling relationships

Parameter	*Scaling Relationship*
Acceleration	N
Seepage velocity	N
Length	1/N
Stress	1
Strain	1
Force	$1/N^2$
Time (diffusion)	$1/N^2$
Time (creep)	1

2.3 *Centrifuge testing procedures*

A schematic representation of the centrifuge test package is shown in Figure 2. The essential elements of the test package are as follows: (i) an aluminium strong box of internal dimensions 650 mm long x 390 mm wide x 325 mm high; (ii) a standpipe and valve arrangement to allow changes in the water table during the test; (iii) two dual-axis servo-electric actuator units placed over the strong box to allow in-flight strength measurements to be made and stress controlled footing tests to be performed; (iv) settlement (laser transducer and target arrangement) and pore pressure transducers are placed to monitor surface settlements and pore pressures respectively.

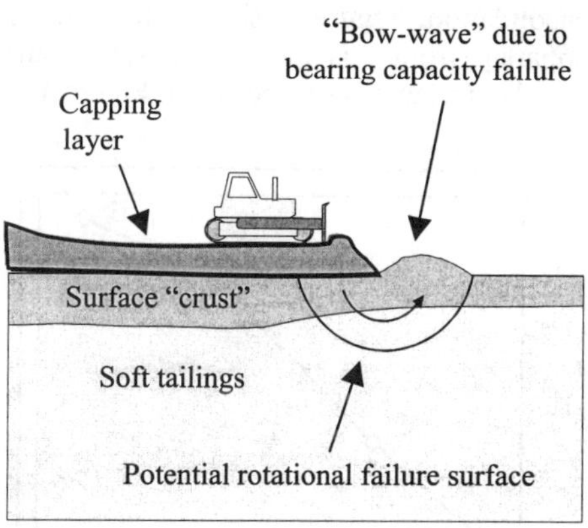

Figure 1: Bearing capacity failure during capping

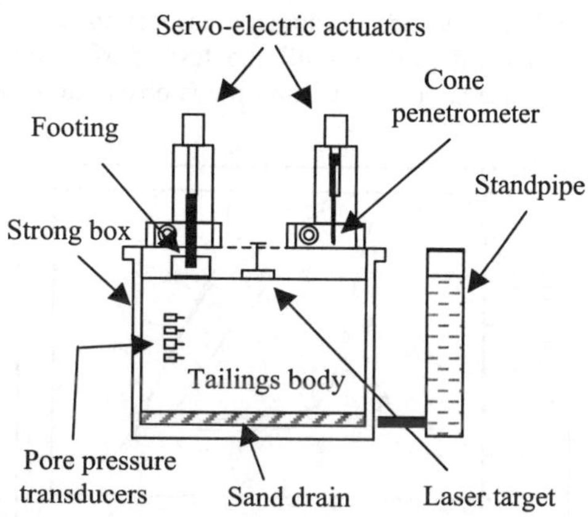

Figure 2: Schematic of centrifuge apparatus

Five centrifuge tests were performed in the laboratory programme reported herein; the test procedure remaining essentially the same for all of the tests. However, there were differences in the water table positions, which are described below. For all of the tests top and bottom drainage conditions were allowed. A summary of the main elements of the test procedures is given below:

(a) The strong box containing the slurried tailings (to a depth of approximately 300mm) was placed on the centrifuge platform and accelerated to a speed equivalent to 100 standard gravities and an initial cone penetrometer test was performed after approximately 1 year of prototype time.

(b) After completion of primary consolidation, a cone penetrometer and a vertical footing test were performed. The centrifuge was then stopped and surface water decanted off and tube samples taken to establish solid and moisture content profiles. The centrifuge was then re-started and a further cone penetration test performed after reconsolidation.

(c) For test A (16% clay) the initial water table was set to 6m below the level of the tailings after completion of primary consolidation. For test B (16% clay) the water table was set at the base of model for the duration of the test. For tests C (20 % clay), D (24 % clay) and E (29 % clay) the water table was initially set at or above the tailings surface, and dropped to 5m below the tailings surface after completion of primary consolidation.

(d) Once equilibrium conditions had been established with a lowered water table, both a cone penetration test and footing test was performed. After a footing settlement of 10% of the footing diameter, footing control was set to a stress hold condition and the sample was resaturated by surface inundation, increasing the base water head or both.

(e) A further cone penetration test was performed after model resaturation had occurred.

(f) After stopping the centrifuge tube samples were taken to record moisture contents.

3 TEST RESULTS

3.1 *Consolidation and settlement*

For each centrifuge test the surface settlement and pore pressures were continuously monitored. From these plots it is possible to estimate the time required to achieve primary consolidation and the surface settlement associated with that consolidation. Figure 3 shows the surface settlement plots for all the tests performed. It is apparent from this plot that a significant increase in consolidation time is observed for the higher clay mixtures as indicated in Figure 4.

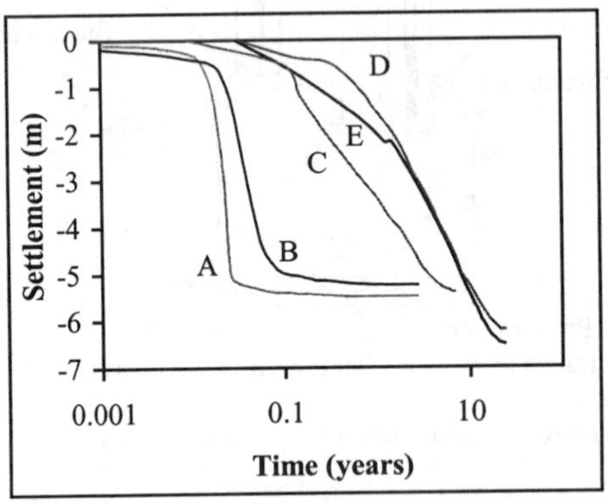

Figure 3: Settlement v. time for all of the tests

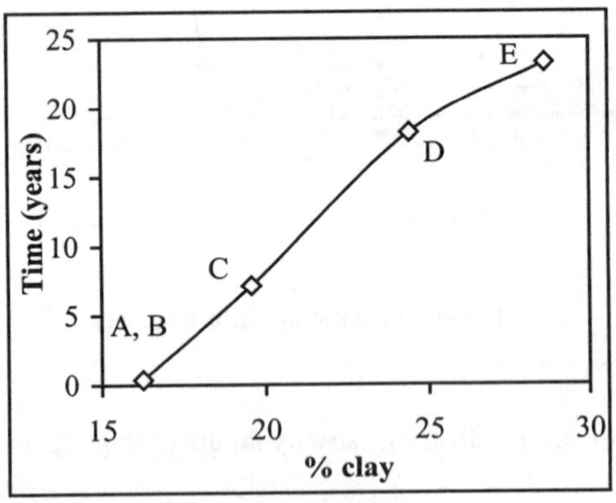

Figure 4: Time to end of primary consolidation

However, the overall settlements during primary consolidation are relatively unaffected by the mix proportions and are about 20-30% of the initial depth. A summary of primary consolidation settlement and times are presented below in Table 4, along with measured additional settlements resulting from lowering of the water table below the model surface. This suggests that for these materials, given a high disposal rate, tailings clay proportions in excess of 17-20% would be unsuitable for valley disposal methods (or for in-pit disposal), because of the excessively long time that would be required for the thick layers of poorly consolidated clayey material to settle.

Table 4: Summary of consolidation and settlement data

Test No.	Mix ratio clay:sand (% clay)	Time to primary consolidation	Settlement during primary consolidation (m)	Additional settlement from water table lowering (m)
A	1:5.16 (16)	100-150 days	5.4	-
B	1:5.16 (16)	100-150 days	5.2	-
C	1:9.12 (20)	7 yrs	5.5	0.8-1.0
D	1:3.1 (24)	18 yrs	6-6.5	0.6-0.8
E	1:2.5 (29)	23 yrs	6-6.5	-

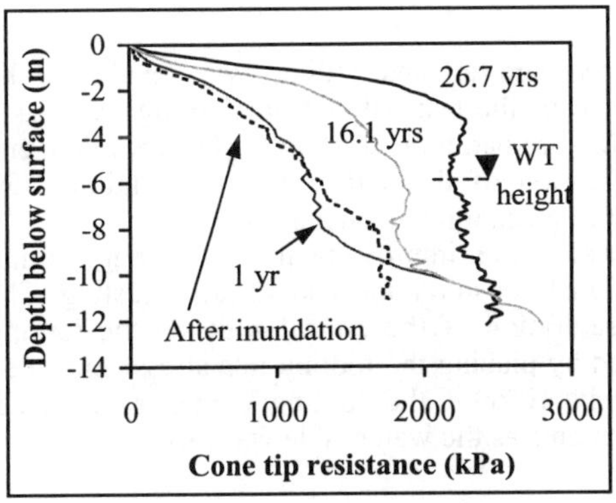

Figure 5a: Cone test sequence for test A

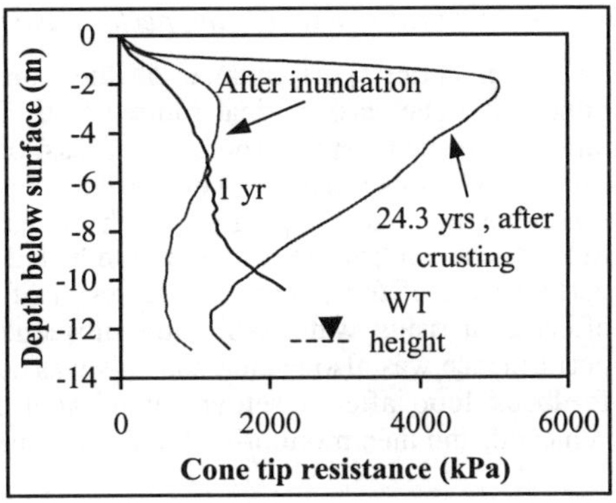

Figure 5b: Cone test sequence for test B

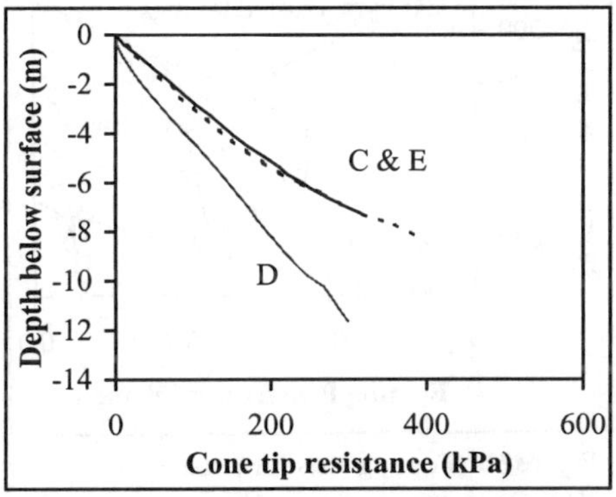

Figure 5c: Cone tests after primary consolidation

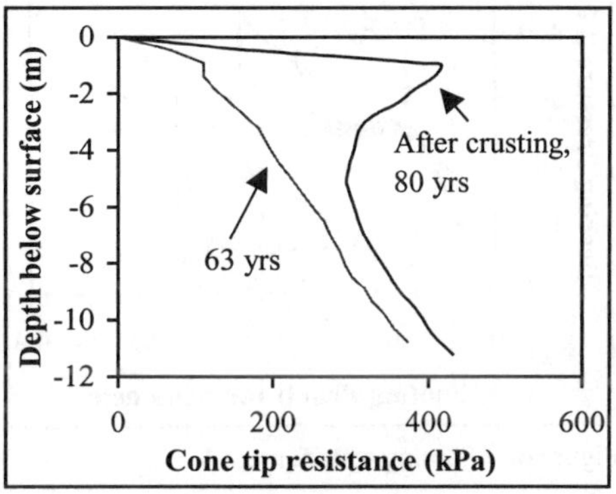

Figure 5d: Cone tests for test D pre/post crusting

3.2 *Strength*

The strength of the tailings was assessed at various stages during the centrifuge test using a cone penetrometer mounted one of the dual axis actuators. Figure 5a shows a typical sequence of cone penetration tests recorded during centrifuge test A (16% clay).

It is apparent that the strength gain in the consolidating tailings results from two mechanisms, namely: the strength associated with self-weight consolidation (compare the cone profiles at 1 and 16.1 years) and secondly the strength gain associated with the lowering of the water table leading to further self-weight consolidation, but more significantly crusting of the surface material (compare the cone tests at 16.1 and 26.7 years). A more dramatic effect of crusting is apparent in Figure 5b where the water table is maintained at the base of the model. It is also observed in Figures 5a and 5b that the effect of raising the water table either by surface inundation (Test A) or from increasing the hydraulic head at the base of the model (Test B) is to significantly reduce the strength of the surface crust.

It is apparent from the test results that the proportion of clay has a significant effect on the cone penetration profiles. For the low clay mixture (Tests A and B) the cone tip resistance at the end of primary consolidation is significantly greater than that for the high clay mixes (Tests C, D and E), compare Figures 5a and 5b and Figure 5c. For all of the tailings mixtures significant improvements in the cone resistance is observed as the result of lowering the water table. There is a large difference in the measured cone tip resistances for the high and low clay materials, which is evident by comparing the results in Figures 5a and 5b, with 5c and 5d. Inspection of this data suggests that for the tailings mixtures with a low clay content (Tests A and B) the cone tip resistance is best interpreted as if penetrating a sand deposit, whereas for the higher clay mixtures the cone data is best interpreted as a cohesive deposit.

3.3 *Assessment of engineering performance*

The engineering performance of the consolidated and crusted tailings was evaluated by conducting standard vertical footing tests. For each of the tests, the initial footing test was a strain controlled test incorporating at least one unload-reload loop and was performed at the end of primary consolidation. This provided an indication of the suitability of the tailings for rehabilitation at an early stage, which can have definite economic advantages.
Subsequent footing tests were conducted to assess the improvement in the engineering performance of the tailings as the result of additional consolidation and surface crusting. The effect of a rising water table and subsequent resaturation of the crustal zone on the footing performance was also evaluated. This was achieved by placing the footing in a stress controlled feedback loop after a settlement of approximately 10% of the footing diameter had been achieved, and then monitoring the footing displacements as the water table was raised.

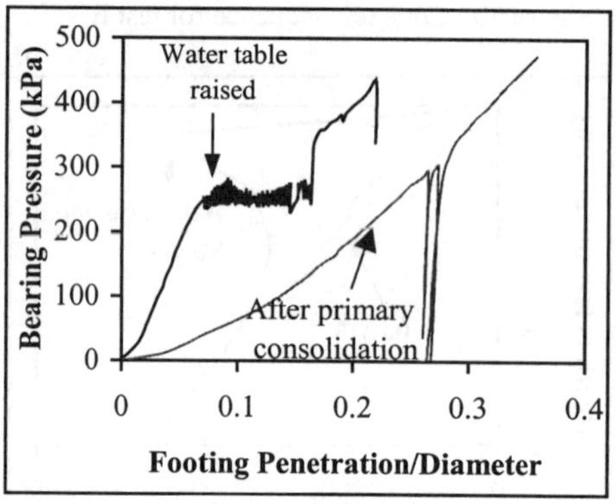

Figure 6a: Footing tests for test 1

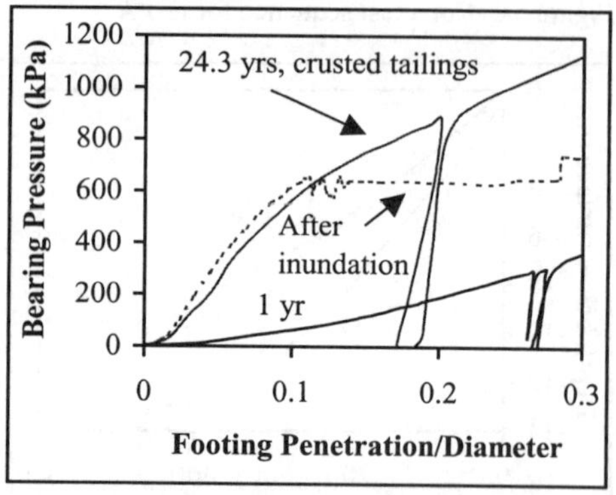

Figure 6b: Footing tests for test 2

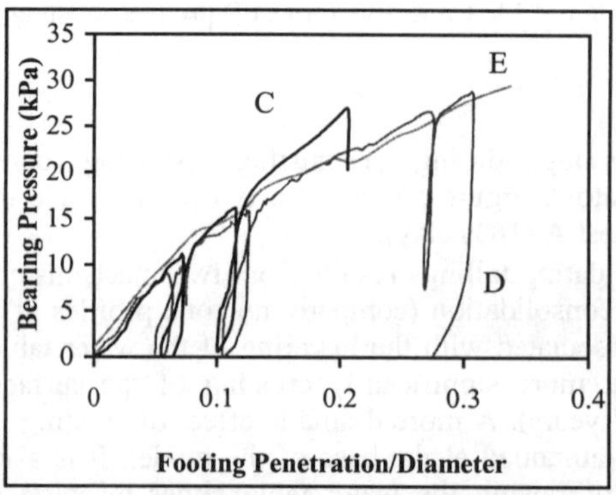

Figure 6c: Footing tests after primary consolidation

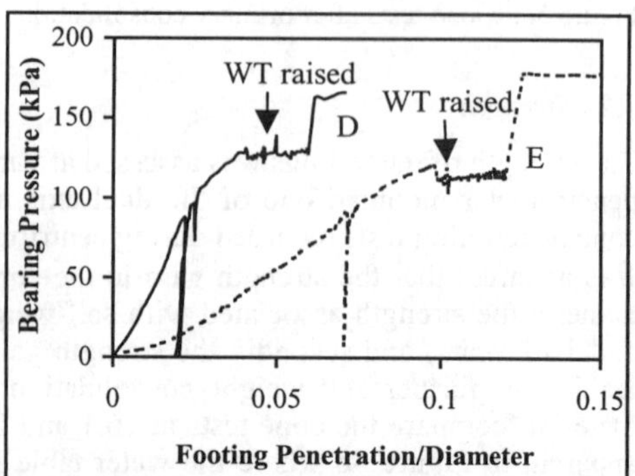

Figure 6d: Footing tests pre/post inundation

The results of the footing tests indicate that there is a significant effect on the footing response depending on the clay content of the tailings. Figure 6a and 6c show the results of footing tests conducted after completion of primary consolidation for the low and high clay content tailings respectively. It is apparent that for all the tailings mixtures with a clay content greater than 17% (Tests C, D, and E) very similar, but significantly weaker bearing responses are observed than for those obtained for the low clay content (Test A, Figure 6a).

For a penetration of about 10% of the footing diameter a bearing pressure of 70 kPa is obtained for the low clay mixture (Test A, Figure 6a) compared to a corresponding value of about 15 kPa for the higher clay mixtures (Tests C, D and E Figure 6c). It is interesting to note that in none of the footing tests is a distinct yield point observed in the bearing response of the material. It can be concluded, that whereas high clay mixtures are not capable of supporting significant loads after completion of primary consolidation, the low clay mixture does have a limited load carrying capacity. However, the performance of any footings placed on this weak material are likely to be governed by settlement considerations.

Figures 6a, 6b and 6d show the footing responses after lowering the water table and during subsequent resaturation. It is evident that for all the tailings mixtures a significantly improved bearing response is produced if some degree of crusting has occurred. The effect of the crustal strength is well illustrated in Figures 6a and 6b for the low clay content tailings. For Test B (Figure 6b) a significant crustal strength builds up as the result of drainage occurring through the base of the model (i.e the water table is at the model base). In this case the bearing pressure of the footing at a settlement of 10% of its diameter is increased from 75 kPa to 600 kPa. Similarly the improvement for Test A where the water table was lowered to 6 m below the tailings surface is also significant, increasing to about 250 kPa. Similarly for the high clay tailings the footing performance is considerably improved as the result of crusting, see Figure 6d. In this case, comparison of the bearing pressures at a footing settlement of 5% of the footing diameter shows an increase from 60 to 140 kPa. It should be noted that for Test E (29% clay) insufficient test time was available to allow any significant crustal strength to develop through drainage and evaporation.

Comparison of the cone data and footing responses indicate that the extent of improvement is clearly related to the degree and depth of desaturation and the generation of suction pressures in the tailings. The development of these suctions is dependent on the permeability of the consolidated tailings and the rate of evaporation from the surface of the model. Since the evaporation rate was not controlled during the test, it is not possible to compare the crustal strengths for different tests at corresponding times (either model or prototype) once the water table has fallen below the model surface. Thus the response of the footings on the desaturated tailings is best related to the cone resistance profiles which provide an indication of the generation of pore suctions through increased cone resistance.

As previously noted, in none of the footing tests performed prior to lowering the water table was a distinct yield in the bearing response of the footing observed. For the low clay tailings this was also the seen to be the case for the footing tests performed after crusting, see Figure 6b.

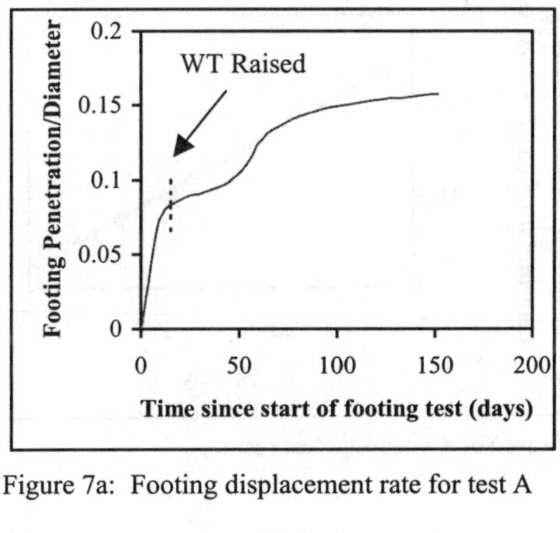

Figure 7a: Footing displacement rate for test A

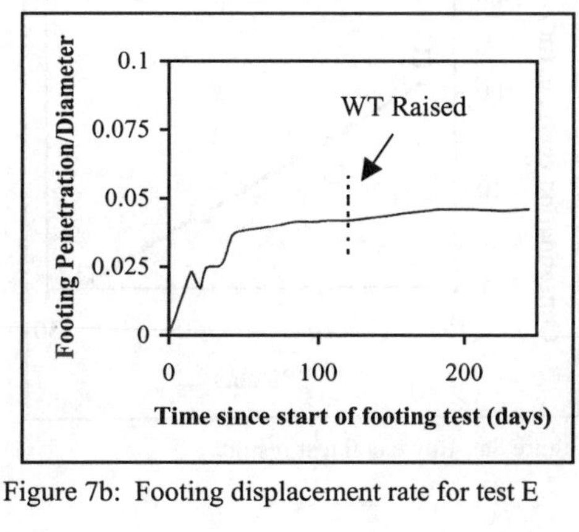

Figure 7b: Footing displacement rate for test E

However, for the higher clay tailings a distinct yield point was observed, see Figure 6d. Figures 7a and 7b show the footing responses associated with inundation of the model surface and raising the groundwater table. For clay contents less than 17%, the effect of inundation of the tailings induced additional footing settlements, see Figure 7a. However, for the higher clay contents there was little or no effect of surface inundation on the footing responses, see Figure 7b. However, since the permeability of the high clay tailings is controlled by the clay fraction it is likely that insufficient time had elapsed such that softening of the high clay tailings can occur before the test was stopping. On the other hand the higher permeability of the low clay tailings means that the surface inundation has an almost immediate effect on the footing response.

4 DISCUSSION

The main conclusion from this data is that the degree of primary consolidation settlement is not greatly influenced by the mix proportions (for this range of materials). In general, settlements of around 5-6 m were observed for the primary consolidation phase, with further settlements of about 1 m being induced by a 6 m lowering of the water table.

Some degree of prediction can be made on the likely primary consolidation times required for various sand:clay mixtures. It should be noted that consolidation times will approach an upper limit governed by the clay permeability and a lower limit governed by the permeability of the sand. For the 20 m thick deposit considered here the respective upper and lower limits appear to be around 25 years and 100 days respectively. It should also be noted that the consolidation times derived from the centrifuge tests refer to the consolidation of an instantaneously formed 20 m thick deposit. In the field, however, the tailings are deposited over a long period of time and will experience fairly rapid initial self-weight consolidation since the initial drainage path will be small - i.e one way drainage to the free surface. Overall consolidation times can be accelerated by the placement of horizontal layers of higher permeability material, which would effectively reduce the drainage paths for the consolidating tailings mixtures.

It is possible to make some estimations of the "bulk" coefficient of consolidation, c_V, from the centrifuge model tests. Using the pore pressure and settlement data to establish the times for primary consolidation to occur, the relevant values of c_V can be found from the following expression

$$c_v = \frac{T_v d^2}{t} \qquad (1)$$

where, T_V= the time factor, d = half the initial depth of tailings and t = the time to the degree of consolidation (refer to Figure 3).

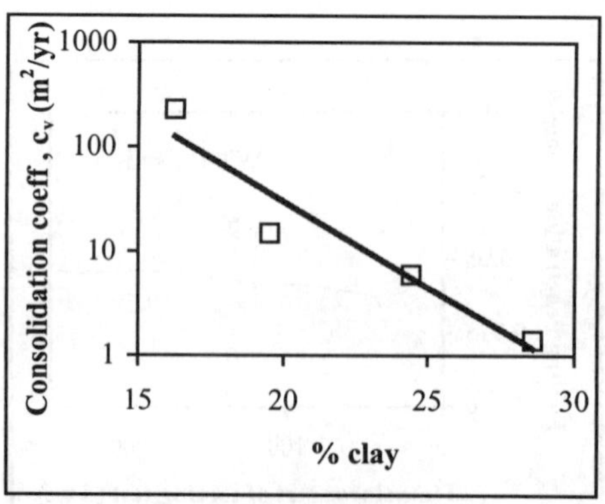

Figure 8a: Rowe cell test results

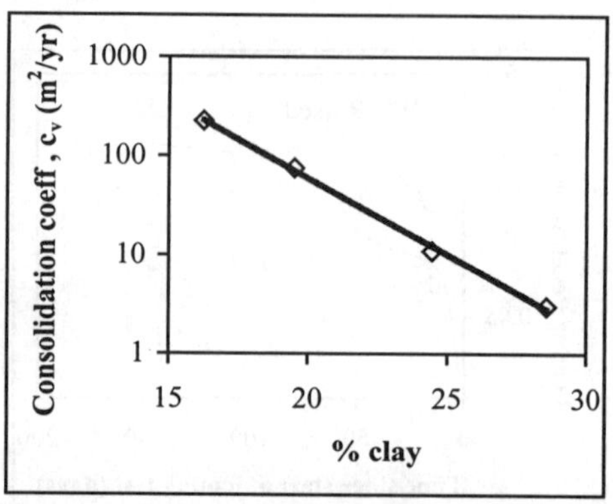

Figure 8b: Centrifuge test results

For all of the tests two-way drainage conditions were enforced in the model. The relevant time factor, T_V, for a 90 % degree of consolidation is 0.94. Values of c_V calculated from the centrifuge model tests have been plotted in Figures 8b and can be compared with the values derived at an effective stress of 150 kPa in the Rowe cell tests in Figure 8a. The agreement is reasonable and indicates that a fairly significant increase of c_V values occurs at clay contents of about twenty %.

As was the case for the primary consolidation times, it must be assumed that values of c_V will be bounded by the dominating influence of either the clay or the sand. From the data presented in Figure 8 it appears that a lower bound value for c_V, of about 1-2 m^2/year (at an effective stress of 150 kPa) occurs with clay:sand mixtures of around 1:3.5. From the laboratory study a corresponding upper bound value for c_V, at an applied effective stress of 150 kPa, of about 250 m^2/year was derived.

For all the tailings mixtures tested, it is apparent that little surface strength is obtained after primary consolidation, when the water table remains at or above the surface of the tailings. The low clay content mixture (17% clay) exhibits a greater surface strength than the higher clay content mixes. Furthermore, significant suctions develop during crusting of the upper layers leading to a considerable increase in strength of the tailings surface. For the mixtures containing greater proportions of clay (Tests C, D and E), a less pronounced strength increase due to crusting is observed. For all the tests, the strength gain due to crusting is significantly reduced following inundation of the tailings surface.

These footing tests confirm that a significant increase in bearing capacity is associated with crusting of the tailings surface. For all the footing responses very low bearing strengths were obtained at the end of primary consolidation such that the material could not be considered capable of supporting any significant loads. The bearing strength of the higher clay tailings (Tests C, D and E) was only about 10% of that achieved for the low clay tailings. It seems reasonable to infer that there is a threshold clay content below which the material response is predominantly frictional, and above which it is cohesive. The significant influence of the clay content is again reflected in the footing response to inundation where the permeability of the tailings is dictated by the clay fraction. For clay contents less than 17%, inundation of the tailings induced additional, and accelerated, footing settlements, see Figure 7a, whereas for the higher clay contents there was little or no effect on the footing response (Figure 7b) over a similar timescale.

5 CONCLUSIONS

The centrifuge model tests reported have shown the influence of clay and sand percentages on the engineering behaviour of tailings. Different mixtures of sand and kaolin clay were used to represent a range of typical tailings materials. The performance of shallow footings and cone penetrometer tests placed on the consolidated and crusted tailings mixtures were evaluated. The effects of changes in groundwater elevation on the footing behaviour and tailings strength were also assessed during the testing programme.

The data indicated that the degree of primary consolidation settlement was not greatly influenced by these particular mix proportions. In general, settlements of around 5-6 m were observed for the primary consolidation phase, with further settlements of about 1 m being induced by a 6 m lowering of the water table. The crusting developed from the changes in water table elevation lead to significant increases in strength and improved footing performance. Subsequent inundation of the crusted surfaces lead to reductions in strength and reduced footing performance, similar to that observed after primary consolidation. The % of clay affected the time for changes to occur, but not the overall settlement (over this range of mixes) and there appears to be a threshold percentage where the behaviour changes from a frictional to cohesive material at around 17% clay.

The study presented has provided well-controlled and highly instrumented test data on the engineering performance of different materials representative of typical crusted tailings. This data would be extremely difficult and expensive to achieve at full scale, and this will allow development and validation of analytical and numerical methods for the prediction of bearing capacity and settlement of crusted storages. Dealing with the rehabilitation of poorly consolidated storages can present considerable problems and severe economic penalties to mine operators. Studies of this type can help to provide suitable analytical tools to assist in the assessment of both tailings disposal and rehabilitation strategies for storages.

6 ACKNOWLEDGEMENTS

Special thanks are due to the staff of Westralian Sands mineral sands mine in Capel, Western Australia for their help during this study and the Geomechanics Group of the University of Western Australia.

7 REFERENCES

Toh S. H, Fahey M and R Kitamura (1991) "The effect of water table lowering on the behaviour of soft clay." Proc. Int. Conf. on Geotech. Eng. for Coastal Development., Geo-Coast '91, PHRI, Japan.

Stone K J L, Randolph M F, Toh S and A A Sales (1999) "Evaluation of the consolidation behavour of mine tailings." ASCE Jour of Geotechnical Engineering, Vol 120, No 3, 973-990.

Fahey M and S. H Toh (1992) "A methodology for predicting the consolidation behaviour of mine tailings." Proc. Western Australian Conf. on Mining Geomechanics, Kalgoorlie, Australia.

Toh S.H and M Fahey (1991) "Numerical and centrifuge modelling of large-strain consolidation". Computer Methods and Advances in Geomechanics, G. Beer, JR Booker and JP Carter (eds), Vol. 1, 279-289, Balkema, Rotterdam.

Fahey M and Y Fujiyasu (1999) "The influence of evaporation on the consolidation behaviour of mine tailings." Proc. 1st International Congress on Environmental Geotechnics, ISSMFE, Edmonton, Canada, Vol.1, 981-986, BiTech Publisheres, Canada.

Rassam D W and D J Williams (1997) "Shear strength of unsaturated gold tailings." Proc., 8th Australia-New Zealand Conf. on Geomech.

Rassam D W and D J Williams (1999) "Bearing Capacity of Desiccated Tailings." ASCE Journal of Geot. And Geoenv. Eng., Vol 125, No 7, pp 600-609.

Newson T A, Fujiyasu Y and M Fahey (1997) "The evaporation behaviour of tailings." Proc of 9th international conference on tailings and mine waste, Fort Collins, Colarado, USA, 137-196.

Arulanandan K, Thompson P Y, Kutter B L, Meegods N J, Muraleetharan K K and C Yogachandran (1988) "Centrifuge Modelling of Transport Processes for Pollutants in Soils." J. Geotech. Eng., ASCE, 119(2), 185-205.

Hensley P J (1989) "Accelerated Physical Modelling of Transport Processed in Soil." PhD Thesis, Cambridge University.

Schofield A N (1980) "Cambridge Geotechnical Centirfuge Operations." Geotechnique, 20, 227-268.

Tailings and Mine Waste '01 © 2001 Balkema, Rotterdam, ISBN 90 5809 182 1

Modeling of a tailings impoundment raising

M. Barrientos & S. Barrera
Geotécnica Consultores S.A., Santiago, Chile

ABSTRACT: The modeling or simulation of a tailings impoundment's growth depends mainly on the adopted tailings deposition scheme, the size of the tailings pond, and the geotechnical and reological characteristics of the tailings. The first two factors correspond to design decisions and may be modified according to the study's characteristics. On the other hand, the behavior of the tailings with regards to the deposition slope and density are intrinsic characteristics that only depend on the nature of the tailings. This means that the estimates of these values are key factors in the impoundment's growth simulation. The range of variability of these parameters, along with the geometric complexity of the topography, makes the study of a tailings impoundment's growth requires geometric calculations of complex surfaces and also requires sensibility analysis of the tailings parameters. This makes hand calculations of the model impractical due to the large amount of human resources needed.It is for these reasons that one of the main objectives of the engineering stage, where the impoundment's behavior is studied in detail, is to be able to count on tools that facilitate the simulations of a tailings impoundment's growth. These tools should be fast, reliable and also allow for the variation of the parameters that govern the growth to obtain a series of results that anticipate the behavior in time of the impoundment's growth.

1 INTRODUCTION

One of the important activities that constitutes a mining operation is the safe and economical disposition of its wastes. In the particular case of a mining operation with a concentrator plant, the main waste is tailings (milled rock slurry without any ore). The tailings disposition translates into the management, in many cases, of large volumes of material.

Historically, Chile has been known as a mining country. The mountainous location of its main sites, the unfavorable topographical characteristics near the concentrator plant, the increasing volume of tailings to be deposited, environmental protection and the costs associated with this part of the mining process condition the characteristics of the tailings impoundments.

Mainly, the cost of a tailings impoundment centers in the constructions of the dams that limit the area where the tailings will be deposited. The dimensioning and location of these, the tailings transport system and the filling strategy condition the potential capacity of the impoundment and, therefore, the investment.

Within the engineering studies that comprise the development of a tailings impoundment, the simulation of its growth is a relevant aspect to consider since the establishment of an adequate impoundment filling strategy orients the tailings deposition, the design of containment structures (dams), the location of the tailings pond, and the design of the tailings transport facilities. The simulation also allows for optimization that implies significant reduction of the costs involved in the facilities and, in some cases, the deferment of investment periods.

Because of this, it necessary to have tools that allow for the development of a growth analysis in a logical, reliable and fast manner to have an array of alternatives for the decision making that is to come. Moreover, these tools should be capable of diverse simulations in an impoundment, avoiding the application of simplifying assumptions.

2 VARIABLES THAT INFLUENCE THE IMPOUNDMENT'S CAPACITY

The capacity of a tailings impoundment depends mainly on the following variables and characteristics:

- Density of deposited tailings
- Average slope of deposited tailings
- Location and volume of the tailings pond
- Tailings discharge points locations

These correspond to two types of factors of different nature:

1) Intrinsic or inherent characteristics of the tailings that obey natural laws and that may only be estimated based on laboratory tests, and
2) Tailings management criteria that are freely adopted during the modeling process and whose variation ranges.

2.1 Density of deposited tailings

The density of deposited tailings is variable within the tailings impoundment with significant differences according to its location: in areas of dry beach; in areas of wet beach; in the area below the tailings pond; in depth; and in areas in contact with permeable and impermeable surfaces. For this reason and to determine the impoundment's capacity in a practical manner, it is necessary to estimate the density of all the tailings deposited as an average.

The dry density of deposited tailings depends on two geotechnical characteristics:

1) Specific gravity of solids *(G)*, which is a constitutive characteristic of the tailings and is always a known parameters with little variability;
2) Void ratio *(e)*, whose range of variability for expected average values fluctuates between 0.8 and 1.2 for copper tailings (Ref 2, 6).

The estimate of expected average density is based strongly on the consultant's experience with other tailings impoundments of similar characteristics supported or backed up by the tailings grain size analysis, sedimentation tests and permeability characteristics.

2.2 Slope of deposited tailings

The deposited slope depends on the following parameters:

1) Tailings grain size
   In general, the finer tailings are deposited in gentler slopes than the coarser tailings. In reality, the deposited tailings slope varies as it moves away from the discharge location due to the segregation process that takes place in the tailings slurry.

2) Concentration of solids (by weight) (Cw)
   The profile of deposited tailings depends also on the concentration of solids (by weight), with the tailings being deposited in gentler slopes at low concentrations and in very steep slopes at high concentrations. (See Figure 1) (Ref 1, 3).

3) Tailings flow at the discharge location
   In general, we observe that gentler deposition slopes are obtained for the discharge location with large flows and the opposite for discharge locations with small flows (See Figure 2) (Ref 3).

4) Underwater tailings discharge

The deposition slope in the tailings pond is significantly larger than on the tailings beach above the water level (Ref 1). It is commonly adopted that the underwater slope is between 5 to 10 times larger than the slope above water.

In general, the profile of tailings deposition may be modeled by equation (1) (Ref 1), that basically depends on the intrinsic properties of each mineral (See Table 1 and Figure 3, Ref 1).

Table 1
Dimensionless Melent'ev master profiles (Ref. 1)

Tailings	N
Copper	1.4
Diamond	1.5
Platinum	2.0
Gold	4.0

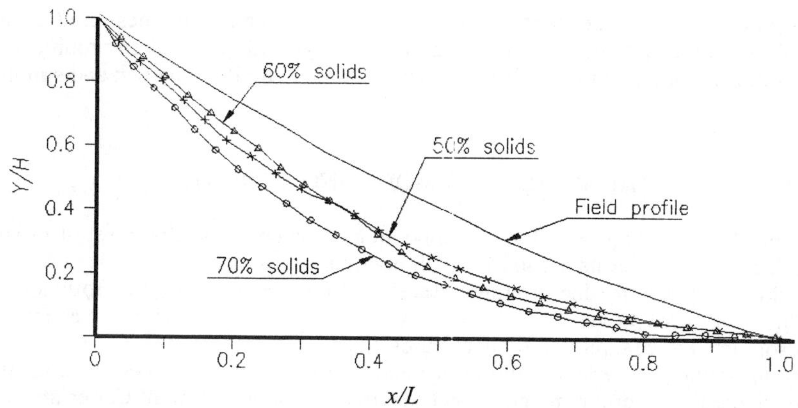

Figure 1
Comparison of Melent'ev master profiles to model hydraulic fill beaches of gold tailings (Cw=50% solids concentration) with field profile for the same material (Ref. 1, 3)

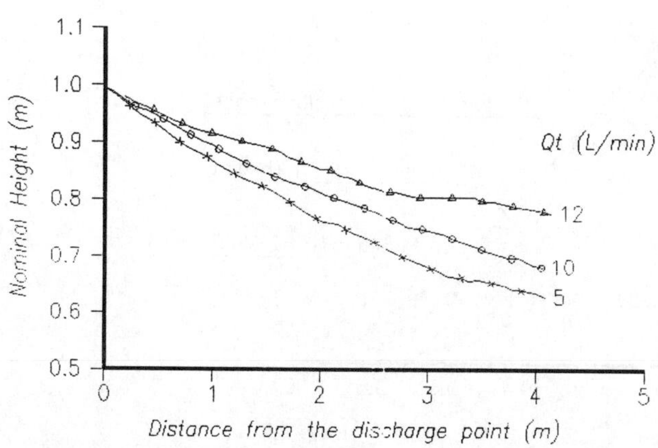

Figure 2
Deposition slope as a function of the discharge flow (Cw=20%) (Ref. 3)

$$y/H = [1-(x/L)]^n \qquad (1)$$

Alternatively, this profile may be approximated by a combination of two straight lines.

In general, the value of the deposition slope strongly influences the impoundment's modeling. It is for this reason that it is common to develop sensibility analysis varying this parameter to limit the impact of its variation.

2.3 Location and volume of the tailings pond

The location and volume of the tailings pond as well as the deposition scheme (dealt with in the following section iv) are variables that usually work in tandem, strongly influencing on the impoundment's capacity. The location of the pond depends, among other factors, on the type of dam, on the topography and on the hydrogeological characteristics of the area. The volume of the pond depends on the reclaimed water rate, on the tailings deposition slope and the water reclaim system.

2.4 Tailings discharge locations

The design of the discharge locations has, as its main objective, the design of an impoundment that allows for a safe and economical tailings storage with minimal environmental impact. Since this means reconciling many different factors, tools that allow the fast and simple simula-

3 DESCRIPTION OF THE SYSTEM FOR MODELING GROWTH

The system for modeling growth of an impoundment is comprised by a set of programs developed in Turbo Pascal for its use in MS-DOS operating systems.

The working surfaces for the programs developed are generated by the Software SURFER (Ref 5). The results are provided in files with ASCII format, which may serve as input data for graphics programs (screen, printer, plotter, AutoCAD).

In order to define the variables and nomenclature used, it is convenient to limit the global system considered. The programs have been developed using a straight Cartesian system and the geometric calculations are done within the tailings impoundment area, therefore, the variables are defined within this framework. The global system is comprised of: the base surface, the impoundment's basin limited by the dams, the deposited tailings, tailings pond, and the discharge locations (Ref 4).

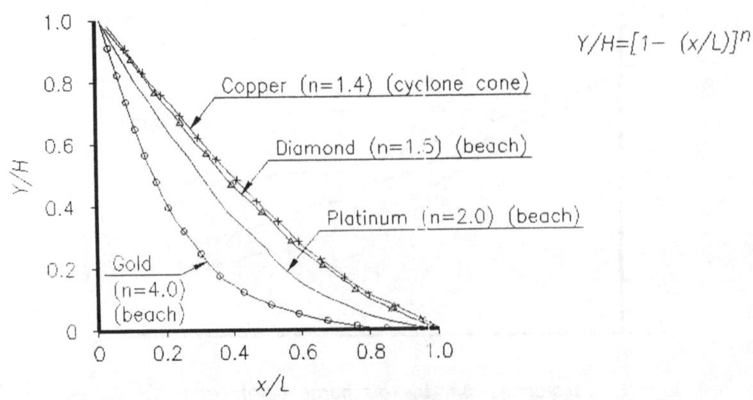

Figure 3
Dimensionless Melent'ev master profiles (Ref. 1)

3.1 Mathematical Models

In order to develop the growth models, three basic elements must be generated at a minimum, which are: (1) base surface (topography), (2) limiting planes, and (3) discharge cone.

3.2 Base Surface

The base surface corresponds to the topographic base of the tailings impoundment area. This is generated by the application of GRID of the SURFER software and it is on this surface that the programs do all the geometric calculations.

At the present time, the programs permit the breakdown of the surface into rectangular or square, triangular or other irregular elements. The SURFER software generates rectangular breakdowns equidistant in each direction. The rectangular breakdown is based on nodes that make up a mesh or surface. Each node is defined by its Easting, Northing and its elevation value.

3.3 Limiting Planes

In order to model the surroundings in which the tailings are deposited (structures such as dams or dikes), it is necessary to generate the impoundment's limiting planes, whose Cartesian equation is:

$$Z = K + A \times X + B \times Y \quad \text{or}$$

$$Z = K + A \times Easting + B \times Northing$$

If we know three points in a plane (x1,y1,z1), (x2,y2,z2) y (x3,y3,z3), the value of the coefficients are calculated as:

$$K = \frac{x1 \times (y2 \times z3 - y3 \times z2) - y1 \times (x2 \times z3 - x3 \times z2) + z1 \times (x2 \times y3 - x3 \times y2)}{x1 \times (y2 - y3) - y1 \times (x2 - x3) + (x2 \times y3 - x3 \times y2)}$$

$$A = \frac{z1 \times (y2 - y3) - y1 \times (z2 - z3) + (z2 \times y3 - z3 \times y2)}{x1(y2 - y3) - y1 \times (x2 - x3) + (x2 \times y3 - x3 \times y2)}$$

$$B = \frac{x1 \times (z2 - z3) - z1 \times (x2 - x3) + (x2 \times z3 - x3 \times z2)}{x1 \times (y2 - y3) - y1 \times (x2 - x3) + (x2 \times y3 - x3 \times x2)}$$

3.4 Discharge Cone

The behavior of deposited tailings from a single discharge is modeled satisfactorily as a cone. The cone's vertex is the discharge location and the tailings adopted deposition slope are used as their directrices.

The equation of the discharge cone is the following:

$$Z(x, y) = Z_d - i\sqrt{(X - X_d)^2 + (Y - Y_d)^2}$$

With :

Z(x,y): Elevation at the point (X,Y)

(Xd, Yd, Zd): Coordinate and elevation of the discharge point

i: Average slope of deposited tailings

By the application in different sectors of the discharge cone equation, it is possible to model the variation of the deposition slope in the beach and the break in slope underwater. (Ref 4)

4 BASIC INFORMATION FOR RUNNING THE GROWTH MODEL

In order to run the tailings impoundment growth simulation, it is necessary to have the following basic background information:

1) General criteria of tailings and tailings pond management within the impoundment

In general and during the study stages of a tailings impoundment, the criteria of tailings management and tailings pond location are modified as a function of the design of the dams that limit the impoundment, the design of the tailings transport and deposition system, and the costs associated with these facilities. The programs allow for the modeling of the impoundment's growth by freely changing the variables related to the tailings discharge locations. These locations condition the tailings management and location of the tailings pond that allow for the maximization of the impoundment's capacity, which is an important aspect for impoundments with limited space which must contain great volumes of tailings.

2) Topographical base

The topography of the area where the tailings will be deposited is the base for the generation of working surfaces, on which the calculation of volume and surfaces are made. The precision of the topographical base required will depend on the level of engineering that is being done and the level of precision that is looked for with the growth model. When the topographical surface are generated with SURFER, one must take precaution and verify that the surfaces generated by the program have a good approximation to the original topography. Errors in the coordinates and elevations of the calculated nodes may lead to errors in the calculations of the growth model. This is particularly relevant in the initial stages of growth.

5 APPLICATION TO A REAL CASE

5.1 Output

As was previously indicated, the objective of the modeling or simulation of a tailings impoundment growth is to anticipate, in a realistic manner, the behavior in time of the impoundment's growth. As a result of the simulation, the following information is obtained:

- Tailings elevation at each discharge location
- Distribution in mass of the tailings at each discharge location
- Elevation of the tailings pond for a given operational volume
- Elevation of the tailings pond for any hydrological event
- Raising rate of the dams and deposited tailings

In addition to that previously indicated, graphic information is obtained (SURFER, AutoCAD), for any given moment in time of the impoundment's growth.

5.2 Application

Once operation of the tailings impoundment has started, verifying the growth model with the real impoundment's growth is ideal to check the adopted parameters and calibrate the model to better estimate future growth.

Currently in Chile's I Region, at 200 kilometers south-east of the city of Iquique (See Figure 4), is located Pampa Pabellon Tailings Impoundment, property of Cía Minera Doña Inés de Collahuasi. A growth model was developed for this impoundment during the design stages of the impoundment with the objectives of: (1) predict the elevation of the tailings in contact with the dam where the freeboard would be minimum (critical), (2) maximize the impoundment's capacity, and (3) estimate the production of waste rock necessary for the construction of the im-

Table 2
Comparison of the Growth Model with Real Behavior

Measurement date	Dam elevation (1) (m a.s.l.)	Tailings elevation (m a.s.l.)		Elevation Difference (B)-(A) (m)
		(1) Site (A)	Model (B)	
31-Dec-99	4137.51	4133.72	4133.71	-0.01
31-Jan-00	4137.51	4134.58	4134.00	-0.58
29-Feb-00	4137.51	4134.60	4134.19	-0.41
31-Mar-00	4141.00	4134.64	4134.61	-0.03
30-Apr-00	4140.92	4134.64	4135.02	0.38
30-May-00	4140.96	4134.60	4135.43	0.83
30-Jun-00	4140.68	4135.23	4135.82	0.59
31-Jul-00	4140.68	4136.04	4136.20	0.16

(1): Average value of three measurement points, located on the left abutment of the dam (See Figure 4)

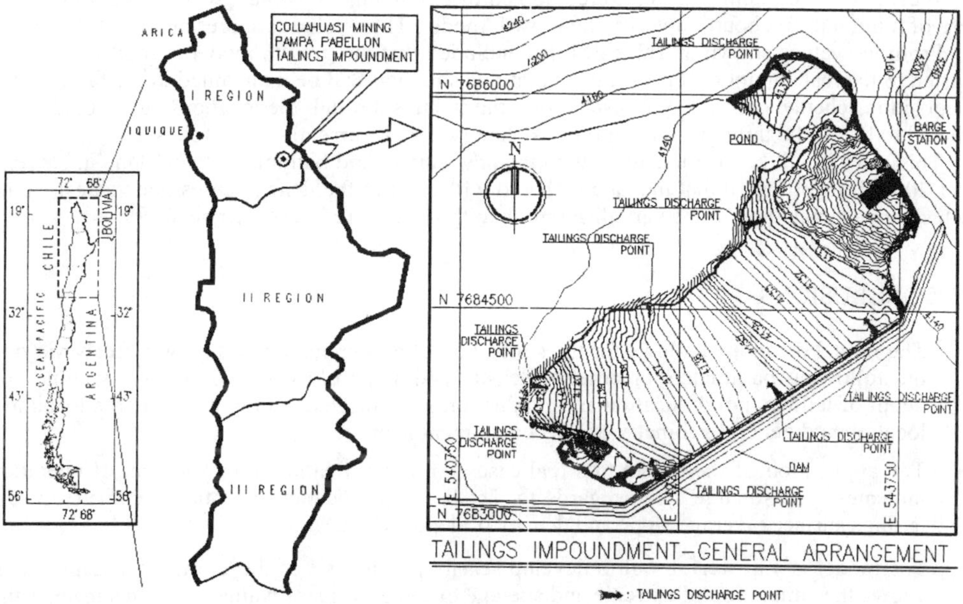

Figure 4
Location of the Pampa Pabellón Tailings Impoundment and
Situation of the Tailings Impoundment in November 1999

poundment's main dam to maintain a minimum operational freeboard based on the impoundment's growth rate.

The basic following parameters were adopted in modeling the impoundment's growth:

- Average slope of tailings deposition of 0.5% (constant)
- Dry density of deposited tailings of 1.35 t/m^3 (constant) (void ratio of 1.0)
- Minimum volume of the tailings pond of 1 Mm^3

Additionally, a series of tailings discharge locations were identified, which maximize the impoundment's capacity and contain the tailings pond on the east side, location where the water reclaim system is located (See Figure 4).

The verification of the adopted parameters was done approximately one year after the impoundment's operation started (approximately in November of 1999), by the topographical survey of the tailings and a bathimetry of the tailings pond. With this information, the following was determined:

1) The volume of deposited tailings. This data together with the tons of production of processed tailings in the same time period informed by the concentrator plant allowed for the determination of the average density of deposited tailings, and
2) The average deposition slope for the most representative discharge locations.

From analyzing the results, it was concluded that the density of deposited tailings was very close to the one assumed (1.35 t/m^3), and that the average deposition slope was close to 0.4% for the discharge location father away from the tailings pond.

A new simulation was done with the adjustment of the adopted parameters and taking the new surface of deposited tailings to November 1999 as base surface (See Figure 4). This was done to better predict the elevation of the tailings in the areas of minimum freeboard (control areas) since the requirement of waste rock for dam construction is dependent on this situation.

Weekly measurement controls were adopted of the tailings in contact with the dam in the areas of minimum freeboard to verify the growth model. Their values are presented in Table 2.

From the values shown in Table 2, it is concluded that the growth model is satisfactorily adjusted to the real situation. It is estimated that this situation will be maintained in the future if the deposition scheme, consisting in discharging the tailings through the locations considered in the model to locate the tailings pond on the east side (where it currently is), is followed.

This growth will be verified again with the new survey and bathimetry to be done in November 2000. It is estimated that the data gathered will confirm the density and slope values; values which are estimated not to be very different than those gathered in November 1999.

6 CONCLUSIONS

- The model developed allows for the simulation of an impoundment's growth freely varying the following parameters or criteria: density and slope of deposited tailings, volume and depth of tailings pond, location and production to be discharged by the discharge locations, location and general geometry of the containment dams.
- The application of the model to a real case and its verification after one year of operation indicates a reasonable fit in regards to the discharge location elevation, geometry of the tailings surface, average slope and density of deposited tailings.
- The model is a powerful tool to develop sensibility analysis of those critical variables as it allows the simulation of extreme and average expected values, evaluating in this manner the impact of this variation in the works and capacity of the impoundment.

REFERENCES

1 Blight G., *The master profile for hydraulic fill tailings beaches, 1994 Proc. Instn Civ. Engrs., Geotech. Engng, 1994, 107, Jan., 27-40*

2 Barrera S., *Deposition densities of tailings in Chilean deposit. Tailings and Mine Waste'98, Fort Collins-Colorado-USA*

3 Küpper A., Morgenstern R.N., Sego C.D., *Laboratory test to study hydraulic fill, 1992 Can. Geotech. J. 29, 405-417 (1992)*

4 Pinto M., *Herramientas para el manejo de modelos topográficos aplicados a depósitos de relaves, 1998, Geotécnica Consultores, Santiago, Chile*

5 SURFER 7.0 Software Manual, *Golden Software, Inc. 1999, Colorado USA.*

6 Vick S., *Planning, Design and Analysis of Tailings Dams, Colorado 1983*

Tailings and Mine Waste '01 © 2001 Balkema, Rotterdam, ISBN 90 5809 182 1

Reduction of surface subsidence and brine inflow prevention in potash mines by subsequent backfilling

Joerg Bodenstein & Henry A. M. Rauche
ERCOSPLAN Ingenieurgesellschaft Geotechnik und Bergbau mbH, Erfurt, Germany
Wolfgang Schreiner
Institut fuer Gebirgsmechanik GmbH, Leipzig, Germany
Karl-Heinz Eulenberger
Institute of Mining, Technical University Bergakademie Freiberg, Germany

ABSTRACT: The slurry backfill technology has been successfully used in potash mines since 1908. The presentation summarises the results of several investigations on slurry backfill in German potash mining during the last 40 years. The positive effect of subsequent slurry backfill on the reduction of dangerous rock movements and surface subsidence is shown with examples.

1 INTRODUCTION

The worldwide annual potash extraction of approx. 23 million tons K_2O (e.g. as in 1999) by conventional mining methods creates mine openings of millions cubic meters. Because of the most ductile mechanical behaviour of the potash ore and the surrounding host rocks, these mine openings close themselves by time-dependent convergence processes. The related mass movement leads to subsidence at the Earth´ s surface, above the mine openings, though the amount and rate of subsidence is controlled by geological (e.g. thickness and tectonic styles of cover rocks) and mining parameters (e.g. depth, overall extraction ratio, room and pillar geometry, extraction rate). The dynamic of this subsidence process influences the land surface (topo-graphy) and its natural and/or artificial dewatering systems as well as the buildings and infrastructure of the towns and villages by the slope gradient and ground deformations.

Inside the brittle cover rocks, this subsidence is mostly realised by frictional sliding on pre-existing faults and joints and occasionally by fracture propagation. Both mechanisms influence the hydraulic regime inside the groundwater aquifers.

On the other hand, the amount of residues from potash ore processing, compared with another mining branches, is relatively high. Therefore, the potash industry traditionally uses these residues as a backfill material.

In the German potash industry, backfill was used to reduce the surface subsidence above the mine fields, as well as for the reduction of mining losses. During the more than 140 years of production, different backfill technologies were developed due to the technical challenge.

In general, two different materials - rock salt from the development of long-life drifts and galleries or, as previously mentioned, residues from the potash ore processing - are used as backfill material. The processing residues are transported in a dry or wet form by downward shaft hoisting and/or conveyor belts ("dry backfill technology") or as a suspension by pipelines (hydromechanical transport, the so-called "slurry backfill technology").

2 ROCK MECHANICAL PROPERTIES OF DIFFERENT BACKFILL MATERIALS USED IN POTASH MINES

The effect of subsequent backfill on the reduction of the surface subsidence as well as on the increase of the exploitation rate, is controlled by its rock mechanical properties. The infill of solid material into the stopes or rooms after the excavation leads to a fast stabilisation of the pillars and the roof. So, the mining induced rock movements can be slowed down and limited by their absolute amount. As a result, the surface subsidence will decrease.

How can we influence the rock mechanical properties? They are controlled by three main factors:

- Mechanical strength,
- Grade of filling,
- Initial density of backfill.

The mechanical strength of backfill is determined by the compaction and solidification of the in-filled material. As the rock masses exert onto the backfill massive by convergence, the near-contour zone will stop its dilatation. The grade of refilling limits the extent of the roof convergence, whilst the initial backfill density limits the movement of the pillars.

During the history of the German potash industry, the following materials were used for subsequent backfill after the ore extraction:

- Residues of the potash processing and
- Rock salt from the drift development.

Both materials are of good mechanical properties to achieve a positive effect on the pillar and roof stabilisation. However, the differences can arise from the initial consistency of the material, as well as from the transport and infill technology. The advantages are on the processing resi-

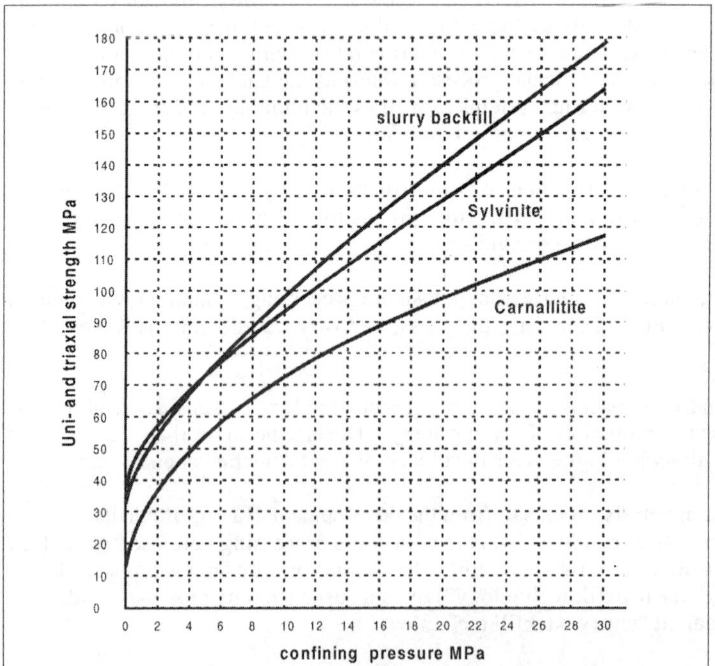

Figure 1: Triaxial strength of solidified slurry backfill in comparison with potash ore

dues side, because of the moisture and better granularity of the material during the infilling, which leads to a fast compaction and consolidation.

As a result, a compact backfill massive, with a strength, comparable to the surrounding natural rock massive, is created. Figure 1 shows the uni- and triaxial strength, depending on the confining pressure, for solidified slurry backfill from German potash mines and for natural Silvinite and Carnallitite, following laboratory tests by Fulda, Jaeger et al. (1988).

Further investigations and laboratory tests on long-term consolidated slurry backfill material from several potash mines during the last years have shown the effect of a higher (triaxial) strength on the backfill material. Figure 2 shows the typical stress-strain-behaviour during a multistage triaxial test, following Sitz et al. (1999).

The following proven technologies have been used for subsequent backfill in the German potash mines:

- Slurry backfill,
- Dry backfill by slinger, pneumatic and drop stowing infill methods,
- Improved dry backfill by the moisturisation and compression of the material during the infill operations.

The latter method has been especially developed for the delayed backfill operations in the abandoned potash mines in Germany, because residues from the potash ore processing (needed for an effective stabilisation of the roofs and pillars) were not available since the shut-down of the mines.

The slurry backfill technology operates with the processing residues using two possible methods of transport:

- Shaft hoisting and haulage by conveyer belts or
- Pipeline systems from the surface to the stopes.

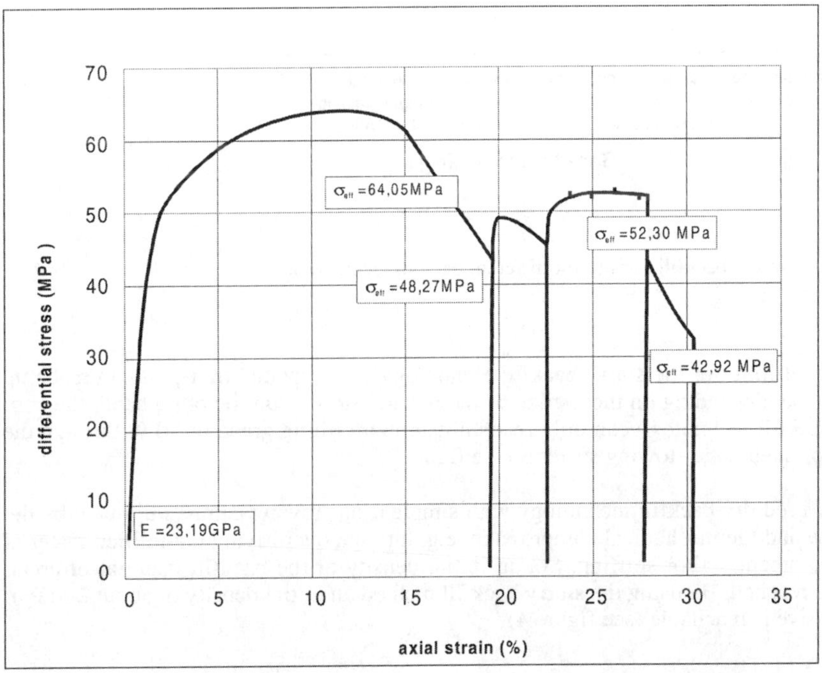

Figure 2: Differential stress and axial strain of solidified slurry backfill in a multistage triaxial test

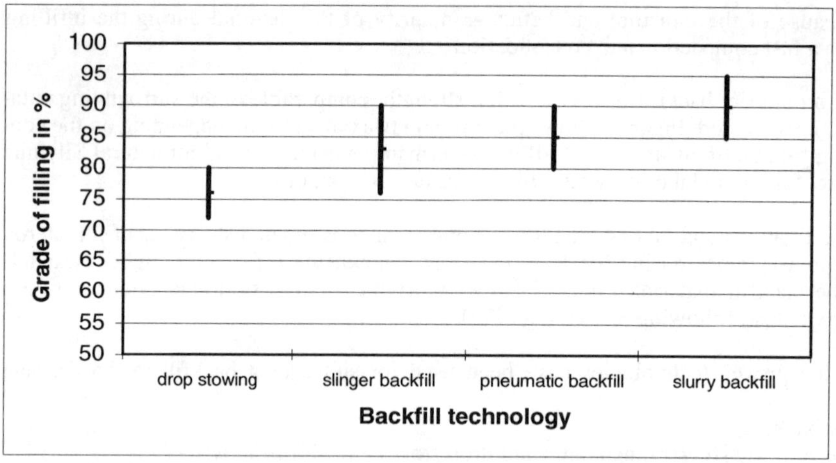

Figure 3: Reachable grade of filling for different backfill technologies

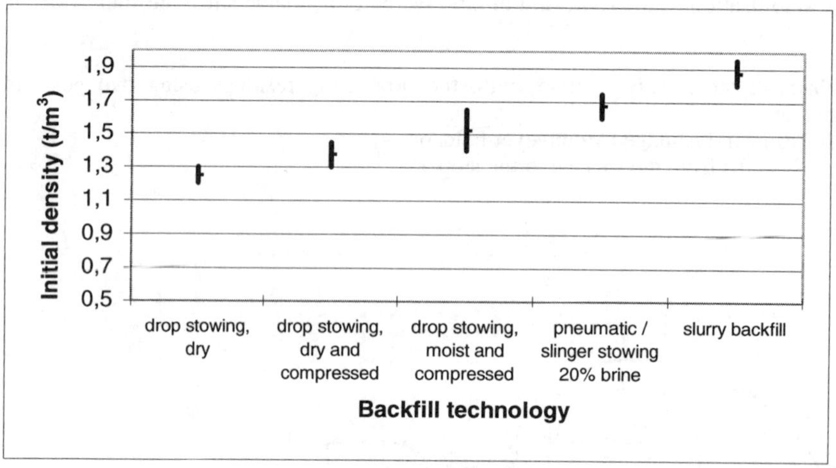

Figure 4: Reachable initial backfill density for different backfill technologies

The grade of refilling for the slurry backfill technology with pipeline transport can reach up to 95 % by volume, depending on the incline of the refilled stopes. On the other hand, the conventional dry backfill technology can only reach a maximum filling grade of 90 % through the use of slingers or pneumatic stowing systems (see figure 3).

The formerly used dry backfill technology with simple drop stowing can be improved by the addition of brine and the mechanical compression, e.g. through overdriving the infilled material with mobile equipment and/or stuffing. So, an initial density of the backfill material of about 1,65 t/m^3 can be reached. By using the slurry backfill method an initial density of about 2 t/m^3 in the backfill massive is reachable (see figure 4).

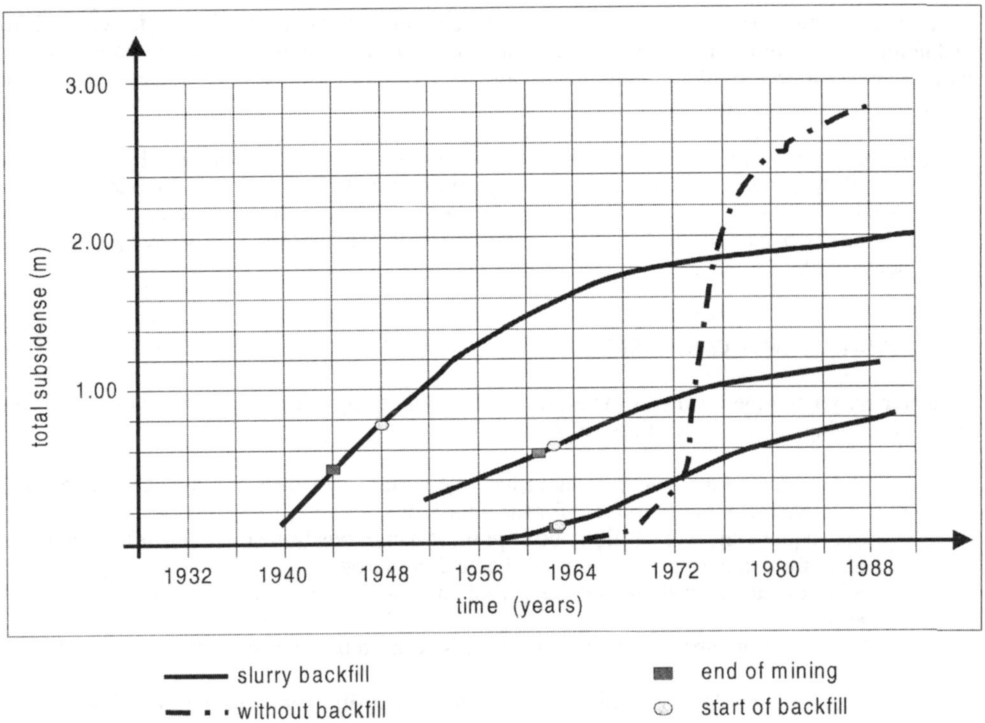

Figure 5: Time-dependent subsidence above potash mine fields

3 LONG-TERM MINE SITE INVESTIGATIONS AS PRACTICAL CASE STUDIES

Which factors can be influenced by the rock mechanical properties of backfill? According to the general protective aims of mining operations, they are:

- Surface subsidence,
- Amount of seismic events or rock bursts,
- Damage of (hydrological) protective horizons by subsidence, leading to brine inflow.

The surface subsidence above potash mine fields is a time-dependent process, following the convergence of the underground stopes. Its horizontal expansion exceeds the edges of the mine fields, and depends on the strata and behaviour of the overburdening rock. The total amount of subsidence is controlled by the parameters (height, width) of the underground stopes, occurring gradually over several hundred years.

Until around 1968 in the Southern Harz potash district it was required by mining legislation to refill the potash stopes after mining. Therefore, since 1908 in most cases the slurry backfill technology was used. After this time the stopes, due to economical reasons, were left open.

Long-term measurements above the German potash mine fields show the effect of subsequent backfill and its influence on the stability of the overburdening rock and surface. Particularly slurry backfill has shown good results, in relation to the surface subsidence behaviour. Figure 5 shows the influence of slurry backfill on the time-dependent subsidence above potash mine fields as an example, following Bodenstein & Schreiner (1999). Particularly the effect of the time-span between the end of mining and the beginning of backfill on the amount of total subsidence is visible.

Another safety aspect for the using of subsequent slurry backfill as a high-effected backfill technology is the possibility to reduce the number and particularly the amount of seismic events, which can be induced by the mining operations. An effective refilling of the stopes will prevent the inducing of dangerous rock bursts.

Reducing the rock movements above potash mine fields by subsequent backfill can prevent the failure of the protective layers, because they often loose their integrity by breaking up under high amounts of subsidence, which leads to brine inflows into the mine. These are the results of modelling on the behaviour of the protective layers for potash mines in Germany, e.g. following Schreiner (1996).

4 SUMMARY AND CONCLUSIONS

In summary from the results of the aforementioned investigations on subsequent slurry backfill the following conclusions can be drawn:

- Subsequent slurry backfill can reduce the surface subsidence and, at the same time, improve the exploitation rate for the potash ore,
- Subsequent slurry backfill helps to prevent dangerous brine or water inflows into the mine field, due to its excellent mechanical properties,
- Subsequent backfill methods are well known from the history of German potash mines,
- The effect of subsequent slurry backfill is verified by the results of long-term investigations,
- Proven technologies for high-effected slurry backfill operations are available.

REFERENCES

Bodenstein, J. & Schreiner, W. 1999. Reduzierung der Bergsenkungen durch Einbringen von Spülversatz. Presentation in Freiberg, Germany. 12. February 1999

Fulda, D.; Jaeger, G. et al. 1988. Ist der eingebrachte Versatz im Kalibergbau in der Lage, die geomechanischen Aufgaben der Abbaupfeiler zu übernehmen ?. In *Neue Bergbautechnik* 18 (1988): 261-264

Schreiner, W. 1996. Modelltechnische Bewertung der geologischen Barrieren im Bereich des Westfeldes Bleicherode. Leipzig, Germany, 05. November 1996

Sitz, P. et al. 1999. Untersuchung und Bewertung alter Spülversatzkörper aus bergbaueigenen Rückständen der Kaliindustrie hinsichtlich ihrer Eignung für Verschlussarbeiten in Untertagedeponien. Final report. TU Bergakademie Freiberg. 1999

Liners, covers and barriers

Tailings and Mine Waste '01 © 2001 Balkema, Rotterdam, ISBN 90 5809 182 1

Slope effect study on the capillary cover design for a spent leach pad

G.Zhan & A.B.Mayer
Hydrology Department, Barrick Goldstrike Mines Incorporated, Elko, Nev., USA
J.McMullen
Barrick Gold Corporation, Toronto, Ont., Canada
M.Aubertin
Department of Civil, Geological and Mining Engineering, Ecole Polytechnique, Montreal, Que., Canada

ABSTRACT: The performance of a cover with capillary barrier effects (CCBE) depends on the hydrological properties of the CCBE materials, layer thickness, weather conditions, vegetation, and the slope geometry of the cover. However, the effect of slope geometry has been commonly ignored in past investigations. A sloping surface may lead to saturated local areas and lateral movement of water that would not predicted in an one-dimensional, unsaturated domain. The concept of the CCBE is applied for the cover design for the AA Leach Pad at the Barrick Goldstrike Mine in order to prevent meteoric water percolation. One-dimensional analysis demonstrated that a cover consisting of 0.91 m of Carlin Formation siltstone (TCS) placed over the leach pad material could effectively eliminate meteoric water percolating through the reclaimed leach pad. This paper further evaluates the performance of the sloping cover under both normal and storm event conditions using a two-dimensional model – HYDRUS2D.

1 INTRODUCTION

The AA Leach Pad is located at the Barrick Goldstrike Mine site in the Little Boulder Basin in Elko and Eureka Counties, Nevada, USA. The total area of the pad is approximately 9×10^5 m^2, and the pad height is between 38 and 76 m. The climate at the site is semi-arid. Average annual precipitation ranges between 0.20 and 0.38 m. Average annual pan evaporation is about 1.50 m.

Using a one-dimensional numerical modeling approach, it was predicted that the meteoric water percolation through the leach pad might be limited or eliminated by a cover with capillary barrier effects (CCBE) (Zhan et al. 2000). In general, a CCBE is created by placing a layer of fine-grained material above a coarse-grained material to induce a capillary barrier effect. Provided sufficient contrasts exist between the unsaturated hydraulic conductivities and water retention curve (WRC) of the coarse- and fine-grained materials, flow through a CCBE is minimized until the upper fine-grained material becomes nearly saturated. In the CCBE configuration of the AA Leach Pad, the coarse material layer is made from the leach pad itself, while the fine material is locally available Carlin Formation siltstone (TCS). In reality, the CCBE of AA Leach Pad is inclined; therefore, the performances of the cover would be different from that represented by an idealized one-dimensional model. The moisture that builds up above the interface between the coarse- and fine-grained materials could flow along the inclined interface. At a certain downgrading point, the fine-grained material could become so saturated and allow water to infiltrate to lower coarse-grained material. This point at which water would flow into the lower material is called the Down Dip Limit or DDL point (Ross 1990).

Over the last few decades, one-dimension studies of CCBE have been performed by numerous researchers (e.g. Frind et al. 1976; Fayer et al. 1992; Benson et al. 1994; Morris and Stormont 1997). However, studies on slope effect on the performance of CCBE are rare. Most analyses on the slope effect have been conducted using analytical approaches (Ross 1990; Morel-Seytoux 1994; Warrick et al. 1997). These solutions are generally based on steady-state conditions, and assume infinite thickness and no evapotranspiration (ET). Following the preliminary modeling results from Aubertin et al. (1997), Bussière et al. (1998) conducted a laboratory study on the cover inclination behaviors which showed significant differences compared with those from analytical solutions and from one-dimensional modeling results. Furthermore, Bussière et al. (2000) used the SEEP/W (GEOSLOPE International 1994) and HELP (Schroeder et al.1994) models to evaluate the influence of slope inclination and material properties on the hydrologic behavior of a CCBE. In order to evaluate the slope effects on the performances of the designed CCBE of the AA Leach Pad and to confirm the one-dimensional simulated results, two-dimensional simulations were conducted in this study using the commercial software HYDRUS2D (Simunek et al. 1999). Some of the main results are presented below.

2 HYDROLOGIC PROPERTIES OF COVER AND LEACH PAD MATERIALS

Hydrologic property tests of leach pad material and TCS material were conducted in 1999 at Daniel B. Stephens & Associates, Inc. in Albuquerque, New Mexico. The laboratory test results were fitted to van Genuchten's equations using RETC code (van Genuchten et al. 1991). Based on van Genuchten's theory, the WRC functions and unsaturated hydraulic conductivity functions have the following expressions respectively:

$$S_e = \frac{\theta - \theta_r}{\theta_s - \theta_r} = (1 + |\alpha h|^n)^{-m} \qquad h < 0, \qquad\qquad (1)$$
$$\theta(h) = \theta_s \qquad h \geq 0$$

$$K(h) = K_s * S_e^{1/2}[1 - (1 - S_e^{1/m})^m]^2 \qquad h < 0 \qquad\qquad (2)$$
$$K(h) = K_s \qquad h \geq 0$$

where θ_s and θ_r denote the saturated and residual volumetric water contents respectively; S_e is the effective water content; K_s is the saturated hydraulic conductivity; m, n and α are empirical constants and $m = 1 - 1/n$ (n>1); θ is water volumetric content; h is matric suction; and $K(h)$ is unsaturated hydraulic conductivity.

The fitted WRC parameters and saturated hydraulic conductivities of these two materials are listed in Table 1. Both the WRC and unsaturated hydraulic conductivity functions are presented in Figure 1.

Table 1. Fitted WRC parameters and saturated hydraulic conductivities of leach pad and TCS materials.

Parameter	θ_r	θ_s	M (1-1/N)	N	α	K_s
					1/cm	cm/s
Leach pad	0	0.19	0.175	1.213	0.084	4.5E-4
TCS	0.044	0.39	0.204	1.257	0.007	8.9E-6

Based on the method proposed by Aubertin et al. (1998), the air entry value (AEV) of a porous medium can be estimated by the suction corresponding to θ equal to $0.9\theta_s$. This approach typically provides an intermediate AEV between Brooks and Corey's (1966) approach and van Genuchten's et al. (1991) approach. Using this methodology, the AEV of the leach pad and TCS materials are estimated to be 10 cm and 120 cm respectively. The measured WRC points of the leach pad material indicate that the water content change is insignificant when suction is larger than 80-200 cm. However, the water content increases rapidly when suction is lower than this range. Baker and Hillel (1990) suggested that water entry value (WEV) can be estimated from this lower inflection point in the WRC. Using this evaluation method, the WEV of the leach pad material is estimated to be 150 cm. If suction at the interface between the fine and coarse layers is greater than the WEV of the coarse layer, very little water will seep into the coarse layer. This complementary approach, which is not shown here, has been used to further evaluate whether or not water could seep below the cover.

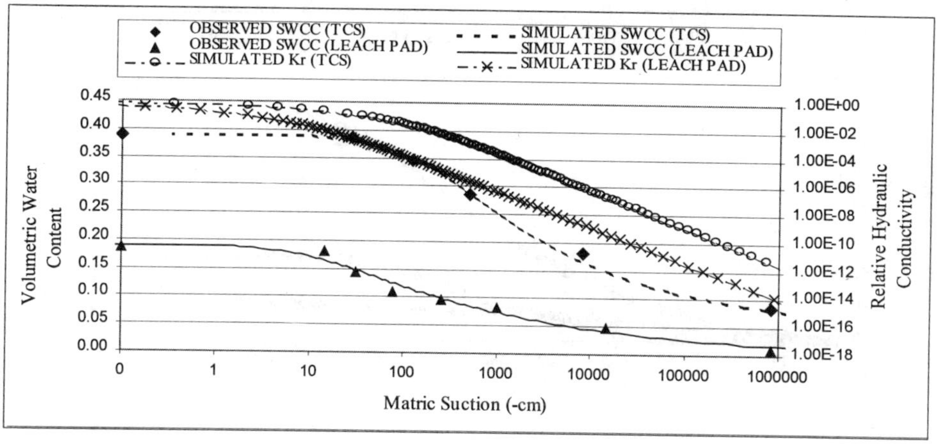

Figure 1. Observed and simulated WRC and relative hydraulic conductivity for the leach pad and TCS materials.

During the construction of the leach pad, a 91 cm thick drainage layer consisting of 2.5 cm diameter material was placed over the synthetic liner made of a geomembrane (considered impervious here). Saturated hydraulic conductivity of this coarse material is estimated to be 1.4E-3 cm/s; saturated water content to be 0.36; and van Genuchten's parameters m, n and α are estimated to be 0.52, 2.08, 0.078 respectively.

3 MODEL INPUTS

In addition to the hydrologic parameters of the TCS, leach pad, and drainage materials, other model inputs include the following:

3.1 *Climatic Data*

A weather station at the mine site has a complete daily data set for the year 1998, which includes air temperature, wind speed, relative humidity, precipitation and radiation. This data set is used for the weather inputs with the following modifications:

(a). The 1998 precipitation data was scaled to match the long term annual average precipitation from a USGS weather station, located about 8 km southeast of the mine site. This state weather station has 22 years of precipitation records with an average annual precipitation of about 33 cm. The shorter record from the mine site weather station is consistent with data from this station.

(b) The daily potential pan evaporation (PE) rates used in this model were obtained from the results of the one-dimensional simulations using SoilCover code (Geo-Analysis 2000 Ltd. 1997). The simulated annual PE of the site is approximately 156 cm. The measured PE rates in the summer seasons agree with the input data.

The cumulative precipitation and PE rates for the simulated period are represented in Figure 2.

3.2 *Vegetation*

The potential transpiration rates of the vegetation were also obtained from the output of SoilCover. In this study, the vegetation is conservatively characterized as "poor", having a leaf area index (LAI) from 0 to 1. The growth period of vegetation is estimated from March 16 to October 15. The root zone depth is specified same as the cover depth. The water uptake reduction

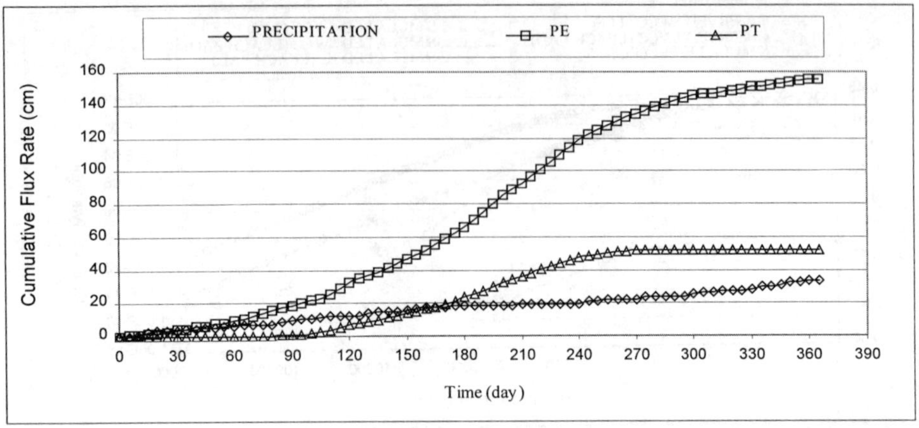

Figure 2. Cumulative precipitation, potential evaporation (PE), and potential transpiration (PT) of the simulation.

model defined by Feddes et al. (1978) was applied. The total annual potential transpiration (PT) was 52 cm.

The cumulative PT rate for the simulated period is represented in Figure 2.

3.3 *Boundary Conditions and Discretization*

The geometry of the leach pad cross-section was assumed to be symmetrical, and only half of it was considered as the study domain. The cover depth was assumed as 1 m.

Theoretically, smaller grid sizes will result in a better numerical presentation. Howeverthe most dense grid system that can effectively be used for our domain, using the HYDRUS2D attached mesh generator, is about 10 cm for the finite element side on the slope surface. Figure 3 illustrates the studied domain. The total number of elements used for the simulated domain is approximately 90,000.

As illustrated in Figure 3, the right boundary is assumed as a no flow boundary because of the symmetry of the pad geometry. The bottom boundary, which represents the synthetic liner beneath the drainage layer, is also defined as a no flow boundary. The cover is represented as the

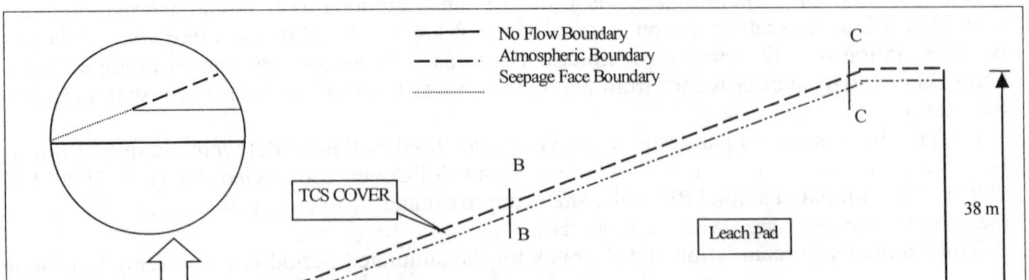

Figure 3. Study domain and boundary conditions of the simulation.

atmospheric boundary and the left side of the drainage layer on the slope face is assumed to be a seepage face boundary. The water flux exchanges only occur in the atmospheric and seepage face boundaries. Meanwhile, the water can also be transferred by transpiration, which is defined as sink terms in HYDRUS2D.

3.4 *Initial Conditions*

The simulated period is one year. To present the model conservatively, a relative initial wet (low suction heads) condition was assumed. The heads at the end of the simulation were used as the new initial condition and the simulation was repeated. This kind of iteration was performed for successive annual periods in order to represent dynamic pseudo steady state conditions.

4 SIMULATED RESULTS

The simulated cumulative fluxes for the entire cover with unit width are presented in Figure 4. In this figure, a positive flux represents flow leaving the domain, while a negative indicates flow entering the domain. Atmospheric flux is the flux leaving (+) or entering (-) the cover surface. Simulated results indicate that there is a small amount of water entering the cover layer because the cumulative atmospheric flux is negative at day 365. Meanwhile, the cumulative plant transpiration is positive and it is much larger than the absolute value of the atmospheric flux. These results indicate that vegetation is capable to uptake the water which enters the cover layer and meteoric water percolation is eliminated with selected cover configuration.

If a DDL condition is reached in the simulation, it should be noticed that the velocity vectors at the nodes delineating the cover bottom have a significant downward component. This could happen if the water flows diversely along the interface from upper slope position to lower slope position; the water content of the cover bottom will then become greater and eventually reach the maximum water content so a downward flow could occur.

The animation of the velocity vector flow obtained from these calculations with HYDRUS2D doesn't show any water percolation through the cover for the recorded time steps. Similarly, the water content profiles at different sections along the slope show no significant differences. Typical simulated water content profiles are presented in Figure 5. Three sections are shown in this figure. The horizontal distances of these sections to the left boundary of the studied domain are 110, 60, and 15 m respectively. These sections are labeled section A-A, B-B, C-C and il-

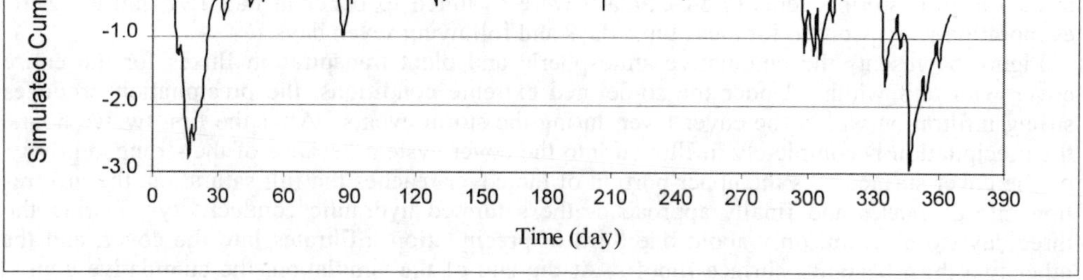

Figure 4. Simulated cumulative atmospheric flux and plant transpiration for normal condition.

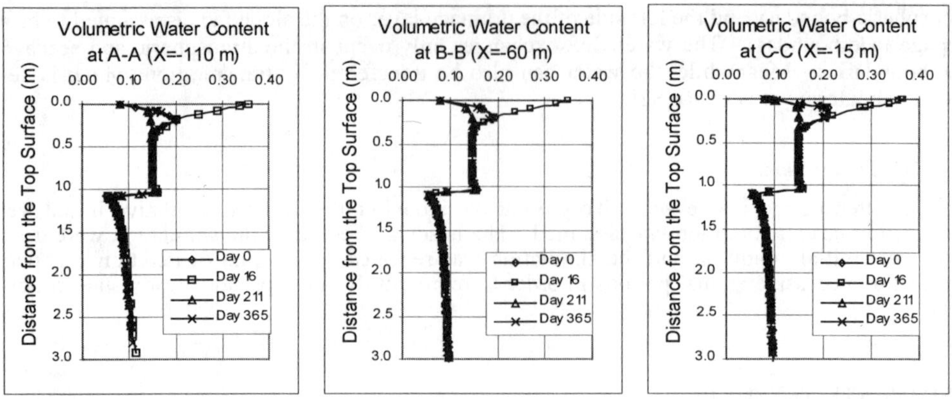

Figure 5. Simulated water content profiles at three sections along the slope for normal condition.

lustrated in Figure 3. The results from days 0, 16, 211, and 365 are presented for illustrative purposes. Day 16 is a wet winter day during a high precipitation period, and day 211 is a dry day in the summer. These results indicate that a strong capillary barrier effect is established. The fine grain material in the cover layer is much wetter than the underlying leach pad material, which then becomes less pervious, and the meteoric water is stored and transferred in the cover layer. In addition, the simulated water content profiles at different sections along the slope show no significant differences. No significant downward flow (DDL) occurs through the sloping cover.

5 SIMULATED COVER RESPONSES TO STORM EVENTS

The previous simulation was conducted with average weather conditions. In order to investigate cover behaviors during extreme events, storm events were considered in the following simulation. The Western Regional Climate Center in Reno, Nevada has the 24-hour storm event information for the mine site. Table 2 lists the expected storm magnitude for different return periods.

Table 2. The 24-hour storm events for the Goldstrike mine site.

Return period (years)	2	5	25	50	100
Storm magnitude (cm)	2.54	3.54	6.32	6.83	7.84

The same model inputs as shown above were used in this simulation, with the exception that three 100-year storm events (7.84 cm/day) were assumed to occur at day 1, 2 and 3. Zero evaporation was specified for these three days and following seven days.

Figure 6 presents the cumulative atmospheric and plant transpiration fluxes for the entire cover with unit width. Under the so-defined extreme conditions, the precipitation produces strong infiltration within the cover layer during the storm events. After the first twelve hours, the precipitation is completely infiltrated into the cover system because of the strongsorptivity on the cover surface. As the upper portion of the cover reaches the full saturation, the infiltration rate decreases and finally approaches the saturated hydraulic conductivity. During the three-day storm event, only about one third of precipitation infiltrates into the cover, and the other two third becomes surface runoff. At the end of the simulation, the cumulative atmospheric flux rate is negative, which means the cover does receive net water flux through the at-

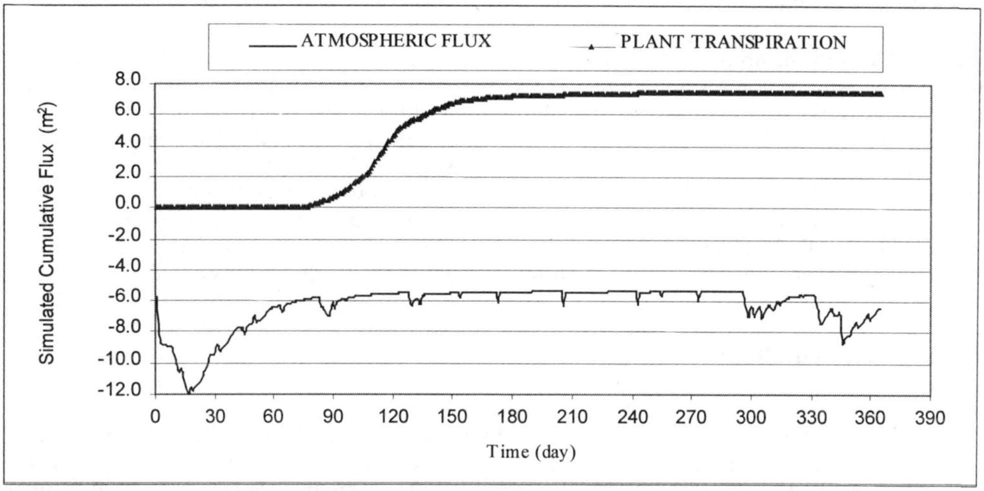

Figure 6. Simulated cumulative atmospheric flux and plant transpiration for storm event condition.

mospheric boundary. On the other hand, the simulated actual transpiration rate is still greater than the atmospheric flux which enters the cover system, so there will be the no infiltration through the bottom of the cover. Compared with the results from the standard simulation, it was found that both evaporation and transpiration rates are much greater in the storm event condition because more water is available in the cover layer. Due to the strong water holding capacity of the cover layer, the water entering the cover system during the storm events is transferred into the atmosphere through evaporation and plant transpiration.

The simulated volumetric water content profiles at sections A-A, B-B and C-C are shown in Figure 7. Storm events cause the wet fronts to travel downward initially. However, the water in the cover is eventually removed through the evaporation and transpiration. Comparing the water content profiles at day 211 and day 365 with the previous standard simulation, the water contents show no differences for all three sections. In addition, the storm events do not affect the water contents at leach pad layer at all, which means that suction at the interface is still higher than the WEV of the leach pad material. Hence, it can be expected that the unsaturated hydraulic conductivity of the pad material will remain low and will restrain water from flowing into the pad. Evaporation and lateral movement will then combine to evacuate the humidity so that no deep infiltration can be generated even under such extreme conditions.

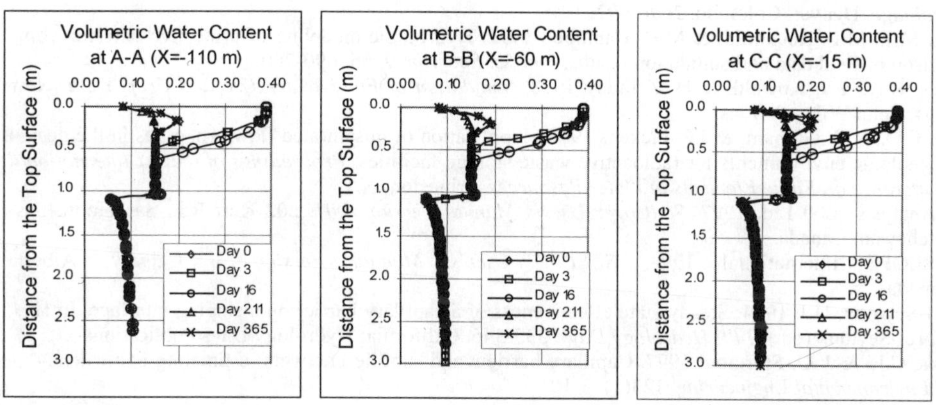

Figure 7. Simulated water content profiles at three sections along the slope for storm event condition.

6 CONCLUSIONS

The results from the above two-dimensional simulations confirm the findings derived from the previous one-dimensional simulation, i.e. a vegetated TCS cover with approximate 1 m thickness will effectively limit meteoric water infiltration to the leach pad during normal and severe weather conditions. Downward flow corresponding to the DDL phenomenon is not expected to occur on the slope of the cover due to the strong water holding capacity of the TCS material. However, the effects of the cover geometry of the CCBE could be significant to other sites and other cover configurations. This is especially true if the slope angle is large (> 20^0) and the saturated hydraulic conductivity of the cover material is high. It is believed that the slope effects can not be ignored in the cover design and severe conditions should be considered.

ACKNOWLEDGEMENTS

This paper is a result of AA Leach Pad reclamation program of Barrick Goldstrike Mines Inc. The inputs from the team members are gratefully acknowledged. Authors thank Dr. Mark Ankeney and Dr. Kuo-Chin Hsu from Daniel B. Stephens & Associates, Inc. for their technical support. Ms. Mayana Kissiova from Golder Associates is also acknowledged for her valuable input. We are especially grateful to Mr. Karl Burke from Barrick for his comments in reviewing this manuscript; however, the authors take full responsibility for any errors or omissions that may occur in this paper.

REFERENCES

Aubertin, M., R.P. Chapuis, A. Bouchentouf & B. Bussière 1997. Unsaturated flow modeling of inclined layers for the analysis of covers. *Proceedings of the fourth International Conference on Acid Rock Drainage*. Vancouver, Canada. 2: 731-746.

Aubertin, M., J.F. Ricard. & R.P. Chapuis 1998. A predictive model for the water retention curve: application to tailings from hard-rock mines. *Can. Geotech J.* 35: 55-69.

Baker, R.S. & D. Hillel 1990. Laboratory tests of a theory of fingering during infiltration into layered soils. *Soil Sci. Soc. Am. J.* 54: 20-30.

Benson, C.H., P.J. Bosscher, D.T. Lane, & R.J. Pliska 1994. Monitoring system for hydrologic evaluation of landfill covers. *Geotech. Testing J.* 17(2): 138-149.

Brookes, R.H. & A.T. Corcy 1966. Properties of porous media affecting fluid flow. *J. of irrigation and drainage division.* 92: 61-89.

Bussière, B., M. Aubertin, H.J. Morel-Seytoux & R.P. Chapuis 1998. A laboratory investigation of slope influence on the behavior of capillary barriers. *Proceedings of the 51st Canadian Geotechnical Conference*. 2: 831-836.

Bussière, B., M. Aubertin & R.P. Chapuis 2000. An investigation of slope effects on the efficiency of capillary barriers to control AMD. *Proceedings of the Fifth International Conference on Acid Rock Drainage*. Denver, Colorado. 2: 969-977.

Fayer, M.J., M.L. Rockhold & M.D. Campbell 1992. Hydrologic modeling of protective barriers: comparison of field data and simulation results. *Soil Sc. Soc. Am. J.* 56: 690-700.

Feddes, R.A., P.J. Kowalik & H. Zaradny 1978. *Simulation of Field Water Use and Crop Yield.* New York: John Wiley & Sons.

Frind, E.O., R.W.Gillham, & J.F. Pickens 1976. Application of unsaturated flow properties in the design of geologic environments for radioactive waste storage facilities. *Proceedings of the 1st International Conference on Finite Elements in Water Resources*. Princeton, N.J.

Geo-Analysis 2000 Ltd. 1997. *SoilCover User's Manual Version 4.01*. 202 Kerr Rd., Saskatoon, Saskatchewan, Canada.

GEOSLOPE International 1994. *SEEP/W User's Manual Version 3.* Calgary, Alberta Canada.

Morel-Seytoux, H.J. 1994. Steady-state effectiveness of a capillary barrier on a sloping interface. In H. J. Morel-Seytoux (ed.), *14th Hydrology Days*: 335-346. California: Hydrology Days Publications.

Morris, C.E. & J. C. Stormont 1997. Capillary barriers and subtitle D covers: estimating equivalency. *J. of Environmental Engineering*. 123(1): 3-10.

Ross, B. 1990. The diviersion capacity of capillary barriers. *Water Resour Res*. 26: 2625-2629.
Schroeder, P.R., T.S. Dozier, P.A. Zappi, B.M. McEnroe, J.W. Sjostrom & R.L. Peyton 1994. *The Hydrologic Evaluation of Landfill Performance (HELP) Model: User's Guide for Version 3*. EPA/600/9-94/168b. U.S. Environmental Agency Risk Reduction Engineering Laboratory, Cincinnati, Ohio.
Simunek, J., M. Sejna & M. Th. Van Genuchten 1999. Simulating water flow and solute transport in two-dimensional variable saturated media, version 2.0. *HYDRUS2D/MESHGEN2D*. U.S. Salinity Laboratory, Riverside, California.
van Genuchten, M. Th., F.J. Leij & S.R. Yates 1991. *The RETC code for qualifying the hydraulic function of unsaturated soils*: 1-85. EPA Rep. EPA/600/2-91/065, U.S. Salinity Laboratory, Riverside, California.
Warrick, A.W., P.J. Wierenga & L. Pan 1997. Downward water flow through sloping layers in the vadose zone: analytical solutions for diversions. *J. of Hydrology*. 192: 321-337.
Zhan, G., A.B. Mayer, J. Mcmullen & M. Aubertin 2000. Capillary cover design for a spent leach pad. *Proceedings of the International Symposium on Hydrology and the Environment*. Wuhan, Hubei, P.R.China.

Tailings and Mine Waste '01 © 2001 Balkema, Rotterdam, ISBN 90 5809 182 1

On the mechanics of wind erosion from tailings dams

G.E. Blight & F. Amponsah Da Costa
Department of Civil Engineering, University of the Witwatersrand, Johannesburg, South Africa

ABSTRACT: Wind erosion can be a major cause of the loss and dispersion of tailings material from a tailings impoundment into its surrounding environment. Such dust dispersion can be a major health hazard to inhabitants in nearby settlements and can also damage the health of animals, degrade crops and cause soil and water pollution. The problem of wind erosion can affect tailings dams in all types of climate, but becomes worse as climatic aridity increases.

The paper will investigate the mechanics of wind erosion from tailings dams. It will report on in situ measurements of wind velocity close to the surface of tailings dams, as well as wind-tunnel tests on model dams to illustrate the surface effects of features such as crest-walls, berms, etc. It will also deal with wind-velocity/particle pick-up relationships and the quantification of wind erosion losses.

1 INTRODUCTION - RING-DYKE TAILINGS DAM PROFILES AND WIND EROSION

Measurements in the South African gold fields area which has a semi-arid climate with a five month-long dry season, (Blight 1989, 1991, 1999), have shown that erosion losses from unprotected tailings dam surfaces are roughly evenly divided between water and wind erosion. Obviously, as a climate gets drier, wind erosion will dominate more and more, while in wetter climates it may be less important. Because sand-dunes are shaped and formed by the wind, it is particularly interesting to study their shape, in relation to the velocity of the wind that has formed them. Figure 1 shows the profile of a wind-formed sand dune, with the windward side to the left. This windward slope of the dune (9.5°) is clearly related to the predominant speed of the wind, and its height to the available supply of sand. The leeward slope or slip face forms at the angle of repose of the sand, namely 33° in this case. The profile of the wind parallel to the slope surface and at a height of 200 mm above the sand has been shown superimposed on the elevational profile. This profile, typical of several measured on sand dunes, is similar in shape to the physical profile. The wind speed increases up the dune and then abruptly decreases shortly before the crest of the dune is reached. Once the crest starts to provide shelter from the wind, the wind speed falls abruptly.

Why should the wind speed increase with height up the dune? Wind represents a movement of the atmosphere relative to the earth. Viscous resistance and turbulence between the earth's surface and the moving air decelerates air movement, resulting in the formation of a retarded boundary layer and a relative acceleration of the air with increasing height above the surface. The vertical velocity gradient of the wind over a uniform surface can be expressed by the well-known equation (e.g. Mulligan 1988):

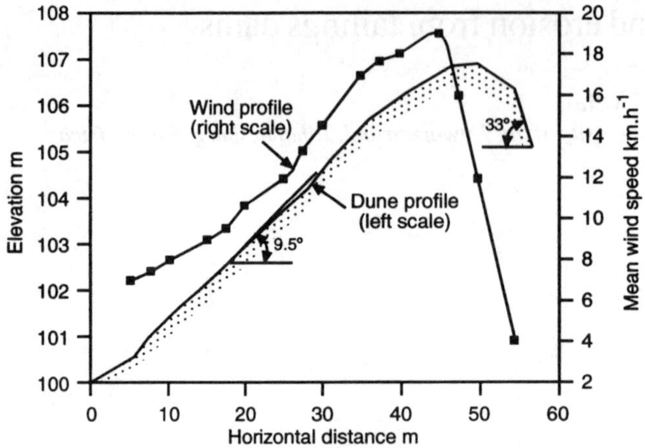

Figure 1. Topographic and wind velocity profiles for a natural sand dune.

$$dv/dz = K(\log_e z/k).v_z \quad (1)$$

where dv/dz is the vertical wind velocity gradient relating wind velocity v (in ms^{-1}) to height z(m), and has units of s^{-1}

- K is von Karman's constant (units m^{-1})
- k is a roughness length (m)
- v_z is the wind velocity at height z

A dune or a tailings dam causes an additional resistance to air flow and modifies the ideal boundary layer wind profile described by equation (1).

Figure 2 shows similar physical and wind profiles for a tailings dam. The wind profile was also measured parallel to the surface at a height of 200 mm. The wind profile is also fairly similar to the physical profile of the tailings dam. It is interesting to see the localized drop in wind speed at the berm half-way up the height of the dam. Peak wind speed occurs at the leading edge of the crest wall with a considerable drop in speed over the 2 m width of the crest wall, and a further drop to a reduced, more or less constant wind speed over the flat top surface of the dam.

From the wind profile one would expect the worst wind erosion to occur over the upper portion of the windward slope, with deposition occurring on the berm and just downwind of the crest wall. Because of the relatively low wind speed over the top of the dam, there should be little net erosion off the top. It is fair to say that if one observes a plume of wind-blown dust coming off the leeward side of a tailings dam, most of the dust has eroded from the upper slopes of the windward side of the dam, and less has come off the top of the impoundment.

Wind velocities do not always increase appreciably with height above the base of a dam, however. Figure 3 shows wind profiles measured when the wind was blowing normally, and parallel to a 10 m high tailings dam slope. Again, the sheltering effects of relatively small features, like small ridges and trenches, berms and berm and crest walls are illustrated. It is obvious that excrescences above the general surface of a dam can be used to provide shelter from wind erosion, and that it is mainly these excrescences that need to be protected against wind erosion, and not the entire tailings surface.

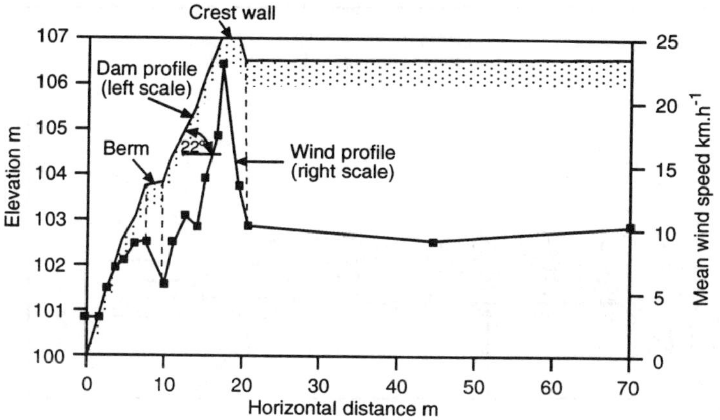

Figure 2. Topographic and wind velocity profiles for a tailings dam slope.

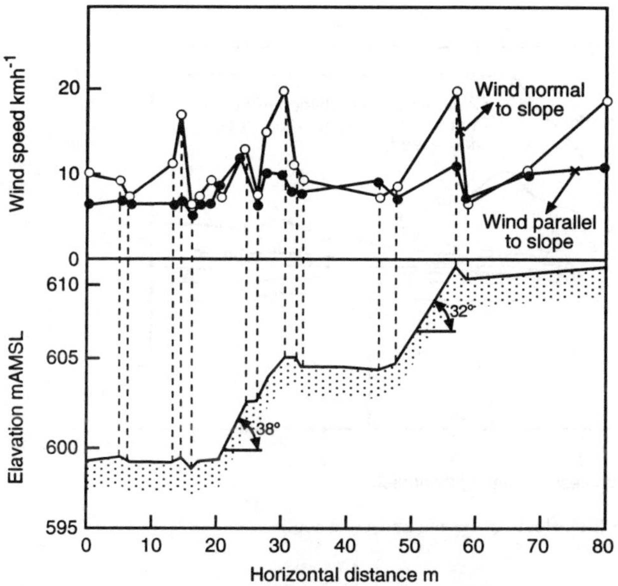

Figure 3. Topographic and wind velocity profiles for a tailings slope with wind normal and parallel to section.

2 WIND TUNNEL TESTS OF MODEL TAILINGS DAM PROFILES

To confirm and possibly extend the conclusions drawn from wind velocity measurements on full size tailings dams, a series of measurements has been made on model tailings dams in a wind tunnel. Figure 4 shows the results of wind velocity measurements made over a simple dam profile. In the wind tunnel, the air approaching the model has a uniform velocity (i.e. an amplification factor of 1.0). The effect of the boundary layer in the field occurs because the model slope partly blocks the wind tunnel, resulting in a venturi effect. As the amplification

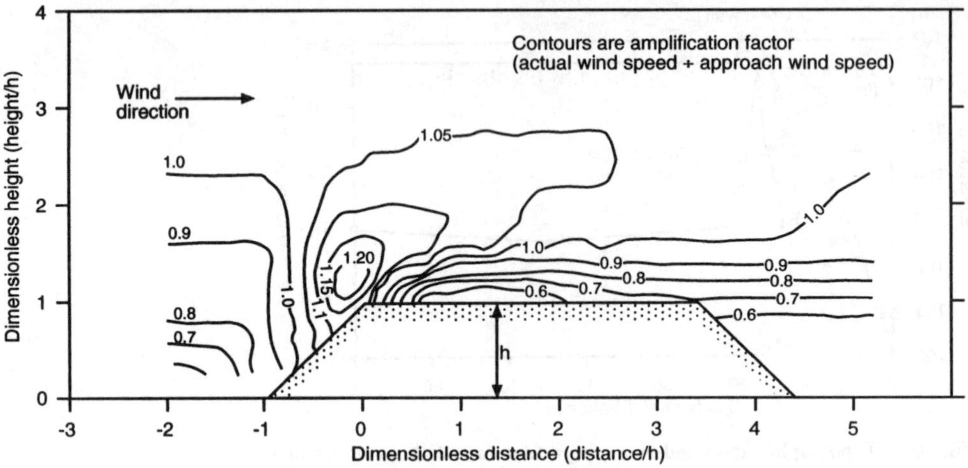

Figure 4. Wind tunnel test on a model tailings dam.

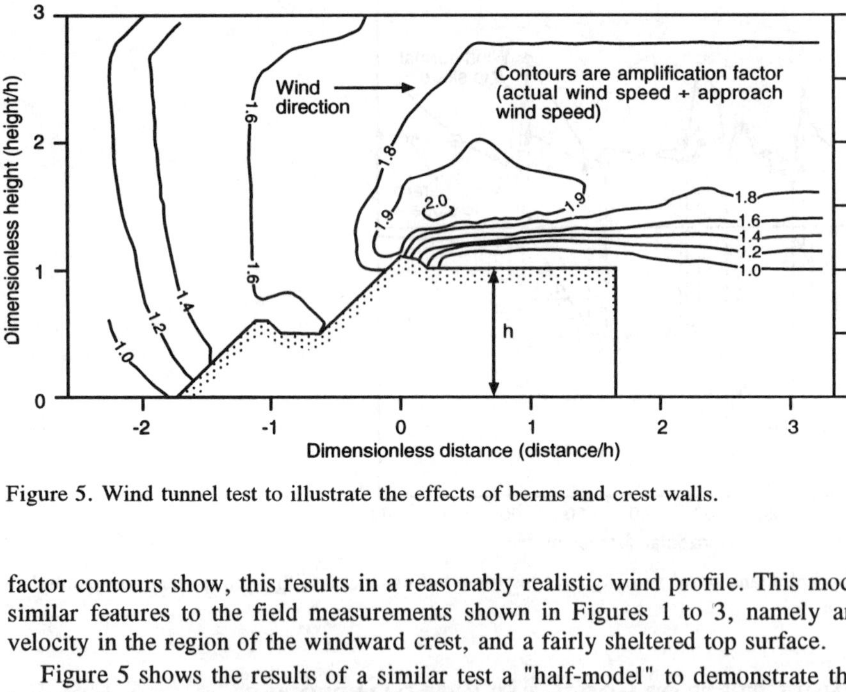

Figure 5. Wind tunnel test to illustrate the effects of berms and crest walls.

factor contours show, this results in a reasonably realistic wind profile. This model test shows similar features to the field measurements shown in Figures 1 to 3, namely an accentuated velocity in the region of the windward crest, and a fairly sheltered top surface.

Figure 5 shows the results of a similar test a "half-model" to demonstrate the effects of a berm and crest-walls. Here again, it is evident that the berm provides some wind shelter, that the highest wind velocities impinge on the crest-wall and that the top-surface of the dam is relatively sheltered. Thus the wind tunnel tests confirm the field wind velocity measurements.

3 TAILINGS PICK-UP VELOCITIES

According to Bagnold's (1941) classical work on wind erosion, erosion of a particulate material does not commence until a threshold or "pick-up" wind velocity has been exceeded. His equation for the transport of material is

$$q = K_t(v - v_t)^3 \tag{2}$$

where q is the rate of material movement (mass per unit area per unit time),
v is the wind velocity at the surface,
v_t is the threshold wind velocity and
K_t is a constant of proportionality.

The threshold velocity v_t is a very important parameter. A knowledge of v_t together with recorded wind data (gust velocities, frequency and duration and direction) can enable wind erosion losses from a tailings dam to be estimated.

The parameters in equation (2) depend on the particle size distribution of the tailings, and a series of different tailings materials has been investigated in the wind tunnel to establish their threshold or wind pick-up velocities and the character of the transport equation. The tests were done by spreading a layer of tailings of known mass on a board measuring 300 mm square and placing this on the floor of the wind tunnel. The wind velocity was then gradually increased until the first grains of tailings left the board at the pick-up or threshold velocity. The pick-up velocity is not well defined, as the first visible movement of particles consists of a rolling motion that results in particle re-arrangement without any particle loss. The first particles lost from the board are lost by rolling, not pick-up.

The wind velocity was then increased a stage and once equilibrium prevailed, the wind velocity was reduced to zero, the board and contents were removed from the tunnel and weighed, replaced in the tunnel and the velocity was then further increased. In this way the particle transport curves for materials with the three particle size distributions shown in Figure 6 were constructed. These curves are shown in Figure 7.

The function most closely fitting the experimental points was exponential, and of the form

$$R = e^{Av} \tag{3}$$

where R = sand transport rate in $kg.m^{-2}.h^{-1}$, and
v = wind velocity in $km.h^{-1}$.

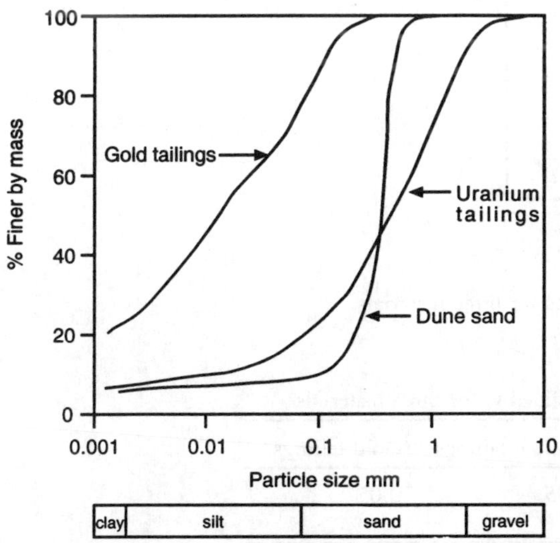

Figure 6. Particle size distributions for three materials investigated for particle pick-up velocity.

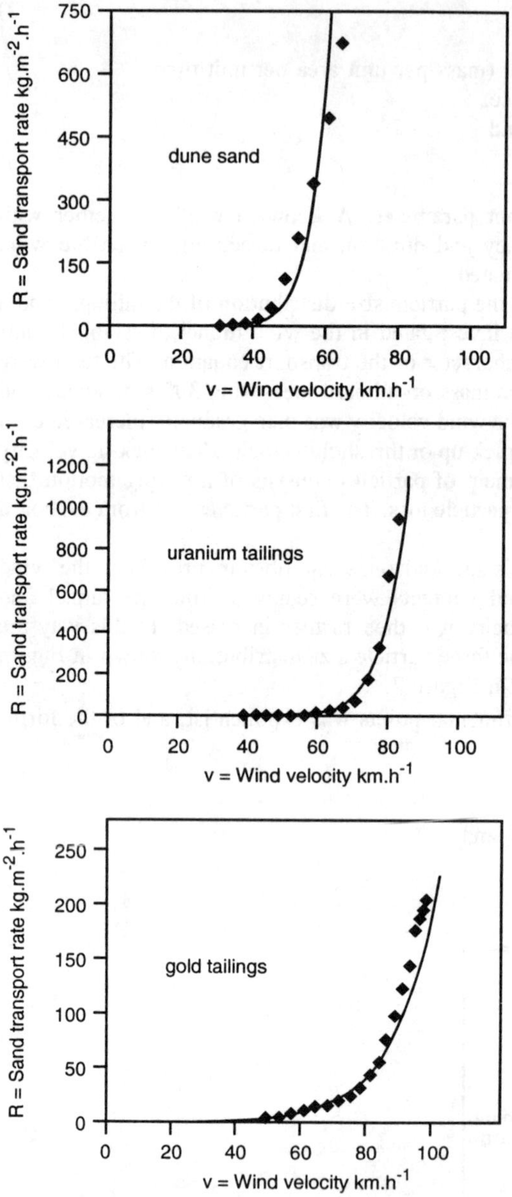

Figure 7. Wind velocity transport characteristics for three materials.

Table 1. Parameter A in equation (3) and re-defined v_t for three materials

Material	Dune sand	Uranium tailings	Gold tailings
A (for R in $kg.m^{-2}.h^{-1}$) and v in $km.h^{-1}$	0.10	0.075	0.05
v_t (in $km.h^{-1}$ for $R = 20\ kg.m^{-2}.h^{-1}$)	30	40	60

The disadvantage of this formulation is that it does not define the pick-up velocity, but it fits the data better than equation (2). If the pick-up wind velocity is redefined as "that velocity that will transport material at a nominal small rate" (e.g. 20 $kg.m^{-2}.h^{-1}$), then pick-up velocities can be calculated. Values for A and v_t in terms of the above re-definition are given in Table 1 for the three materials referred to in Figures 6 and 7.

4 FIELD EVIDENCE

The observations recorded for a dune (Figure 1) where sand movement was taking place at a wind velocity of between 8 and 20 $km.h^{-1}$ do not agree very well with the value of v_t in Table 1 of 30 $km.h^{-1}$. Only one set of field measurements is available to compare with the wind tunnel data given in Figure 7 and Table 1. Observations of dust generation on the impoundment from which the uranium tailings were sampled (over a period of five years) showed that material pick-up is negligible unless the one hour average wind velocity exceeds about 45 $km.h^{-1}$, which does agree approximately with the wind tunnel data (Table 1). Estimates of annual tailings loss caused by wind erosion are shown in Figure 8. There appears to be a linear relationship between erosion E and the square of one hour maximum wind velocity (not $(v_1 - v_t)^3$ as predicted by Bagnold's equation). Although E has been expressed in $T.ha^{-1}$ per year, it is important to note that most or all of the losses took place in one (or a few) severe windstorms, each lasting some hours (or a few days at most).

Severe wind storms at this location, in which the one hour maximum wind velocity exceeds 35 $km.h^{-1}$ are uncommon, even though gust speeds as high a 130 $km.h^{-1}$ are regularly recorded. For example, in a six month period (the "windy season") analysed in detail, the highest one hour maximum wind speed recorded was 33.5 $km.h^{-1}$ and 27 $km.h^{-1}$ was exceeded only twice in 180 days.

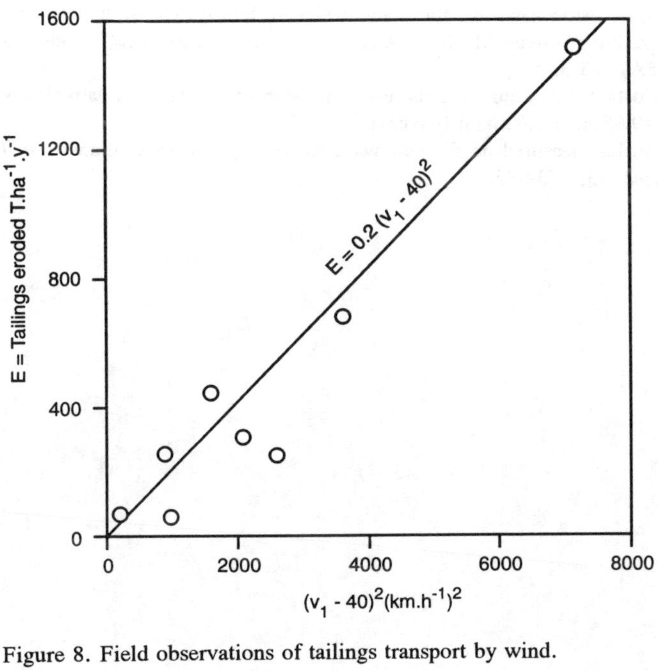

Figure 8. Field observations of tailings transport by wind.

5 CONCLUSIONS

There is obviously a lot still to learn concerning the mechanics of wind erosion from tailings dams. The following observations (first published more than a decade ago, Blight 1989, 1991) concerning the distribution of erosion losses from tailings impoundment have been confirmed:

.1 The area of a tailings dam that is most vulnerable to wind erosion is the vicinity of the slope crest or crests. If the tailings surface in this zone is protected, erosion losses will be greatly reduced.

.2 Because they are sheltered from the worst effects of the wind, the beaches or top of a impoundment are not as subject to wind erosion as the outer slopes.

Both these conclusions have been supported by field observations of the distributions of erosion losses, by field measurements of near-surface wind profiles, and by wind tunnel tests on models.

.3 Wind tunnel tests have also been used to study the threshold or pick-up wind velocity at which the wind will start to transport particles along the surface. This velocity is not well defined, as the first movement of particles takes place by rolling and rearrangement of the surface, and actual particle pick-up only takes place at higher velocities. A modified pick-up velocity, at which a nominal small rate of particle pick-up occurs, established by wind tunnel tests, agrees roughly with particle pick-up velocities observed in the field.

.4 Even though severe wind storms may occur only a few times in a year, the loss of tailings by wind erosion may be as high as $1500T.ha^{-1}.y^{-1}$ or a depth of tailings of about 100 mm (Figure 8).

REFERENCES

Bagnold, R.A 1941. *The physics of blown sand and desert dunes.* London: Methuen.

Blight, G.E. 1989. Erosion losses from the surfaces of gold-tailings dams. *Journal South African Inst. Mining and Metallurgy*, 89(1):23-29.

Blight, G.E. 1991. Erosion and anti-erosion measures for abandoned gold tailings dams. *Reclamation 2000: Technologies for Success*. 8th National Meeting, American Society for Surface Mining and Reclamation, Durango, Co, USA, 323-330.

Blight, G.E. & F. Amponsa-Da Costa 1999. Improving the erosional stability of tailings dam slopes. *Tailings and Mine Waste '99*. 197-206, Rotterdam: Balkema.

Mulligan, K.R. 1998. Velocity profiles measured on the windward slope of a transverse dune. *Earth Surface Processes and Landforms*, 13, 573-582.

Tailings and Mine Waste '01 © 2001 Balkema, Rotterdam, ISBN 90 5809 182 1

Internal shear strength of a geosynthetic clay liner at high normal loads

J.T.Olsta
CETCO, Arlington Heights, Ill., USA

R.H.Swan, Jr
GeoSyntec Consultants-SGI\R\, Atlanta, Ga., USA

ABSTRACT: Geosynthetic clay liners (GCLs) have seen increasing usage of GCLs in the mining industry. Potentially, GCLs could cut costs in the liner design of heap leach pads and tailings piles. However, little is known about the behavior of GCLs under the extremely high normal loads encountered in some of these applications.

To study the internal shear behavior of a needlepunched GCL under high normal loads two sets of shear tests were performed. The first set of tests was conducted in a standard 300 mm by 300 mm (12 in. by 12 in.) direct shear box at normal loads of 350, 700, and 1050 kPa (50, 100 and 150 psi). The second set of tests was conducted in a smaller 150 mm by 150 mm (6 in. by 6 in.) direct shear box at normal loads of 1050, 1750, and 2800 kPa (150, 250 and 400 psi). The smaller shear box was used at the higher normal loads due to limitations in the application of the higher normal stresses to the larger shear box. A common normal load of 1050 kPa (150 psi) allowed for comparing results of the two shear boxes.

Test results indicated good correlation between the small and large shear boxes. At 1050 kPa (150-psi) normal load, the peak internal strength of the GCL for the large and small shear boxes was 332 kPa (48.2 psi) and 334 kPa (48.4 psi), respectively. At the highest normal load, 2880 kPa (400 psi), the GCL had a peak strength of 668 kPa (96.9 psi).

In addition to the direct shear box test results, the peel strength test result for the same GCL sample is indicated. Peel strength tests have been correlated to direct shear strength and used as an index test in MQC and CQA test programs.

1 INTRODUCTION

A geosynthetic clay liner (GCL) is a man-made product typically consisting of a layer of sodium bentonite clay placed between two geosynthetics. GCLs contain an industry standard minimum of 3.65 kilograms per square meter of sodium bentonite clay as measured using ASTM D5993. The interlayer sodium ions on the montmorillonite clay platelets make it hydrophilic and high swelling. Sodium bentonite used in GCLs also has excellent filter cake properties resulting in maximum fluid loss of 18 ml per ASTM D5891 and a maximum hydraulic conductivity of 5×10^{-9} cm/s per ASTM D5887.

There are unreinforced and reinforced GCLs available for use. Unreinforced GCLs are typically manufactured using water-soluble adhesive and/or pressure to bond the bentonite to geotextiles or a geomembrane. Since the clay mineral montmorillonite is typically the main component of bentonite, it is well-known to have a low shear strength (Olson, 1974), these products are typically used in man-made earthen projects which are relatively flat in nature that do not present significant shear stresses or differential bearing pressures.

Reinforced GCLs are typically manufactured by needlepunching the top and bottom geotextiles together to encapsulate the bentonite layer. The physical bonding of the geotextiles enhances the internal shear strength of the GCL. Peel strength tests have been correlated to internal shear strength and are used as an index test in Manufacturer's Quality Control (MQC) and Construction Quality Assurance (CQA) test programs (Richardson, 1997). All US manufacturers of reinforced GCL currently certify to a minimum peel strength of 66 N (15 lbs.) per modified ASTM D4632.

GCLs have been manufactured since the early 1980's and have been used in landfill liner systems since 1986 (Koerner, 1996). More recently GCLs have been used in numerous mining applications, such as liners in tailings ponds and in closure (cap) applications (Miller, 1999). GCLs have a potential use in liner systems for heap leach pad applications and some GCLs have been proposed for use in current heap leach pad designs. GCLs have been shown to be compatible with the typical dilute cyanide solution which is used on heap leach pads to break down the spent ore to recover the desired mineral. In laboratory testing (Wallace, 1991), a hydraulic conductivity of <1 x 10^{-9} cm/s was reported after two pore volumes of permeation with a 600 ppm sodium cyanide solution at confining pressures greater than 210 kPa (30 psi). However, in general, geosynthetics are subjected to much higher stresses in heap leach pads than in other civil engineering applications. Heap heights can reach 125 m (Van Zyl, 1997) with vertical loads of up to 2,800 kPa. Consequently, there is concern for the internal shear strength of GCL under these conditions.

2 EXPERIMENTAL SETUP

To study the internal shear behavior of a needlepunched GCL under these extremely high normal loads two sets of shear tests were performed. The first set of tests was conducted in a standard 300 mm by 300 mm (12 in. by 12 in.) direct shear box at normal loads of 350, 700 and 1050 kPa (50, 100 and 150 psi). A schematic of the direct shear box test configuration is presented in Figure 1.
The second set of tests was conducted in a smaller 150 mm by 150 mm (6 in. by 6 in.) direct shear box at normal loads of 1050, 1750 and 2800 kPa (150, 250 and 400 psi). The smaller shear box was used at the higher normal loads due to limitations in the application of the higher normal stresses to the larger shear box. A common normal load of 1050 kPa (150 psi) allowed for comparing results of the two shear boxes. Samples used in both size shear boxes incorporated the use of textured steel gripping surfaces affixed to the upper and lower plates to achieve complete and uniform gripping of the GCL geotextiles. The reinforced GCL selected had a nonwoven geotextile on one side and a woven geotextile on the other side. Testing was conducted in accordance with ASTM D 6243 under the following test conditions. Samples were hydrated for 48 hours under each normal load and then sheared at a constant displacement rate of 1 mm/min (0.04 in./min).

3 RESULTS

Test results are summarized in Table 1. The results indicate good correlation between the small and large shear boxes. At 1050 kPa (150 psi) normal load, the peak internal strength of the GCL

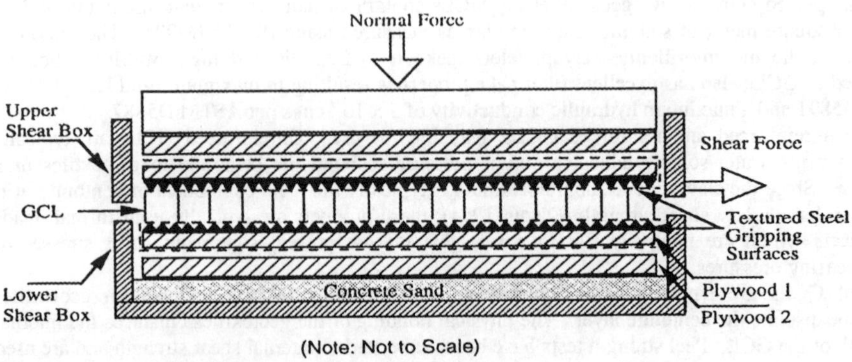

Figure 1. Direct Shear Box Test Configuration

Table 1. Internal shear strength of Bentomat ST reinforced GCL under hydrated conditions.

Shear Box Size (mm x mm)	Normal Stress (kPa)	Peak Load (N)	Large Displacement* Load (N)	Peak Strength (kPa)	Large Displacement* Strength (kPa)
300 x 300	350	13,690	3483	148	37
300 x 300	700	22,910	7259	247	78
300 x 300	1050	30,870	12,670	332	137
150 x 150	1050	7744	3367	334	145
150 x 150	1750	11,090	5048	478	217
150 x 150	2800	15,510	7117	668	306

* Large displacement measured at 50 mm for 300 by 300 shear box and 35 mm for 150 by 150 shear box.

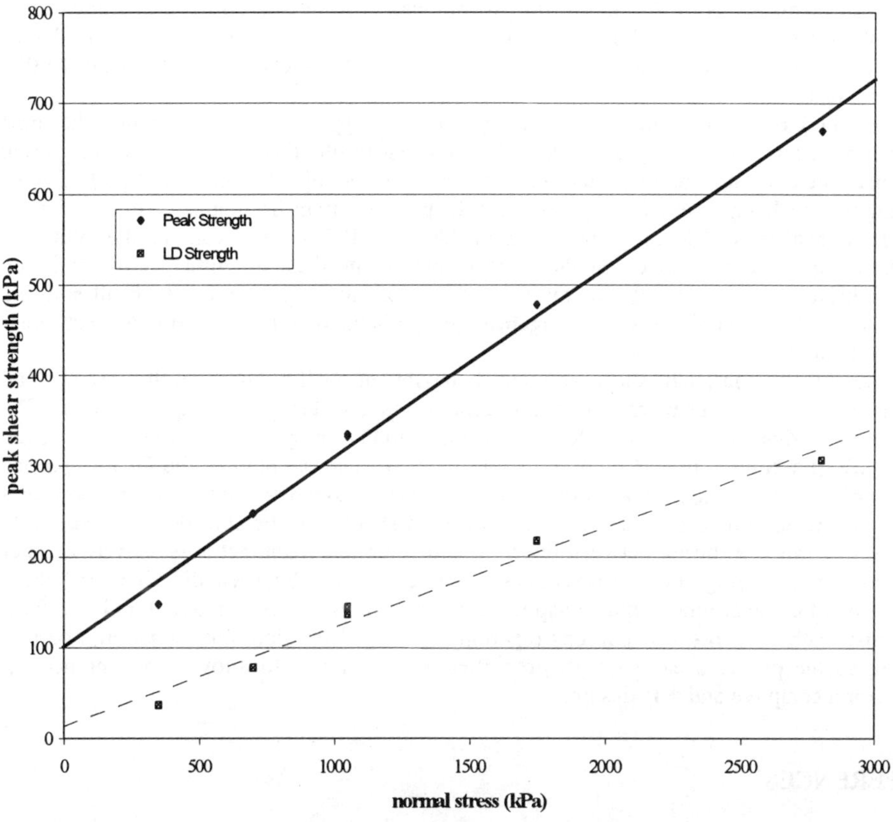

Figure 2. Internal shear strength of Bentomat ST reinforced GCL under hydrated conditions.

for the large and small shear boxes were 332 kPa (48.2 psi) and 334 kPa (48.4 psi), respectively. At the highest normal load, 2800 kPa (400 psi), the GCL had a peak strength of 668 kPa (96.9 psi). At 1050 kPa (150 psi) normal load, the large displacement (LD) strength of the GCL for the large and small shear boxes were 137 kPa (19.8 psi) and 145 kPa (21.0 psi), respectively. At the highest normal load, 2800 kPa (400 psi), the GCL had a large displacement strength of 306 kPa (44.4 psi).

Figure 2 depicts the data graphically using a traditional best-fit line for illustrative purposes. It yields a 12° peak friction angle with a peak adhesion of 100 kPa (2095 psf) and an 6° large displacement friction angle with a large displacement adhesion of 13 kPa (275 psf). The data is actually curvilinear with a point of maximum curvature at approximately 1050 kPa normal stress.

The GCL used in the above internal shear strength testing had a MQC peel strength of 110 N (25 lbs.) using a modified ASTM D4632 test method. This is a typical peel strength for reinforced GCL with a specified minimum peel strength of 66 N (15 lbs.).

4 SUMMARY AND CONCLUSIONS

It can be seen from the test results that the GCL has a high apparent peak friction angle at the highest normal stress of 2880 kPa (400 psi) of 14°. Although when comparing this peak apparent friction angle with the peak apparent friction angle of 20° for the GCL under typical landfill loading conditions of 700 kPa (100 psi) one can see the true effect of the curvilinear behavior of the GCL. This curvilinear behavior occurs in the peak internal strength of all soil-like materials as well as interfaces of soil against geosynthetic and geosynthetic against other geosynthetic materials.

However, the large displacement shear strength behavior typically does not exhibit the same dramatic curvilinear behavior. It can be seen from the test results that the GCL has an apparent large displacement friction angle at the highest normal stress of 2880 kPa (400 psi) of 6.3°. Comparing this peak apparent friction angle with the peak apparent friction angle of 6.5° for the GCL under typical landfill loading conditions of 700 kPa (100 psi) one can see the minimum effect of the curvilinear behavior on the large displacement shear strength. This minimum curvilinear behavior is also similar in the large displacement internal strength of all soil-like materials as well as interfaces of soil against geosynthetic and geosynthetic against other geosynthetic materials.

Understanding the internal shear strength behavior of GCLs under high normal load conditions is very important when one is designing a lining system for a heap leach pad. The data presented in this paper demonstrates the potential strength properties for a reinforced GCL under specific test conditions. In addition to the internal shear strength of the GCL, interface shear strength of the GCL against a geomembrane or other geosynthetic materials and soils that may be encountered in a specific design/site application should also be considered. Because of differences in surface texturing between different geomembrane manufacturers as well as other geosynthetics and varying soil characteristics, interface friction between the GCL and these materials should be determined on a site-specific basis. All testing, both internal and interface, should consider the actual site conditions regarding wetting, hydration, consolidation, and rate of loading so the proper shear strength properties are developed to allow good engineering judgment for a complete and safe design.

5 REFERENCES

Koerner, R.M., 1996. Geosynthetic clay liners, part one: an overview. *Geosynthetic Fabrics Report,* 14(3): 22-25.

Miller, B., 1999. *Proc. of Sixth Int'l Conf. on Tailings and Mine Waste, Ft. Collins, CO, USA, 24-27 Jan. 1999*: 353-358. Rotterdam, Balkema.

Olson, R.E., 1974. Shearing strength of kaolinite, illite and montmorillonite. *Journal of Geotechnical Engineering.* 100 (10).

Richardson, G.N., 1997, GCL internal shear strength requirements. *Geosynthetic Fabrics Report,* 15(2):20-25.

Van Zyl, D., 1997. *Proc. of Geosynthetics '97 Conf., Long Beach, CA, USA, 11-13 March, 1997*:1079-1090. St. Paul, IFAI.

Wallace, J.F., 1991. Final report permeability testing of Bentomat liner. Correspondence from Dames & Moore to American Colloid Company. 15 July, 1991.

Tailings and Mine Waste '01 © 2001 Balkema, Rotterdam, ISBN 90 5809 182 1

Tensile stresses and the durability of PE pipes

R.W.I. Brachman
Department of Civil and Environmental Engineering, University of Alberta, Edmonton, Alb., Canada

ABSTRACT: Polyethylene (PE) drainage pipes are extensively used in tailings and mine waste applications because of their excellent resistance to chemical degradation, relative ease of construction and overall good performance. Limiting the long term tensile stresses in PE pipes is necessary to obtain adequate long term durability. If tensile stresses are too large and in the presence of a flaw or defect, PE pipes may be susceptible to stress cracking–a brittle failure occurring at stresses well below yield. The tensile stresses that may occur in tailings and mine waste applications are discussed. Results from analytical solutions, large-scale laboratory tests and numerical analyses are presented to provide a measure of the tensile stresses from large earth pressures, coarse backfill materials, and perforations. Sources of flaws or defects in PE pipes are also described. Recommendations are given for design and construction procedures that promote greater durability of polyethylene pipes.

1 INTRODUCTION

Polyethylene drainage (PE) pipes are extensively used in tailings and mine waste applications because of their excellent resistance to chemical degradation, relative ease of construction and overall good performance. However, unique conditions may arise for polyethylene pipes in tailings and mine waste applications (e.g., large tailings dams, gold mining heap leach pads) that are not typically encountered by pipes used in more conventional gravity-flow sewer applications. Even if a pipe is properly designed to withstand the external stresses from the earth pressures (e.g., see Moore 2000), the pipe must have sufficient long term durability to perform its intended function.

A great amount research has been conducted on the long term durability of polyethylene for both natural gas distribution pipes and geomembranes (e.g., see summaries by Mruk 1990 and Rowe 1998) and in general the findings from this research are applicable to buried polyethylene pipes. Essentially, the long term durability of polyethylene is related to both chemical and physical aging of the polymer. Chemical aging involves the breakdown of the molecular structure, and eventually a reduction in engineering properties of the material, whereas physical aging involves an adverse change in the crystallinity of the material. Mruk (1990) has identified: (a) photodegradation, (b) oxidative degradation, and (c) stress cracking as the mechanisms that may limit the life of polyethylene piping, and noted that modern polyethylene resins generally have excellent resistance to these forms of aging. Current standards require that polyethylene piping contain adequate ultraviolet stabilizers and antioxidants to limit the chemical degradation under normal conditions. Test procedures exist to evaluate the stress cracking resistance of polyethylene pipes (ASTM D2837).

The objective of this paper is first, to clarify the importance of tensile stresses on the long term durability of polyethylene pipes. Limiting tensile stresses in PE pipes is necessary to obtain adequate long term durability. If tensile stresses are too large, PE pipes may be susceptible to stress cracking. This is a brittle failure mechanism that occurs at stresses well below yield. If allowed to propagate, cracking may lead to collapse of the pipe. Second, the tensile stresses that may occur for buried polyethylene pipes in tailings and mine waste applications arising from enormous overburden pressures, coarse gravel backfill, and perforations are discussed and quantified using results from analytical solutions, large-scale laboratory tests and numerical analyses. Third, the sources of flaws and defects in PE pipes are described and the potential failure mechanisms are discussed.

2 STRESS CRACKING

Three conditions must exist for stress cracking to occur. The first requirement is a polymer microstructure that will allow the propagation of the crack. Polyethylene is a polymer that may be susceptible to this failure mechanism because of its semicrystalline structure. The second requirement is a flaw or defect in the material that serves to initiate the crack. Common sources of defects include flaws induced by the seaming process and construction damage. The third requirement is the presence of long term tensile stresses. Tensile stresses can promote crack propagation to the stage of failure.

Although modern polymer formulations generally have excellent resistance to stress cracking, it is very important to limit long term tensile stresses and, to the maximum practical extent possible, limit the number of defects in the pipe to obtain long term durability. The next sections detail and quantify the nature of tensile stresses in PE pipes. This is followed by a discussion on the sources of defects in PE pipes.

3 PIPE STRESSES UNDER DEEP BURIAL

The principle source of tensile stresses in PE pipes occurs from the external stresses applied to the pipe from the overburden material. Figure 1a shows an idealized installation of a deeply buried pipe. For example, these conditions could represent a pipe buried under a tailings dam, or within a leachate collection system in a deep heap leach pad. A region of soil around the pipe is isolated in Figure 1b showing idealized earth pressures acting distant from the pipe. These pressures have a vertical component σ_v arising from the weight of the overlying materials above the pipe and a horizontal component $K\sigma_v$ associated with the restraint against lateral soil movement within the embankment, where K is the coefficient of lateral earth pressure.

The buried pipe acts as a component of the soil-pipe system where the stresses in the pipe are a function of both the soil and pipe stiffness, and the resulting soil-structure interaction. Maximum compressive hoop stresses occur at the interior springline of the pipe, and typically maximum tensile hoop stresses occur at the interior crown and invert. The magnitude of tensile stresses for PE pipes predominantly depends on overburden pressures σ_v, soil modulus E_s, and

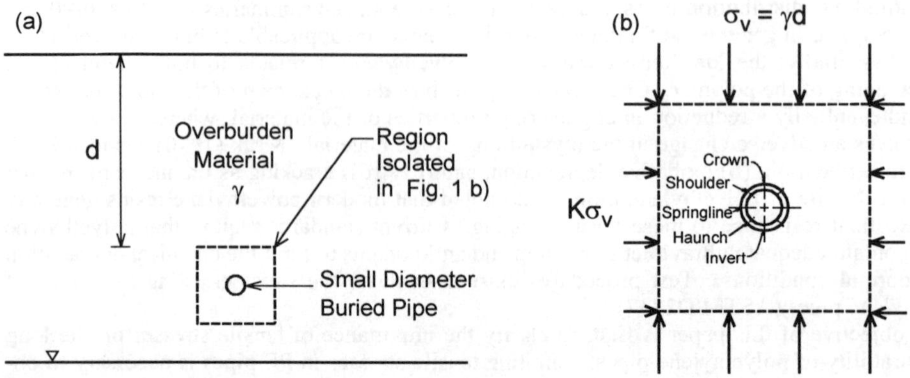

Figure 1. (a) Pipe under deep burial. (b) Earth pressures acting on soil-pipe system.

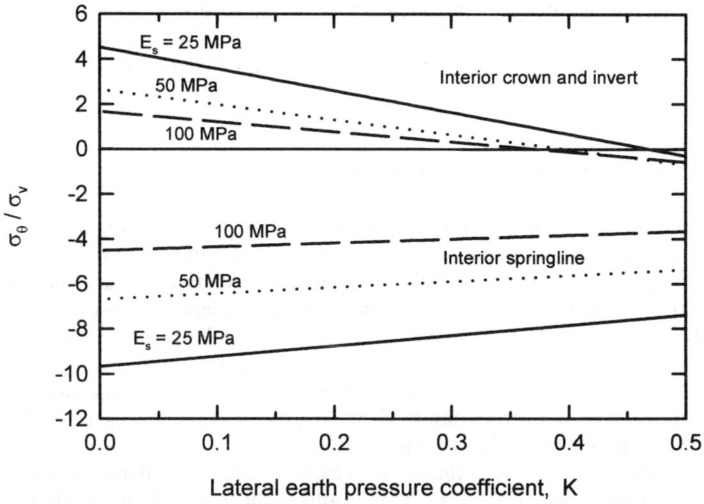

Figure 2. Hoop stresses normalized by vertical overburden stress at the interior crown, invert and springline for a plane pipe with DR of 11 calculated for different soil modulus E_s and K. (Tension positive).

lateral earth pressure coefficient K. Figure 2 shows normalized hoop stresses (σ_θ/σ_v) at the interior springline and crown calculated from the equations of Höeg (1968) for a typical PE pipe with DR=11 (DR is the ratio between the outside diameter of the pipe OD and the pipe thickness). Pipe stresses are shown for three values of E_s and also for a range of K values. In general, the lower E_s or K, the greater the tensile stresses at interior crown and invert. Smaller tensions occur at interior crown and invert for a thinner pipe (ie. larger DR), however, it may be prudent for the pipe to have sufficient thickness to limit local bending stresses from coarse backfill, as discussed in the next section.

Tensile stresses can also exist at the exterior springline if the pipe is thick enough (DR < 11) and the soil has a low K and E_s. Laboratory measurements by Brachman (1999) on a 220 mm OD, DR 9 PE pipe buried in 50 mm coarse gravel showed tensile hoop stresses at both the interior crown and invert, and at the exterior springlines. Pipe design is still controlled by limiting the tensile stresses at the interior invert location, since these tensile stresses for this case were nearly three times greater than those at the exterior springlines. However, the presence of tensile stresses at the exterior springline may be important for establishing limits of allowable cracks or damage at this location.

The value of K used during pipe design impacts the magnitude of tensile stresses and depends primarily on the ground conditions and on the lateral restraint provided to the soil around the pipe. For some narrow heap leach pads there may be little or poor lateral restraint, thus K may be closer to the active earth pressure limit K_a, normally calculated viz:

$$K_a = (1 - \sin \phi') \div (1 + \sin \phi') \qquad (1)$$

where ϕ' is the angle of internal friction of the ground. For burial under an extensive tailings dam, K is probably closer to the at-rest value K_o, which is normally estimated using the simple expression:

$$K_o \approx (1 - \sin \phi') \qquad (2)$$

Selection of a lower value of K for pipe design would provide a greater estimate of the tensile stresses and thus be conservative for the long term durability of the pipe.

The stiffness of the backfill material is very important for PE pipes since the pipe obtains support from the backfill. Soil stiffness, often characterised by soil modulus E_s, is a nonlinear function of confining stresses. The greater the confinement provided to the backfill soil, the stiffer it will be, and hence result in smaller pipe deflections and stresses. Guidance for the selection of soil modulus during pipe design can be obtained from the data base reported by Selig (1990).

4 LOCAL BENDING STRESSES FROM COARSE BACKFILL

Several cases arise where coarse gravel backfill is required to be placed directly against the pipe, for example, coarse crushed ore against a collection pipe in a heap leach pad, or coarse gravel backfill around a pipe for seepage control beneath a tailings dam. Also, relatively coarse gravel backfill (nominal 50 mm coarse gravel) is used in municipal solid waste landfills to minimize the biological clogging of the leachate collection system. In these cases, the pipe is supported and loaded at discrete points around the circumference rather than the more continuous support provided by other backfill materials (e.g., sand, well-graded gravel). Local bending stresses arising from the discontinuous support could potentially lead to greater tensile stresses in the pipe and thus affect the long term durability of the drainage pipes.

Two important factors influencing the local bending effects from coarse backfill are the spacing between contacts and differences between contact forces. Both of these factors are related to the size (relative to the pipe), shape and arrangement of the gravel particles and interactions between individual gravel particles.

To illustrate the effects of discrete backfill contacts, consider the hypothetical case of a pipe subject to discrete contact forces of equal magnitude and equally spaced around the exterior surface of the pipe. In this case, the contact forces would induce local bending stresses resulting in incremental tension opposite the contact and incremental compression half way between two contacts on the interior surface of the pipe. On the exterior surface of the pipe, local bending would produce incremental compression beneath the contact and incremental tension half way between two contacts. When added to the stresses in the pipe from the earth pressures (as discussed in the previous section), the incremental bending stresses will either increase or decrease the tensile or compressive stresses in the pipe. Any differences in contact forces around the pipe would further increase any local bending effects.

A recent study (Brachman 1999) was conducted to quantify the effect of coarse gravel backfill on PE pipes used in municipal solid waste landfills. This work was conducted with crushed dolomitic limestone (50 mm nominal size) consisting of large angular particles with 70% finer than 51 mm and only 8% finer than 38 mm. Although the type of backfill used in landfills will probably be coarser than that used in most other buried pipe applications (possibly with the exception of very coarse material in a heap leach pad), it is useful to examine the principal finding of the work on landfill pipes.

Two types of test were conducted. The first series of tests were conducted under hoop compressive loading where the soil is subjected to a uniform radial pressure (ie. K=1). These tests were primarily used to study the nature of the coarse gravel contacts and to quantify the effects of local bending stresses under simplified loading conditions. The second series of tests were conducted under the biaxial earth pressures expected to prevail under field conditions for a deeply buried pipe. The principal objective of these tests was to measure the effect of coarse drainage stone on the mechanical response of the pipe.

4.1 *Hoop compression tests*

The hoop compression tests showed large variation of strain measured over small portion of the inside of a 320 mm OD DR 11 pipe. The variations in hoop strains shown in Figure 3a can be directly attributed to the coarse gravel since similar strain measurements were nearly identical when tested with medium sand backfill. These measurements were made directly opposite a region of hand placed contacts. Figure 3b shows the location of the strain gauges as well as the location of gravel contact points (obtained with carbon paper) and the outline of each stone in this region (obtained from measurements after the test).

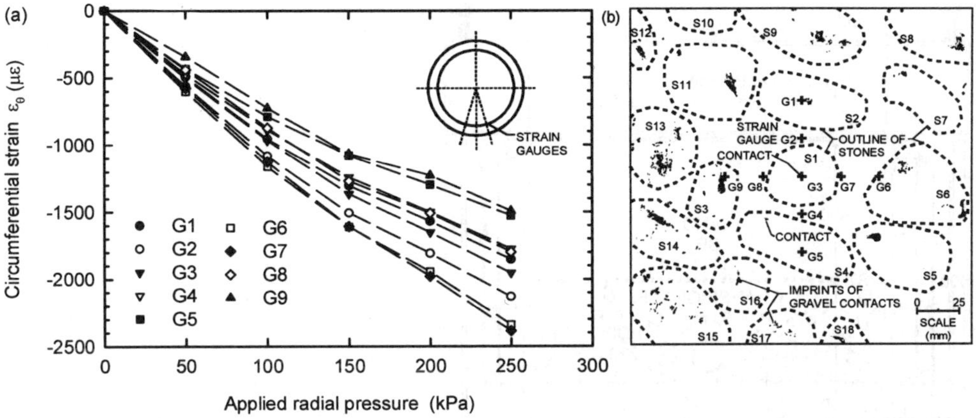

Figure 3. (a) Circumferential strains measured for a 320 mm OD DR 11 pipe buried in 50 mm coarse gravel and subject to uniform radial pressure. (b) Location of strain gauges and gravel contact points.

The largest compressive strains were measured at locations G6 and G7. Since these two points were located in between two widely spaced gravel contacts (Fig. 3b), the large compressive strains are consistent with large incremental compression on the inside surface of the pipe from local bending. The smallest compressive hoop strain was found at location G9 – directly opposite a gravel contact. The large incremental tension from local bending at this location produces a small compressive strain measurement. The close proximity of other gravel contacts near G9 (see Fig. 3b) may also contribute to smaller compression at this point. At the other locations, the strain measurements are not solely related to contact location. It is believed that variations in contact force at each gravel contact produces additional local bending.

Despite the large variations in strain, no circumferential tension was measured. Thus, for the 50 mm coarse gravel tested and the uniform radially compressive pressures, this particular pipe was thick enough such that the local bending effects are not sufficiently large to lead to circumferential tensile strains in the pipe. The local bending effects, however, result in both increases and deceases in circumferential compression.

4.2 *Biaxial compression tests*

Large scale laboratory tests were conducted under biaxial earth pressures to assess the influence of nominal 50 mm coarse gravel backfill on the structural response of 220 mm OD DR 9 leachate collection pipes (Brachman 1999). One set of specific conditions representative of a medium-size municipal solid waste landfill in Ontario, Canada were examined and are shown in Figure 4. The pipe was placed within a 600 mm deep layer of coarse gravel. The drainage layer was underlain by a 200 mm thick clay layer to include the effect of a more compressible layer beneath the gravel and pipe. Poorly graded medium sand was used to fill the remainder of the test cell. Overburden pressures were simulated with a pressurised air bladder. Lateral earth pressures were developed by limiting the lateral soil strains with stiff steel walls.

The coarse gravel backfill was found to dominate the structural response of the pipe. The coarse gravel backfill results in larger pipe deflections and greater variations in deflections than those obtained for a medium-dense sand backfill. For the pipes tested with coarse gravel backfill, the average vertical and horizontal diameter changes were approximately -1% and 0.8%, of the mean diameter at an applied pressure of 250 kPa. However, the measured variations of deflection do not preclude the use of coarse (50 mm) gravel in direct contact with leachate collection pipes. In fact, 220 mm OD DR 9 PE pipes would be expected to perform well in a medium-size landfill (vertical pressures up to 250 kPa) for the gravel, perforation size and perforation pattern tested (Brachman 1999).

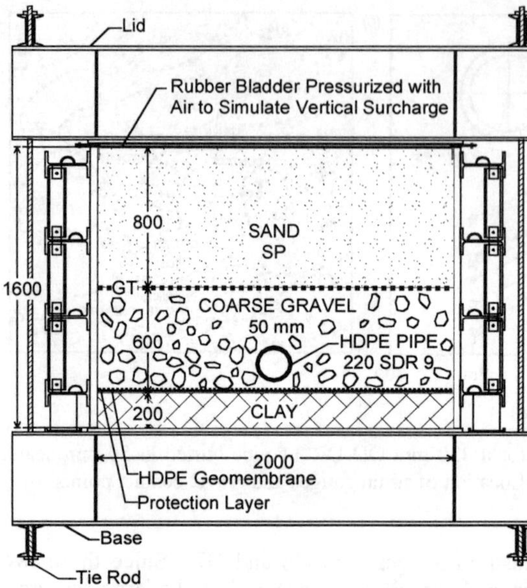

Figure 4. Laboratory facility used for biaxial compression tests on landfill leachate collection pipes.

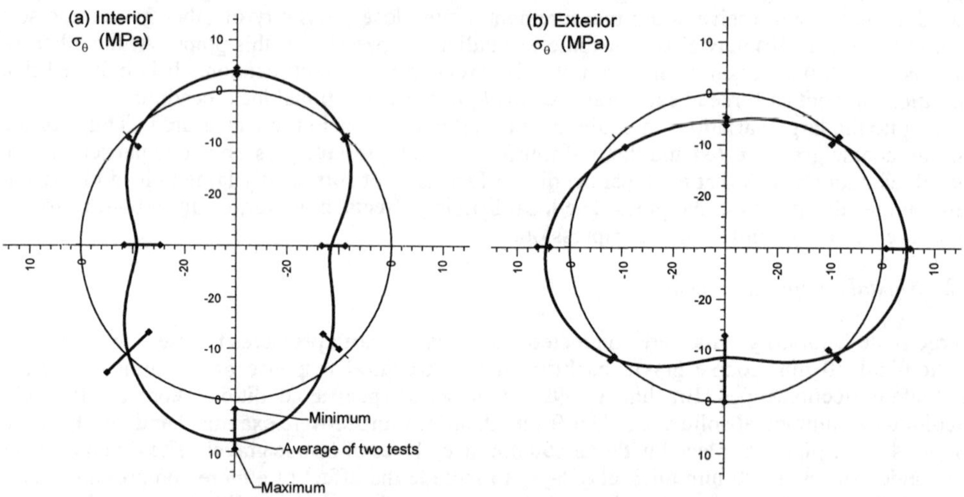

Figure 5. Hoop stresses for a 220 mm OD DR 9 pipe buried in 50 mm coarse gravel for (a) interior and (b) exterior surfaces of the pipe at an applied vertical surcharge of 900 kPa. (Tension positive).

Average hoop stresses from two biaxial compression tests are shown in Figure 5 for an applied vertical pressure of 900 kPa. The pipe stresses were greatly influenced by local bending effects caused by the coarse gravel backfill as shown by the large difference between maximum and minium vales at each location around the pipe. With coarse gravel backfill the maximum strains were found to be roughly twice the average value (Brachman 1999). Based on the measured results it appears that the estimates of stress obtained from the average of measured strains should be multiplied by factor of two to account for the increases due to the coarse gravel backfill.

5 PERFORATIONS

Pipes used in tailings and mine waste applications often contain perforations to permit collection of fluid. Perforations cause stress concentrations in the pipe and may produce tensile stresses even in a purely compressive stress field depending on axial restraint conditions. Thus, it is important to carefully consider the stresses near perforations during pipe design to obtain long term durability of the pipe.

Laboratory testing of PE pipes with large perforations has also been conduced to study the influence of perforations on pipe response (Brachman 1999). It was found that the principal effect that the perforation has on the structural performance of the pipe is a local perturbation in the pipe stresses with its influence concentrated around the hole. Numerical analyses (three-dimensional finite element analysis) of circular perforations in the pipes has shown that for a constant applied pressure, the maximum stresses around the hole depend on: (1) the circumferential location of the hole, (2) the size of the hole, (3) the thickness of the pipe, and (4) axial spacing of the hole.

The author is currently preparing design charts based on the laboratory testing and numerical analysis for perforated pipes; the following suggestions are based on the work conducted to date. First, perforations should be located at the shoulders and haunches (ie. quarter points) of the pipe, since this is where hoop stresses are smallest (see Fig. 5). It is probably best not to place perforations at the either the crown or invert since this will result in large magnifications of tensile stresses at these locations. Second, whenever possible, the holes should not be greater than 1/20 of the pipe diameter to minimize the stress concentrations. Larger holes can be used, however, careful consideration must be given to stresses in the pipe. If larger holes are required, thicker pipes should be used as the stress concentrations are smaller for thicker pipes. Third, an axial centre-to-centre spacing of 6 times the perforation diameter or greater between perforations is recommended, to ensure adjacent rows of perforations do not influence one another, and thus minimize the stress concentration around the hole.

Another factor influencing the stresses around a perforation are the axial restraint conditions for the pipe. Axial plane strain conditions (ie. full axial restraint) are normally assumed to occur under deep burial, when conditions of long and prismatic geometry exist along the pipe axis. However, some situations may arise where the pipe can experience only minimal axial restraint (e.g., at a thermal expansion joint or where the pipe enters a manhole) and result in stress concentrations around the hole worse than for conditions of full axial restraint.

For example, consider a pipe with outside diameter OD and DR 11 that has four circular holes of diameter a located at the shoulder and haunches, Figure 6. The holes have size ratio of OD/a

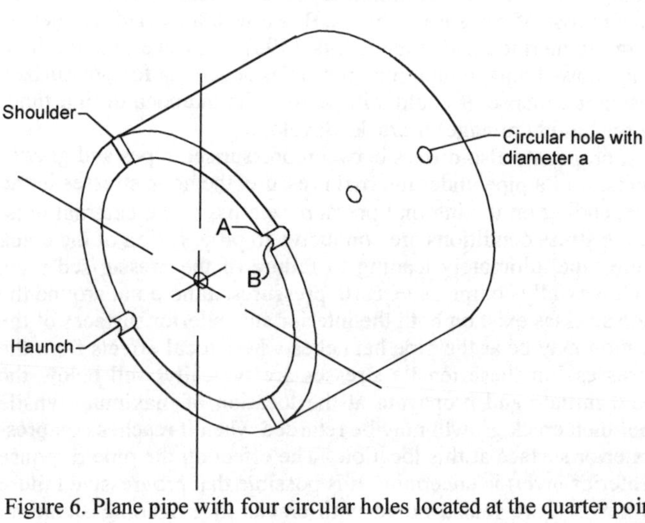

Figure 6. Plane pipe with four circular holes located at the quarter points.

equal to 20, and an axial spacing of 6a. This geometry has been analysed for two cases with different axial end conditions: one with full axial restraint, and the other with no axial restraint at the ends of the pipe. The maximum compression stress concentration factor (ie. stress at perforation ÷ stress without perforation) occurs at point A for both cases, but increases from 2.3 to 2.5 for conditions of zero end restraint. More important, however, tension is produced at point B as the stress concentrations go from compression of 0.75 with full axial restraint to tension of 0.4 for zero end restraint. Thus tensile stresses can be produced around perforations for pipes with only minimal axial restraint and should be considered during pipe design, especially for pipes under very large overburden stresses.

6 MAXIMUM ALLOWABLE LONG TERM TENSILE STRESSES

The hydrostatic design basis (HDB) is often used to limit the long-term tensile stresses in thermoplastic pipes subjected to internal pressure (e.g., gas distribution pipes). The HDB value is found by testing the pipe subject to uniform internal pressure for various times-to-failure. The hoop stress extrapolated to 10^5 hours on a logarithmic plot of stress versus time-to-failure is defined as the HDB stress (see ASTM D 2837). It is common practice to limit the working stresses to the hydrostatic design stress (HDS), which is one-half of the HDB. The hydrostatic design stress can be conservatively used to limit the tensile stresses for plane pipes subjected to external earth pressures arising from deep burial. It is recommended that the long term tensile stresses in polyethylene piping be less than the hydrostatic design stress.

7 FLAWS AND DEFECTS IN PE PIPES

Conditions conducive to stress cracking exist for buried PE pipes. Axial cracking of the pipe at locations distant from butt fusion joints, and circumferential cracking at the butt fusion joints are the likely modes of failure for leachate collection pipes. This assertion is based on the work of Parmar and Bowman (1989), who reported on the brittle failure of polyethylene pipes with butt-fusion joints subject to internal pressure. Tensile hoop stresses may lead to axial cracking, with the crack often initiating at a local defect in the material. The fusion process involved in joining sections of pipe together may induce circumferential cracks that could propagate under tensile axial stresses.

The consequences of stress cracking may be quite different for pressurized pipes and gravity-flow drainage pipes. Stress cracking in pressure pipes ultimately leads to failure of the pipe involving the development of leaks and loss of pressure. The HDB is a well accepted and generally successful basis for pressure rating thermoplastic pipe (Mruk 1990). However, the implications of stress cracking in a gravity-flow drainage pipe may not be as severe as for pressurized pipes. Provided that the pipe does not collapse, it could still perform its intended design function (that is to provide a pathway for liquid drainage) if cracks develop.

The extent to which a crack can propagate also differs between pressurised pipes and gravity flow pipes subject to earth pressures. For a pipe under internal pressure, the hoop stresses in the pipe wall are generally tensile (depending on the internal pressure relative to the external pressures). Once a crack is initiated, the stress conditions are conducive to propagation of the crack through the entire thickness of the pipe, ultimately leading to failure of the pressurised pipe. However, when subject to external, biaxially compressive earth pressures, at no point around the pipe circumference do tensile hoop stresses exist on both the interior and exterior surfaces of the pipe (see Fig. 5). The only exception may be at the pipe haunches where local effects from the gravel backfill lead to tensile stresses–but these tensile stresses are typically well below the HDS. Conceivably, a crack could initiate and propagate at the location of maximum tensile hoop stress (ie. interior invert), but then crack growth may be retarded when it reaches compressive stresses that prevail at the exterior surface at this location. The effect on the pipe response if an axial crack develops at the interior invert is uncertain. It is possible that progressive failure of the pipe could result. Since stresses are redistributed with cracks in a bending situation, propagation of a crack through the wall may or may not be possible.

Stress conditions in PE pipes may also be conducive to propagation of circumferential cracks induced by the butt fusion seaming. Again, tensile stresses do not prevail through the entire wall thickness, which may limit the extent to which a crack can propagate. Parmar and Bowman (1989) indicated that removal of the internal weld bead (heated polyethylene material that "rolls-into" the pipe under pressure and subsequently cools–where cracks may initiate) relocated the failure away from the butt fusion joint and increased the fatigue life of the pipe. For critical applications, one may wish to undertake the potentially expensive removal of the interior weld beads to reduce the potential for circumferential cracking, provided that during the process of bead removal additional cracks or defects are not induced.

Another source of flaws in PE pipes is from damage that occurs during construction. Damage can occur from handling of bundles of pipes and individual pipes. Excessive damage can occur if the pipe is dragged on the ground surface during pipe seaming and installation. This can be easily remedied by sliding the pipe on top of closely spaced rubber tires to prevent the pipe from touching the ground surface. A properly implemented construction QA/QC program is required especially for critical projects. Good communication is also required between the pipe designer and contractors.

8 SUMMARY AND CONCLUSIONS

Polyethylene drainage pipes are extensively used in tailings and mine waste applications because of their excellent resistance to chemical degradation, relative ease of construction and overall good performance. Preventing premature failure from stress cracking is important to obtain adequate long term durability of PE pipes. Long term tensile stresses and flaws or defects may lead to stress cracking, and potentially, collapse of the pipe.

Stress conditions conducive to stress cracking may exist for buried polyethylene pipes and can be worsened when backfilled in coarse materials from local bending induced by discrete contacts. Cracks are most likely to propagate in the axial direction from tensile hoop stresses, and at butt fusion joints in the circumferential direction from tensile axial stresses.

Tensile stresses from the earth pressures under deep burial can be quantified using the existing analytical solution of Höeg (1968) with careful selection of ground stiffness and lateral earth pressure coefficient. Coarse backfill directly against the pipe may lead to local bending stresses depending on the size of the particles relative to the pipe and the pipe thickness. Large scale laboratory testing provides the best means of quantifying the local effects from backfill materials. Tensile stresses may result around perforations, even in a purely compressive stress field if the pipe has only minimal axial restraint. Three-dimensional finite element analysis can be used to calculate the stress concentrations around perforations for use during pipe design. Other construction related issues like differential settlements, or shear stresses from slopes may result in tensile stresses in pipes and need to be considered on a case-by-case basis.

Essential requirements to obtaining adequate long term physical durability of the pipe are: (1) sources of tensile stresses should be identified and quantified, (2) tensile stresses should be limited to values below the hydrostatic design stress, and (3) damage to the pipe should be minimized to the maximum practical extent possible.

ACKNOWLEDGEMENTS

The research reported in this paper has been funded by the Natural Sciences and Engineering Research Council of Canada. The large scale laboratory testing was conducted at the University of Western Ontario under the direction of Drs. I.D. Moore and R.K. Rowe.

REFERENCES

Brachman, R.W.I. 1999. *Mechanical performance of landfill leachate collection pipes*. Ph.D. Thesis, The University of Western Ontario, London, Canada.

Höeg, K. 1968. Stresses against underground structural cylinders. *Journal of the Soil Mechanics and Foundations Division*, Proceedings of the American Society of Civil Engineers, SM4: 833-858.
Moore, I.D. 2000. Buried pipes and culverts. *Geotechnical and Geoenvironmental Engineering Handbook*, R.K. Rowe (Ed.), Kluwer Academic Publishing, Norwell, U.S.A. In press.
Mruk, S.A. 1990. The durability of polyethylene piping. *Buried Plastic Pipe Technology*. ASTM STP 1093, G.S. Buczala and M.J. Cassady (Eds.), ASCE, pp. 21-39.
Parmar, R., & Bowman, J. 1989. Crack initiation and propagation paths for brittle failures in aligned and misaligned pipe butt fusion joints. *Polymer Engineering and Science*, 29(19):1396-1405.
Rowe, R.K. 1998. Geosynthetics and the minimization of contaminant migration through barrier systems beneath solid waste. Proceedings of the 6th International Conference on Geosynthetics, Atlanta GA, USA, pp. 27-103.
Selig, E.T. 1990. Soil properties for plastic pipe installations. *Buried Plastic Pipe Technology*. ASTM STP 1093, G.S. Buczala and M.J. Cassady (Eds.), ASCE, pp. 141-158.

Hydrology and geochemistry

Tailings and Mine Waste '01 © 2001 Balkema, Rotterdam, ISBN 90 5809 182 1

Groundwater characterization and alternative evaluation for the Split Rock uranium mill tailings project

L.(Toby) Wright
Shepherd Miller Incorporated, Fort Collins, Colo., USA

ABSTRACT: This paper presents a case study of the groundwater characterization and development of ground water corrective action alternatives for the Western Nuclear, Inc. Split Rock Site in central Wyoming. A comprehensive multi-year site characterization was performed that included geologic, hydrologic and geochemical studies. These data were integrated with historical site data into a revised site conceptual model that provided the basis for groundwater flow and transport models. A broad range of technologies were screened and combined to develop four potential site-wide alternatives to provide the requisite protection of public health and safety and the environment. A preferred plan was selected based on the evaluation of these alternatives on their respective benefits and costs and using the flow and transport models.

1 INTRODUCTION

This paper presents a brief case study of the groundwater characterization and evaluation of groundwater corrective action alternatives for the Western Nuclear Inc. (WNI) Split Rock Site in Central Wyoming.

Western Nuclear Inc. processed uranium ores at the Split Rock Site from 1957 through 1981. During this period, the facility processed a total of approximately 7.7 million tons of ore with mill tailings from the process deposited into several contiguous areas that ultimately covered a total of approximately 180 acres. Following a decline in the world uranium market, the mill went on standby status in 1981 and moved to reclamation status in 1986.

The Nuclear Regulatory Commission (NRC) regulates uranium mill tailings in the United States. Reclamation of tailing sites is based on site specific conditions with general requirements to provide protection of human health and the environment for 1,000 years. There are three elements of reclaiming at uranium mill tailing sites:

1. Mill decommissioning and remediation of contaminated soils,
2. Surface reclamation, and
3. Groundwater

Mill decommissioning at the Split Rock site occurred in 1989 and consisted of dismantling the mill and burying the contaminated materials in the tailings impoundment. The mill decommissioning completion was approved by the NRC in 1990. Surface reclamation and remediation of contaminated soils consisted of identifying soils that were contaminated in the mill area and the soils near the tailings impoundment that were impacted by the wind-blown tailings. The contaminated soils were excavated and placed in the tailings impoundment. The NRC approved the soil clean up in 1999.

2 GROUNDWATER

Historically, the NRC considered groundwater reclamation activities independently of the mill decommissioning and surface reclamation activities. Groundwater monitoring has occurred on site since the 1960s. Changing regulations caused the monitoring program to expand from a few wells and a few constituents to the current monitoring program, which consists of over 30 wells that are monitored for 16 constituents.

In 1987, WNI reported to the NRC that monitoring wells showed groundwater quality concentrations near the tailings had increased and had become statistically greater than background concentrations. This initiated a process that resulted in the construction of a groundwater Corrective Action Program (CAP). This groundwater CAP, which began operation in 1990, called for pumping groundwater from wells in the Northwest and Southwest Valley to lined ponds constructed in the Southwest Valley (see Figure 1). These waters were subsequently spray-evaporated over the un-reclaimed tailings.

As surface reclamation of the tailings progressed, the surface area available for spray evaporation decreased. As a result of the decreasing area available for spray evaporation, the total volume of ground water pumped annually decreased. Subsequent pumping was reduced to the point where evaporation could be achieved using just the lined CAP ponds. The decreased volume being pumped was approved by the NRC and authorized by subsequent amendments to WNI's license. At the present time, the NRC mandated that CAP continue to operate with 6 to 15 million gallons of ground water being pumped and evaporated annually.

3 GROUNDWATER CHARACTERIZATION

In order to complete site reclamation, groundwater corrective action efforts had to be complete. However, the existing site groundwater conceptual model was not complete and existing corrective action efforts had not proved to be effective in returning groundwater concentrations to background or drinking water standard levels. Therefore, in 1995 WNI initiated a comprehensive plan to characterize the hydrologic and geochemical aspects of the site

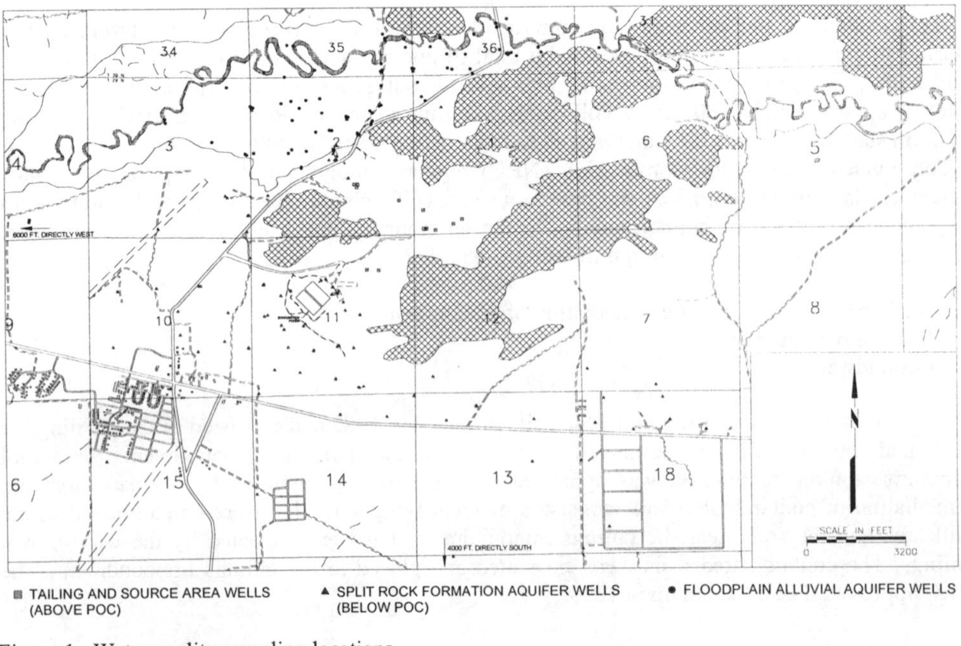

Figure 1. Water quality sampling locations

groundwater system. The first step of this characterization consisted of the development of a comprehensive scoping and quality assurance document. This document was developed to provide comprehensive strategic direction and quality objectives for the project to maximize the efficiency and effectiveness of the characterization effort. The characterization consisted of more than 100 borings, installation of more than 100 wells and more than 100 drive point piezometers, and analysis of 4,700 water samples and 245 solids samples.

4 CONCEPTUAL SITE MODEL

The results of the characterization studies, along with a review of the historic conditions, led to the development of a conceptual site model that incorporated both the hydrologic and geochemical conditions at the site.

5 HYDROLOGY

The groundwater and transport of site-derived constituents occurs primarily in the Split Rock Formation aquifer, and the Sweetwater River Alluvium, also called the floodplain alluvial aquifer. The floodplain alluvial aquifer is a shallow aquifer (15 to 30 feet thick) of river sediments that overlies the Split Rock Formation. This aquifer is hydrologically connected to the underlying Split Rock Formation. This shallow and highly permeable floodplain alluvial aquifer was formed where the river cut and meandered across the Split Rock Formation, which fills the alluvial basins between the Green Mountains to the south and the Granite Hills.

At the Split Rock Site, ground water flows from areas high in elevation surrounding the Main Tailings Impoundment, down the Northwest and Southwest Valleys, and then merges with the regional groundwater flow. Ground water flows northwest out of the Northwest Valley and merges with northeastward regional flow. Ground water also flows southwest out of the Southwest Valley, meets regional flow, and diverges into two flow paths around the granite outcrops, one to the north and one to the east. Areas where the granite bedrock is near to the ground surface, beneath the Sweetwater River floodplain, force ground water to flow upward into shallower portions of the flow system, primarily in the Sweetwater River alluvium.

Local ground water in the Split Rock Site area is recharged from direct precipitation on the valley floor and from precipitation run-off from the surrounding granite hillsides. Approximately 0.6 inches per year of precipitation infiltrates the valley floor to deep recharge, while approximately 6 inches per year of runoff from the surrounding granite hillsides recharge the alluvial aquifer. In addition, drainage of the tailings has historically input up to 1,400 gpm to the upper valley ground water system though current seepage rates are approximately 150 gpm.

Since tailings and water disposal in the tailings impoundment ceased in 1986, tailings drainage and consolidation have greatly diminished, and the elevated ground water levels beneath the tailings caused by tailings drainage have largely dissipated. Tailings seepage rates are presently estimated to be approximately 150 gpm and are expected to reach long-term, steady-state rates of less than 5 gpm in the next 30 years.

6 CHEMISTRY

Constituents derived from tailings have been identified in ground water at concentrations above background at the Split Rock Site. The identification of the constituents for which license conditions must be established is a step-wise process that identifies the constituents that are potentially hazardous to human and ecological health related to former mill site activities and that exist at concentrations above naturally existing (i.e., background) concentrations.

The concentrations of ammonia, manganese, molybdenum, nitrate, combined radium (Ra-226+228), and uranium, were identified to be above the protective standards beyond the edge of the tailings reclamation cover. Although no existing domestic uses were impacted by the constituents at that time, the relative abundance and potential future transport of the constituents

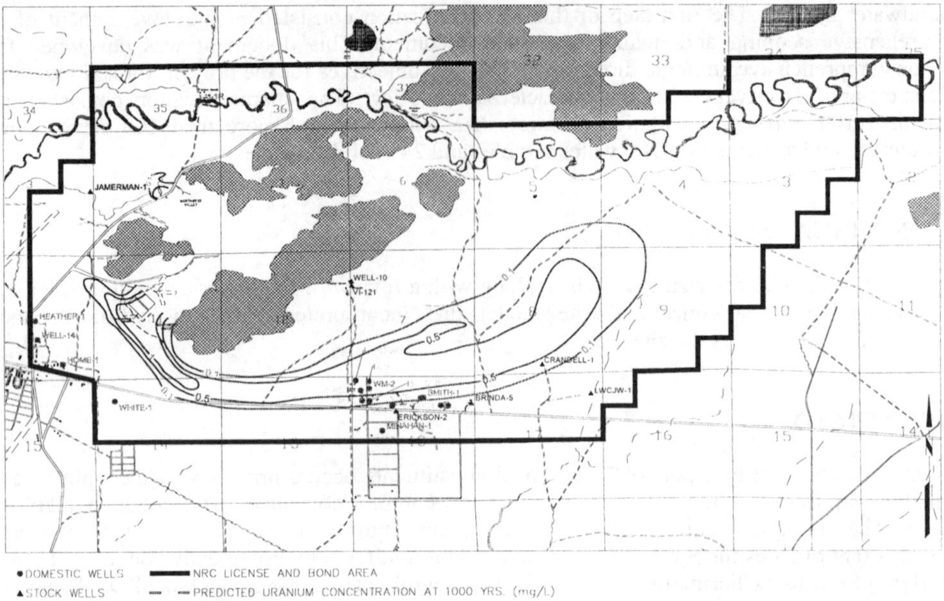

Figure 2. Domestic and stock wells

may have affected the future if no protective actions were taken. Therefore, corrective action was considered to be required for the future protection of public health and safety and the environment.

The six constituents of concern (COC) mentioned above are the focus of the various potential corrective action alternatives developed for this program. Specifically, uranium, which is the most prevalent and mobile constituent of the six COC, with the largest long-term source, was the principal focus of the corrective action selection process. Any potential alternative that adequately addressed uranium risks also adequately addressed concerns relating to all other COC.

Figure 2, illustrating maximum current uranium concentrations at each ground water sampling point, shows that uranium has migrated from the impoundment area down the Northwest and Southwest Valleys. Concentrations in the upper valleys have decreased over the last 10 years as the seepage flux has diminished, and the portion of runoff from the granite outcrops that infiltrates to ground water beneath the tailings impoundment continues to dilute the remaining seepage.

Uranium, and the other COC, continued to be introduced to the ground water system through residual tailings seepage and through slow disassociation of uranium from aquifer solid materials. Because the majority of uranium in the aquifers at the Split Rock Site was associated with aquifer solids, and because tailings seepage rates are declining, future concentrations of uranium in ground water will depend largely on how uranium is released from solids. Factors that affect the way uranium is released include the chemistry of the water that passes through the aquifer, the form of uranium associated with the solids, and the fraction of uranium on the solids that is available for release. Several studies on the aquifer materials were conducted to characterize these factors. The test included mineralogical analyses, bulk chemical analyses, distribution coefficient testing, batch and column leach testing, and sequential extraction testing.

Based on the test results, an empirical relationship between the concentration of uranium on solids and dissolved in the ground water was developed. This relationship involves a variable partitioning coefficient (K_d) to represent the sum of all chemical interactions under equilibrium conditions. An analytical model based on this variable K_d was developed and applied using site specific input values of porosity, K_d, the initial concentration in water, the concentration of inflowing water, and the bulk density of solids. The results of this analytical model were incorporated as the source of uranium to the calibrated flow model (MODFLOW) and transport model (RAND3D).

7 ALTERNATIVE EVALUATION

Using the chemical and hydrologic characterization data integrated with the calibrated flow and transport models, four corrective action alternatives were developed to address long-term protection of human health and environment from future potential exposure pathways. These alternatives were developed after screening a comprehensive suite of potentially applicable technologies. The screened technologies were then configured such that all alternatives would provide the necessary protection of human health and the environment.

These four alternatives were evaluated relative to the costs (economic and non-economic) and benefits (restoration of areas where groundwater could be accessed). A summary of the costs and benefits for each alternative is presented on Table 1.

Table 1 Corrective action alternatives detailed analysis summary

Alternative Criteria	Institutional Control Only	Hydraulic Diversion With Institutional Control	Focused Pumping with Institutional Control	Perpetual Containment With Institutional Control
Protective	Yes	Yes	Yes	Yes
ALARA				
Costs:				
Monetary Costs:	$114,000	$17,910,000	$107,850,000	$117,350,000
Non-monetary costs:				
Aesthetic Impacts	None	Low (perpetual)	High (25 + yrs)	Very High (perpetual)
Worker Health Impacts	None	Low	Moderate	High
Environmental Impacts	None	Low	High	Moderate – High
Resource Value:				
Ground Water Consumption	None	Evap. Loss = None Extract. Loss = 250 gpm (1,000 yrs)	Evap. Loss = 1,876 gpm (25 yrs) Extract. Loss = None	Evap. Loss = 265 gpm (25 yrs) Evap. Loss = 35 gpm (1,000 yrs) Extract. Loss = None
Total Control Area	5,275 acres	2,795 acres	3,818 acres	2,965 acres
Benefit:				
Socioeconomic Benefits	None	Low	Moderate	Moderate
[1]Restored Ground Water Resource Area	0	2,480 acres	1,457 acres	2,310 acres
[2]Restored Ground Water Resource Value	$0	$302,560	$177,750	$281,820

Notes: [1]Restored Ground Water Area is the difference in Ground Water Resource Control Area between Baseline Institutional Control Alternative and other alternatives.

[2]Ground Water Resource Value is assumed to be equal to land values of Restored Ground Water Resource Area, based on present average land values of $122/acre.

7.1 *Alternative No. 1 - Institutional Control with Alternate Water Supply*

The Institutional Control Alternative allowed natural flushing to attenuate, disperse, and dilute site-derived constituents to meet protective standards at the points of exposure (POEs) with no active treatment or mitigation measures. The compliance strategy of this alternative consisted of removing the drinking water pathway for exposure via institutional control(s), including providing a replacement drinking water supply, when and if it becomes necessary, for existing residents within the area requiring control.

7.2 *Alternative No. 2 - Hydraulic Diversion With Institutional Control*

The Hydraulic Diversion Alternative incorporated a perpetual hydraulic control of flow to prevent eastward constituent migration from the Southwest Valley and alternate compliance standards with institutional controls to meet protective standards at the POEs with no active treatment measures. The compliance strategy of the Hydraulic Diversion Alternative consisted of limiting the area potentially impacted by site-derived constituents through perpetual hydraulic control and removing future drinking water pathways for exposure via institutional control(s). Hydraulic control would be achieved by pumping fresh groundwater from outside the plume area into 16 injection wells at a combined rate of approximately 500 gpm to create a groundwater divide, deflecting flow from the Southwest Valley toward the north. The natural dilution and attenuation characteristics of the ground water and surface water systems would allow this alternative to meet protective levels at the POEs with no active treatment. The POEs consisted of the boundary of controlled area, small ephemeral surface water ponds or seeps in the Sweetwater River floodplain, and the Sweetwater River.

7.3 *Alternative No. 3 - Southwest Valley Focused Pumping With Institutional Control*

Alternative No. 3 incorporated active treatment and targeted pumping of a selected area in the Southwest Valley and institutional controls to meet protective standards at the POEs. The compliance strategy of Alternative No. 3 consisted of removing the dissolved and sorbed constituents in the Southwest Valley from the focused pumping area by controlled flushing of the aquifer using injection and pumping wells. This would reduce the amount of site-derived constituents in the ground water system, control the short-term and intermediate-term amount of uranium flowing from the Southwest Valley through interim hydraulic control and ex-situ treatment, and reduc the long-term concentration and distribution of site-derived constituents relative to the Alternative No. 1. Pumping would consist of pumping roughly 1,900 gpm from 29 wells and injecting 1,700 gpm of non-impacted ground water from an up gradient area into 34 wells located in the focused pumping area of the SW Valley. Pumped water would be evaporated from approximately 1,000 acres of lined evaporation ponds. The future drinking water pathway for exposure would be eliminated via institutional controls and no active treatment of the Northwest Valley or floodplain are proposed. Natural flushing would occur in the Northwest Valley and reduce concentrations over time. The maximum extent of ground water contamination out the Northwest Valley had already occurred and is protective; therefore, additional action was not proposed for this alternative.

7.4 *Alternative No. 4 - Perpetual Containment Pumping With Institutional Control*

Alternative No. 4 incorporated immediate and perpetual containment pumping of the Northwest Valley and cleanup of an area outside the Southwest Valley beyond the area influenced by containment pumping. Following cleanup outside the Southwest Valley, cleanup pumping in this area would be discontinued and perpetual containment pumping at a lower pumping rate would be initiated. The initial Southwest Valley pumping would consist of pumping 960 gpm from 19 wells and injecting 800 gpm into 16 wells located in areas of elevated concentration in the Southwest Valley. Containment pumping in the NW Valley would consist of pumping 5 wells at a combined rate of 100 gpm. Initial Southwest Valley pumping would be terminated after approximately 25 years at which time the 19 Southwest Valley extraction wells and all 16 injection wells will be abandoned. Three Southwest Valley pumping wells in the valley mouth would then be pumped at a combined rate of 40 gpm, while the 5 NW Valley wells would continue to be pumped at a combined rate of 100 gpm to provide perpetual containment of both valleys. At that time, the treatment facility capacity requirements would decrease resulting in decommissioning and reclamation of a large portion of the treatment facility.

Pumped water would be processed through a conventional water treatment plant using pH adjustment and reverse osmosis (RO) membrane technology. Evaporation of treatment brines from the RO system for the initial pumping period (25 years) would require approximately 143 acres of evaporation ponds. Approximately 23 acres of sludge disposal cells would be required to handle the treatment sludge from the water treatment plant. The clean effluent from the water treatment plant would be re-injected to the aquifer as part of the fresh water injection system. However, additional fresh water would be required (approximately 160 gpm) to meet the fresh water injection water requirements (800 gpm). The water to supplement the fresh water injection would be drawn from a new supply well installed approximately 1 mile south (up gradient) of the control area. An estimated 20 acres of evaporation ponds would be required to evaporate the brine from the treatment facility for the long-term containment pumping. An estimated 140 acres of sludge disposal cells would be required to accommodate the solid treatment waste generated over the 1,000-year design life although, conceptually, treatment and disposal would be required in perpetuity.

The compliance strategy of Alternative No. 4 consisted of removing the dissolved and sorbed constituents in the Southwest Valley aquifer area by controlled flushing of the aquifer with injection and pumping wells. This would reduce the amount of site-derived constituents in the ground water system and control the short-term and intermediate-term amount of uranium flowing from the Southwest Valley through interim hydraulic control. Perpetual containment pumping of both valleys, at relatively low pumping rates following cleanup outside the Southwest Valley, would prevent all future migration of site-derived constituents from the valleys. The drinking water pathway for exposure in the floodplain area and areas directly outside the valley mouths, which were impacted, would be eliminated via institutional control.

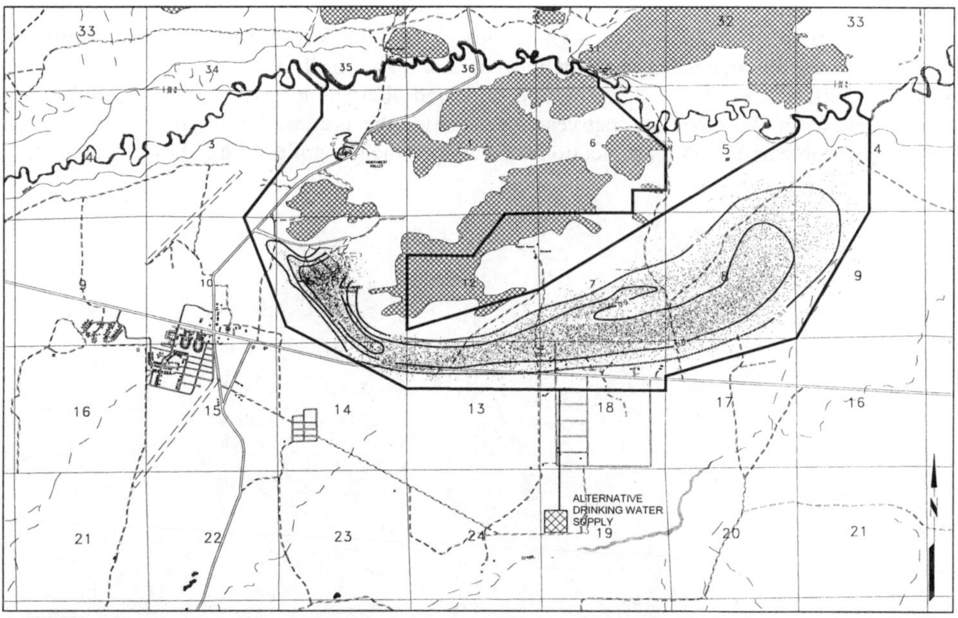

Figure 3. Proposed correction action alternative no. 1, institutional control

8 SELECTED ALTERNATIVES

From the four alternatives described above, the proposed ground water corrective action alternative for the WNI Split Rock Site is Alternative No. 1 – Institutional Control with Alternate Water Supply (Figure 3). This alternative eliminates the potential future human exposure pathway with institutional controls and eliminates access to ground water for human consumption within the proposed long-term care boundary. In addition, it provides for an alternate drinking water supply in the future, should it be required, for existing or future residents within the proposed long-term care boundary who use local ground water for their drinking water supply.

All ground water constituent concentrations beyond the proposed long-term care boundary will remain at background levels under all conditions. Furthermore, site-derived constituents will remain within the range of natural background in the river except during possible brief periods of extremely low flow. In the relatively near future, site-derived constituents in the river will not exceed concentrations within the range of natural background concentrations even during the most extreme low flow conditions.

Evaluation of the entire aquatic ecosystem during the period of peak loading (1995-1996) indicated that there were no adverse effects to the system from the combined range of site-derived constituents in the river. Further, decreased future loading to the river and the self-flushing nature of the river system assures that there will be no long-term cumulative impacts to the river. Over time, the concentration of site-derived constituents in the flood plain will return to protective values for all receptors and concentrations in the river will remain within the range of natural background even during the extreme worst-case low flow events. Therefore, a reasonable assurance of present and future protection to all receptors is provided for the surface water system.

The proposed preferred alternative has no impacts on surface lands, traditional land use, wildlife habitat or on the aesthetics of the area. No changes in ground water use beyond the long-term care boundary will be required. In addition, all traditional land use within the long-term care boundary should be allowed with the exception of use of ground water for domestic drinking water supply.

All other alternatives evaluated had significantly greater impacts on surface lands, traditional land use, wildlife habitat, aesthetics of the local area and significantly greater occupational risks to workers. Further, the cost of the proposed preferred alternative is 100 times to 1,000 times less expensive than the other alternatives. As a result, the preferred alternative provides the requisite reasonable assurance of protection and satisfies the principles of the ALARA process.

Tailings and Mine Waste '01 © 2001 Balkema, Rotterdam. ISBN 90 5809 182 1

Rabbit Lake groundwater monitoring optimization program

D. Pritchard
Golder Associates Limited, Calgary, Alb., Canada
D. Rezansoff & P. Landine
Cameco Corporation, Saskatoon, Sask., Canada

ABSTRACT: The Rabbit Lake Uranium Mine, located in Northern Saskatchewan, Canada, has been in operation since 1975, and has undergone major changes and expansions during the past 23 years, including the Eagle Point expansion, and mining of the A and D Zones. As a consequence of these changes and expansions, the environmental monitoring program has become increasingly complex. Such changes usually result in an increase in the amount of environmental data to be collected, analyzed, comprehended, and reported. The increasing complexity of information makes it difficult to assess the conditions being monitored, and the original purpose(s) of the monitoring program tend to get "lost". In order to address this situation, a review of the monitoring program has been completed, with the objective of optimizing the program to provide only that information required to meet the monitoring objectives.

1 INTRODUCTION

1.1 *History*

The Rabbit Lake Uranium Mine, located in Northern Saskatchewan, Canada, has been in operation since 1975, and has undergone major changes and expansions during the past 23 years, including the Eagle Point expansion, and mining of the A and D Zones. As a consequence of these changes and expansions, the environmental monitoring program has become increasingly complex. Such changes usually result in an increase in the amount of environmental data to be collected, analyzed, comprehended, and reported. The increasing complexity of information makes it difficult to assess the conditions being monitored, and the original purpose(s) of the monitoring program tend to get "lost". When this situation arises, a review of the monitoring program is necessary, with the objective of optimizing the program to provide only that information required to meet the monitoring objectives.

This report describes the methods and results of the review and optimization of two groundwater-monitoring programs conducted at the Rabbit Lake facility: the RP-Series Groundwater Monitoring Program, and the 81-Series Groundwater Monitoring Program. Both of these programs comprise a network of groundwater monitoring wells installed adjacent to potential sources of groundwater contamination. The wells are sampled at a regular frequency (typically once per year), and the samples are analyzed for a variety of chemicals, including major ions, nutrients, metals, and radionuclides.

These groundwater monitoring well networks monitor groundwater quality adjacent to three principle source areas: 1) B-Zone Waste Rock and Ore Stockpiles; 2) Rabbit Lake In-Pit Tailings Management Facility (RLITMF) and associated waste rock stockpiles, and 3) Above-Ground Tailings Management Facility (AGTMF). The locations of these source areas are presented on Figure 1.

Figure 1

MONITORING PROGRAM REVIEW AND OPTIMIZATION OVERVIEW

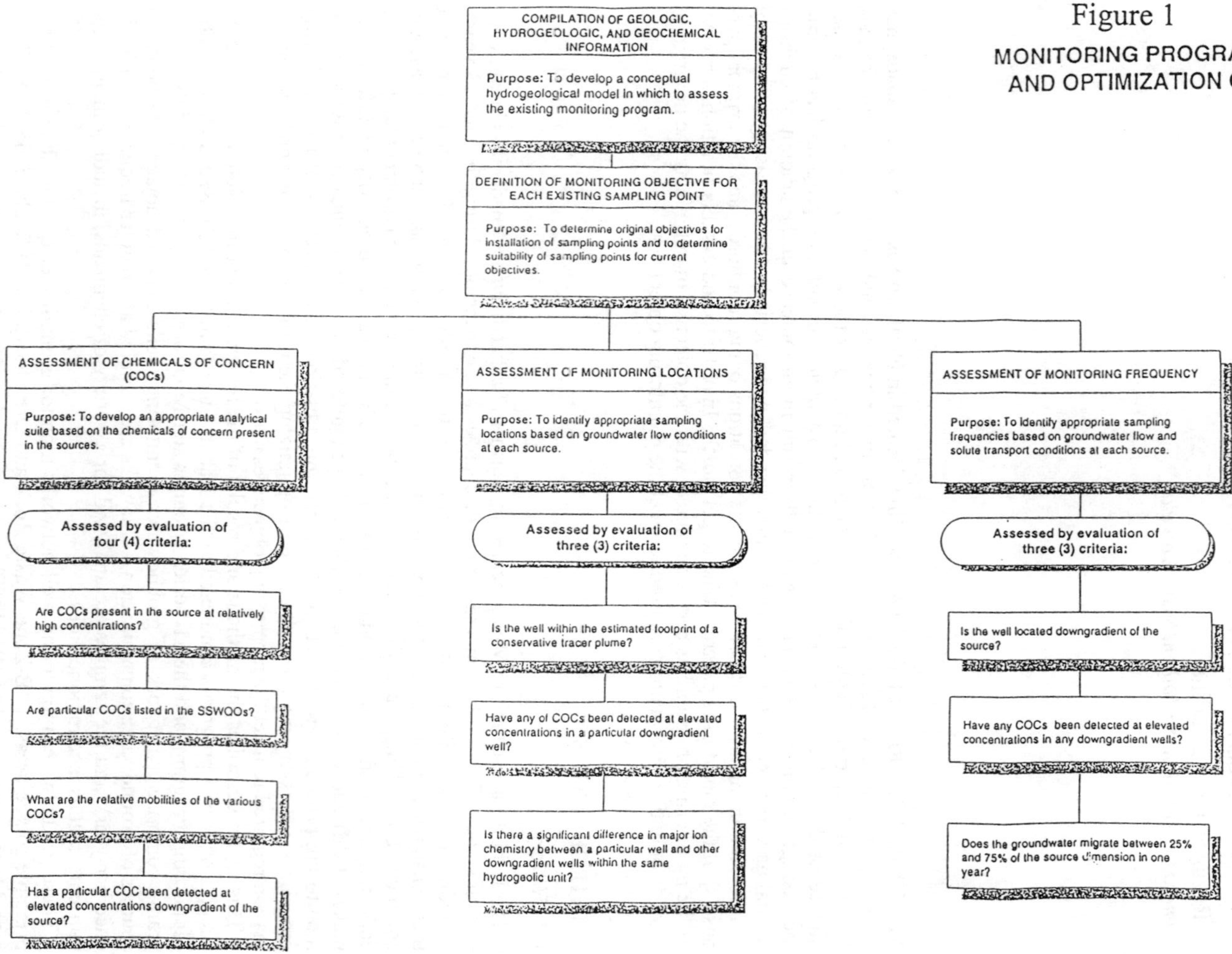

2 TECHNICAL APPROACH

2.1 Review of the Monitoring Objectives

A summary of the Technical Approach used for the Review and Optimization Program is presented on Figure 1. The monitoring program review began with an assessment of the monitoring objectives for each well nest. Company reports were compiled and reviewed to determine why groundwater monitoring was being conducted at each well, what information was being collected, and what the information was to be used for.

2.2 Review of the Geological and Hydrogeological Environment

Available geological and hydrogeological information was reviewed to establish the principal hydrogeologic units, to establish the groundwater flow directions and rates, and to identify the extent of potential contaminant plumes migrating from each of the source areas. Information on the geochemistry of the principal sources of potential groundwater contamination was also compiled and summarized, and chemicals of concern (COCs) were identified. Sources of information included Golder, 1995; Golder, 1999; Ruhrmann and Holl, 1990; Cameco, 1992; and SENES, 1993.

2.3 Assessment of the Groundwater Monitoring Well Locations

The objectives of this assessment were to:

- evaluate the locations of the existing wells with respect to the orientation of predicted contaminant migration directions associated with each source;
- evaluate the locations of the existing wells with respect to differences in groundwater chemistry between hydrogeological units; and
- evaluate the concentrations of chemicals of concern in each well to identify potential impacts that require continued monitoring.

Selection of appropriate well locations was based on three factors:

<u>1. Is the well within the estimated footprint of a conservative tracer plume?</u>

Information on predicted conservative tracer plumes from solute transport modelling conducted by Cameco for the three principal source areas was assessed, and the locations of the monitoring wells with respect to the predicted tracer plumes were evaluated. Sources of information included Golder, 1995; Golder, 1999; Ruhrmann and Holl, 1990; and internal Cameco studies.

<u>2. Have any Chemicals of Concern been detected at elevated concentrations in the well?</u>

Elevated concentrations were defined using three criteria: i) exceedences of Saskatchewan Environment and Resources Management (SERM), Saskatchewan Surface Water Quality Objectives (SSWQO), and/or Canadian Council of Ministers of the Environment (CCME) guidelines, ii) concentrations that appear to be above background concentration ranges as defined by SENES 1993, and iii) concentrations that appear to have increasing trends during the monitoring period. Sources of this information included time series graphs of groundwater analytical data produced by Golder Associates from Cameco's Rabbit Lake environmental monitoring database; Cameco, 1997; the Saskatchewan Surface Water Quality Objectives; the Metal Mining Liquid Effluent Regulations (MMLER); the SERM Permit; and CCME, 1991.

It is important to note that the SERM, SSWQO and CCME criteria were used only for comparison purposes to identify chemicals that might be considered "elevated" with respect to conditions elsewhere in Canada. Comparison to these criteria does not imply that these criteria are applicable to the Rabbit Lake Facility.

3. Is there a significant difference in major ion chemistry between this well and other downgradient wells within the same hydrogeologic unit?

The major ion chemistry of each well was reviewed to identify potential differences in groundwater type-chemistry between well locations and between hydrogeologic units (i.e. bedrock, interface, and overburden wells). The purpose of this review was to make sure that all hydrogeologic units were being monitored, and to identify potential redundancy in monitoring locations. Sources of this information included piper plots of groundwater analytical data produced by Golder Associates from Cameco's Rabbit Lake environmental monitoring database.

A weighting system was used to rank each well against the key questions listed above. If the answer to the particular question was yes, a score of 1 was given to the well. If the answer was no, a score of 0 was given. The scores were totalled, and wells with total scores ≥ 2 were included on the required monitoring list, while wells with total scores <2 were removed from the required monitoring list.

For example, if a well had elevated concentrations of a chemical (score=1), but the well was not within the footprint of a tracer plume (score=0) and there was no significant difference between the major ion chemistry of this well and the others (score=0), the well would receive a total score of 1.

2.4 Assessment of the Chemical Parameters

A "long-list" of chemicals possibly associated with the Rabbit Lake facilities was compiled by SENES, 1993, and comprised arsenic, copper, iron, lead, nickel, zinc, uranium, radium226, thorium230, and lead210. An assessment of this long-list of chemicals was conducted, and included:

- A review the current analytical suite;
- selection of appropriate chemical parameters for monitoring based on the source chemistries; and
- revision of the parameters analyzed to reflect the appropriate parameters.

Selection of appropriate chemical parameters was based on four factors, again in the form of questions:

1. Is the chemical in the source at relatively high concentrations, such as concentrations above the relevant regulatory criteria (SSWQO, MMEL, and SERM Permit)?

A review of the source zone geochemistry was conducted to identify chemicals of concern (COCs). Chemicals associated with mining and milling operations at Rabbit Lake, with concentrations above regulatory criteria, or with elevated concentrations with respect to background conditions (in the case where no regulatory criteria have been established) were identified as COCs. Source zone information included tailings porewater and seepage chemistry data from Cameco's Rabbit Lake environmental monitoring database; the SSWQO; the MMLER; the SERM and AECB Operation Approvals; and CCME, 1991.

2. Is the chemical listed in the relevant regulatory criteria?

Relevant regulatory criteria were reviewed to determine which chemicals of concern have had criteria established. Sources of information included the Saskatchewan Surface Water Quality Objectives; the MMLER; the SERM Permit; and CCME, 1991.

3. Does the chemical have a relatively high mobility in groundwater?

A review of the relative mobility of the principle metals and radionuclides present in the sources was conducted to determine which chemicals are most likely to appear in the groundwater monitoring wells. Sources of information for this review included Sheppard and Thibault, 1990; Golder Associates Ltd., 1995; Golder Associates Inc., 1997; Golder Associates, 1999; and AECL, 1994.

4. Has the chemical been detected at elevated concentrations in downgradient wells?
A weighting system was used to rank the chemicals against the key questions listed above. If the answer to the particular question was yes, a score of 1 was given to the chemical. If the answer was no, a score of 0 was given. The scores were totalled, and chemicals with total scores ≥3 were included on the monitored parameter list, while chemicals with total scores <3 were removed from the monitored parameter list.

2.5 Assessment of the Monitoring Frequency

By default, all RP-Series and 81-Series wells have been sampled on an annual basis. The objectives of this assessment were to:

- review the well locations with respect to groundwater flow rates;
- select an appropriate monitoring frequency that would adequately detect and characterize contaminant migration; and
- revise the sampling frequency for each well network as required.

Selection of the appropriate monitoring frequency was based on three factors:

1. Is the location of the well downgradient of the source?
Information on groundwater flow directions for the three principal source areas was assessed, and the locations of the monitoring wells with respect to the direction of groundwater flow were evaluated. Sources of information included Golder, 1995; Golder 1999; Ruhrman and Holl, 1990, and internal Cameco documents.

2. Have COCs been detected at elevated concentrations in the well?
The approach, rationale and sources of information to address this question were presented in Section 2.3.(3) above.

3. Is the approximate groundwater velocity between 25% and 75% of the source zone dimension in the direction of groundwater flow?
The frequency of sampling should be sufficient to detect contaminant release as early as possible, with a low likelihood of missing the release, and to provide enough data points to adequately describe the concentrations of the release. In order to assess the frequency of sampling, we compiled information on the approximate groundwater velocity in the vicinity of each of the three sources. The hypothetical migration was then assessed for a conservative chemical contaminant away from the source to evaluate the appropriate frequency of monitoring required. Because the assessment of migration did not involve computer modelling, we used several conservative assumptions. First, it was assumed that a release from the source would be in the form of a slug, or discrete release. A slug release was assessed to make sure that our monitoring frequency did not miss a potential short-term release from the source. This assumption is considered to be conservative because in reality releases from the principal sources are likely to be continuous, as the sources have been in place for some time, and will remain in place for the foreseeable future. Second, it was assumed that contaminant migration would not undergo dispersion, retardation or decay (i.e. a non-reactive tracer). A non-reactive tracer was assessed because it would provide the most conservative estimate of contaminant migration rate and extent.

The assessment of the hypothetical contaminant migration indicated that:

- if groundwater migrates downgradient a distance of greater than 75% of the length of the contaminant source (in the direction of groundwater flow) in one year, samples collected at a frequency of one per year would have a relatively high probability of miss-

Figure 2

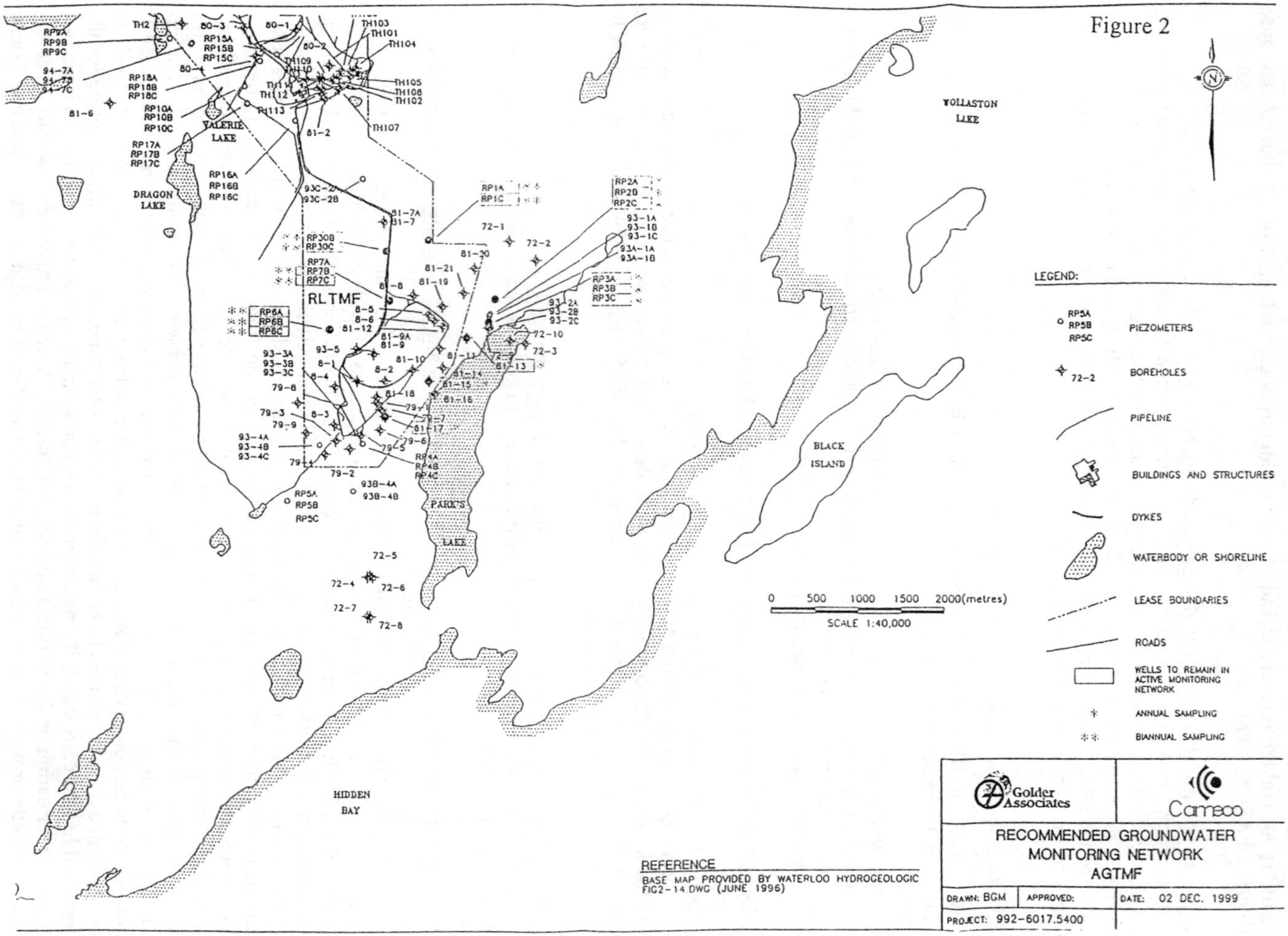
WOLLASTON LAKE
BLACK ISLAND
HIDDEN BAY
PARK'S LAKE
VALERIE LAKE
DRAGON LAKE
RLTMF
0 500 1000 1500 2000(metres)
SCALE 1:40,000
LEGEND:
PIEZOMETERS
BOREHOLES
PIPELINE
BUILDINGS AND STRUCTURES
DYKES
WATERBODY OR SHORELINE
LEASE BOUNDARIES
ROADS
WELLS TO REMAIN IN ACTIVE MONITORING NETWORK
ANNUAL SAMPLING
BIANNUAL SAMPLING
REFERENCE
BASE MAP PROVIDED BY WATERLOO HYDROGEOLOGIC FIG2-14.DWG (JUNE 1996)
Golder Associates
Cameco
RECOMMENDED GROUNDWATER MONITORING NETWORK AGTMF
DRAWN: BGM
APPROVED:
DATE: 02 DEC. 1999
PROJECT: 992-6017.5400

ing the detection of the contaminant plume, and that the sampling should be increased to a semi-annual frequency (twice per year);

- if groundwater migrates downgradient a distance of between 25% and 75% of the length of the contaminant source (in the direction of groundwater flow) in one year, samples collected at a frequency of one per year would be sufficient to detect the contaminant plume; and
- if groundwater migrates downgradient a distance of less than 25% of the length of the contaminant source (in the direction of groundwater flow) in one year, the sampling frequency can be increased to a biannual event (every two years), with a sufficient likelihood of detecting the contaminant plume.

- A weighting system was used to rank each monitoring well against the key questions listed above. If the answer to the particular question was yes, a score of 1 was given to the well. If the answer was no, a score of 0 was given. The scores were totalled, and wells with total scores >2 were included on the annual monitoring frequency list, while wells with total scores <2 were recommended for monitoring every two years.

3 REVIEW OF SOURCE ZONE MONITORING PROGRAMS

The method of monitoring program review was described in Section 2 above. Below is an example of the method applied to the wells in the vicinity of the Rabbit Lake Above-Ground Tailings Management Facility (AGTMF) (See Figure 2).

3.1 Above-Ground Tailings Management Facility

3.1.1 Groundwater Monitoring Objectives

A total of twenty-four RP-Series wells have been installed at eight locations adjacent to the Above-Ground TMF. All twenty-four of these wells were installed in 1991 for the purpose of monitoring groundwater flow and chemical characteristics associated with the AGTMF (Ruhrmann and Holl, 1990). In addition, 81-Series wells were installed at six locations in 1981 in the vicinity of the AGTMF for the purpose of establishing groundwater flow conditions in the bedrock in support of engineering design (Ruhrmann and Holl, 1990).

Because the RP-Series wells were installed for the purpose of monitoring groundwater flow and quality adjacent to potential contaminant sources in the vicinity of the In-Pit TMF, they are considered suitable for continued groundwater monitoring use in this area. Therefore, there is no justification for removing or replacing the RP-Series wells on the basis of monitoring objectives. However, removal of some of the 81-Series wells from active monitoring is justified for the following reasons:

- The wells were originally installed for the purpose of geotechnical assessment of bedrock, unrelated to contaminant transport;
- The lateral and vertical locations of the sampling points are not related to specific sources of potential groundwater contamination, or to specific pathways of contaminant migration (i.e. porous flow through overburden, flow through specific bedrock features such as faults, etc.); and
- Some of the wells have been designated as "less reliable", which brings into question the reliability of the remainder of the wells.

Therefore, the majority of the 81-Series wells are not required for an active monitoring program of groundwater quality related to the potential sources of contamination associated with the In-Pit TMF. However, due to the absence of RP-Series monitoring wells immediately to the east and downgradient of the AGTMF, wells 81-13, 81-15, and 81-17 have been retained for further assessment as to their suitability for inclusion in an active monitoring program.

Table 1

Rabbit Lake Above-Ground Tailings Management Facility (AGTMF)
Groundwater Monitoring Program Optimization
Summary

Well #	Pre-Optimization		Post Optimization	
	Monitoring Frequency	Parameters	Monitoring Frequency	Parameters *
RP-1A	Annual	**	Bi-annual	S^{-2}, As, Fe, SO_4, U, Ra^{226}
RP-1B	Annual	**	-	-
RP-1C	Annual	**	Bi-annual	S^{-2}, As, Fe, SO_4, U, Ra^{226}
RP-2A	Annual	**	Annual	S^{-2}, As, Fe, SO_4, U, Ra^{226}
RP-2B	Annual	**	Annual	S^{-2}, As, Fe, SO_4, U, Ra^{226}
RP-2C	Annual	**	Annual	S^{-2}, As, Fe, SO_4, U, Ra^{226}
RP-3A	Annual	**	Annual	S^{-2}, As, Fe, SO_4, U, Ra^{226}
RP-3B	Annual	**	Bi-annual	S^{-2}, As, Fe, SO_4, U, Ra^{226}
RP-3C	Annual	**	Annual	S^{-2}, As, Fe, SO_4, U, Ra^{226}
RP-4A	Annual	**	-	-
RP-4B	Annual	**	-	-
RP-4C	Annual	**	-	-
RP-5A	Annual	**	-	-
RP-5B	Annual	**	-	-
RP-5C	Annual	**	-	-
RP-6A	Annual	**	Bi-annual	S^{-2}, As, Fe, SO_4, U, Ra^{226}
RP-6B	Annual	**	Bi-annual	S^{-2}, As, Fe, SO_4, U, Ra^{226}
RP-6C	Annual	**	Bi-annual	S^{-2}, As, Fe, SO_4, U, Ra^{226}
RP-7A	Annual	**	-	-
RP-7B	Annual	**	Bi-annual	S^{-2}, As, Fe, SO_4, U, Ra^{226}
RP-7C	Annual	**	Bi-annual	S^{-2}, As, Fe, SO_4, U, Ra^{226}
RP-30A	Annual	**	-	-
RP-30B	Annual	**	Bi-annual	S^{-2}, As, Fe, SO_4, U, Ra^{226}
RP-30C	Annual	**	Bi-annual	S^{-2}, As, Fe, SO_4, U, Ra^{226}
81-07A-01	Annual	***	-	
81-07A-02	Annual	***	-	-
81-09A-01	Annual	***	-	-
81-09A-02	Annual	***	-	-
81-10	Annual	***	-	-
81-11	Annual	***	-	-
81-12	Annual	***	-	-
81-13	Annual	***	Annual	S^{-2}, As, Fe, SO_4, U, Ra^{226}
81-14	Annual	***	-	-
81-15	Annual	***	Annual	S^{-2}, As, Fe, SO_4, U, Ra^{226}
81-16	Annual	***	-	-
81-17	Annual	***	Annual	S^{-2}, As, Fe, SO_4, U, Ra^{226}
81-18	Annual	***	-	-
81-19	Annual	***	-	-
81-20	Annual	***	-	-
81-21	Annual	***	-	-

* including field measurements of pH, electrical conductivity, temperature, and water level
** Ca, Cl, HCO3, K, Mg, Na,SO4, T-Hard, T-Alk, Bo, Fl, Hg, Se, Ag, Al, As, Ba, Be, Cd, Co, Cr, Cu, Fe, Mn, Mo, Ni, Pb, Vn, Zn, TDS, Conductivity, pH, Ra, U, field-temp, field pH, field conductivity
*** Ca, Cl, HCO3, K, Mg, Na,SO4, NO3, NH4, Fe, TDS, conductivity, pH, Ra, U, field-temp, field pH, field conductivity

3.1.2 Geological and Hydrogeological Environment

The overburden in the vicinity of the AGTMF consists of silty sand tills with variable amounts of gravel, cobbles, and boulders. The tills are described as being dense to very dense. Overburden thickness is about 10 metres. The overburden is underlain by metamorphic bedrock, including biotite paragneiss and quartzite. The RP-Series wells have been installed in the overburden, the overburden-bedrock interface, and within the upper weathered bedrock. The 81-Series wells have been installed at various depths within the bedrock. Groundwater flow within the overburden and upper weathered bedrock is generally towards the east and southeast.

3.1.3 Assessment of the Groundwater Monitoring Locations

Monitoring well locations were evaluated and scored as discussed in Section 2.3.

RP-Series Wells

Three sets of RP-Series wells were located in positions hydraulically upgradient of the AGTMF. These wells were RP-1A, B, and C, RP-6A, B, and C, and RP-30A, B, and C. Of the remaining RP wells installed adjacent to the AGTMF, only three out of five of the downgradient well locations were actually located within the estimated footprint of the conservative tracer plume. In fact, no RP-Series wells were installed immediately downgradient (to the east) of the AGTMF in an area where the highest concentrations of a predicted plume would be found. Elevated concentrations of chemicals of concern had been detected in seven of the fifteen downgradient wells, and in all five of the downgradient well locations. This indicates that the monitoring wells are detecting chemicals of concern migrating from the AGTMF. Finally, the evaluation shows that five of the fifteen downgradient wells, at three of the five downgradient locations, had significantly different major ion chemistries. Again, the difference in major ion chemistry is a result of elevated sulphate concentrations in the five wells, which has been caused by migration of sulphate from the AGTMF. This suggests that there is potential redundancy in the number of downgradient wells, as the non-impacted major ion chemistry of the groundwater is relatively consistent.

The results of the evaluation and scoring indicate that eight of the fifteen downgradient monitoring wells, at three of the five downgradient locations, should remain in the active monitoring program. The upgradient wells (RP-1, RP-6, and RP-30 nests) should also remain in the active monitoring program.

81-Series Wells

As was stated in Section 3.3.1, only three of the nineteen 81-Series wells were retained for consideration as active monitoring locations. These wells are 81-13, 81-15, and 81-17, and are located immediately east of the AGTMF, between the toe of the AGTMF and Park Lake. An assessment of time series graphs of iron, sulphate, and total uranium indicates elevated and in some cases increasing concentrations of some or all of these chemicals of concern. Data on arsenic concentrations in these wells was not available. Because these wells are located immediately downgradient of the AGTMF, are within the predicted footprint of a contaminant plume, and have elevated and increasing concentrations of chemicals of concern, these wells should be included in an active monitoring program. However, because of the age of the wells and the installation method and objectives (for geotechnical purposes), consideration should be given to replacing these wells with properly designed and constructed groundwater monitoring wells.

3.1.4 Assessment of the Chemical Parameters

The chemical parameters were assessed against four factors to determine their appropriateness for consideration as chemicals of concern. These factors included relatively high concentrations in the source, established Surface Water Quality Objectives, relatively high mobility, and detection at elevated concentrations downgradient of the source.

The results of the assessment indicate that, of the long-listed chemical parameters, arsenic, iron, sulphate and uranium should be retained as monitored chemicals of concern for the AGTMF monitoring network.

3.1.5 Assessment of the Monitoring Frequency

The monitoring frequency for wells installed adjacent to the AGTMF was evaluated and scored by assessing the wells position relative to the source, by identifying elevated concentrations of chemicals of concern in the wells located downgradient of the AGTMF, and by assessing the conservative groundwater velocity in the vicinity of the AGTMF.

<u>RP-Series Wells</u>

Three sets of RP-Series wells were located in positions hydraulically upgradient of the AGTMF, and the remaining five sets of wells can be considered downgradient, although only three sets of wells were located within the estimated footprint of the conservative tracer plume. Elevated concentrations of chemicals of concern had been detected in seven of the fifteen downgradient wells, and in all five of the downgradient well locations. This indicates that the monitoring wells are detecting chemicals of concern migrating from the AGTMF. The evaluation also shows that groundwater adjacent to and beneath the AGTMF migrates at a rate of between 8 and 23 m/year in the overburden, and between 4 and 255 m/year in the bedrock. These rates result in groundwater migrating generally less than 75% of the average length of the AGTMF in one year in the bedrock, and often less than 25% of the average length of the AGTMF in one year in the overburden. As with the In-Pit TMF, the relatively high rate of migration calculated for the bedrock (221 m/year) represents flow along a discrete fracture. Again, because the upper bedrock is highly fractured, and has been considered in previous studies to be an equivalent porous medium, this high rate of migration is not considered representative of Darcy flow conditions.

The results of the evaluation and scoring indicate that ten of the fifteen downgradient RP-Series wells, at all five of the downgradient locations, should remain at an annual monitoring frequency. The upgradient wells and the remaining downgradient wells can be monitored at a biannual frequency (every two years). The evaluation did not identify wells that should have a monitoring frequency of more than once per year.

<u>81-Series Wells</u>

Because wells 81-13, 81-15, and 81-17 are located downgradient of the AGTMF and because elevated concentrations of chemicals of concern have been detected in these wells, the wells should be monitored on an annual basis. This sampling frequency is consistent with the RP-Series wells (RP-2 and RP-3) located in the vicinity of 81-13, 81-15, and 81-17.

4. RECOMMENDATIONS

4.1 Above-Ground Tailings Management Facility

The revised program comprises eighteen wells at nine locations, as compared to forty wells initially. Five of the wells are to be sampled annually, with the remaining ten wells (including the upgradient wells) to be sampled biannually. The recommended analytical suite comprises arsenic, iron, sulphate, and total uranium, along with field measured parameters of pH, temperature, and electrical conductivity. Groundwater depth/elevation measurements must also be collected during the sampling events. It is recommended that this monitoring program be reviewed every five years, and revisions to monitoring location, frequency, and parameters be made as required by changes in chemical concentrations, monitoring objectives, etc. We also recommend that the construction and physical condition of wells 81-13, 81-15, and 81-17 be assessed. If required, these wells should be substituted with suitable nearby wells or replaced with appropriately designed and constructed groundwater monitoring wells. Decisions on well upgrades, substitutions and/or replacement will have to be made after field inspection of the wells in this area.

5 SUMMARY

A structured review of two mature groundwater monitoring programs at a uranium mine in Northern Saskatchewan was undertaken. Existing wells were reviewed in terms of local geology, hydrology, well location, chemical parameters and monitoring frequency. The result was a significant reduction to the required parameter lists and frequency of monitoring, which was approved by the mine's federal and provincial regulators.

This form of monitoring program review may be applied to other programs, particularly where an abundance of historical data exists for several monitoring wells within one or more monitoring programs.

REFERENCES

Cameco Corporation. 1992. Environmental Impact Statement: Collins Bay A-Zone, D-Zone, and Eagle Point Development; Revised 1992.

Golder Associates Ltd. 1995. TMF Hydrogeologic Modelling Study. Prepared for Cameco Corporation. (Golder #942-6236).

Golder Associates Ltd. 1997. Design Assessment Task: Generic Geosphere and Biosphere Conceptual Models. Prepared for Chem-Nuclear Systems, Inc. (Golder #963-1576).

McLean, Joan E. and Bert E. Bledsoe, . 1992. Behavior of Metals in Soils. Ground Water Issue. United States Environmental Protection Agency.

Pendias, Henryk and Alina Kabata-Pendias. Trace Elements in Soils and Plants. CRC Press. Second Edition.

Ruhrmann, G. and N. Holl. 1990. Review of the Groundwater Monitoring Strategy in the Rabbit Lake Area.

SENES Consultants Limited. 1993. Background Concentrations, Environmental Transport, and Toxicity of Selected Metals and Radionuclides.

Sheppard, Marsha I. and D.H. Thibault. 1990. Default Soil Solid/Liquid Partition Coefficients, KdS, for Four Major Soil Types: A Compendium. Health Physics Vol. 59. No. 4 (October). pp. 421-482.

Waterloo Hydrogeologic Inc. 1999. AquaChem Aqueous Geochemical Data Analysis and Plotting. Software.

Tailings and Mine Waste '01 © 2001 Balkema, Rotterdam, ISBN 90 5809 182 1

Numerical simulation of solute transport in a spent heap leach pad

G. Danko
Mining Engineering Department, Mackay School of Mines, University of Nevada-Reno, Nev., USA

ABSTRACT: A mathematical model based on a Set of Partial Differential Equations (SPDE) is used for the simulation of a conservative solute transport in a spent heap leach pad. The model is solved using the MULTIFLUX code configured for the solution of an SPDE. Four equations in the SPDE are found sufficient for adequate modeling solute transport in a heterogeneous rock formation based on comparison to experimental results of a large-scale in situ field test. After calibration of the hydrotransport properties and coefficients of the SPDE model, it is used to predict temporal and spatial concentration variations for an assumed conservative solute in four different, characteristic regimes of a heterogeneous leach pad. The results indicate that the concentration differences between the regimes rapidly diminish with time. The variation of concentration with time is described with an attenuation function and/or coefficient both of which can conveniently be used for the prediction of rinsing time necessary to achieve a given target concentration.

1 INTRODUCTION

The coupled convection-dispersion transport of a conservative solute can be modeled based on a set of partial differential equations. For a formation of fractured and porous rock with residual and moving water not in equilibrium with each other, dual-porosity models are used consisting of a system of two equations. Such a model is applied by Toride (Toride et. al, 1995) in CXTFIT, also referred to as a mobile-immobile (MIM), dual porosity model. The balance of the mobile and stagnant concentrations, c_m and c_s is written as follows as a function of time, t, and flow direction, x:

$$\theta_m \frac{\partial c_m}{\partial t} = \theta_m D_m \frac{\partial^2 c_m}{\partial x^2} - q\frac{\partial c_m}{\partial x} - \alpha(c_m - c_s) \tag{1}$$

$$\theta_s \frac{\partial c_s}{\partial t} = \theta_s D_s \frac{\partial^2 c_s}{\partial x^2} + \alpha(c_m - c_s) \tag{2}$$

where θ_m and θ_s are the mobile and stagnant volumetric water contents, D_m is the dispersion coefficient in the mobile phase; D_s is the diffusion coefficient in the stagnant phase and is assumed to be zero in CXTFIT; q is the specific discharge for the mobile phase; and α is a mass transfer coefficient. The set of equations (1) and (2) simplifies the spatially distributed, three-dimensional transport system by using only two distinct regimes. This simplification does not take into account the effect of preferential fracture flows; therefore, the predicting power of the model involving a large spatial domain with field inhomogeneities is limited.

In order to improve the mathematical model, a four-equation, mobile-stagnant-preferential-lysimeter SPDE configuration is developed. The model is used for the evaluation of concentration variations in a spent leach pad during rinsing (Zhan and Danko, 2000).

2 NUMERICAL MODEL CONFIGURATION

MULTIFLUX, a coupled hydrothermal (HT)-computational fluid dynamic (CFD) model (Danko, 1999), is configured for the solution of the convection-diffusion/dispersion problem associated with solute transport in a spent leach pad. Only the CFD module of MULTIFLUX is used for the application.

2.1 *Balance Equations*

First, equations (1) and (2) are expanded by including more terms and combined into equations (3) and (4), to express more complex interactions. Second, a new balance equation (5) is employed to model a fast-tract, loosely connected preferential flow and transport phase with c_p concentration, θ_p water content, and D_p dispersion coefficient. Third, a mass balance equation (6) is applied to govern the concentration variations, c_l, in the concentration lysimeter samplers with a representative, relative water content θ_l, and mass transport connections to the mobile, stagnant, and preferential phases, represented by β_m, β_s, and β_p, respectively. The set of equations is as follows:

$$\theta_m \frac{\partial c_m}{\partial t} = \theta_m D_m \frac{\partial c_m}{\partial x^2} - q_m \frac{\partial c_m}{\partial x} - \alpha_m (c_m - c_s) - \beta_m (x)(c_m - c_l) \tag{3}$$

$$\theta_s \frac{\partial c_s}{\partial t} = \theta_s D_s \frac{\partial^2 c_s}{\partial x^2} + \alpha_m (c_m - c_s) + \alpha_p (c_p - c_s) - \beta_s (x)(c_s - c_l) \tag{4}$$

$$\theta_p \frac{\partial c_p}{\partial t} = \theta_p D_p \frac{\partial^2 c_p}{\partial x^2} - q_p \frac{\partial c_p}{\partial x} - \alpha_p (c_p - c_s) - \beta_p (x)(c_p - c_l) \tag{5}$$

$$\theta_l \frac{\partial c_l}{\partial t} = \beta_m (x)(c_m - c_l) + \beta_s (x)(c_s - c_l) + \beta_p (x)(c_p - c_l) \tag{6}$$

2.2 *Boundary and Initial Conditions*

A complex set of boundary conditions is applied to model the specific solute transport mechanisms on the upper and lower boundaries of the spent leach pad domain as follows:
At x = 0

$$q_m c_m (0,t) - \theta_m D_m \frac{\partial c_m (x,t)}{\partial x}\Big|_{x=0} = q_m c_{in} (t) + \gamma_m (c_{in} (t) - c_m (0,t)) \tag{7}$$

$$-\theta_s D_s \frac{\partial c_s(x,t)}{\partial x}\Big|_{x=0} = \gamma_s (c_{in}(t) - c_s(0,t)) \tag{8}$$

$$q_p c_p(0,t) - \theta_p D_p \frac{\partial c_p(x,t)}{\partial x}\Big|_{x=0} = q_p c_{in}(t) + \gamma_p (c_{in}(t) - c_p(0,t)) \tag{9}$$

where γ_m, γ_s and γ_p are surface mass transport coefficients.

At x = L (lower boundary)

$$-\theta_s D_s \frac{\partial c_s(x,t)}{\partial x}\Big|_{x=L} = \gamma_s (c_m(L,t) - c_s(L,t)) \tag{10}$$

where γ_s represents mass transport between the mobile efflux and the stagnant regime over the pad liner.

Homogeneous initial condition is assumed:

At $t = 0$

$$c_m = c_s = c_p = 0 \tag{11}$$

2.3 *Model Parameter Optimization*

A simplified parameter optimization is used to demonstrate the application of the four-equation MULTIFLUX model configuration to a large-scale field test. In contrast to a CXTFIT evaluation with $D_S = 0$ in equation (2), (Zhan and Danko, 2000), $D_m = D_S = D_p = D$ constraint is applied.

Step 1. $q_m, q_p, \theta_m, \theta_s$, and θ_p are optimized with $\alpha_m = \alpha_p = \alpha$ and $q_m + q_p = q$, the known water flux density constrained and β_m, β_s, and β_p mass transport coefficients defined specifically for each sampler as an exclusive connection to one regime only:

$$\begin{aligned} &\beta_m(x_i) = \beta, \quad \beta_s(x_i) = 0, \quad \beta_p(x_i) = 0 \quad \text{if } e_m = \min(e_m, e_s, e_p) \\ &\beta_m(x_i) = 0, \quad \beta_s(x_i) = \beta, \quad \beta_p(x_i) = 0 \quad \text{if } e_s = \min(e_m, e_s, e_p) \\ &\beta_m(x_i) = 0, \quad \beta_s(x_i) = 0, \quad \beta_p(x_i) = \beta \quad \text{if } e_p = \min(e_m, e_s, e_p) \end{aligned} \tag{12}$$

where e_m, e_s, and e_p are squared error-of-fit values for sampler i summed for N samples taken over the sampling period T:

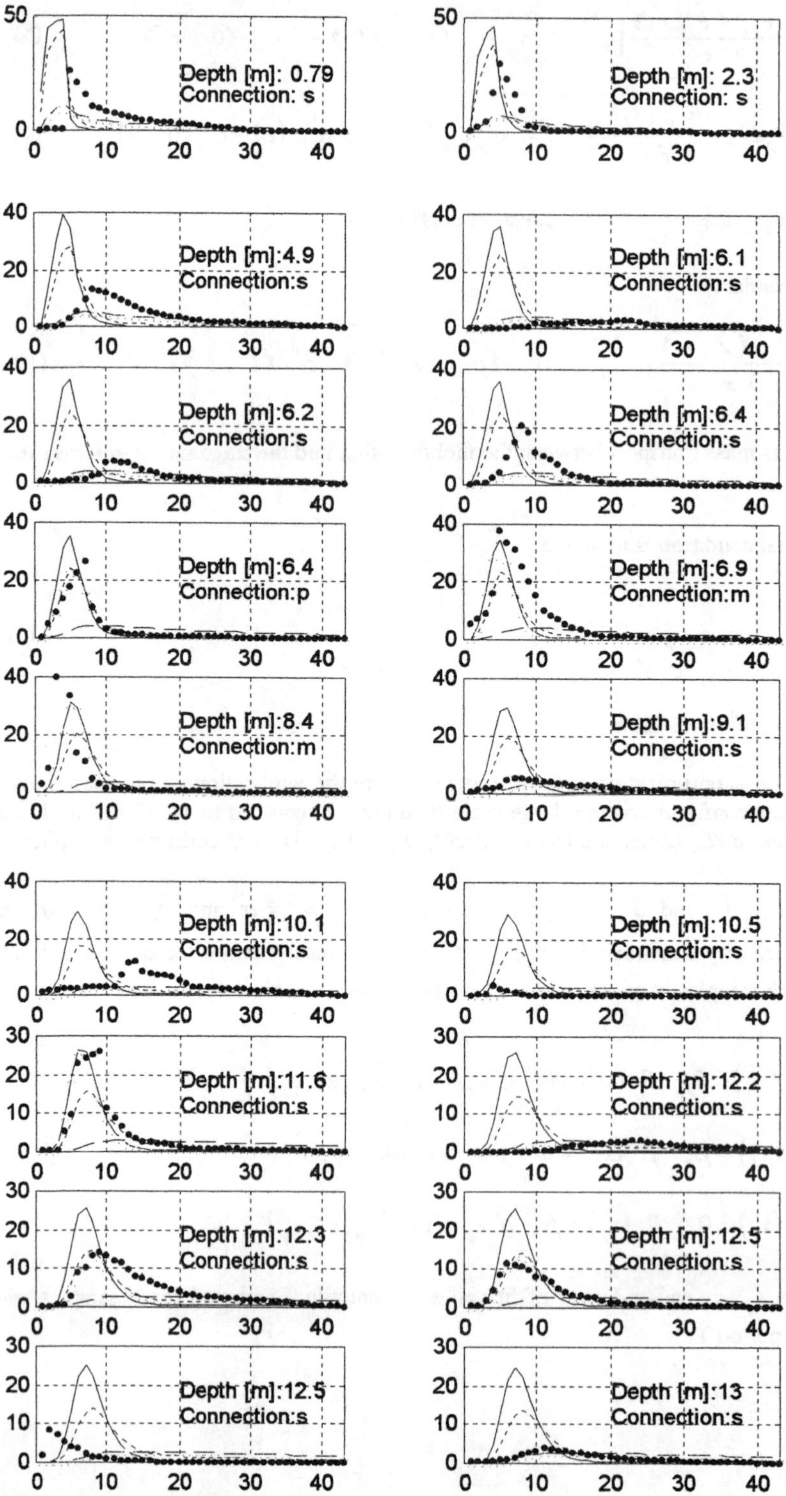
Depth [m]: 0.79
Connection: s
Depth [m]: 2.3
Connection: s
Depth [m]:4.9
Connection:s
Depth [m]:6.1
Connection:s
Depth [m]:6.2
Connection:s
Depth [m]:6.4
Connection:s
Depth [m]:6.4
Connection:p
Depth [m]:6.9
Connection:m
Depth [m]:8.4
Connection:m
Depth [m]:9.1
Connection:s
Depth [m]:10.1
Connection:s
Depth [m]:10.5
Connection:s
Depth [m]:11.6
Connection:s
Depth [m]:12.2
Connection:s
Depth [m]:12.3
Connection:s
Depth [m]:12.5
Connection:s
Depth [m]:12.5
Connection:s
Depth [m]:13
Connection:s

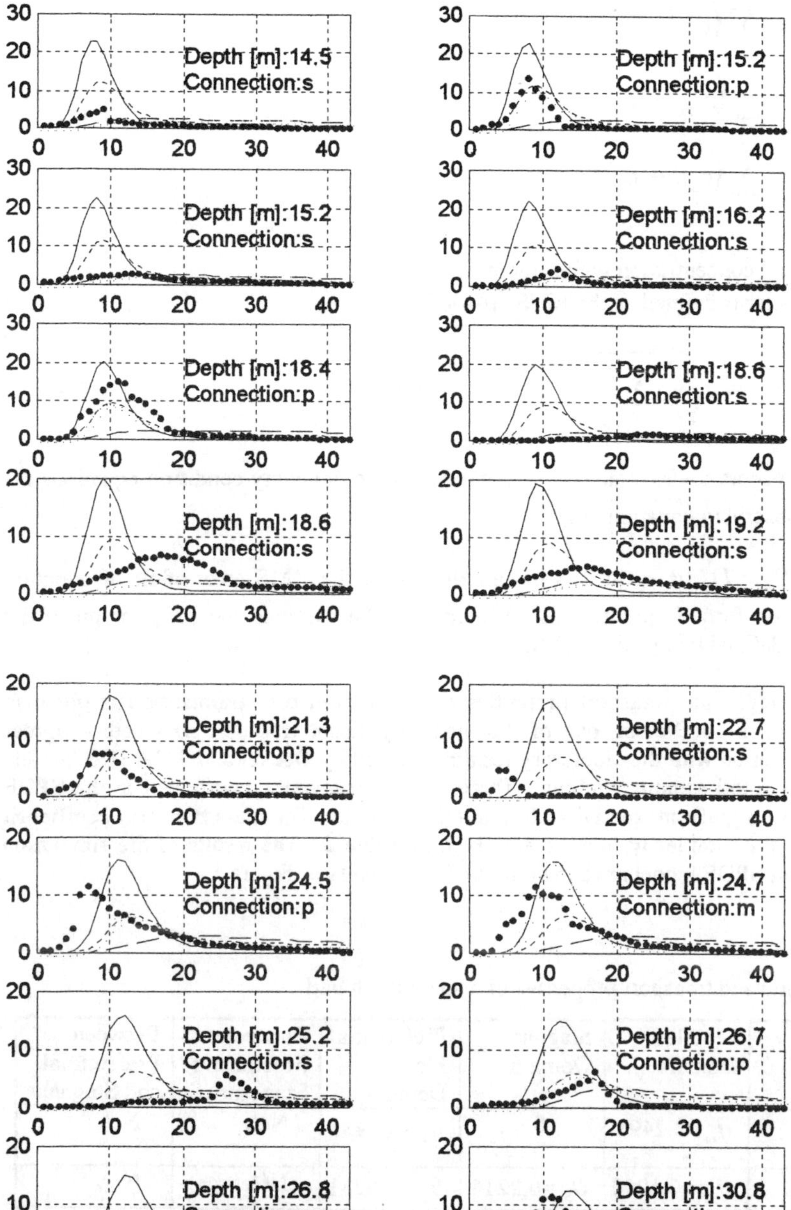

Figure 1. Simulation and measurement results for 34 lysimeters. Solid, dotted and dashed lines indicate the mobile, stagnant and preferential flow domain simulated concentrations. The bold dots depict the measurement results while the 'x' symbol shows the simulated concentrations of the lysimeters.

$$e_m = \sum_N (c_M - c_c)^2$$
$$e_s = \sum_N (c_M - c_s)^2 \qquad (13)$$
$$e_p = \sum_N (c_M - c_p)^2$$

and c_M is the measured concentration with sampler i.
The optimized parameter is defined as the RMS error-of-fit:

$$RMS = \sqrt{\frac{1}{N}\sum_N (c_M - c_\ell)^2} = \min \qquad (14)$$

Step 2. The mass transport coefficients γ_m, γ_s, and γ_p in the boundary condition equations (7) – (9) are optimized observing equation (14).

Step 3. $D = D_m = D_s = D_p$ is optimized minimizing again the RMS error-of-fit of equation (14). The optimization from Step 1 to 3 is repeated with the dynamic re-configuration of the sampler connections defined in equations (12).

The set of hydrologic and transport properties obtained from the optimization is given in Table 1. The dispersion coefficient, one of the most significant transport parameters, agrees within 20 percent deviation with the laboratory results obtained by Decker and Tyler (Decker and Tyler, 1998) using 6 m tall sample columns. Therefore, the dispersion coefficient in the SPDE model does not require significant spatial scaling in this example. The mass transport coefficient and water content of the sampler lysimeters are given in Table 2. The results of the simulation with the four-equation SPDE model are shown in the 34 sub-plots of Figure 1.

TABLE 1. Hydrologic and transport properties of a spent leach pad

	Mobile Domain	Stagnant Domain	Preferential Flow Domain	Between Mobile and Stagnant	Between Preferential and Stagnant
Rinsing water flux density [m/day]	q_m=0.2491	0	q_s=0.0471	N/A	N/A
Water content	θ_M=0.1000	θ_S=0.2214	θ_P=0.0231	N/A	N/A
Dispersion coefficient [m^2/day]	D_M=1.886 9	D_S=1.8869	D_P=1.8869	N/A	N/A
Mass transport coefficient within the pad [1/day]	N/A	N/A	N/A	α_M=0.0031	α_P=0.0024
Surface mass transpor coefficient [1/day]	γ_M=2.035	γ_S=0.1759	γ_P=0.1546	N/A	N/A

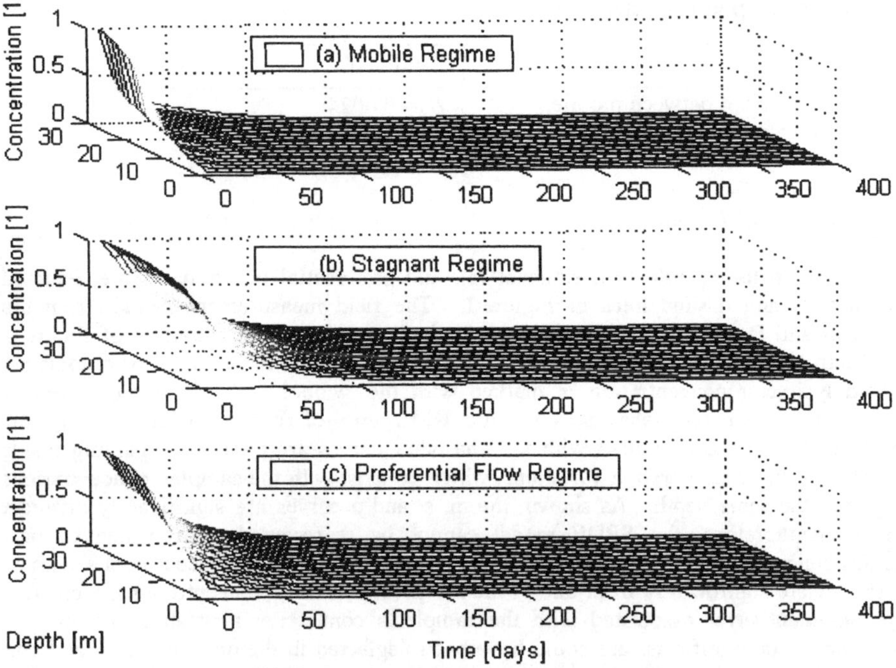

Figure 2. Concentration variations with depth and time during rinsing in the mobile (a), stagnant (b), and preferential flow regimes (c).

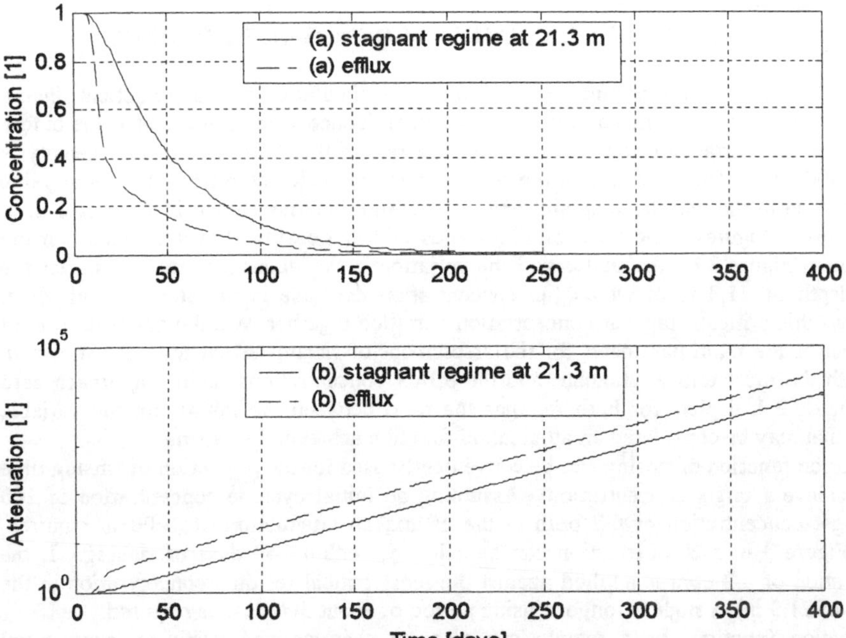

Figure 3. The concentration variation (a) and the attenuation function (b) with respect to the rinsing time in the stagnant regime at 21.3 m (solid line) and in the efflux (dashed line).

Table 2. Lysimeter transport characteristics

Water content	θ_l =0.0002
Mass transport coefficient between mobile, stagnant, or preferential domain [1/day]	β = 0.0024

2.4 *Experimental vs. model results*

The concentrations in the mobile (m), stagnant (s), and preferential (p) domains are depicted using solid, dotted, and dashed lines in Figure 1. The field measurement results from the lysimeters (Zhan and Danko, 2000) are shown with bold squares. The sub-plots of Figure 1 indicate the depth and the individual lysimeter connection to any one of the m, s, or p domains. The simulated lysimeter concentration is marked with the symbol 'x.' Only the lysimeter concentrations are considered relevant since the RMS error-of-fit is calculated using the measurement points and those marked with x. The other curves are plotted for gaining inside information about other concentration variations, which co-exist with the sampled concentrations within the pad at the same depth. As shown, the m, s, and p curves are significantly different from each other; therefore, the SPDE model cannot be represented with a conventional convection-dispersion equation (CDE) model. The sampler's simulated concentrations marked with x do not deviate significantly from one of the sampled m, s, or p concentrations, indicating that the measurement error associated with the sampler's connection resistance and internal chamber volume is not significant and could have been neglected in the present study, and was neglected in another evaluation (Zhan and Danko, 2000) using CXTFIT. The RMS error-of-fit for the whole data set is 2.79 ppm, significantly better than the 3.6 obtained from the MIM model (Zhan and Danko, 2000).

3 NUMERICAL SIMULATION OF CONTAMINANT REMOVAL BY RINSING

The SPDE model with optimized parameters is used for the simulation of a hypothetical rinsing process. The pad is assumed to have a uniform contaminant concentration of a unit value before rinsing commences. Constant rinsing water application rate of 0.2962 m/day in flux density is considered for 400 days. The decreases in the concentration are calculated from the initial value in the pad with time in the mobile, stagnant, and preferential domains shown in Figures 2 a, b, and c, respectively. Figures 2 a-c show similar trends with noticeably slower decrease for the stagnant domain in Figure 2 b. An analysis of the variation in the stagnant regime identifies the most critical depth of 21.3 m at which the concentration decrease is the slowest with time. Figure 3 a shows this critical, stagnant concentration variation together with the concentration of the efflux, which is the main parameter for the assessment of rinsing efficiency. As shown in Figure 3 a, both the most critical stagnant and the efflux concentrations rapidly approach zero with time. Figure 3 b depicts for both regimes the ratio between the initial and the variable concentrations that may be considered an attenuation function achieved by rinsing.

The attenuation function of rinsing can be conveniently used for the prediction of rinsing time necessary to achieve a target concentration. Assuming an initial cyanide concentration of 200 ppm and a target concentration of 0.2 ppm in the efflux, an attenuation of 1000 is required. According to Figure 3 b, such attenuation can be achieved within 260 days of rinsing. If the target concentration of 0.2 ppm is applied against the most critical residual concentration in the pad (at a depth of 21.3 m), a slightly longer rinsing period of about 300 days is required.

The attenuation function for a regime can also be characterized using an exponential approximation for the numerical results represented by the curves shown in Figure 3 b. The exponential approximation, applicable for large time periods (greater than 30 days), defines a constant attenuation coefficient by Equation (15) as follows:

$$C(t) = C_o * \exp[-a * (t - t_o)] \quad (15)$$

where C(t) and C_o are the time-dependent and initial concentrations, a is the attenuation coefficient for rinsing, t is the rinsing time in days, and t_o is a constant time shift. Using Eqn(15), the attenuation coefficients for both the mobile and the stagnant regimes are a = 0.0241 1/day, and t_o is 26 and –16 days for the mobile and the stagnant regimes, respectively.

4 CONCLUSIONS

The hydrologic and transport properties of a four-equation SPDE model based on the MULTIFLUX code can be obtained for the explanation of in situ measurements of a spent leach pad. The dispersion coefficients of the model are in close agreement with the values obtained from laboratory column tests. The RMS error-of-fit is lower than that obtained from CXTFIT for the same measurement data set.

The SPDE model with identified coefficients can be conveniently used to simulate a hypothetical rinsing process. As expected, the concentration decrease during rinsing is faster in the mobile and preferential flow domains than in the stagnant regime.

A non-dimensional rinsing attenuation function can be constructed based on the numerical solution of the SPDE model. The attenuation curve can be used for the prediction of rinsing time for a given initial and a target concentration. A slightly longer rinsing time is required to decrease the concentration to the same ratio in the stagnant regime than that in the mobile phase.

A numerical example is provided to illustrate the rinsing time prediction based on the SPDE model. The results of 260-300 days of rinsing time are consistent with current, industrial considerations.

5 ACKNOWLEDGEMENT

The contribution of the University of Nevada, Reno, Nevada, U.S.A. and the University of Technology, Helsinki, Finland, in the form of faculty sabbatical development support are gratefully recognized and highly appreciated.

REFERENCES

Danko, G., 1999. "MULTIFLUX code description," CRWMS M80 MOL. 19990409.0096, DOE, Yucca Mountain Project Record Center, Nevada, U.S.A.

Decker, D.L. & S.W. Tyler, 1998. "Evaluation of flow and solute transport parameters for heap leach recovery material," *J. of Environmental Quality*. 28:2; 543-555.

Toride, N., F.J. Leij, & M.Th. van Genuchten, 1995. "The CXTFIT code for estimating transport parameters from laboratory or field tracer experiments, version 2.0", Res. Rep. No. 137, U.S. Salinity Laboratory, USDA, Riverside, CA.

Zhan, G., & G. Danko, 2000. "Experimental and Numerical Study of In-Situ Leach Pad Hydrology," *Proceedings of the International Symposium on Hydrogeology and the Environment*. Wuhan, Hubei, P.R. China.

Tailings and Mine Waste '01 © 2001 Balkema, Rotterdam, ISBN 90 5809 182 1

The use of natural stable isotopes of water to delineate pond leakage in a semiarid environment

R.T. Leigh
Leigh Geological Services Incorporated, Green River, Wyo., USA

ABSTRACT: In semiarid and arid environments, isotopically light water molecules are preferentially vaporized, enriching remaining water with heavier isotopes of hydrogen and oxygen. Water composition of tailings ponds in this climate have an elevated concentration of these isotopes, which are sufficiently distinct from ground water to be used as a tracer. Delineation of pond leakage utilizing the heavier water isotopes is useful to define plume extents and magnitude.

In a fifteen-year case study, the use of stable isotopes of hydrogen and oxygen was preferable to water quality monitoring for accurately mapping pond water migration. Natural ground water composition at this location was indistinguishable from tailings pond salts. This was the result of dissolution of similar natural salts as ground water elevations rise and buffering reactions of the tailings pond brines with natural minerals in the surrounding rocks. These compounding factors obscured the delineation of tailings pond leakage.

In semiarid and arid environments, the use of natural heavier isotopes of hydrogen and oxygen in water complements the use of monitoring ground water quality and provides better definition for mapping pond leakage.

1 INTRODUCTION

Utilizing evaporation ponds in semi arid climates is an approach applied to manage tailings. Tailings are composed of mining and process waste, including solids and dissolved salts from inefficiencies of the refining process. Slurry pipelines are utilized to transport the tailings to the evaporation ponds for disposal. Tailings pond sizes are dependent upon solids generation, storage life and evaporation and often exceed 640 acres (one square mile) in size. Pond fluid elevations are controlled by the high evaporation rates that also concentrate the dissolved salts.

Tailings ponds constructed prior to the use of membrane liners leak into the surrounding shallow ground water system resulting in rising ground water levels and apparent degradation of the ground water quality and often form surface brine crust during high evaporation of the summer months.

It was necessary to determine whether the rising ground water levels were due to migration of tailings pond water or simply due to increase hydraulic head as a result of rising pond water elevations as solids in-filling occur. Also, were the brine crust resulted from transportation of pond solutes?

Plume delineation from traditional analysis of inorganic constituents fail to define the lateral extents of pond leakage. The unique isotopic composition of the pond water was sufficiently distinct from natural ground water to be used as a tracer for plume delineation.

2 REGIONAL SETTING

2.1 *Climate*

The study area is located in the Green River Basin of southwestern Wyoming, which has a continental dry, cold temperate boreal climate at an elevation of 6500 ft. The average annual precipitation is approximately 8 inches with approximately 40 inches of evaporation. The region has relatively cool climate with an average annual temperature of 44° F with extremes ranging from -40° to 100° F. The study is subject to strong gusty winds from the west which average 7 mph.

2.2 *Geology*

The Eocene Bridger Formation underlies the study area to a depth of approximately 300 ft. The Bridger depositional environment was predominately fluvial and lacustrine and is composed of four lithofacies: laminated carbonates/calcareous claystones, evaporites, claystones/mudstones, and sandstones/siltstones.

Weathering of the near surface strata has created an upper zone of permeability that extends locally to a depth of 60 ft. Laminations of gypsum up to three-quarters-of-an-inch thick are present in the near surface weathered claystones. Occurrences above the piezometric surface indicate gypsum deposition occurred prior to tailings pond usage and is probably related to playa geomorphology.

The underling non-weathered strata have permeability's less than 0.01 millidarcey. Higher permeability's, in the 10 millidarcey range, have been identified in the non-weathered sandstone lens creating isolated perched aquifers.

2.3 *Regional Hydrogeology*

The study area tailings pond was constructed in a pre-existing playa lake. Depocenter of the lake contains clay-rich soils, with low permeability, surrounded by fluvial silts and clays. The playa is located on a hydrological divide draining predominately to the northeast and secondarily south. Highland areas are located to the east and west.

Seasonally the playa collected run-off during the spring that completely evaporated by August. Historical monitoring information indicates the piesometric surface to be at a depth of 45 ft. Ground water samples collected from the weather zone aquifer, prior to development of the tailings pond, contained 12,000 mg/l Total Dissolved Solids (TDS) near the playa and decreasing to 2,000 mg/l TDS at greater distances. The major constituents include sodium, calcium, sulfate, carbonate, and chloride ions. pH values range 7 to 9.

3 DEUTERIUM AND OXYGEN-18 AS POND WATER TRACER

3.1 *Stable Isotope Systematics*

The use of stable isotopes of water to delineate tailings pond water leakage was developed from research published by Knauth and Beeuas, 1985. Their investigation focused on identifying the source of water inclusions in salt horizons in the Palo Duro Basin, Texas, by determining the isotopic composition. A similar approach was used for this study.

Natural water molecules are a mixture of stable hydrogen and oxygen isotopes (excluding tritium). Hydrogen exist as two stable states, (H hydrogen-1 and D deuterium/hydrogen-2) and oxygen exist as three stable isotopes (oxygen-16, oxygen-17, oxygen-18). During evaporation and precipitation the relative proportions of these isotopes change. Water composed of $HH^{16}O$ has a higher vapor pressure than water composed of $HD^{16}O$ or $HH^{18}O$ or $HD^{18}O$. Lighter molecules are preferentially partitioned into the vapor phase during evaporation. The reverse occurs during precipitation. The first water droplets to form are isotopically heavy.

The meteoric water cycle involves repeated evaporation/precipitation steps leaving natural water with a wide variation of isotopic compositions. Produced primarily by evaporation of ocean water, meteoric water isotopic fractionalization is temperature dependent. Therefore ocean air masses at different latitudes have different composition.

Precipitation at coastal areas are isotopically heavy. As air masses move inland, the remaining water vapor becomes progressively lighter. Seasonal isotopic variation at one location occurs as a result of temperature, air mass source location and flow path distance.

Stable isotope analysis of oxygen and hydrogen are reported as variations in isotope ratios with respect to SMOW (Standard Mean Ocean Water) ratios. Specifically isotope compositions of deuterium and oxygen-18 are reported as ratios defined by:

$$\delta = \left(\frac{R_{sample}}{R_{SMOW}} - 1\right) \times 1000$$

R= Ratio of deuterium to hydrogen-1
Or oxygen-18 to oxygen-16
For the sample or SMOW

The meteoric Water Line MWL has been defined by the isotopic composition of global precipitation (Craig, 1961), Figure 1. Negative values indicate lighter isotopic composition relative to SMOW. In general, the cooler the climate or the greater distance an air mass travels from the ocean, the further down the MWL a sample will plot.

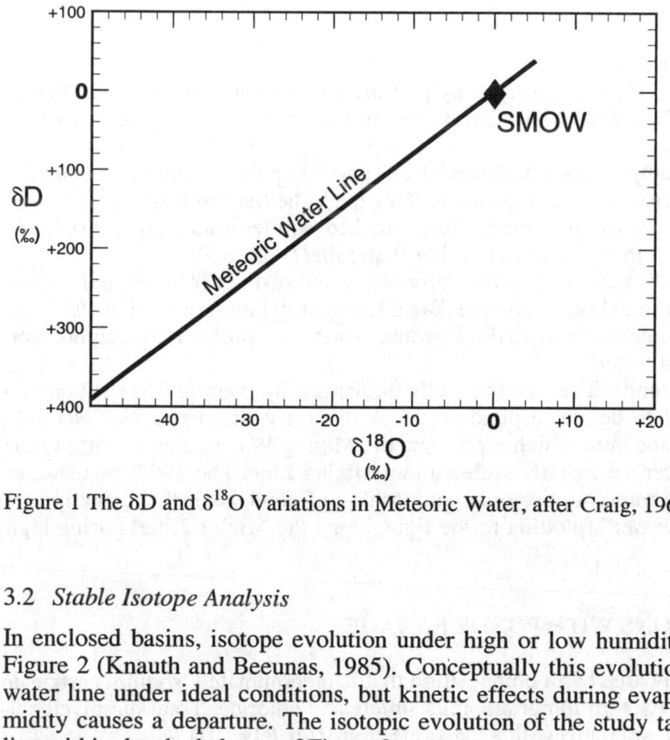

Figure 1 The δD and $\delta^{18}O$ Variations in Meteoric Water, after Craig, 1961

3.2 *Stable Isotope Analysis*

In enclosed basins, isotope evolution under high or low humidity climates varies as illustrate, Figure 2 (Knauth and Beeunas, 1985). Conceptually this evolution should parallel the meteoric water line under ideal conditions, but kinetic effects during evaporation in the presence of humidity causes a departure. The isotopic evolution of the study tailings pond water should fit a line within the shaded area of Figure 2.

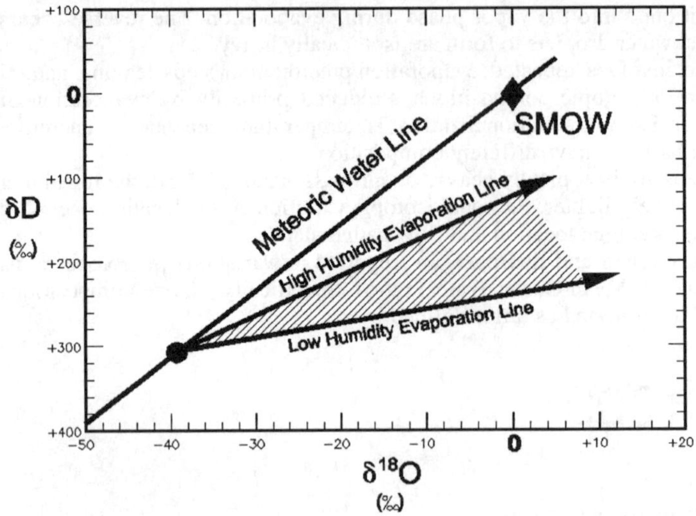

Figure 2. Isotopic Evolution of Evaporating Meteoric Water, after Knauth and Beeunas, 1985

4 MIXING MODEL

Initial isotopic analysis of the study area were completed in August 1985, and expanded during 1986. Samples collected include:

- Precipitation, rain and snow events
- Pond Water
- Observation Wells
  - Shallow weather zone wells
  - Deep perched aquifer wells

A unique fingerprint for the study area emerges by plotting the isotopic compositions (Figure 3). Like the global Meteoric Water Line MWL illustrated on Figure 2, a MWL develops which is characteristic for the study area.

"Natural" formation water samples collected away from possible pond leakage plots parallel to the MWL but contains lighter isotopic composition. This might be the result of reaction with soils and rocks as infiltration occurs over time. Deep perched aquifer water are isotopically lighter still and plot further down the Natural Formation Water line (Figure 3)

Tailings pond water was confirmed to be isotopically heavy and distinct from natural formation water to be used as a tracer to delineate pond leakage. Lateral migration of pond water leakage is evident in the isotopic composition of shallow ground water, as a progressive composition change due to mixing with pond water.

Plotting the isotopic water compositions of the wells influenced by pond leakage (Figure 3) develops a mixing model used to delineate pond water. A divergent line from the "Natural" Formation Water line NFW is apparent which represents the Mixing Water Line. Mixing ratios of formation water to pond water are linearly scaled along Mixing Line. The 100% pond water indicator represents the average minimum isotopic composition of pond water that actually varies seasonally and becomes "heavier" (plotting to the right, along the Mixing Line) during high evaporation months.

5 GEOCHEMICAL REACTIONS WITH POND LEAKAGE

Tailings pond water of the study area has a composition that is predominately sodium carbonate (reported as Total Alkalinity TA) with minor amounts sulfate and chloride. Total dissolved solids TDS seasonally exceed 100,000 mg/L with a pH value approximately 10.0.

Background ground water quality information is limited.

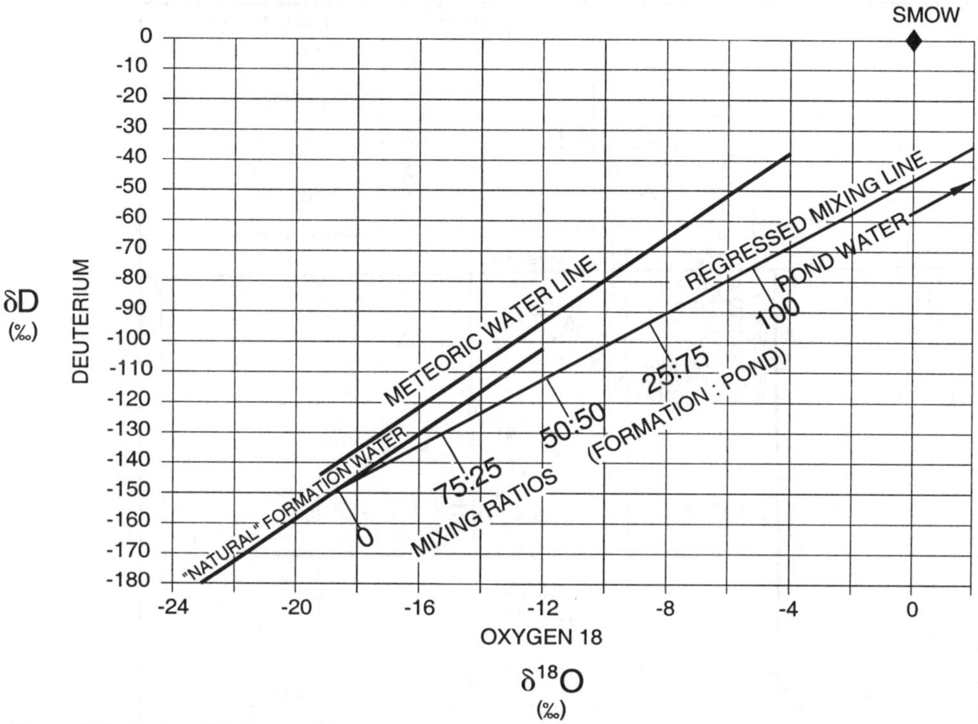

Figure 3 Study Area Mixing model

5.1 *Reaction Series*

A chain of events occurs in response to tailings pond water leakage

1. Rising ground water elevations occur in the immediate vacinity of the pond. This propagates a hydraulic response to increasing heads.
2. Dissolution of natural (sodium sulfate) salts occur as fluid levels rise.
3. Buffering reactions of gypsum with pond water (sodium carbonates) increase sodium sulfate concentrations. $CaSO_4 + Na_2CO_3 => Na2So_4 + CaCO_3$
4. Gypsum depletion and rising TA concentrations

5.2 *Ground Water Monitoring*

The study area contains over 100 observation wells completed at various depths. Some wells were clustered at key locations and screened at specific zones. Single wells in the shallow weathered zone were located to monitor changes downgradient of the tailings pond (Figure 4).

5.3 *Graphical Presentation*

Inorganic and isotopic water samples were collected quarterly through 1998 and then semi-annually. The following graphics illustrate TDS, sulfate, and TA changes with fluid elevation changes and pond water concentration estimates indicated. Geochemical reactions are apparent.

- Study Area Tailings Pond Hydrograph
  Seasonal pond water elevation changes result from evaporation cycles
- Observation Wells A and D
  Well A illustrates fluid level changes in hydraulic response to tailings pond water fluctuations. Note: initial water level rise in Well A has a minimal change in isotopic composition and a small increase in Total Dissolved Solids TDS and sulfates SO_4. This mi-

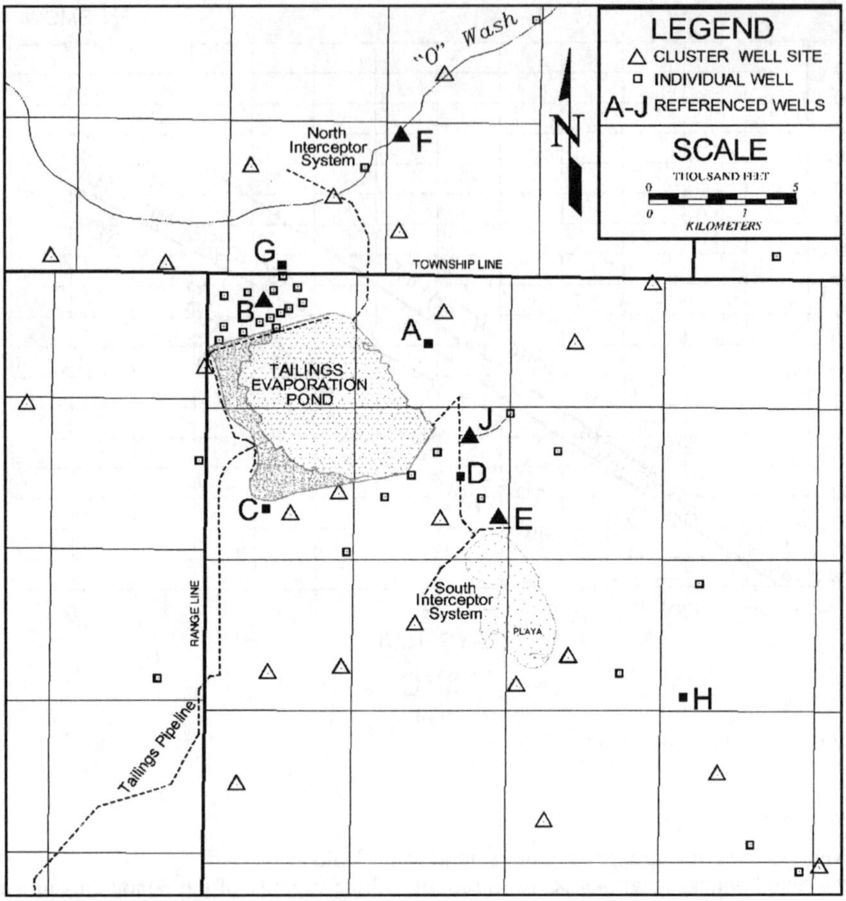

Figure 4 Study Area Observation Well Location Map

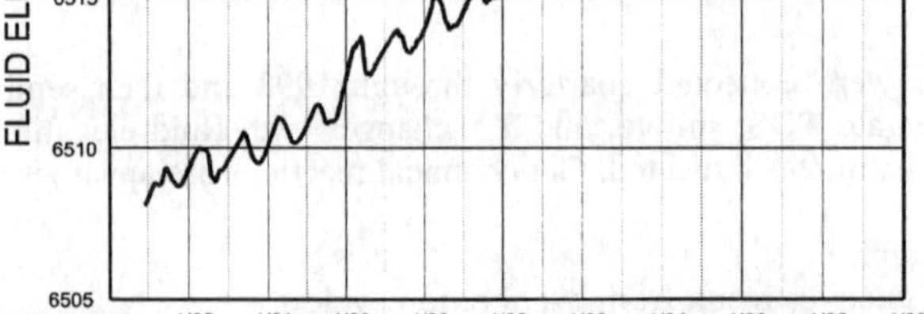

nor change probably resulted from dissolution of limited amounts of natural sodium sulfate salts occurring in the vadous zone. Continual migration of pond water containing sodium carbonates eventually react with natural gypsum, increasing sodium sulfate concentrations. As the buffering effects of the gypsum is depleted, sodium carbonate TA concentrations will increase with continual pond water leakage.

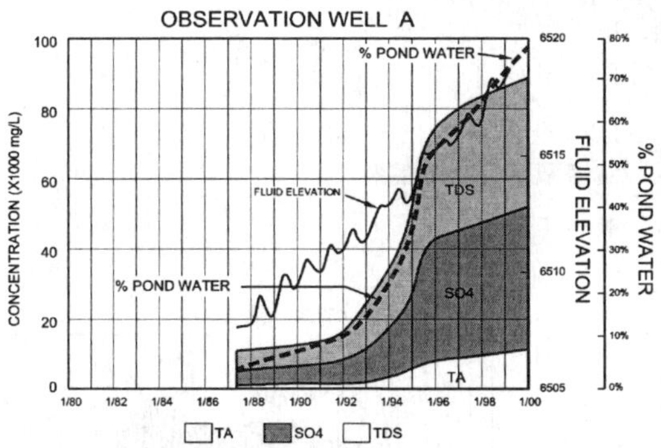

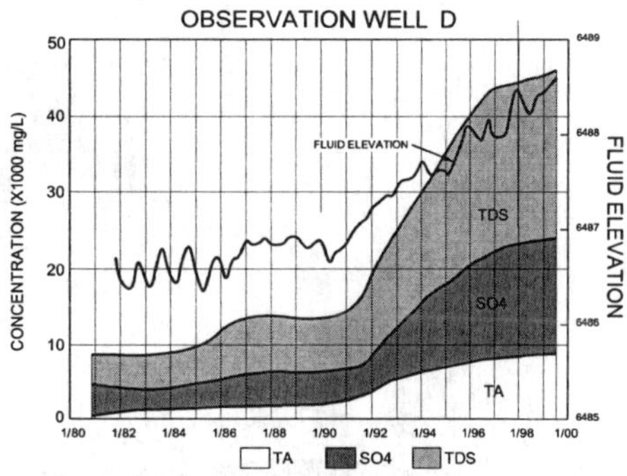

- Observation Well C
  Well C exhibits rising fluid elevations in response to rising hydraulic head of the tailings pond. No TDS change is evident. A lack of natural "salts" and minimal gypsum laminations are responsible.
- Observation Well J
  Well J contains relatively high concentrations of pond water and increasing TDS concentrations. This illustrates continual buffering effects of gypsum.

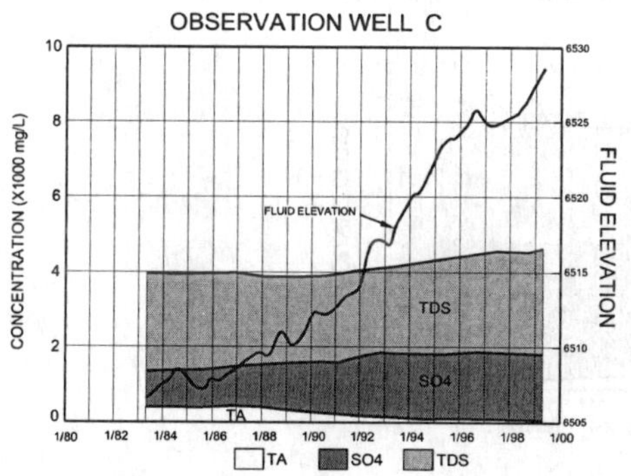

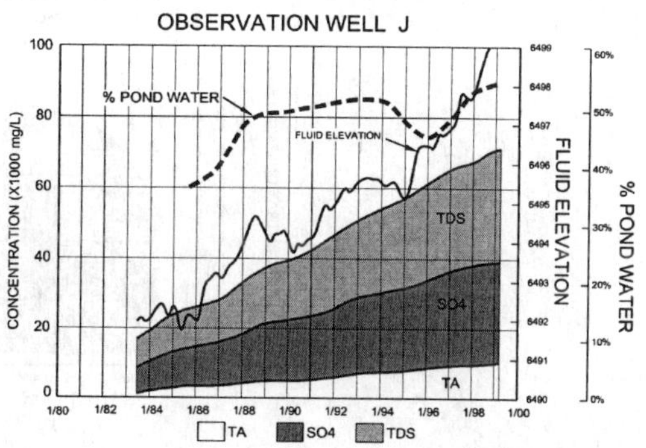

- Observation Well B
  Well B is located immediately downgradient of the tailings pond. The buffering effects of gypsum have been completed and high TA concentrations are observed.
- Observation Wells F, G, and H
  A series of recovery wells were installed in 1987 to reduce fluid elevations downgradient of the tailings pond and to create a containment zone with the help of the regional topography (Figure 4). Composite graphs for Wells F, G, and H, illustrate responses to the interceptor systems. Well F located downgradient of the North Interceptor System exhibits improving isotopic composition and water quality.

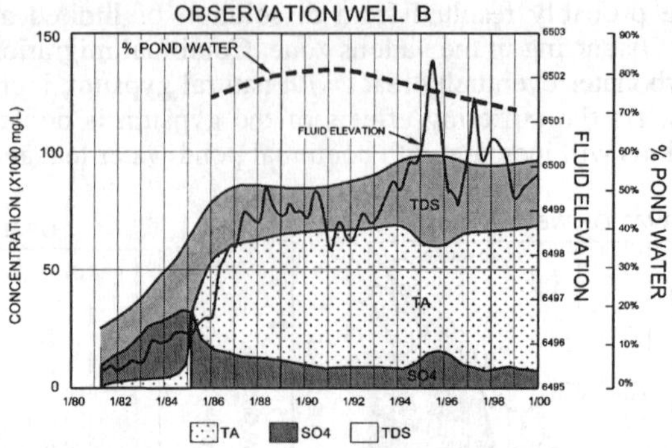

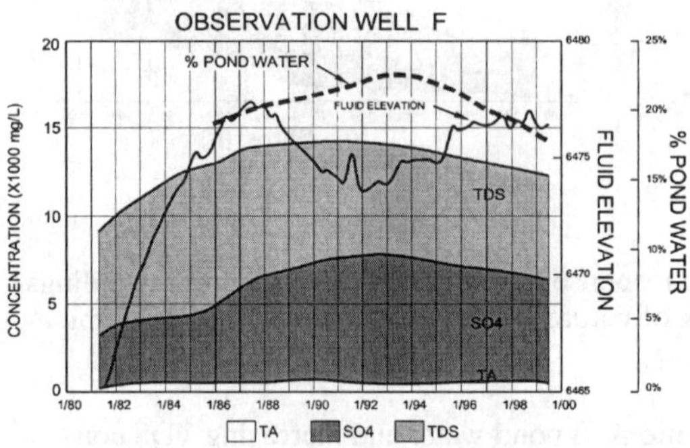

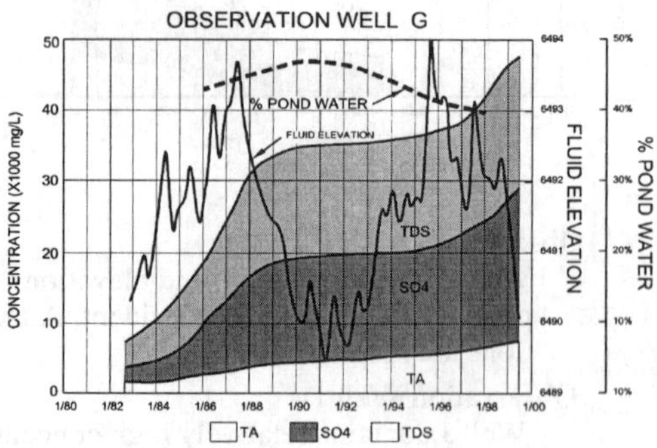

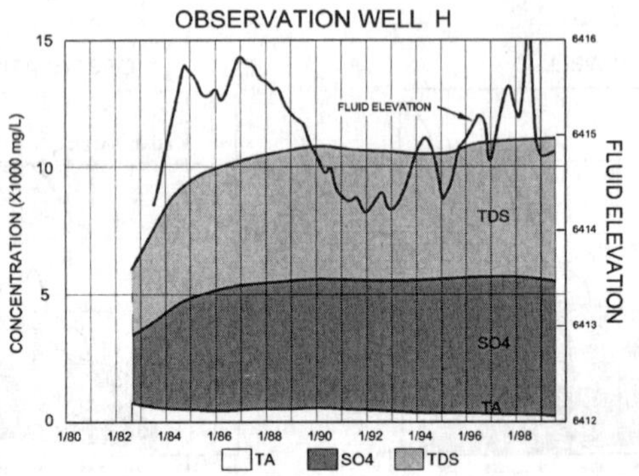

6 SUMMARY

Stable isotopes of hydrogen (deuterium) and oxygen (oxygen-18) have been shown to be useful in delineate tailings pond water leakage in a semi arid environment. These isotopes were distinct from natural ground water to be used as a tracer and provided better plume definition than inorganic analysis. Mapping lateral pond water migration is not presented in this paper because of confidentiality issues. This delineation was defined. Observations conclude that pond water leakage utilizing isotope mapping had a smaller areal distribution than what was indicated by inorganic analysis.

REFERENCES:

Craig, H. 1961. Isotopic variation in meteoric waters. *Science.* 133: 1702-1703

Knauth, L. P., and M. A. Beeunas. 1985. *Origin of fluid inclusion water in bedded salt deposits, Palo Duro Basin, Texas.* Office of Nuclear Waste Isolation, Battelle Memorial Institute, Columbus, OH. BMI/ONWI-569

Tailings and Mine Waste '01 © 2001 Balkema, Rotterdam, ISBN 90 5809 182 1

Use of reverse osmosis product water to decrease restoration time

R.R.Cellan & R.A.Waterland
Homestake Mining Company, Grants, N.Mex., USA
G.L.Hoffman
Hydro-Engineering L.L.C., Casper, Wyo., USA

ABSTRACT: A 300 GPM reverse osmosis (R.O.) treatment plant has been added to the Grants Project to produce approximately 270 GPM of low total dissolved solids (TDS) water for injection into the alluvial aquifer. The R.O. treatment is being used to decrease the time of restoration and reduce the required evaporation capacity. The R.O. treatment allows the restoration system to be operated at a larger collection rate without additional evaporation capacity. This permits the restoration to be completed quicker. The low TDS R.O. product water is also very aggressive in removing dissolvable constituents from the alluvial material. Analyses have shown that replacement of fresh water injection with R.O. product injection will also reduce the amount of time required for restoration of the aquifer. Restoration timing for TDS, uranium, selenium and molybdenum with the use of R.O. product injection was evaluated.

1 BACKGROUND

The uranium industry began in the Grants, New Mexico area in the early 50's after the discovery of uranium bearing minerals by an early prospector at Haystack Butte in 1949, which is located at the west end of the Grants mineral belt.

Homestake Mining Company, in partnership with two other companies, was one of the original pioneers in the uranium business. Homestake exploration began in the Grants area in the early 50's. Homestake mining operations began in 1956. The Homestake ores came mainly from mines located in the Ambrosia Lake Mining District of the Grants mineral belt.

The Homestake - Grants milling complex started full time production of uranium concentrate or yellow cake in late 1957. The mills were constructed in 1956 in conjunction with the U.S. Atomic Energy Commission to supply their demands of uranium to the government.

Over the years, the organization and ownership of the Grants operations went through several changes. The earlier Homestake partnership groups became United Nuclear - Homestake Partners in 1968. This partnership was subsequently dissolved in 1981 and the operation became known as Homestake Mining Company - Grants.

At the time that operations began in the mid 50's there were no State or Federal regulations defining the design or construction of tailing facilities used by the mining industry. What constituted best practice at that time was defined by economic decisions such as the site location or the availability of construction materials. In the case of the Homestake Grants site, water for mill processing defined the location of the milling site and thus the tailings storage sites which are almost always located as close as possible to the milling facilities.

The tailings impoundments at the Grants site were built using a starter dike for a dam. The dike was simply bulldozed from the local sandy soil and the tailings deposited within. As the requirement for more storage became necessary, the tailing embankments were raised by building the dam from cycloned sand from the tailings slurry. The sand became the impoundment dam while the remaining portions of the tailings slurry or the slimes were collected and stored within the dam.

In the beginning of the project there were two separate tailings impoundments in use. Both were center-line constructed dams. After several years of operation one of the two impoundments was stopped from being used. At the time of closing that dam, referred to as the Small Tailing Impoundment, it contained about 2 million tons of tailings. The other, referred to as the Large Tailing Impoundment, contained slightly over 20 million tons of tailings at closing in 1990.

It should be obvious from the construction techniques when the milling facilities were first constructed that there wasn't any of the environmental safeguards required by today's standards built into any of the tailings structures. So it was not surprising that in later years, when new ground-water regulations were enacted and sites were required to perform site reviews, contamination of ground water was discovered.

The ground-water contamination at the Grants uranium tailings site was initially defined in 1975 and 1976 resulting in the site having the distinction of being selected as a CERCLA site. A more thorough and complete site review was completed in 1995 to define the complete extent of all the ground-water systems at this site.

A large-scale ground-water restoration program has existed at Homestake's Grants Project site since 1977. That program was based on pump and treat technology. Disposal of the collected water has been in two lined evaporation ponds. Each of the ponds has a fine mist spray systems installed to enhance evaporation. This method of disposal is controlled by the weather conditions that affect the evaporation rate and volume. This, also in turn, severely limits the collection rate during a wet cycle and controls the restoration rate.

The restoration of the ground water will govern the time required for eventual closure of this site. The rate at which the ground-water restoration standards are met is critical to the time required to complete the restoration project.

In order to have better control over the restoration rate, a program was initiated in 1998 to use reverse osmosis (R.O.) to supplement the existing evaporation systems. The R.O. program has enabled the collection rate to be increased significantly without having to expand the less controllable lined pond capacity.

A 300 GPM low pressure R.O. unit followed by a 90 GPM high pressure R.O. unit was installed at the site. These units were designed to produce approximately 270 GPM of R.O. product water. The product water has been used to replace some fresh water injection in the restoration program.

2 HISTORICAL GROUND-WATER RETORATION

A ground-water containment program that was initiated in 1977 consisted of fresh water injection and collection of the contaminated water near the tailings. The injection in conjunction with the collection has developed a hydraulic barrier, which has reversed the ground-water flow from the injection well line back to the line of collection wells. Since the program began, a volume of 2.8 billion gallons (average rate of 246 GPM) of contaminated ground water has been collected at the Grants site while a volume of 4.1 billion gallons (average rate of 339 GPM) of fresh water has been injected into the alluvial aquifer.

The restoration program has been greatly accelerated since milling operation ceased in 1990 and the main goal has been restoration rather than ground-water management to meet milling operation goals. Since that time significant progress has been made towards total restoration of the aquifer.

2.1 *Concentrations*

The concentrations at the time of the CERCLA listing in 1981 exceeded 20,000 mg/l for TDS, 40 mg/l for uranium, 100 mg/l for molybdenum and 10 mg/l for selenium in the ground water.

Over time a part of the aquifer has been restored as restoration retreats towards the Large Tailings Impoundment. Presently, the major portion of the remaining high concentrations are concentrated around and beneath the Large Tailings Impoundment. Ground-water concentrations near the southern and eastern sides of the Small Tailings Impoundment are also part of current ground-water restoration efforts.

2.2 *Fresh Water Injection Restoration*

Fresh water injection had progressed to a line of injection wells extending from the east side of the Small Tailings to the west of the Large Tailings by 1993 (see Figure 1 for location of the injection wells with a solid dot). Collection has been maintained near the tailings piles during this period and is shown in Figure 1 with a line of wells represented by connecting open circles. Well KM was used to monitor the restoration due to the fresh water injection. Figure 1 also shows the location of well KM relative to the fresh water injection lines.

Table 1 presents the pre-injection concentrations at well KM and the concentrations after 9.4 pore volumes had moved to well KM from the fresh water injection line. A pore volume was defined as the 50% change in concentration for the conservative constituents of chloride, TDS and sulfate. Table 1 presents the concentrations of these three parameters and the key restoration constituents of uranium, molybdenum and selenium.

Adequate restoration of selenium resulted from the fresh water injection because this parameter behaves similar to a conservative constituent at this site during restoration. Uranium and molybdenum concentrations restored to a level near 2 mg/l but became fairly steady due to the leaching of these constituents off of the alluvial material. Figure 2 shows a plot of the uranium and sulfate concentrations versus time on the bottom x-axis for monitoring well KM. Sulfate concentrations restored to a level near the injection concentration of 700 mg/l in 1995 while the uranium concentrations became fairly steady at slightly less than 2 mg/l. The fresh water concentration for uranium is less than 0.02 mg/l (see Table 1). Figure 2 also shows the uranium and sulfate changes in concentration versus pore volume on the upper x-axis and shows that approximately three pore volumes were needed for the sulfate concentrations to approach the injection concentration. With fresh water the uranium concentrations were still elevated at greater than 9 pore volumes.

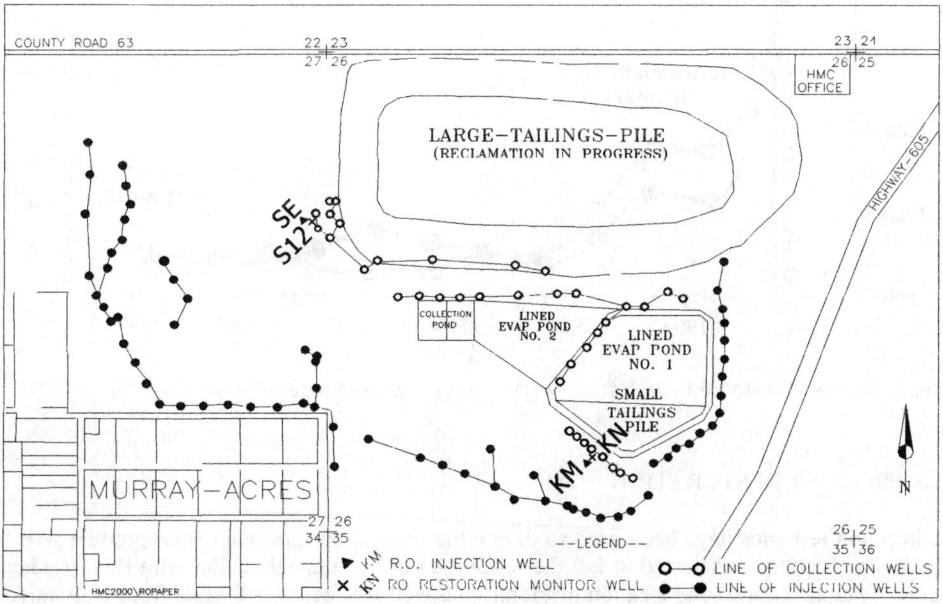

Figure 1. Homestake's grants tailings and R.O. product injection sites

Table 1. Concentrations of consituents.

WELL	DATE	Pore Volume	Cl (mg/l)	TDS (mg/l)	SO_4 (mg/l)	U (mg/l)	Mo (mg/l)	Se (mg/l)
KM	#	0	746	8000	3900	22	47	1.7
	04/20/1999	9.4	204	2050	740	1.79	2.29	0.027
KN	#	0	199	2030	755	1.71	2.68	0.076
	06/14/2000	9.28	48	585	142	0.045	0.08	0.007
S12	#	0	173	1980	947	2.26	2.23	0.434
	06/14/2000	2.22	30.8	438	95.2	0.219	0.43	0.055
FRESH WATER		---	198	1810	656	0.018	<0.03	0.006
R.O. PRODUCT		---	12	43	12	0.039	0.08	0.008

NOTE: # = PRE-INJECTION CONCENTRATION

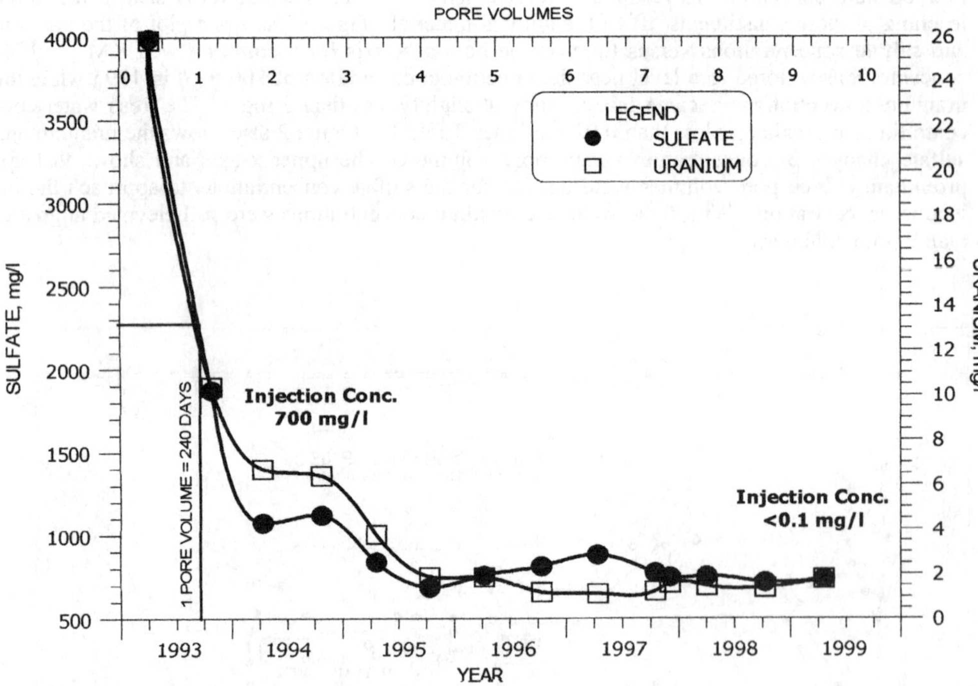

Figure 2. Well KM restoration with fresh water versus time and pore volume

3 R.O. PRODUCT RESTORATION

Two injection test sites have been used to define the restoration rates with R.O. product water. Wells KN and S12 were selected to test the restoration of the alluvial aquifer with R.O. product injection. Figure 1 shows the location of monitoring well KN which is 80 feet from R.O. injection well KM and monitoring well S12 which is located 90 feet from injection well SE.

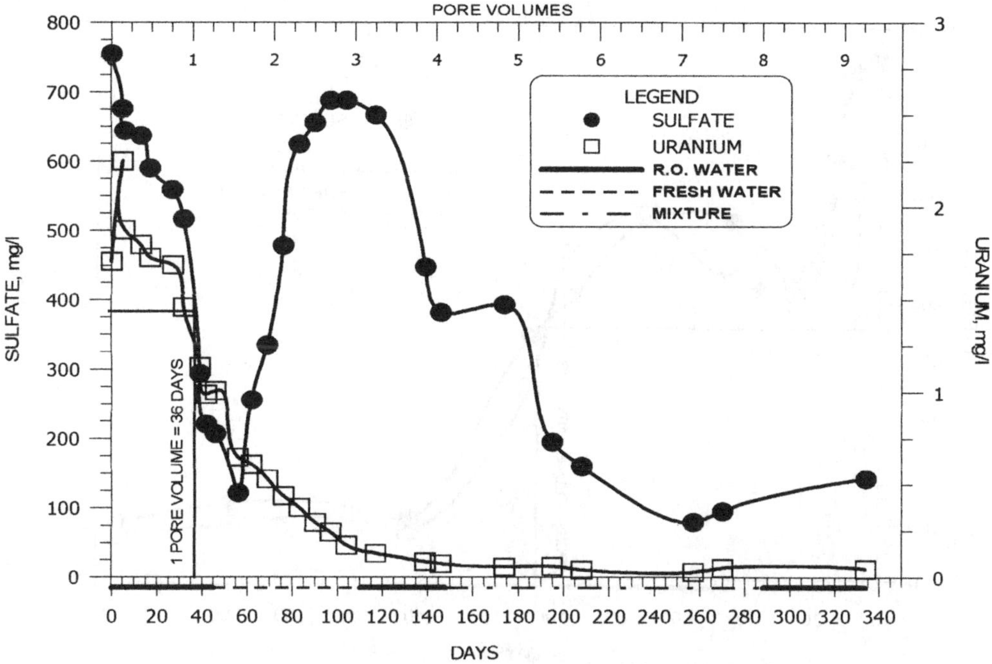

Figure 3. Well KM restoration with R.O. product water versus time and pore volume

3.1 *Concentrations*

Table 1 presents the concentration of the key parameters for the R.O. product water, pre-injection monitoring well concentrations, and one additional set of concentrations after restoration for the two monitoring wells. The R.O. product water has a very low TDS and is, therefore, very aggressive in stripping dissolvable constituents off of the alluvial material.

3.2 *Well KN Restoration*

Injection into well KM was initiated to test the restoration of the alluvial aquifer at well KN which is located 80 feet northeast of well KM, toward the collection wells near the south side of the Small Tailings. The R.O. injection testing was conducted during the start up of the R.O. plant and therefore some interruption in the R.O. product supply was experienced due to adjustments in the plant. The pattern of the line shown on the x-axis shows when R.O. product, fresh water, or a mixture of the two was injected into well KM.

Figure 3 shows the changes in sulfate and uranium concentrations for monitoring well KN versus time on the bottom x-axis. A pore volume for this test was estimated to be 36 days based on 50 % change in the conservative constituent concentration. This travel time fits the groundwater velocity based on the aquifer properties and hydraulic gradient for this area of the alluvial aquifer. The sulfate concentrations showed a typical restoration curve until the switch to fresh water caused the sulfate level to increase. The uranium concentrations continued to decline at a reasonable restoration rate but may have been lagged slightly, because fresh water was used during portions of the test. Figure 3 also shows the plot of the same data versus pore volume that shows adequate restoration by 3 pore volumes with uranium concentrations stabilizing at less than 0.1 mg/l. The R.O. product restoration of well KN resulted in a stable uranium concentration that is less than one tenth of the stable concentration with fresh water restoration.

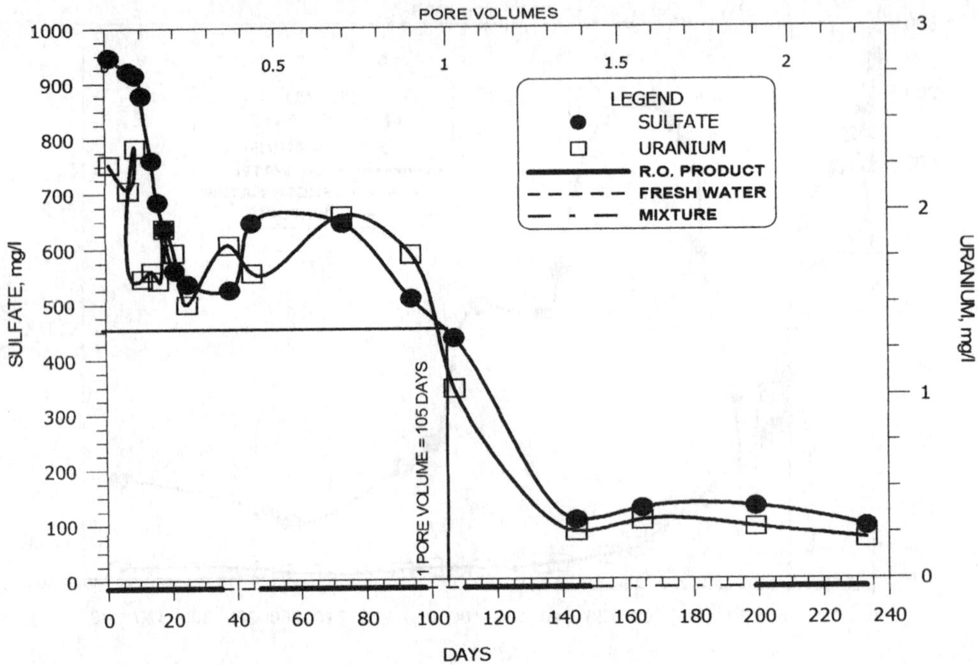

Figure 4. Well S12 restoration with R.O. product water versus time and pore volume

3.3 *Well S12 Restoration*

R.O. product water was injected into well SE while well S12 was monitored for changes in concentrations in the alluvial aquifer 90 feet from the injection well. Figure 4 shows the sulfate and uranium concentrations versus time for monitoring well S12. This figure shows that the changes in uranium concentrations mirror the changes in sulfate concentrations. The 50% change in the conservative constituent concentration was used to estimate one pore volume to be 105 days. This amount of time fits the ground-water velocity based on aquifer characteristics for this area of the alluvial aquifer. The concentrations versus pore volume for well S12 are also presented in Figure 4 which shows that the sulfate and uranium concentrations have reached low values between 1 and 2 pore volumes.

4 CONCLUSIONS

R.O. product injection will significantly reduce the amount of time required to restore the ground water aquifer at the Grants site by as much as a year at the present R.O. product output of 270 GPM. As a result the restored concentrations for uranium and molybdenum, using R.O. product water, will likely be about one fifth of the value that would have resulted with only fresh water injection.

At present Homestake Mining is in the process of doubling of the R.O. plant size to 600 GPM to further decrease the amount of time for restoration of this site. The increased rate of R.O. product injection is predicted to further reduce the restoration schedule by as much as two years. The total volume required to be collected is estimated to be approximately 13 % less than the present eleven-year pumping program of 2.5 billion gallons.

Tailings and Mine Waste '01 © 2001 Balkema, Rotterdam, ISBN 90 5809 182 1

Arsenic speciation in natural waters as related to remediation approaches

Anthony J. Bednar, James F. Ranville & Thomas R. Wildeman
Department of Chemistry and Geochemistry, Colorado School of Mines, Golden, Colo., USA
Kathleen S. Smith
US Geological Survey, Denver, Colo., USA

ABSTRACT: Arsenic is a potentially toxic and carcinogenic element that is of environmental concern, even when present in waters at low concentrations. USEPA is considering reducing drinking water standards to as low as 5 ppb. Many regions of the country have serious arsenic contamination problems in soils and also surface and groundwaters, especially those with histories of mining and certain types of agriculture. We have examined existing analytical methods, and developed new methods, for the speciation of both inorganic and organic arsenic compounds. Data from a comparison study of speciation methods, preformed collaboratively with USGS, are presented. These methods are being used in studies of arsenic remediation for various aqueous and soil systems including: acid mine drainage, naturally-contaminated drinking water wells, and agricultural problems. Arsenic remediation experiments using catalytic photo-oxidation, followed by adsorption/coprecipitation with oxyhydroxides, are also being performed.

1 INTRODUCTION

Due to its potential toxicity, USEPA is considering lowering maximum contaminant levels (MCL's) for arsenic in drinking water from the current 50 ppb to between 2 and 10 ppb. Because of this, understanding the chemistry of arsenic in natural systems is vital. Both the redox chemistry and sorption processes of arsenic have been well studied in controlled laboratory conditions, yet much remains unknown in natural systems, and is the cause of debate in the literature (Manning et al. 1998, Bowell 1994, McGeehan & Naylor 1994). A deeper understanding of these topics is important for remediation of sites contaminated by arsenic (Dutre et al. 1998).

Our interest in arsenic originates from a need to remediate acid mine drainage, contaminated drinking water wells, and agricultural sites contaminated with organic and inorganic arsenic compounds. For the purposes of this paper we focus primarily on acid mine drainage (AMD) and naturally contaminated drinking water wells where inorganic arsenic species usually predominate.

The Colorado front range region in particular has a history of mining that has left many areas with serious acid mine drainage problems. The same mineralization that causes acid mine drainages can also be the source of contamination of local drinking water wells. These problems magnify the need to understand the chemistry of arsenic in natural and contaminated waters. To understand arsenic chemistry, accurate speciation methods must be developed.

2 REDOX CHEMISTRY

Arsenic occurs generally in two oxidation states, +III and +V, barring microbial activity which can also produce arsine and methylated species. As (III) occurs usually as a neutral oxyacid (H_3AsO_3), while As (V) is found as an anion at most natural pH's (i.e. pH~3-8; $H_2AsO_4^-$,

$HAsO_4^{2-}$) (Vink 1996). These multiple oxidation states of arsenic, with the possible coupling to iron and manganese redox reactions in samples such as acid mine drainage, provide an interesting array of potential redox reactions (Scott & Morgan 1995, McGeehan 1996).

The need for a better understanding of arsenic chemistry is at least two fold. First, As (III) is more toxic than As (V), so quantification of total arsenic as well as As (III) is important in risk assessment (Tamaki & Frankenberger 1992). Secondly, As (V) is easier to remove from acidic systems by sorption onto iron and manganese oxyhydroxides or coprecipitation with these same oxyhydroxides (Wilkie & Hering 1996, Jain et al. 1999). Many water treatment systems use some type of coagulation-precipitation method for arsenic removal from water (Hering et al. 1997). However, some have postulated that the formation of a ferric arsenite phase is also a possible mechanism (Raven et al. 1998).

3 PHOTOCHEMISTRY

It has been shown that exposure of As (III) to UV light catalyzes its oxidation to As (V) (Brockbank et al. 1988, Johnson & Pilson 1975, Wensheng et al. 2000). It remains to be determined if this is a reliable and reasonably efficient mechanism that could be used in a large scale passive treatment system (Khoe et al. 2000).

Because of the low cost and ease of implementation and maintenance of passive treatment systems, we began preliminary investigations of photooxidation of arsenite as a possible first step in remediation of AMD. Removal of photochemically generated arsenate would then be accomplished by sorption/co-precipitation with iron oxyhydroxides. A list of possible oxidizing agents was considered, including: oxygen, iron (III), and natural organic matter (NOM) [Swannee River]. These would generally be available in a passive AMD photoreactor. Figure 1 shows the preliminary results of our photooxidation experiments. It is evident As(III) is oxidized and that both Fe (III) and NOM oxidize arsenite faster than dissolved atmospheric oxygen alone (DI sample). Dark control samples showed little or no oxidation, indicating the need for photocatalysis. Speciation was performed using an ion exchange method modified from Ficklin 1983 with detection by Graphite Furnace AAS.

Figure 2 shows results from a second experiment conducted under identical experimental parameters as that of the first experiment except for an increased irradiation time (5 hours versus 4 hours previously). The dark controls show almost negligible oxidation, while the Fe (III) solution exposed to light has a greater amount of oxidation than the DI solution.

Because of the limited solubility of Fe (III), experiments were performed at pH~3.5 using hydrochloric acid to lower the pH (instead of nitric acid because of the possible redox reactions). These conditions approximate acid mine waters, although the use of sulfuric acid would make

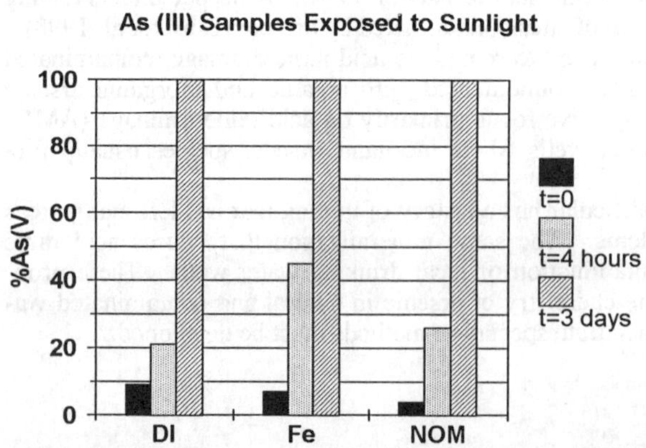

Figure 1. Percent oxidation of As (III) to As(V) at different time intervals.
Samples were aerated and exposed to sunlight in quartz tubes. Initial [As]=250ppb (as As (III)), [Fe^{+3}]=500ppb. DI=deionized water, NOM=natural organic matter.

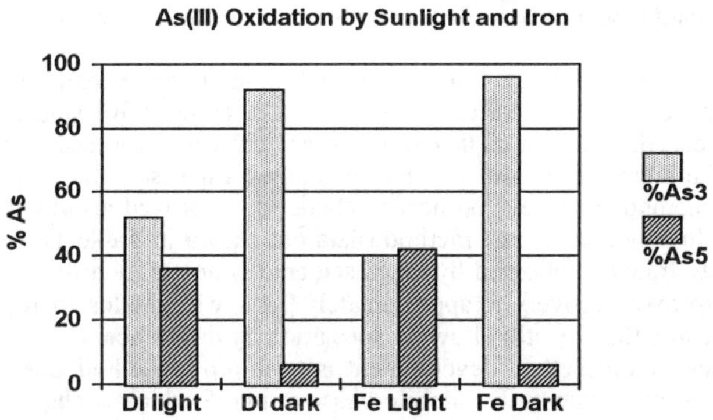

Figure 2. Oxidation of As(III) increased by addition of sunlight and Iron (III).
Samples were acidified to pH~3.5 (HCl) with $[Fe^{+3}]$=500ppb and initial [As]=250ppb. Exposed to sunlight for 5 hours.

the match closer. In future experiments, iron and manganese oxyhydroxide colloids will be added to compare oxidation rates in solution versus surface catalysis and/or reductive dissolution. Colloid surfaces may have a large impact on an engineered treatment system depending on their effect on oxidation rates.

4 SPECIATION METHODS

During the photochemistry experiments we realized that our ability to determine the effectiveness of the oxidation was limited due to the inherent analytical uncertainty in the available speciation technique. This obstacle encouraged us to perform a comparison study of common speciation methods published in the literature. Table 1 shows an abbreviated comparison between anion exchange chromatography (Ficklin 1983), 2 types of HPLC-ICP-MS (Garbarino & Burkhardt 1998, see for example Magnuson et al. 1996, CETAC Technologies), and Hydride Generation Atomic Absorption (HGAAS) (see for example Perkin Elmer), and the calculated values. In general it was noted that in "pure" stock solutions, the methods were somewhat similar in results, except for a general underestimation by the anion exchange method. However, as the sample matrix became more complex or total arsenic concentrations decreased, many methods

Table 1. Arsenic speciation results obtained by several methods for synthetic and natural samples.

Matrix	DI Water		Fe^{+3},Al^{+3}, SO_4^{2-}		Cl^-, SO_4^{2-}		Real AMD	
	As(3)	As(5)	As(3)	As(5)	As(3)	As(5)	As(3)	As(5)
Anion Ex	32	23	37	22	30	30	1354	616
HPLC	41	26			46	29		
CETAC	46	26			49	28		
HGAAS	48	32	47	28	46	35	28	2300
Calc'd	50	25	50	25	50	25	As(T) ~2200	
Matrix	DI Water		Fe^{+3},Al^{+3}, SO_4^{2-}		Cl^-, SO_4^{2-}			
	As(3)	As(5)	As(3)	As(5)	As(3)	As(5)		
Anion Ex	622	496	668	560	608	574		
HPLC	890	530			920	560		
CETAC	983	580			986	588		
HGAAS	1000	690	980	630	950	640		
Calc'd	1000	500	1000	500	1000	500		

$[Al^{+3}]=[Fe^{+3}]$=1000ppb, $[Cl^-]=[SO_4^{2-}]$=10000ppb, all As concentrations given in ppb.

appeared to have biases, even when total arsenic was well above the instrumental detection limits.

We are continuing to investigate and compare these 4 arsenic speciation methods. Because of the variability between the different methods on identical samples, further comparitive studies of arsenic speciation methods are needed. Results of this research are critical if a successful study of the redox chemistry is to be done. The results of the methods comparison study are also being used to modify existing methods and develop new methods for improved accuracy. Our minor modifications to the Ficklin anion exchange method (data not shown in Table 1) involves using DI water for the arsenite fraction followed by increased acid concentration for the arsenate fraction. These changes improved recovery to approximately 90%, with the loss being entirely restricted to the arsenate fraction, thereby still allowing speciation by difference.

A primary effort in our arsenic speciation method development is to improve method flexibility. Many HPLC methods utilize a phosphate buffer mobile phase at pH>5 which is chemically incompatible with AMD. Methods are available which utilize nitric acid mobile phases, yet these can involve complex gradient programs and take up to 20 minutes per sample (Londesborough et al. 1999). We are focusing on eluent simplicity, durability, and speed, without sacrificing accuracy. Chemical treatment of the samples prior to speciation, such as cation complexation with EDTA, is one promising avenue we are persuing to extend the flexibility of current methods.

5 SUMMARY AND DISCUSSION

We are currently investigating the redox chemistry of arsenic as a way of developing and determining the efficiency of remediation techniques for AMD, drinking water wells, and agricultural applications. Because of the different characteristics of the 2 primary inorganic arsenic species, a comparison of analytical methods study was initiated. The results of this study have shown that certain methods may have inherent difficulties with certain sample matrices, indicating the need for further investigations into these techniques, and possible modifications.

Thus far our preliminary results are promising, particularly for AMD. The presence of Fe (III) as an oxidizing agent in these waters, when coupled with sunlight photocatalysis, offers a cheap and reliable method for arsenic oxidation and removal. Modifications of existing speciation methods continue to be a central theme in this research as it is crucial for quantifying arsenic redox chemistry as well as in risk assessment for arsenic containing waters.

This research is on-going with large areas yet to be studied. Organoarsenicals, especially as used in animal feeds and as cotton herbicides, offer an interesting challenge as they often involve both biotic and abiotic processes. Remediation of areas contaminated by organoarsenicals will most certainly contain at least as many difficulties in determining arsenic species as sites containing only inorganic species.

6 ACKNOWLEDGEMENTS

The authors wish to thank: Paul Lamothe, Geologic Division, USGS, Denver, CO; John Garbarino, Methods Research and Development Program, USGS, Denver, CO; Kirk Nordstrom USGS, Boulder, CO, and David Blowes, Univerity of Waterloo, Ontario, Canada, for participating in the comparison of methods study. We are also grateful for support for this research from the USGS Toxic Substances Hydrology Program, and the Sussman Foundation.

REFERENCES

Bowell, R.J. 1994. Sorption of Arsenic by Iron Oxides and Hydroxides in Soils. *App. Geochem.* 9: 279-86.

Brockbank, C.I., Bailey, G.E., & Low, G.K-C. 1988. Photochemical Decomposition of Arsenic Species in Natural Waters. *Envir. Tech. Letts.* 9: 1361-6.

CETAC Technologies, Inc. Inorganic Arsenic-Selenium Speciation Kit, ANX4605-AS.

Dutre, Veronika, Kestens, Christel, Schaep, Johan & Vandecateele, Carlo. 1998. Study of the Remediation of a Site Contaminated by Arsenic. *Sci. of the Tot. Envir.* 220: 185-94.

Ficklin, Walter H. 1983. Separation of Arsenic (III) and Arsenic (V) in Groundwaters by Ion-Exchange. *Talanta.* 30(5): 371-3.

Garbarino, J.R. & Burkhardt, M.R. 1998. Presentation: Determination of Arsenicals in Natural Water Using Ion Chromatography and ICP-MS. 40th Annual Rocky Mountain Conference on Analytical Chemistry, Denver, CO.

Hering, Janet G., Chen, Pen-Yuan, Wilkie, Jennifer A. & Elimelech, Menachem. 1997. Arsenic Removal From Drinking Water During Coagulation. *J of Envir. Eng.* 123(8): 800-7.

Jain, Amita, Raven, Klaus P., & Loeppert, Richard H. 1999. Arsenite and Arsenate Adsorption on Ferrihydrite: Surface Charge Reduction and Net OH^- Release Stoichiometry. *E. S. & T.* 33: 1179-84.

Johnson, D.L. & Pilson, M.E.Q. 1975. The Oxidation of Arsenite in Seawater. *Envir. Letts.* 8(2): 157-71.

Khoe, G.H., Emett, M.T., Zaw, M. & Prasad, P. *Minor Elements 2000: Processing and Environmental Aspects of As, Sb, Se, Te, and Bi.* Littleton, CO. Society for Mining, Metallurgy, and Exploration.

Londesborough, Susan, Mattusch, Jurgen, & Wennrich, Rainer. 1999. Separation of Organic and Inorganic Arsenic Species by HPLC-ICP-MS. *Fres. J. Anal. Chem.* 363:577-81.

Magnuson, Matthew L, Creed, John T. & Brockhoff, Carol A. 1996. Speciation of Arsenic Compounds by Ion Chromatography with Inductively Coupled Plasma Mass Spectrometry Detection Utilizing Hydride Generation with a Membrane Separator. *J. Anal. Atomic Spec.* 11:893-8.

Manning, Bruce A., Fendorf, Scott E. & Goldberg, Sabine. 1998. Surface Structures and Stability of Arsenic (III) on Geothite, Spectroscopic Evidence for Inner Sphere Complexes. *E. S. & T.* 32:2383-8.

McGeehan, Steven L. 1996. Arsenic Adsorption and Redox Reactions: Relevance to Transport and Remediation. *J. Envir. Sci. Health.* A31(9): 2319-36.

McGeehan, & S.L., Naylor, D.V. 1994. Sorption and Redox Transformation of Arsenite and Arsenate in Two Flooded Soils. *Soil Sci. Soc. Am. J.* 58:337-42

Perkin Elmer, Inc. As(T) by Hydride Generation AAS.

Raven, Klaus P., Jain, Amita & Loeppert, Richard H. 1998. Arsenite and Arsenate Adsorption on Ferrihydrite: Kinetics Equilibrium, and Adsorption Envelopes. *E. S. & T.* 32: 344-9.

Scott, Michael J. & Morgan, James L. 1995. Reactions at Oxide Surfaces: Oxidation of Arsenic (III) by Synthetic Birnessite. *E. S. & T.* 29: 1898-1905.

Tamaki, S. & Frankenberger, W.T. Jr. 1992. Environmental Biochemistry of Arsenic. *Reviews of Envir. Contam. & Tox.* 124: 79-110.

Vink, B.W. 1996. Stability Relations of Antimony and Arsenic Compounds in the Light of Revised and Extended Eh-pH Diagrams. *Chem. Geol.* 130: 21-30.

Wensheng, Zhang, Pritam, Singh, & Muir, David M., 2000. *Minor Elements 2000: Processing and Environmental Aspects of As, Sb, Se, Te, and Bi.* Littleton, CO. Society for Mining, Metallurgy, and Exploration.

Wilkie, Jennifer A. & Hering, Janet G. 1996. Adsorption of Arsenic onto Hydrous Ferric Oxide: Effects of Adsorbate/Adsorbent Ratios and Co-occuring Solutes. *Colloids & Surfaces.* 107: 96-110.

Tailings and Mine Waste '01 © 2001 Balkema, Rotterdam, ISBN 90 5809 182 1

Comparison of kinetic tests for sulfide mine tailings

M. Benzaazoua, B. Bussière & A. M. Dagenais
Unité de Recherche et de Service en Technologie Minérale, Université du Québec en Abitibi-Témiscamingue, Rouyn-Noranda, Qué., Canada

ABSTRACT: Eight different sulfide mine tailings were tested by two types of kinetic prediction tests, the humidity cell (30 to 60 cycles duration) and the column test (11 to 12 cycles duration), for determination of mineral dissolution rates. These tailings had a wide variety of acid generating potentials. The leachates produced by the tailings "Lou", "Lar" and "Cla" were near neutral pH at the start of the test, and became acid after few months, while the five tailings "Jou" stayed at a near neutral pH for the duration of the test. The differences in geochemical behavior are discussed for the two types of tests and for the different samples. Oxidation-neutralization rate of minerals was determined for each of the two types of kinetic tests. The results demonstrate that there is little significant difference between the rates as determined by the water qualities obtained from both the column tests and the humidity cells tests for these cases. This is contrary to results from several previous studies.

1 INTRODUCTION

Environmental considerations have become one of the primary factors in the economic feasibility of any mining project especially those with sulfide rich ores. Acid mine drainage is the main problem caused by the mining of sulfide ore. Under certain conditions (the presence of oxygen and water and the possible presence of bacteria like Thiobacillus), oxidation of sulphide minerals within the tailings (mainly pyrite and pyrrhotite but also ore sulfides lost in the milling process) can occur and generate toxic elements and acid when the neutralization potential of the waste is either absent or exhausted.

The predictive tests used to evaluate the acid generation potential of mine waste are critical for determining the environmental impact of a mine site and for choosing a waste management method to minimize that impact. Static tests, kinetic tests, and geochemical models have been developed and are used on a regular basis, primarily to establish if and when a given material would generate acid. Static tests have been so named because they are usually conducted at a given point in time, and do not account for the rate and evolution of the observed reactions. They generally rely on measuring the sulphur content to predict the Acidic Potential (AP) of the waste and a titration to estimate the Neutralization Potential (NP). When the results of static tests are inconclusive (the test results are in a relatively large uncertainty zone), or when one is interested by the evolution of drainage quality in time, kinetic tests are desirable.

There are different types of kinetic tests to estimate the reaction rates and depletion time for various minerals; the most common tests are column and humidity cell. The main advantage of the humidity cell test is that it returns data rapidly (test durations range from 20 to 52 weeks) on a smaller volume of material than a column test. On the other hand, the column test is considered to be more representative of actual field conditions (Bradham and Caruccio, 1991). Some investigations from previous reports have tried to compare the different kinetic tests. Some of them demonstrated that, depending on the type of test used, the results in term of water quality and reaction rates are different (e.g. Caruccio and Geidel, 1986; Bradham and Caruccio, 1991). To investigate this aspect, eight (8) samples with a wide variety of acid generating potentials were tested by both column and humidity cell tests. The objective was to determine if the kinetic test type influences the results of mineralogical dissolution and water quality. The emphasis was on the dissolution of carbonate minerals and the oxidation of sulfide minerals.

2 EXPERIMENTAL PARAMETERS

In this study, both column and humidity cell tests were used to determine the rate of acid generation and metal release of each individual tailings sample. The tests were conducted in accordance with known established practice in the field (Lawrence & Marchant, 1991; SRK, 1989; Morin & Hutt, 1997).

2.1 *Static tests*

The Sobek ABA test (Sobek *et al.* 1978) consists of an evaluation of the sulphur present in the sample (in the sulfide form) for calculation of the acidity potential (AP) and an acid-base titration for the neutralization potential evaluation (NP). In this study, the modified procedure of the ABA test proposed by Lawrence et Wang, 1997 is used. The net neutralizing potential (NNP) represents the difference between the NP and AP values. One of the most common criteria on the acid generating potential is the following (SRK, 1989):

- if the tailings NNP is higher than +20 kg $CaCO_3$/t, the sample is considered not acid generating,
- if the tailings NNP is lower than -20 kg $CaCO_3$/t, the sample is considered acid generating,
- if the tailings NNP is between -20 and +20 kg $CaCO_3$/t, the sample is in an uncertain zone.

2.2 *Physical characteristics*

Grain size analyses on the tailings samples were done with a Malvern Mastersizer which is a laser based instrument. The relative density (D_r) was determined with an Helium pycnometer (Micromeritics). Specific surface was evaluated with a Micromeritics Gemini Surface area analyzer.

2.3 *Column tests*

The experiments were performed in vertically mounted Plexiglas® columns placed 2 m above the floor. The columns have an inside diameter of 0.106 m and height of 1 m. The bottom of each column is designed to provide water-table control using a ceramic plate with a bubbling pressure of 0.5 bar to prevent air entry. The water-table was fixed 2 m below the bottom of the columns. The columns were filled with 0.6 m of sulfide tailings. The total duration of the experiment was 1 year with 11 to 12 flush-drainage cycles. The flush was done with 2 liters of deionized water. The drained solutions were recovered at the bottom of the column in an oxygen free environment and were analyzed for pH, ORP, conductivity, SO_4^{2-}, Ca, Mg, Mn, Fe, Cu, Zn, Si and Na contents.

2.4 *Humidity cell tests*

Humidity cell tests were performed in a chamber made of Plexiglas® drilled to provide air input and output. The cells have an inside diameter of 0.14 m and 0.35 m height. The cells were filled with 1 kg of fine sulfide tailings sample placed on a punched grid. The total duration of the individual experiments ranged from 30 to 60 cycles depending on the sample. Each cycle was run for seven days. Dry air was passed through the sample container for the first 3 days and humidified air for the next three days. On the seventh day, the sample was rinsed with 50 ml of deionized water. The drained solution were analyzed for pH, ORP, conductivity, $SO4^{--}$, Ca, Mg, Mn, Fe, Cu, Zn, Si and Na contents, the same analysis as the column tests.

2.5 *Tailings chemical analysis and water quality*

The tailings samples and the water collected from tailings samples were analyzed by Inductively Coupled Plasma (ICP-AES) analysis. The water quality was determined with the ICP-AES technique for $SO4^{--}$, Ca, Mg, Mn, Fe, Cu, Zn, Si and Na contents. Pore water pH and Eh measurements were made immediately on fresh sample with a pH/mV meter. The redox potential was measured with a combination platinum electrode and the measurements were corrected by the use of a known calibration standards.

3 TAILINGS CHARACTERISTICS

Four different types of tailings called "Jou", "Lou", "Lar" and "Cla" were placed in columns and humidity cells. Five different samples from the "Jou" mine were studied : BH96-2, BH96-9, BH96-11, BH96-11ox and UQ-8. They were taken at different locations in the "Jou" tailings pond. The sample BH96-11ox represent a prior oxidized material sampled at the surface. The corresponding fresh material is BH96-11. The acid generating tailings (Lou, Lar and Cla) were sampled at the end-circuit of three Canadian mine concentrators and can be considered to be fresh tailings. The chemical analysis of the individual samples are presented in Table 1.

As shown in Table 1, the "Jou" tailings have a sulphur contents ranging from 5 to 10 wt %. They are characterized by significant quantities of Ca, Mg, Mn corresponding to naturally occurring carbonates in the residue (calcite and dolomite mainly). The remaining mineralogy is represented by quartz (the predominant mineral in the tailings samples) and aluminosilicates, as demonstrated by the Si, Al, K, and Na analysis results. The other tailings contain a higher amount of sulphur: 12 % for "Lou" tailings, 15.6 % for "Lar" tailings and 29.1 % for "Cla" tailings. One can also notice from the Ca, Mg, and Mn contents, that the carbonate content is lower when compared to the "Jou" tailings. From all the sites sampled, the sulphur content is primarily in pyrite except for tailings "Cla" where pyrrhotite is also present. Other sulfide minerals are present in negligible amounts depending on the tailings sample (sphalerite, chalcopyrite and nickelite).

Table 1. Chemical analyses of the eight sample studied

	BH96-2	BH96-9	BH96-11	BH96-11ox	UQ-8	Lou	Lar	Cla
Al Wt%	4.82	2.49	4.49	3.24	4.89	4.48	6.64	1.02
Si Wt%	16.2	10.5	13.9	11.4	15.4	17.68	22.03	3.09
K Wt%	0.43	0.19	0.51	0.4	0.5	0.46	1.51	0.16
Na Wt%	2.33	0.83	2.06	1.23	1.92	0.31	1.14	0.21
Ti Wt%	Nd	nd	nd	Nd	nd	0.23	0.40	0.27
Fe Wt%	15.1	23.2	12.8	22.1	15.3	19.92	14.73	40.23
S tot Wt%	4.78	10.6	6.36	6.43	5.99	12.7	15.60	29.1
SO_4^{--} Wt%	0.34	0.48	1.91	6.59	0.5	0.51	1.52	0.4
Ca Wt%	3.89	4.34	4.31	4.4	4.95	1.41	1.27	0.66
Mg Wt%	1.76	1.95	1.58	1.38	2.11	3.51	0.51	0.52
Mn Wt%	0.71	0.93	0.48	1.17	0.47	0.26	0.03	0.05
Cu ppm	20	25	40	20	80	689	690	442
Zn ppm	90	130	110	90	120	1699	13635	118
Pb ppm	40	25	‹25	‹25	‹25	298	705	30
As ppm	220	830	520	340	380	275	62	

nd not determined

The Table 2 summarizes the modified Acid Base Accounting (ABA) static test results. The acidity potential is corrected by deducing the initial sulfate content of each tailings sample. The NP measurements reflect the carbonates content of each tailings sample studied. The NNP balance indicates that one of the "Jou" samples (BH96-9) is potentially acid generating, and the others are close or in the uncertain zone. Tailings "Lou", "Lar" and "Cla", have significant acid drainage potential.

In term of particle size distribution (Figure 1), the eight tailings studied have small size differences, but are typical of Canadian mill tailings (e.g. Bussière, 1993; Aubertin et al., 1996).

Table 2. Static test results of the eight samples studied

	BH96-2	BH96-9	BH96-11	BH96-11ox	UQ-8	Lou	Lar	Cla
PN kg $CaCO_3$/t	120	150	150	115	180	47	13	11
%S	4.78	10.6	6.36	6.43	5.99	12.7	15.6	29.1
%S en SO_4	0.12	0.16	0.64	2.20	0.17	0.51	1.52	0.4
%S en sulfure	4.66	10.44	5.72	4.23	5.82	12.19	14.08	28.7
PA kg $CaCO_3$/t	145.8	326.2	178.8	132.1	182.0	396.9	487.5	909.4
PNN net kg $CaCO_3$/t	-25.8	-176.2	-28.8	-17.1	-2.0	-349.9	-474.5	-898.4

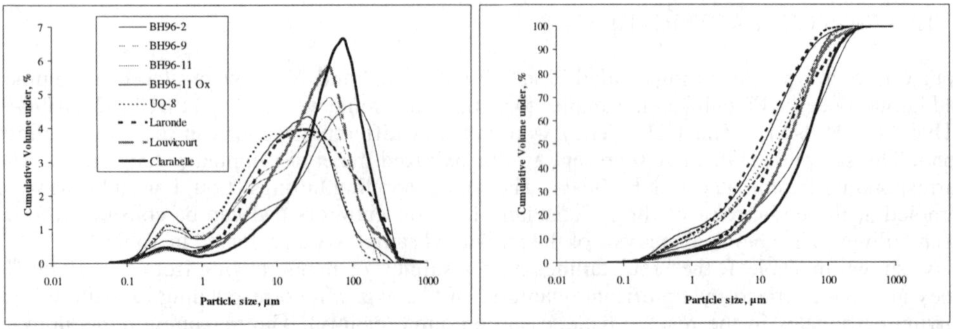

Figure 1 Particle size distributions of the tailings samples studied

Table 3. Physical characteristics of the eight samples studied

	BH96-2	BH96-9	BH96-11	BH96-11ox	UQ-8	Lou	Lar	Cla
D_r	2.98	3.26	3.00	3.17	3.04	3.237	3.196	4.171
S.S.A. m²/g	0.682	0.845	0.820	1.096	0.958	1.4555	1.894	1.487
% under 80 µm	66.6	89.0	89.4	94.4	96.3	87.71	89.38	72.1
D_{10} (µm)	1.9	0.8	1	0.5	0.8	3.32	2.56	5.35
D_{50} (µm)	42.5	16.2	19.2	11.4	9.7	27.57	19.87	47.16
D_{90} (µm)	169.2	80	78.4	54.8	50.6	89.63	126.48	137.13
Mean (µm)	123.2	58.9	61.1	43.1	36.8	42	44.88	62.95

The mean particle sizes range from 40 to 120 µm (BH96-2, has the coarsest tailings). Table 3 summarizes the main physical parameters of the eight tailings studied. As demonstrated by the data of Table 3, the sample densities are indicative of the mineralogy (presence of sulfide minerals) while the specific surface area reflects the particle size distribution and the presence of phyllosilicates.

4 GEOCHEMICAL RESULTS FROM THE KINETIC TESTS

The results of the water quality analysis for both kinetic tests are presented in this section. Because of space constraints, only the evolution of the main elements and parameters involved in the neutralization and oxidation processes are presented (pH, Eh, SO_4^{2-}, Ca, Mg, Mn and Fe). For this study, it was assumed the neutralization is controlled by carbonate dissolution (silicates dissolution is ignored). This assumption is acceptable for the purpose of comparing kinetic tests, but would be unacceptable for the study of the acid generating potential of any given tailings. Figure 2, 3 and 4 represent the measured concentration versus time for all of the major species. The plotted values don't take into account either the volume of the recovered waters or the weight of the solid sample.

4.1 *pH, ORP variations*

The pH remained high (greater than 7) during all of the tests for all of the "Jou" samples while the ORP increased slightly. The type of test (column or humidity cell) doesn't seems to influence either the pH and ORP variations for this material (Figures 2 and 3). The percolated water pH of the other tailings also drops to low values (pH between 3 and 4) by the end of the experiment for both test types. However, this phenomenon appears to be slower for the columns compared to the humidity cell tests, due to the length of the column cycles. The corresponding ORP for the column is usually lower than in the humidity cells which confirms that the environment in the humidity cells is more-oxidative due to more oxygen presence in the humidity cells.

4.2 *Main oxidation and neutralizing products*

Many species from the oxidation-neutralization phenomena are detected in the flushed waters recovered from the column (Figure 2) and humidity cell experiments (Figure 3). The main comments can be summarized as follows:

Calcium, magnesium and manganese : Most of these elements are contained in carbonate minerals which dissolve when pyrite and other sulphides oxidize. Throughout the test, the pH of the "Jou" leachate remained high because of the significant neutralization potential of these carbonates in the waste material (represented by the amount of Ca, Mg and Mn in the flushed waters). However, this neutralization potential seems to decrease in the acidic tailings cases

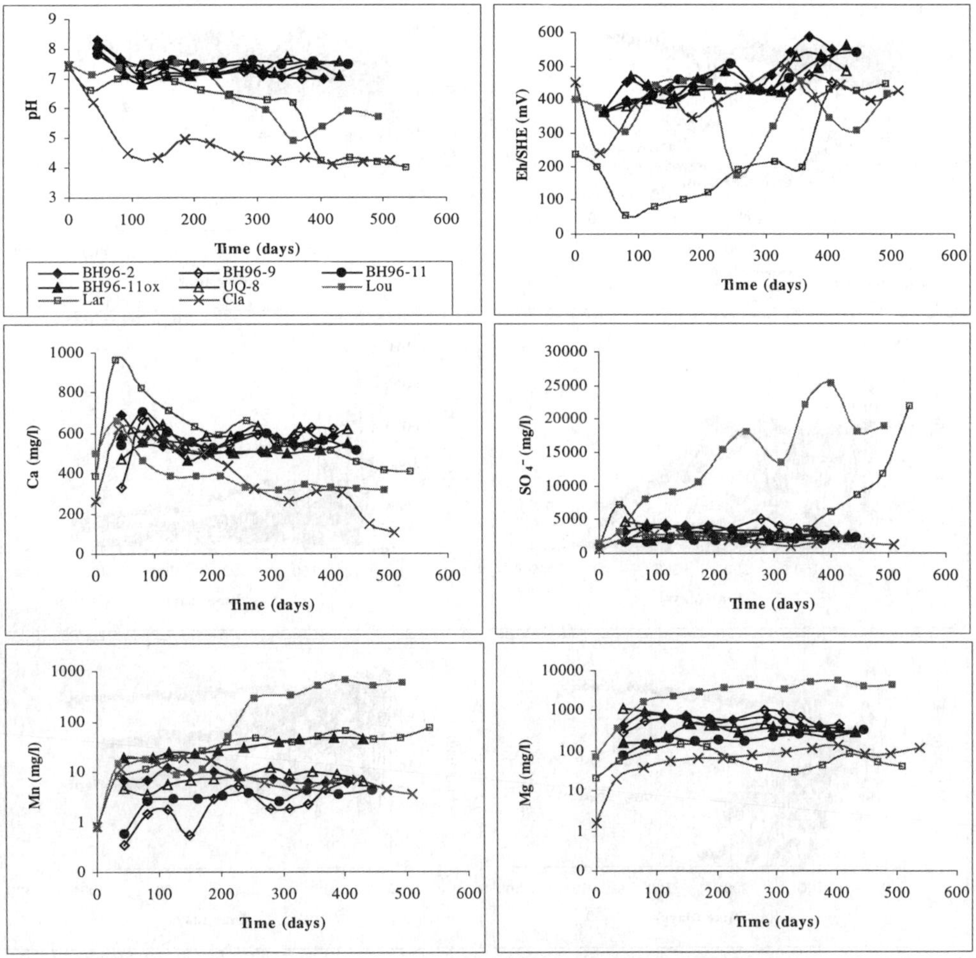

Figure 2 Water quality corresponding to the Column tests done on the eight sample

after depletion of these carbonates. This can be noted by looking at the evolution of Ca, Mg and Mn in the flushed waters. This phenomenon is observed in both kinetic tests.

Sulfates : The sulfates are produced by the oxidation of sulfides. Atmospheric oxygen reacts with the sulfide minerals between the two flush cycles. The reaction starts at the top of the column, and proceeds downward as the oxygen diffuses through the oxidized layers to the fresh material. Near neutrality conditions only involve oxygen as an oxidizing agent. However, when the pH becomes acidic (as is the case with the Lou, Lar and Cla tailings at the end of the experiment), the dissolved ferric iron will contribute to the sulfide oxidation. This may explain some of the differences observed in the production of sulfates for the two types of kinetic test. Also some precipitation in the form of jarosite or gypsum may occur, especially for the "Cla" and "Lar" tailings which are not observed in the humidity cell test and could explain the lower sulfate production. As for the Jou tailings, the production of sulfate is more or less constant through time for both types of test.

Iron : The concentrations of iron in flushed waters from the "Jou" tailings were lower than detection limit even if there is sulphide oxidation as demonstrated by the soluble sulfate presence. Tailings "Lou" released some soluble iron at the end of the column test but negligible amounts during the humidity cell test (Figure 4). The iron detected in the water from the "Lar"

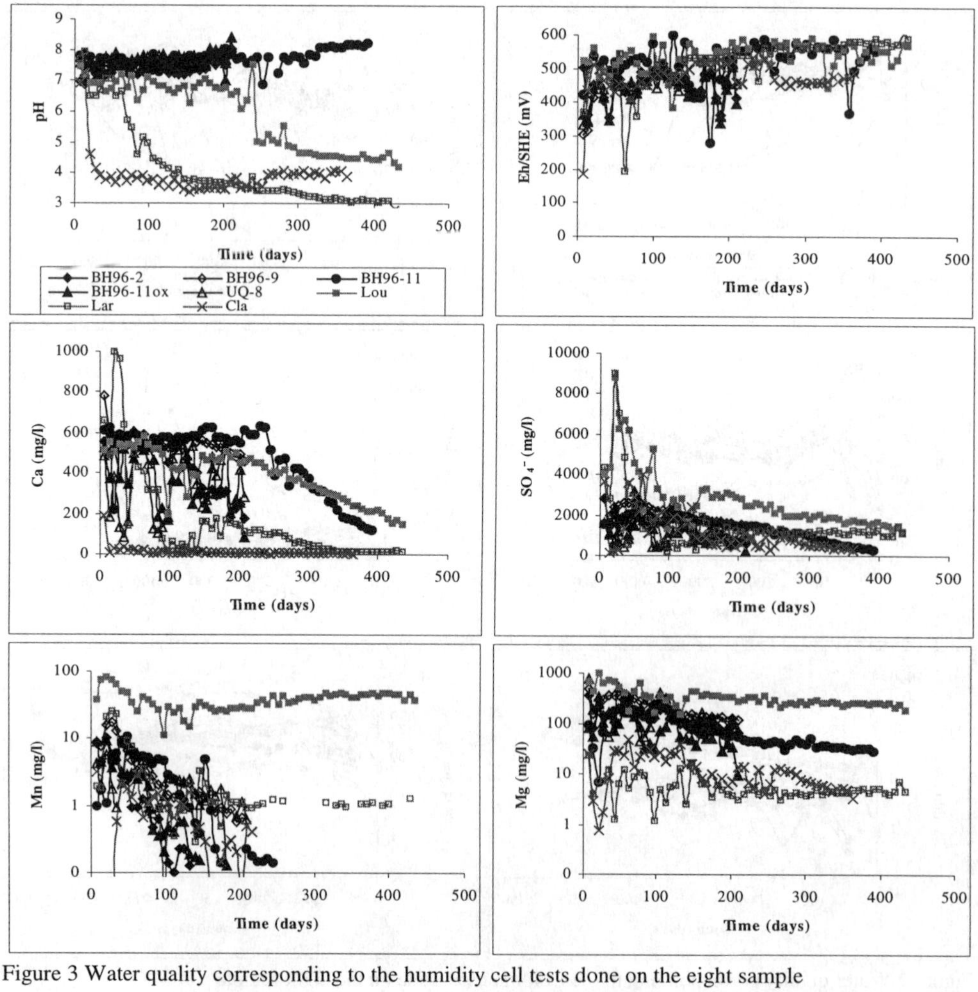

Figure 3 Water quality corresponding to the humidity cell tests done on the eight sample

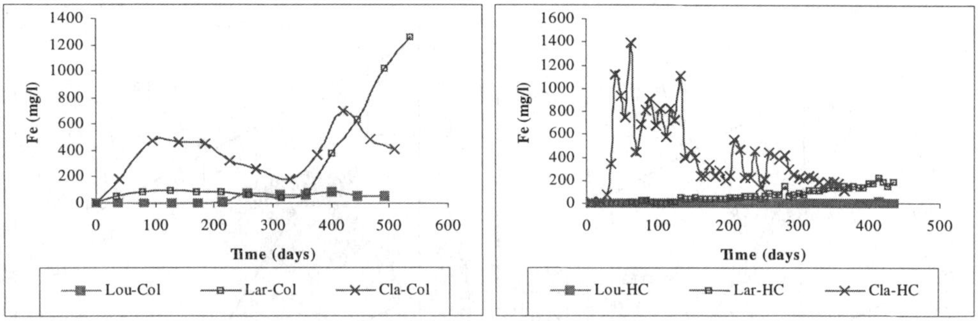

Figure 4 Iron concentration measured in the flushed waters from Lou, Lar and "Cla" tailings
(left : column tests; right : humidity cell tests)

tailings was lower in the humidity cell experiments than in the column tests (Figure 4). The "Cla" tailings released the highest amount of iron in the waters immediately at the start of the kinetic experiments (Figure 4). After 150 days in the humidity cell, the concentrations dropped to values lower than 400 ppm, a phenomenon that was not observed in the column tests.

5 ANALYSIS RESULTS

5.1 *Eh-pH diagram corresponding to the Iron speciation*

The obtained pH and redox potential values were plotted in an iron Eh-pH diagram (Figures 5 and 6). The exercise allows one to understand the pH and Eh variation during the tests and better understand the evolution of iron speciation. All of the pH-Eh projections are localized in the hydroxide stability field for "Jou" tailings (Figure 5). This explains the absence of iron in the flushed waters recovered from both kinetic tests for the "Jou" tailings.

The geochemical behavior of the other three tailings is quite different (Figure 6). The pH decreases from neutral values at the beginning to acidic values (pH bellow 4) at the end of the ex-

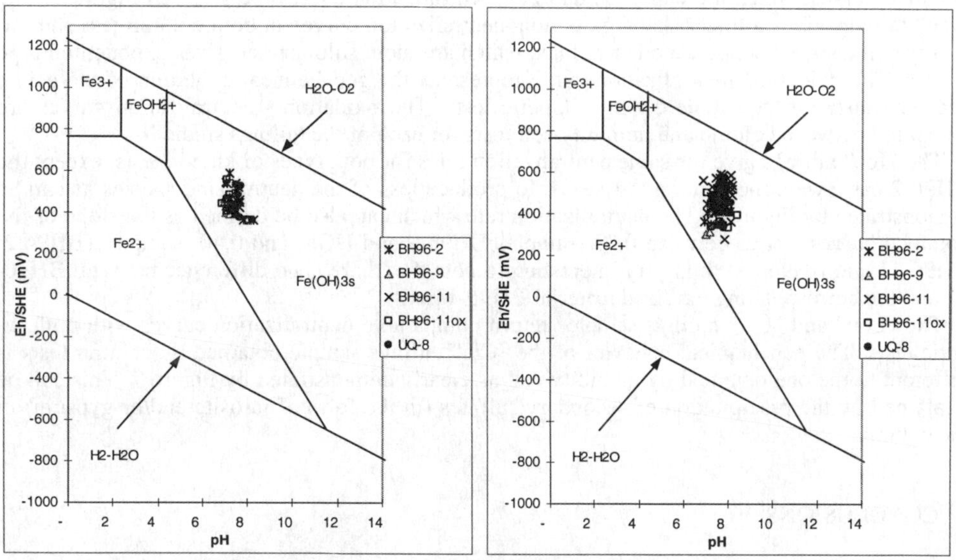

Figure 5 Eh and pH projection into the iron species Eh-pH diagram for Jou tailings
(left: column tests, right : humidity cell tests)

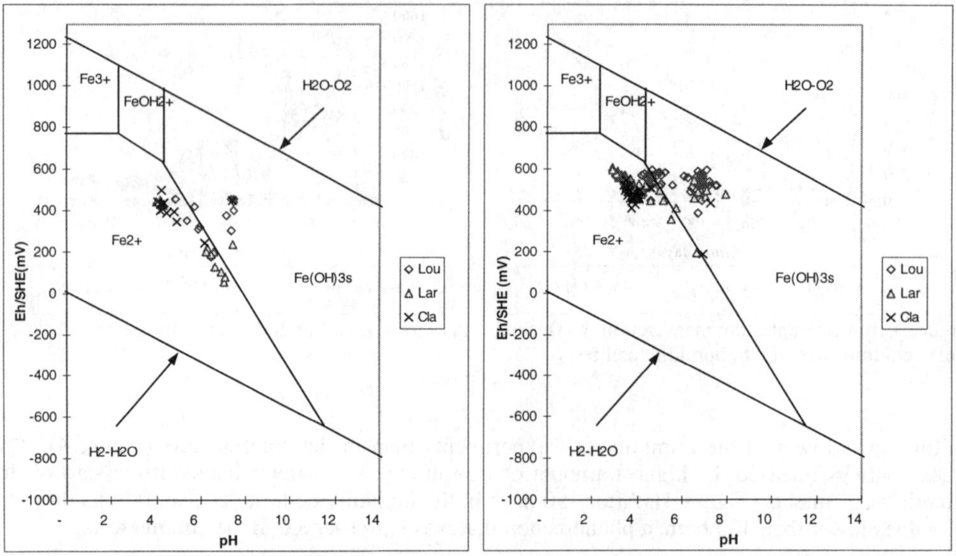

Figure 6 Eh and pH projection into the iron species Eh-pH diagram for Lou Lar and Cla tailings (left: column tests, right : humidity cell tests)

periments. The ORP increase gives evidence of an oxygen reactive medium. All of the points are located near the stability line between Fe^{2+} and $Fe(OH)_{3s}$; this explains why there is a certain amount of iron in the flushed water of these three samples that doesn't amount to the stœchiometry of pyrite oxidation.

5.2 *Oxidation – neutralization rates*

The cumulative mass of sulfates measured in the flushed waters are plotted versus the cumulative amount of calcium, magnesium and manganese. Sulfates represent the main oxidation product while the term Ca+Mg+Mn represents the main carbonate dissolution products. The obtained curves, which are called oxidation – neutralization curves are given in Figure 7. The "Jou" tailings gives a linear shape oxidation–neutralization curves in both column test and humidity cell tests, whereas, the other tailings which are more sulphur rich gives exponential type curves. The oxidation–neutralization curve represents the geochemical evolution of the acidic and neutralization potentials during a kinetic tests. The oxidation - neutralization curves are compared between column and humidity cell tests for each of the tailings studied.

The "Jou" tailings give the same neutralization rates for both types of kinetic tests, except the BH96-2 case, were the column test seems to produce less of the neutralizing species as can be demonstrated by Figure 7. The neutralization rate which can also be defined as the slope of the obtained curves, varies between 0.37 (samples BH96-9 and UQ8) and 0.40 (samples (BH96-2, BH96-11 and BH96-ox). It is very interesting to note that there is no difference between BH96-11 and the corresponding oxidized material BH96-11ox.

The "Lou" and "Lar" tailings samples return comparable neutralization curves with both kinetic tests. The geochemical behavior of the "Cla" tailings sample obtained by column tests is different to the one obtained by humidity cell as clearly demonstrated by Figure 7. This can be explained by the precipitation of secondary sulfates (in the form of jarosite and/or gypsum) in the column.

6 CONCLUSIONS

The main objective of this study was to compare the two types of kinetic tests: the column test and humidity cell test. The main conclusions to be drawn from this study are:

- The relations between oxidation rates and neutralization rates were similar in both test types and for all of the tailings studied, except for the "Cla" tailings;
- The depletion of minerals is faster in humidity cell tests due to the larger number of cycles;
- The concentration in elements are in the same order of magnitude for both tests,
- Both tests can be used as predictive tools for estimating leachate quality and oxidation – neutralization reaction rates.

Bradham and Carrucio (1991) also compared different kinetic tests. They stated in their conclusions, that the type of test influences the reaction rates (oxidation and neutralization). Our findings concur with the fact that the type of test will affect the total production of dissolved species from oxidation and dissolution reactions. This is because humidity cell test is faster than column test leading to a difference in the mineral depletion rate in the samples. The two kinetic tests used in this study differ in the frequency, duration and number of flushing cycles for a given time experience. The sample mass utilized, the oxygen and heat availability are also quit different in the two types of test. This explains why the reaction rates are different from test to test. But when we compare the oxidation products with the neutralization products, the ratio remains similar for both types of test in the conditions used for this study.

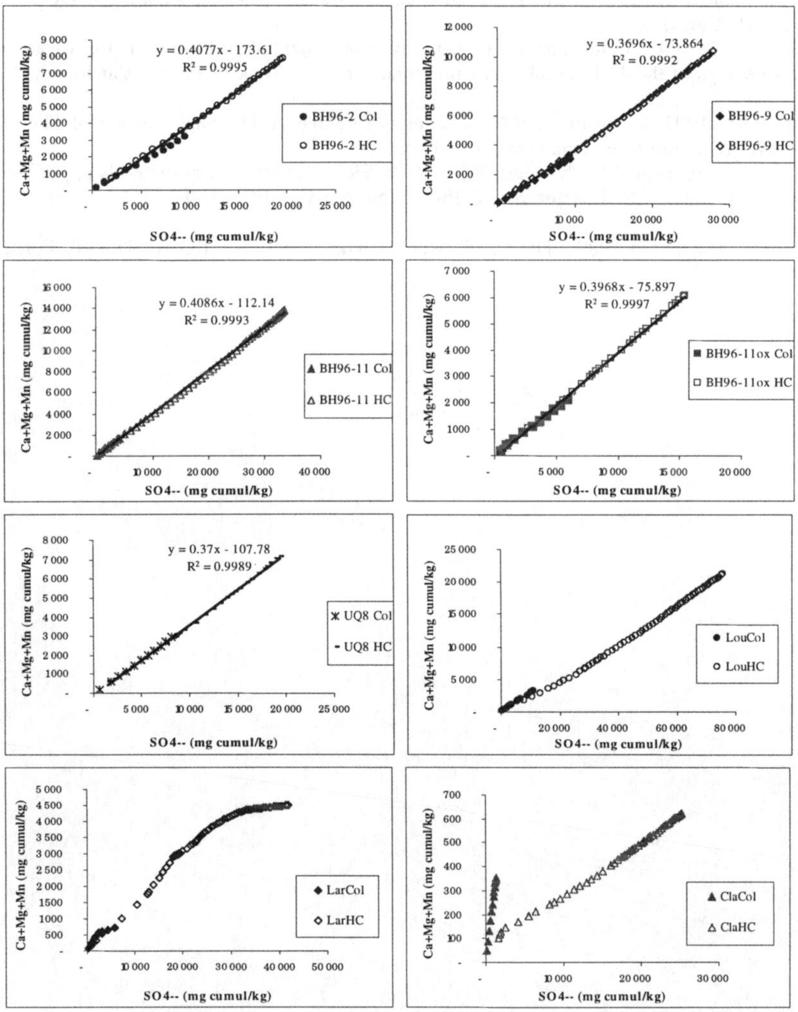

Figure 7 Comparison between the oxidation – neutralization curves obtained for the eight samples with the two type of tests

ACKNOWLEDGEMENTS

The authors would like to acknowledge the "Fonds Institutionnel de Recherche" and the "Unité de Recherche et de Service en Technologie Minérale" for their financial contributions.

REFERENCES

Aubertin, M., Bussière, B. & Chapuis, R.P. (1996). Hydraulic conductivity of homogenized tailings from hard rock mines. Canadian Geotechnical Journal, 33 : 470-482.

Bussière, B. (1993). Évaluation des propriétés hydrogéologiques de résidus miniers utilisés comme barrières de recouvrement. Master thesis, Mineral Engineering Department, École Polytechnique de Montréal, 171 pages.

Bradham, W.S. & Caruccio, F.T. 1991. A comparative study of tailings analysis using acid/base accounting, cells, columns and Soxhlets. 2nd International Conference on the Abatement of Acidic Drainage, Montréal, 1 : 157-173.

Caruccio, F.T. & Geidel, G. 1986. An evaluation of mine waste-overburden analytical techniques. Proceedings of the 1986 National Symposium on Mining, Hydrology, Sedimentology, and Reclamation, Lexington, Kentucky.

Lawrence R.W. & Marchant P.B. 1991. Acid Rock Drainage Prediction Manual, Rapport CANMET, MEND Project 1.16.1b, Canada.

Lawrence, R.W.& Wang, Y. 1997. Determination of neutralization potential in the prediction of acide rock drainage, Proceedings of the 4th International Conference on Acid Rock Drainage, Vancouver, 1 : 451-464.

Morin, K.A. & Hutt, N.M., 1997. Environmental Geochemistry of Minesite Drainage : Practical Theory and Case Studies. MDAG Publishing, Vancouver (Canada).

Sobek, A.A., Schuller, W., Freeman, J.R. & Smith, R.M. 1978. Field and laboratory methods applicable to overburdens and minerals. U.S. Environmental Protection Agency, EPA-600/2-78-054 (PB-280-495).

SRK (Steffen, Robertson & Kirsten) 1989. Draft Acid Rock Technical Guide. BC AMD Task Force, Vol. 1.

Tailings and Mine Waste '01 © 2001 Balkema, Rotterdam, ISBN 90 5809 182 1

GIS-facilitated, hydrogeochemical modeling for mine waste reclamations

Drummond Earley III, Elizabeth A. Parker & Kenny Calhoun
Daniel B. Stephens and Associates, Albuquerque, N.Mex., USA

ABSTRACT: Hard rock mine closure regulations require an assessment of the potential for sulfide-bearing mining wastes to impact groundwater quality through acidic seepage. Data from static and kinetic tests, seepage monitoring, and hydrogeochemical modeling are used for making these assessments. Standard hydrogeochemical models often do not adequately account for complicated mixtures of mining wastes and the size- and site-dependent characteristics of waste rock stockpile facilities. An alternative modeling approach was developed that overcomes these limitations for existing stockpiles and their post-closure regraded configurations processed through custom GIS software (see companion paper by Jordan in this volume). The approach was tested on a conceptual, sulfide-bearing stockpile facility constructed by a U.S. hard rock metal mine located in the arid west. Comparative stockpile modeling results indicate that standard slope regrading and reclamation practices used in this setting do not substantially decrease the potential mass loading of contaminants to the underlying aquifer(s).

1 INTRODUCTION

Open pit metal mining creates large volumes of mining wastes that are now subject to state and federal mine closure regulations. These regulations require pre-closure assessment of these wastes in terms of their potential to affect the environment; degradation of surface and groundwater quality are the primary concerns. Static and kinetic tests, seepage monitoring, and hydrogeochemical modeling are conducted for making these assessments. Based on these findings closure designs must be drafted and, once the facility or the mine itself reaches its design life, implemented.

Alpers & Nordstrom (1999) summarize the general application of geochemical models to the simulation of water-rock interactions in mining environments. They emphasize the need to define clear and testable objectives and hypotheses in the early stages of modeling projects. In this study, a hydrogeochemical mass loading model was developed specifically to simulate infiltration and the accompanying reaction of stockpile materials with descending meteoric water in order to evaluate the relative environmental performance of the preferred closure design alternatives. The primary metric for the evaluation was the annualized mass loading results because they are a more reliable indicator of the potential for groundwater impacts than model-predicted concentrations of constituents in basal seepage (Earley & Parker 1999). Although mass loading was the primary evaluation criterion, estimates of basal seepage quality were also derived from the model's output. The model was designed to conduct predictive (i.e. forward geochemical modeling) using site-specific data commonly collected as part of mine closure studies and general geochemical principles.

To demonstrate the model, selected reclamation alternatives were simulated. Geographic information systems (GIS) computer software was used to render alternative stockpile regrading designs as part of the proposed reclamation plan for a sulfide-bearing stockpile at a hard rock mine site located in the arid west. GIS has been widely used in the analysis of hydrologic and geochemical field data and recently has been interfaced with surface and groundwater flow models and hydrogeochemical modeling (Wells & Pearson 1998). GIS analysis can also be integrated with hydrogeochemical models of mine stockpiles.

2 METHODS

A broad continuum of potential slope configuration designs exist for reclamation of stockpile facilities. In this study three major stockpile outslope configurations were modeled for a mine stockpile located in the western U.S.:

- Stockpiles with existing slope configurations that are now at an approximate angle of repose (base case alternative)
- Conceptual regrading of the base case by pushing down the slopes to a maximum gradient of 4:1 between benches (push-down alternative)
- Conceptual regrading of the base case by pulling back the slopes to a maximum gradient of 4:1 between benches (pull-back alternative)

For simplification, it was assumed that a stable cover and plant community could not be installed on the angle of repose slope. All top surfaces and 4:1 outslopes were assumed to be covered with a 60-cm coarse cover material (non-acid forming) with a thin topsoil dressing and vegetation. This cover design allows vegetation to transpire and remove meteoric water that is stored in the coarse cover material. Currently such dry-barrier or storative cover concepts are being considered and used for reclamation of stockpiles and tailings situated in arid to semiarid localities (ICARD 2000).

2.1 *GIS recontouring*

The conceptual reconfiguration was conducted using GIS software (ArcView) to create a numerical representation of the topography that would result from regrading the stockpile outslopes. The general rendering process is described briefly in this section; the companion paper by Jordan in this volume discusses the process in more detail.

For the push-down and pull-back alternatives, conceptual reconfiguration of the stockpiles involved modifying the base case stockpile configurations to reflect a net 4:1 gradient on the outslopes. Under the push-down alternative, material from the upper portion of the stockpile is redistributed to the lower portion by balancing the cut and fill volumes. Under the pull-back alternative, the footprint of the stockpile remains fixed, and outslope materials are redistributed to the stockpile's top surface. If there is insufficient area on the stockpile to contain the excess materials associated with the pull-back reconfiguration, the material is allocated to a hypothetical stockpile configuration with 4:1 slopes. Excess material from the pull-back reconfiguration is referred to herein as the pull-back cone because its exact location and configuration is not specified, but for the purposes of this study, an idealized right cone geometry will be assigned to the volume. This allows the hypothetical stockpile's infiltration and mass loading to be modeled.

2.2 *Hydrogeochemical model*

2.2.1 *Theoretical basis*

The hydrogeochemical model was developed for a Lagrangian reference frame on the basis of the following equation for advective transport in a column filled with a porous matrix:

$$q_m = q \bullet C \tag{1}$$

[equation 1-25 in Zheng & Bennett 1995]

where q_m = mass of solute transported advectively per unit time across a unit area normal to the flow, q = volumetric fluid discharge per unit area, and C = constituent concentration.

Derivations of this equation are used to calculate uncorrected dissolved mass loading and equivalent seepage compositions for finite stockpile areas of constant depth.

The volumetric stockpile infiltration (calculated by multiplying the volumetric fluid discharge per unit area [q in Equation 1] by the footprint area of the stockpile times the concentration of constituents in the basal seepage [C in Equation 1]) is equal to the total mass loading of a dissolved constituent (M) transported (advectively) per unit time across the stockpile footprint area (mass loading is related to q_m in Equation 1 through multiplication by the stockpile's footprint area). Therefore, the mass loading parameter is a facility-specific extensive measure of the combined environmental impact from meteoric water flow and geochemical reactions occurring in the stockpile. The mass loading results for the reference design will be used to compare the relative performances of the base case, push-down, and pull-back alternatives. The relative differences are emphasized rather than the absolute values because many of the modeling uncertainties cancel when modeling results are normalized to the same reference condition.

The method for estimating seepage quality and mass loading from stockpiles requires three modeling operations that are performed in the following sequence:

- Run variably saturated flow model for prediction of infiltration rates. These rates are manually fed into a spreadsheet model.
- Run spreadsheet application which calculates length-normalized fluxes and equivalent concentrations that represent first order estimated basal seepage compositions.
- Run geochemical model for prediction of mineral precipitation and maximum (constituent) solubility. The output is fed manually into an updated spreadsheet model to calculate corrected seepage concentrations and annual mass loading for attenuation of dissolved constituents.

The spreadsheet application mathematically couples the flux calculations with estimates of infiltration and attenuation obtained from the flow and geochemical models, respectively. The sequential spreadsheet operations were conducted using discretized equations derived from Equation 1.

2.2.2 *Model assumptions*

The most important model assumptions and approximations are listed below:

- Standard temperature and pressure (25°C, 1 bar) occur throughout the stockpile.
- The stockpile's bulk composition is randomly uniform throughout.
- Rainfall reacts instantaneously with the top surface to produce an entry seepage water composition that is similar to runoff from the stockpile.
- Steady-state constituent dissolution occurs from the end-member stockpile materials.
- One-dimensional, steady-state matrix flow but three-dimensional mixing of pore fluids occurs throughout the stockpile.
- The stockpiles are completely aerated by the atmosphere through diffusion and convection.
- Distinct pore waters within the stockpile are perfectly mixed by the time they reach the location where the basal seepage exits the stockpile.
- Mineral precipitation and dissolution occurs along the flow path, but no adsorption or ion exchange is considered in the model.

Some of these assumptions and the mass loading model's conceptual basis are illustrated in Figure 1.

2.2.3 *Infiltration rates*

Infiltration rate estimates for the reclamation alternatives considered in this paper were obtained from UNSAT-H modeling of the interactions that occur in the plant, soil, water, and atmosphere systems at the stockpile surface. These systems were assumed to be mature and well established after closure.

2.2.3.1 UNSAT-H model

UNSAT-H is one of the numerical models that has been widely applied toward modeling variably saturated flow and infiltration at mine sites (ICARD 2000). UNSAT-H is a one-dimensional finite-difference code developed by Battelle Pacific Northwest Laboratory (Fayer & Jones 1990) that simulates infiltration, drainage, redistribution, surface evaporation, and plant water uptake from soils. Infiltration rates are generated for a representative element volume (REV) of the stockpile surface (Fig. 2). UNSAT-H uses soil hydrological, meteorological, and vegetation data to model fluxes of moisture and energy.

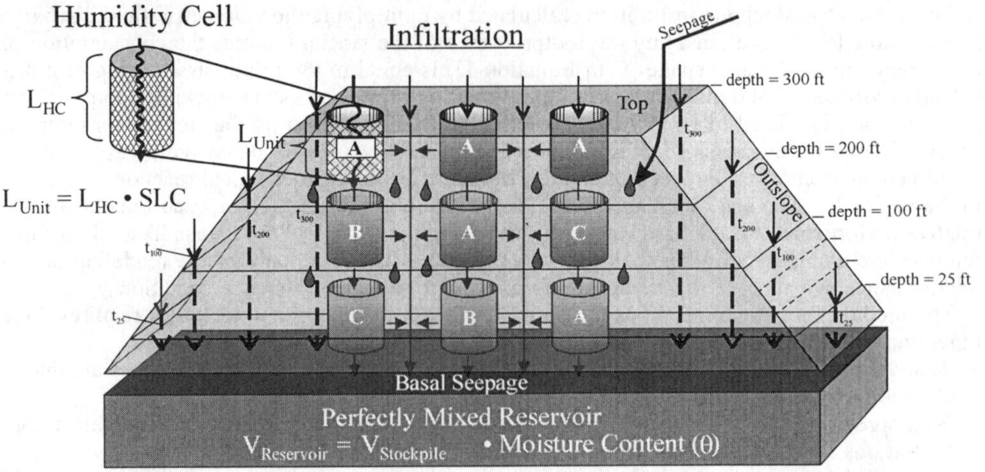

Figure 1. Conceptual stockpile mass loading model; L=length, HC=humidity cell, SLC=scaling law coefficient, A=material type A, B=material type B, C=material type C, V=volume, t_{25}=travel time for 25-foot depth intervals

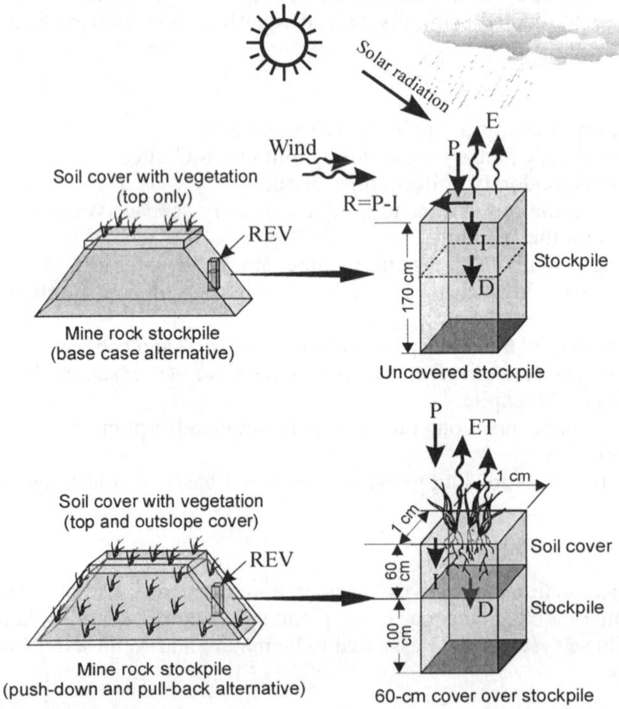

Figure 2. Conceptual diagram of UNSAT-H Model parameters; I=infiltration, D=drainage, P=precipitation, E=evaporation, ET=evapotranspiration, R=runoff, REV=representative element water

Figure 2 shows a conceptual diagram of the model parameters. The upper boundary conditions of the UNSAT-H simulations were calculated from measured daily weather data. The model grid was constructed to be 1.60 meters in depth for the covered case and 1.70 meters in depth for the uncovered case simulations. The lower boundary was specified as a unit gradient, or free-draining, boundary. By assuming a unit gradient, any moisture that propagates to the base of the cover will drain freely out of the system (model domain). The covered stockpile model domain contains 32 nodal points and the uncovered scenario domain contains 37 nodal points.

2.2.3.2 Mine site hydrology data

Precipitation, temperature, wind speed, and other site-specific climatic data are required to conduct the soil hydrologic modeling. Long-term temperature and precipitation records (spanning more than 100 years) were available for a nearby weather station. In general, the regional climate is arid but temperate, with temperatures ranging from –4°C in January to 31°C in July. Long-term data on wind speed and direction, relative humidity, cloud cover, and solar radiation were not available for the local station; these data were therefore derived from the closest climate stations where the data were available. Solar radiation was calculated from published relationships based on latitude (Barbour et al. 1987).

A representative 10-year (1977-1986) data record was selected for the present UNSAT-H infiltration modeling. Figure 3 presents a bar chart of the annual precipitation and shows total annual winter and summer precipitation for the local station from 1977 through 1986. Climatically, the selected simulation period is representative of the historical records for this region, and normal precipitation for the mine area averages approximately 39 centimeters.

One of the more important vegetation parameters, the LAI distribution, describes the ratio of leaf surface area to the soil surface area. To be conservative in terms of potential plant productivity, the annual LAI distribution selected for use in the UNSAT-H modeling assumes a distribution that ranges from nearly 0 in the winter to 0.5 during the peak growing season.

UNSAT-H requires three coefficients to describe the root density function, $R = ae^{-(depth * b)} + c$. For these simulations, parameters were determined assuming that the root length and density decrease nearly linearly with depth. About 40 percent of the root mass is present within the upper quarter of the cover, 30 percent in the next quarter, 20 percent in the third quarter, and 10 percent in the last quarter (Ayers & Westcot 1989). Therefore, the coefficients were put into the model as a = 0.25, b = 0.03, and c = 0.001. The specified maximum rooting depth is equal to the thickness of the cover being modeled. In the simulations, roots were not allowed to penetrate the materials beneath the cover.

The soil hydrologic data that were used in the UNSAT-H modeling are shown in Table 1. The hydraulic conductivity used to model the stockpile materials was 1×10^{-2} cm/s. This value was calculated from the results of field and laboratory tests conducted at the mine site and on samples collected from the mine, respectively.

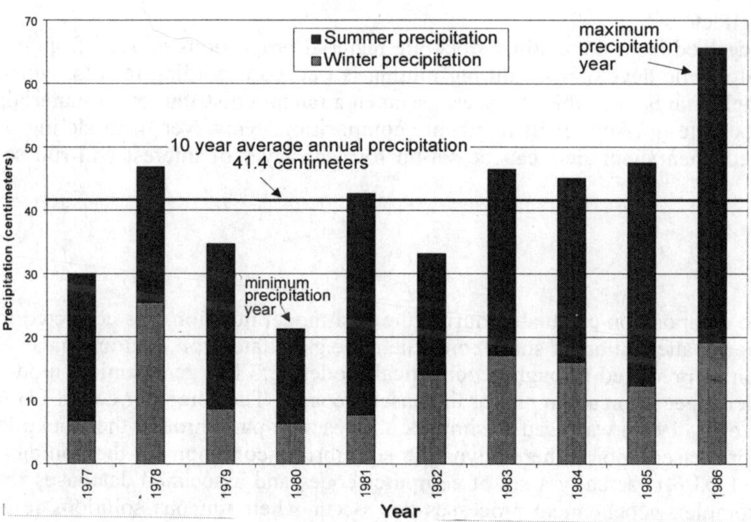

Figure 3. Precipitation record, 1977 through 1986

Table 1. Hydraulic parameters for UNSAT-H model simulations

	Sample Description	
Parameter	High K_{sat}, moderate-storage cover	Covered stockpile material
Sample number	1D–1	NA
K_{sat} (cm/s)	6.2×10^{-3}	1.0×10^{-2}
Available water capacity (volume %)	0.09	NA
USDA soil texture*	Sand	NA
Rock fraction	0.67	NA
Volumetric water content:		
Saturated moisture content θ_s (fraction)	0.38	0.14
Residual moisture content θ_r (fraction)	0.03	0.01
α (cm^{-1})	0.0638	0.0425
N (dimensionless)	1.3356	1.2134

* Explanations of soil types available from the *Soil Survey Manual* (SSDS 1993)
NA = Not applicable α = 1/Air-entry value N = Fitting parameter

The properties for the cover material were selected to be representative of a high-conductivity, moderate-storage material that is similar to a geologic formation that is a potential local source of borrow material. The hydraulic properties of this formation were determined in the laboratory from a sample collected from the mine area.

2.2.4 *Mass loading*

2.2.4.1 Kinetic testing data

The material dissolution rates and length-normalized fluxes were extracted from site-specific humidity cell test (HCT) results, which measure advective transport in a matrix-filled column, and weighted according to the proportion of material types in the stockpiles (Table 2).

2.2.4.2 Rate scaling factors

The rates were also scaled to correct for the effective contact area. The conceptual basis for these operations is shown in Figure 1. The scaling factor values were selected based on a combination of experimental observations and were applied to the fluxes derived from the HCT results. After path length and rinse water volume normalization, the short-term laboratory fluxes were multiplied by the fractional scaling factor to account for the slower specific field dissolution rates of mining wastes (Morin & Hutt 1997).

2.2.4.3 Rate weighting factors

Weighting factors are derived from the distinct stockpile material proportions derived from existing mine records. Stockpile development during mining is tracked according to lifts, which represent time-stratigraphic units. For this study we assumed a random distribution of materials and represented the stockpile in terms of its aggregate composition. However, if modeling of distinct layers is desired, then submodels can be set-up for each layer of interest and run sequentially.

2.2.5 *Reaction paths*

2.2.5.1 EQ3/6 model

The equivalent seepage composition calculated during the first model iteration was corrected in the second iteration for the attenuation of some constituents (e.g. sulfate, iron, and manganese) by mineral precipitation as predicted through geochemical modeling. The geochemical model also calculates constituent speciation and a pH for the basal seepage. The computer code EQ3/6 (Wolery et al. 1990; Wolery 1992) was used to simulate the reaction path through the stockpile to make these corrections based upon a thermodynamic equilibrium condition for the homogeneous aqueous phase. EQ3/6 is actually a set of computer codes and associated databases for use in modeling the complex geochemical processes that occur when aqueous solutions (e.g. rainfall, groundwaters, saline waters, effluent streams) react with soil, rock, or solid waste mate-

Table 2. Static and HCT testing results for stockpile material types (MA) A, B, and C

Parameters	MA-A	MA-B	MA-C
Static testing (Sobek et al. 1978) results ($tons\ CaCO_3$ per 1000 tons):			
ABA	4.51	-46.6	-146
AGP	0.63	46.6	188
ANP	5.14	<0.5	41.8
HCT length-normalized flux ($mg*cm^{-2}*month^{-1}*cm^{-1}$):			
TDS	4.31	91.42	46.11
SO_4	1.1029	59.5057	29.7015
Fe	0.1073	0.6318	0.3768

ABA = Acid-base account = ANP – AGP
AGP = Acid-generation potential
ANP = Acid-neutralization potential
HCT = Humidity cell test
TDS = Total dissolved solids

rials (NCDC 1999). The software was specifically used to simulate the effects of chemical equilibrium, batch fluid mixing, and empirical leaching kinetics. The software handled the relatively high-ionic-strength seepage compositions though the use of the extended Debye-Hückel model (Wolery 1992) and the supporting databases.

2.2.5.2 Reaction rates

The contact lengths of water particles moving through the stockpile materials were estimated from stockpile area thickness probabilities derived from GIS analysis of stockpile surface and under-dump topographies (Fig. 4). The probabilities were based upon 25-foot depth intervals (Fig. 1). The travel times (t_{25}, t_{50}, t_{100}, etc.) were calculated from the integral stockpile area thickness probabilities and the modeled infiltration rates assuming steady-state flow conditions. The infiltration (drainage) estimates (for covered and uncovered stockpile areas) used in the mass loading calculations were derived from the years 1980, 1986, and the average for the 10-year climate period (i.e. 1977 through 1986). The two specific years represent the upper (wet year of 1986) and lower (dry year of 1980) bounds of the 10-year climatic period analyzed in the UNSAT-H simulations. Thus, mass loading estimates are provided for the average, wet, and dry years of the 10-year climatic period analyzed.

3 RESULTS

3.1 *Infiltration*

UNSAT-H treats any precipitation that cannot infiltrate into the surface material as runoff that is lost to the system. Therefore, the total precipitation falling on the surface equals the infiltration plus runoff. The UNSAT-H model predicted no runoff for either scenario; therefore the precipitation equals infiltration (i.e. all precipitation infiltrated into the surface).

The UNSAT-H model results show predicted infiltration (drainage) rates ranging between <0.1 (dry year) and 2.7 centimeters per year (cm/yr) (wet year) for the covered stockpile scenario, and between 0.4 and 13.6 cm/yr for the uncovered stockpile dry year and wet year scenarios, respectively. The average estimated infiltration rates for the 10-year period were approximately 0.82 cm/yr for the covered stockpile and 4.9 cm/yr for the uncovered stockpile. These average infiltration rates equate to approximately 2 percent and 12 percent of the average annual precipitation measured at the weather station.

3.2 *Mass loading*

The mass loading estimates for the stockpile are presented in Table 3. These results show that the estimated mass loading at the base of the stockpile ranges between 313 metric tons per year (push-down alternative) and 735 metric tons per year (base case alternative) for the average in-

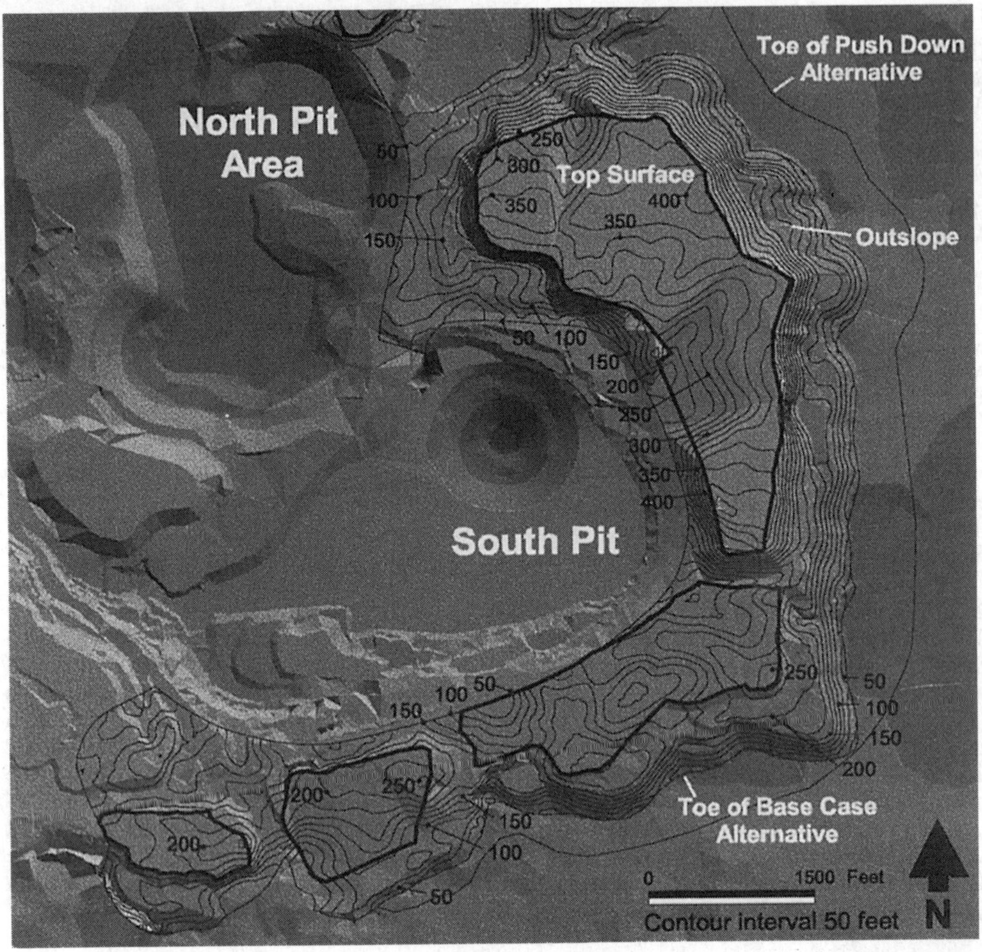

Figure 4. Hillshade TIN image of the mine's topography and depth contours for the base case stockpile configuration. The push-down alternative footprint area is also shown. The TIN image shows that the stockpile is constructed adjacent to an open mine pit (South Pit).

filtration scenario. For the base case alternative, the estimated total mass loading beneath the stockpile for the wet year and dry year are 1833 metric tons per year and 186 metric tons per year, respectively. For the wet year scenario, the push-down alternative mass loading estimate is 717 metric tons per year and for the pull-back alternative (pull-back plus the pull-back cone) the mass loading estimate is 730 metric tons per year. The mass loading estimates for the dry year are 168 metric tons per year for the push-down alternative and 177 metric tons per year for the pull-back alternative (pull-back plus the pull-back cone).

3.3 *Seepage concentrations*

The average equivalent sulfate concentrations in basal seepage for the stockpiles alternatives are also shown in Table 3. These results indicate that the greatest equivalent sulfate concentrations are found beneath the top surface areas for each alternative, with the highest concentrations occurring in the dry year for the pull-back alternative. Additionally, the wet year scenarios for each alternative show equivalent sulfate concentrations that are significantly lower than the equivalent sulfate concentrations for the dry year scenario because of the dilution effect. The estimated maximum concentrations beneath the top surface area in the base case alternative

Table 3. Comparison of stockpile mass loading modeling results, based on total dissolved solids

Facility	Scenario for average precipitation	Planimetric area Top (%)	Planimetric area Slope (%)	Mass loading Top (%)	Mass loading Slope (%)	Mass loading Total (metric tons/yr)	Highest equivalent SO_4^{-2} concentration Top (mg/L)	Highest equivalent SO_4^{-2} concentration Slope (mg/L)
Dry Year	Base case	34.5	65.5	35.4	64.6	186	37,941	9,113
	Push-down	18.1	81.9	22.3	77.7	168	32,866	25,318
	Pull-back	0.0	100.0	0.0	100.0	127	NA	26,951
	Pull-back cone	NA	100.0	NA	100.0	50	NA	30,209
Average Year	Base case	34.5	65.5	14.3	85.7	735	7,497	4,140
	Push-down	18.1	81.9	20.0	80.0	313	6,930	6,173
	Pull-back	0.0	100.0	0.0	100.0	236	NA	6,337
	Pull-back cone	NA	100.0	NA	100.0	90	NA	6,649
Wet Year	Base case	34.5	65.5	11.6	88.4	1833	4,730	3,870
	Push-down	18.1	81.9	18.7	81.3	717	4,578	4,440
	Pull-back	0.0	100.0	0.0	100.0	535	NA	4,470
	Pull-back cone	NA	100.0	NA	100.0	195	NA	4,504

range between 4,730 milligrams per liter (mg/L) [wet year] and 37,941 mg/L (dry year). The estimated maximum concentrations in basal seepage for the pull-back alternative range between 4,470 mg/L (wet year) and 30,209 mg/L (dry year). For the push-down alternative, the estimated average concentrations beneath the top surface area range between 4,578 mg/L (wet year) and 32,866 mg/L (dry year). The thicker material sequences associated with the pull-back alternative produce the highest average equivalent sulfate concentrations (Table 3).

4 DISCUSSION AND CONCLUSIONS

Reductions in infiltration achieved through cover installation do not necessarily translate into direct reductions in mass loading. This is because constituent dissolution is mostly rate-limited as opposed to solubility-limited, and the reduction in the infiltration rate is offset by an increase in the concentration of the constituent in basal seepage such that the mass loading stays constant (see Equation 1). Any gains in environmental performance can be attributed mostly to sulfate precipitation that occurs along the reaction path when travel times are increased as a result of decreased infiltration.

Given the enormous costs associated with regrading, the modeling results for the push-down and pull-back alternatives do not show a significant improvement in stockpile mass loadings as compared to the base case alternatives, especially for the dry and average year results (Table 3). The estimated total mass loading between the individual stockpile alternatives are not substantially different partly because the decrease in infiltration that is achieved through cover installation is offset by the increase in infiltration created by the greater net surface areas of the stockpile(s) after the pull-back and push-down regrading.

REFERENCES

Alpers, C.N. & Nordstrom, D.K. 1999. Modeling of water-rock interactions in mining environments. In Plumlee, G.S. & Logsdon, M.J.. *The Environmental Geochemistry of Mineral Deposits.* Reviews in Economic Geology 6A:289-324.

Ayers, R.S. & Westcot, D.W.. 1989. *Water quality for agriculture.* FAO Irrigation and Drainage Paper No. 29, Rev. 1. Rome.

Barbour, M.G., Burk, J.H., & Pitts, W.D. 1987. *Terrestrial plant ecology,* second edition. Menlo Park, California: The Benjamin/Cummings Publishing Company, Inc.

Earley, D. III & Parker, E.A. 1999. A hybrid probabilistic-deterministic modeling approach for the assessment of seepage quality and quantity from sulfide-bearing mining wastes. *GSA Abstracts with Programs* 31:A-70.

Fayer, M.J & Jones, T.L. 1990. *UNSAT-H Version 2.0: Unsaturated soil water and heat flow model.* PNL-6779, Prepared for the U.S. Department of Energy, Pacific Northwest Laboratory, Richland, Washington. 281p.

ICARD, 2000. Understanding waste rock facilities in dry and wet climates. Short course notes from the 5th International Conference on Acid Rock Drainage, Denver, Colorado, May 2000.

Morin, K.A. & Hutt, N.M. 1997. *Environmental geochemistry of minesite drainage.* Vancouver, British Columbia, Canada: MDAG Publishing.

National Climatic Data Center (NCDC). 1999. Western U.S. Climate historical data summaries. Western Regional Climate Center. <http://www.wrcc.sage.dri.edu> Accessed January 1999.

Sobek, A.A., W.A. Schuller, J.R. Freeman, & R.M. Smith. 1978. Field and laboratory methods applicable to overburdens and minesoils. Report EPA-600/2-78-054, U.S. National Technical Information Report PB-280 495. 403 p.

Soil Survey Division Staff (SSDS). 1993. *Soil survey manual.* Handbook no. 18, 2nd Edition, USDA-Soil Conservation Service. Washington, D.C.: U.S. Government Printing Office.

Wells, M.J.M. & Pearson, R.S. 1998. Environmental applications of geographic information systems. 215th ACS National Meeting, Dallas Texas, March 29-April 2, 1998. American Chemical Society Division of Environmental Chemistry, *Preprints of Extended Abstracts* 38(1).

Wolery, T.J., Jackson, K.J., Bourcier, W.L., Bruton, C.J., Viani, B.E., Knauss, K.G., & Delany, J.M. 1990. Current status of the EQ3/6 software package for geochemical modeling. Chapter 8 In Mechior, D.C. & Bassett, R.L. *Chemical Modeling of Aqueous Systems II.* ACS symposium Series 416, p. 104-116.

Wolery, T.J. 1992. *EQ3/6, A software package for geochemical modeling of aqueous systems: Package overview and installation guide (Version 7.0).* Lawrence Livermore National Laboratory, UCRL-MA-110662.

Zheng, C. & Bennett, G.D. 1995. *Applied Contaminant Transport Modeling.* New York: Van Nostrand Reinhold.

Tailings and Mine Waste '01 © 2001 Balkema, Rotterdam, ISBN 90 5809 182 1

Stochastic modeling of stormwater and receiving stream concentrations

T.J.Cox, R.W.Chappell & R.L.Olsen
Camp Dresser and McKee Incorporated, Denver, Colo., USA

ABSTRACT: A stochastic approach was developed and applied to the Butte, Montana hillside abandoned mining site for modeling stormwater runoff and subsequent receiving stream loadings. This approach enabled capture and quantification of the uncertainty associated with stormwater quality data and allowed the prediction of copper and zinc concentrations caused by runoff from the Butte hillside during storm events. Runoff flows were generated in a spreadsheet model using the rational method and stormwater concentrations were input as probability distribution functions (PDFs). Correlations between sampling sites were also incorporated into the model. The PDFs were combined with runoff hydrographs and stochastically modeled using Monte Carlo simulation. Stream loadings predicted by the model PDFs were combined with ambient stream flow and quality in a mass balance to generate expected stream concentrations in the form of cumulative distributions functions (CDFs). The final stream concentration CDFs were used to evaluate the probabilities of exceeding instream standards at various locations during a specific storm event.

1 INTRODUCTION

The Silver Bow Creek / Butte Area NPL site encompasses the majority of the historic Butte Mining district where metals mining has been conducted on a large scale for over a century. The site consists of former mining, milling, smelting, and related facilities and associated waste rock dumps, tailings impoundments, mill and smelter wastes, and contaminated soils within and surrounding the populated areas of Butte and Walkerville, Montana (CDM 1990). Surface water in Silver Bow Creek (SBC), the receiving stream at the site, is impaired as a result of impacts from mining-related waste materials and from urban discharge (DEQ 1998). Elevated concentrations of metals leached and eroded from mining-impacted soils and waste materials, as well as channel alterations and industrial and municipal point source discharges, have impaired water quality within the creek such that populations of fish and other aquatic species are very low to non-existent.

A preliminary remedial action objective for SBC is to return the creek to its beneficial uses, which includes providing protection of aquatic communities from direct contact with and/or ingestion of site-related contaminants. SBC is greatly impacted by stormwater runoff from the Butte Hillside adjacent to the upstream end of the creek. Therefore, to meet remedial goals, episodic stormwater runoff events will need to be controlled so that acute in-stream water quality exceedances within SBC are prevented to the greatest extent practicable.

This paper describes a modeling approach implemented to predict acute instream copper (Cu) and zinc (Zn) concentrations resulting from stormwater runoff under specified storm conditions and to evaluate the effectiveness of existing and planned Best Management Practices (BMPs). Copper and Zn were selected for modeling because aquatic organisms are sensitive to elevated concentrations of these two metals; however, the modeling approach is applicable for any contaminant in stormwater. The approach involved using stochastic methods to incorporate the uncertainty associated with measured stormwater quality data. Specific objectives of this investigation (CDM 2000) were to (1) model instream concentrations in SBC for three 24-hour design storms (2, 5, and 10 year), (2) compare predicted concentrations with and without existing BMP controls, (3) evaluate and prioritize target areas for future BMPs, and (4) identify and evaluate significant modeling data gaps to guide subsequent sampling plans.

2 SITE BACKGROUND

The Silver Bow Creek / Butte Area NPL site encompasses approximately 85 square miles (mi^2). The area targeted in this study (Figure 1), which covers an area of approximately 5 mi^2, is a subregion of this larger site and encompasses the town of Walkerville, the part of Butte just north of the initial reach of SBC.

SBC is a small mountain stream with low to moderate discharge during normal flow conditions (18 - 23 cubic feet per second [cfs]) relative to the potential volume of stormwater runoff

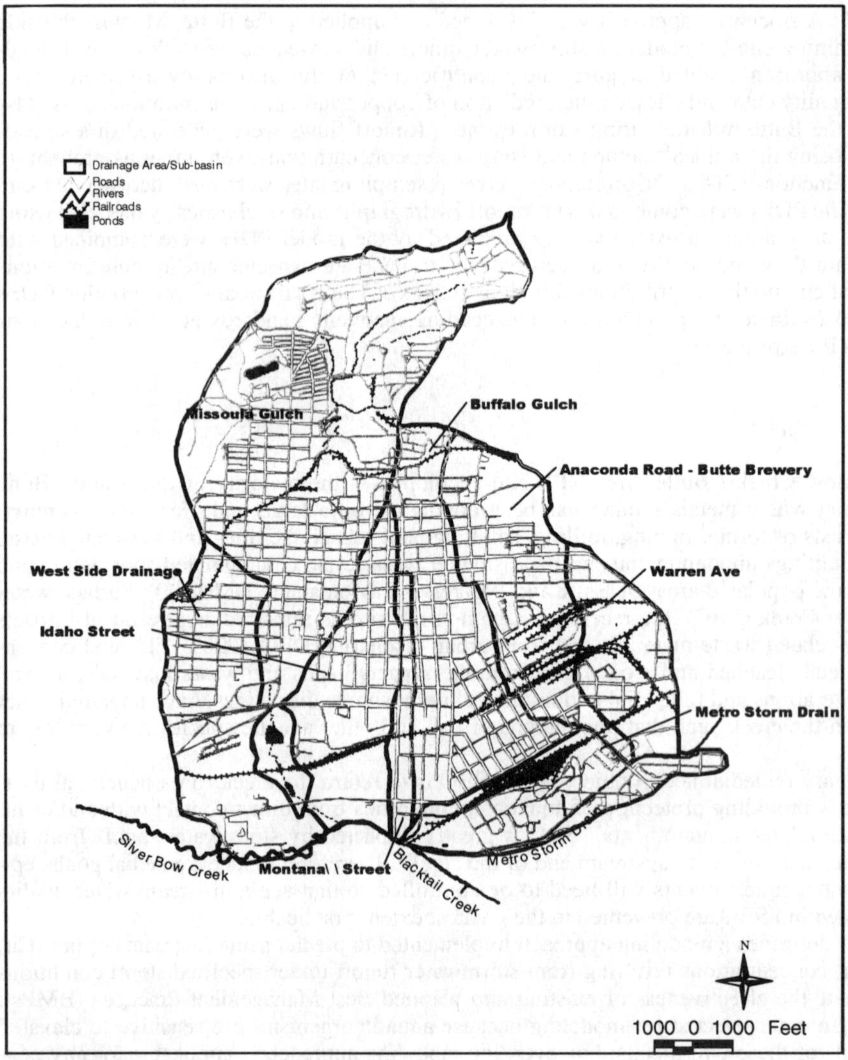

Figure 1. Butte hillside location map.

(e.g., 477 cfs for the 10-year, 24-hour peak flow [ESA 1998]). The effective drainage area of SBC is approximately 103 mi^2. SBC originates at the confluence of the Butte Metro Storm Drain and Blacktail Creek. The Metro Storm Drain is an open channel that was constructed in the early 1930s. The upper portion of this drain is dry except during storm runoff or snowmelt events while the lower portion receives flow via ground water discharge and, during normal flow conditions, contributes between 0.3 and 0.5 cfs to SBC. The primary source of flow in SBC is inflow from Blacktail Creek, which originates in the Highland Mountains and has a drainage basin area of approximately 95 mi^2. Blacktail Creek normally contributes 11 to 15 cfs to SBC. The Metro Storm Drain and SBC receive flow from several sub-drainage basins on the Butte Hillside during stormwater runoff and snowmelt, including Warren Avenue (Warren), Anaconda Road/Butte Brewery (Anaconda), Buffalo Gulch (Buffalo), Missoula Gulch (Missoula), Montana Street (Montana), Idaho Street (Idaho), and West Side (West Side).

In addition to the perennial stream flow and stormwater runoff, SBC receives regulated discharge from the Butte Metro Waste Water Treatment Plant (WWTP) of between 5 and 9 cfs. Additionally, Lower Missoula Gulch intercepts shallow groundwater and maintains a baseflow discharge to SBC of 0.1 to 0.3 cfs (ESA 1999). BMPs have been implemented in the past five years in the area and include a combination of engineered controls (catch basins, channels, culverts, etc.) and reclamation practices (grading, soil covers, erosion control fabric, etc.).

3 STORMWATER RUNOFF MODELING

Stormwater runoff modeling was performed to predict runoff from the Butte Hillside sub-basins discharging to SBC under varying storm conditions. The model was based on the HEC-STORM model algorithm. The model uses the rational method to predict runoff at hourly timesteps from a given watershed for a given storm event, $Q = CIA$, where Q = runoff flow (cfs), C = runoff coefficient (an empirical coefficient that captures the ratio of expected runoff to precipitation and is dependent on watershed characteristics), I = rainfall intensity (in/hour), and A = drainage area (acres). The model also tracks available depression storage (ponding volume from small depressions throughout the drainage area) and subtracts out a corresponding abstraction (as the depression storage fills up) at each timestep. Use of this model to predict runoff hydrographs is generally valid only for small urban watersheds where the time of concentration (the time it takes runoff from the uppermost portion of the watershed to reach the discharge point) is small.

Runoff coefficients (C) and depression storage (inches) for each sub-basin were calculated from land-use characteristics and assumed percent imperviousness values for the various land-use categories. Sub-basins were delineated for both pre-BMP (prior to the start of BMP implementation about five years ago) and post-BMP (existing basin) scenarios to be modeled.

Twenty-four hour design storms at 2, 5, and 10 year recurrence intervals were selected for modeling. The storm hydrographs were calculated using a Type 2 distribution of precipitation totals taken from the Precipitation Intensity Frequency Atlas for Montana (NOAA 1988), as shown in Figure 2. These storms were input to the model at hourly timesteps and show the greatest intensity in the first hour and decrease in intensity in subsequent hours.

The major BMPs were incorporated in the model (for the post-BMP scenarios) through diversions in the runoff flow, reductions in drainage areas, land-use alterations, and explicit modeling of the detention pond system constructed in the Missoula Gulch sub-basin. The model simulates runoff inflow captured by the ponds, and overflow and controlled outflow from each pond. The overflows and controlled outflows add to the uncaptured basin runoff and discharge to SBC.

4 STATISTICAL ANALYSES OF STORMWATER QUALITY DATA

Statistical analyses of Cu and Zn concentrations (dissolved and total) were conducted to determine the input variables for both pre and post-BMP models. Statistical results indicated that Cu and Zn concentrations for the model inputs were lognormally distributed. Therefore, natural log transformed data were used to generate geometric means and geometric standard deviations for use in the stochastic modeling. For sub-basins with insufficient data for statistical analysis, sto-

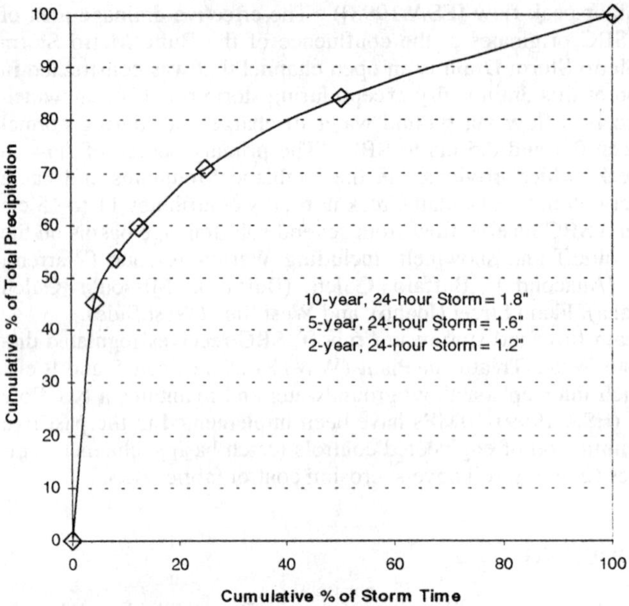

Figure 2. Type 2 design storm.

chastic input parameters were estimated from other sub-basins with similar land-use characteristics.

Statistical analyses were also conducted to identify data correlations between the various sub-basins. This analysis was limited to stormwater data pairs, i.e., samples collected on the same day from two or more sub-basins. The resulting sets of correlation coefficients were averaged to obtain a single correlation coefficient for use in the model. The use of correlations in this manner is based on the assumption that concentrations among the sub-basins for a particular storm event will be related. For example, if concentrations at a particular sub-basin are relatively high, concentrations at all other sub-basins will also be relatively high.

5 STOCHASTIC PREDICTIONS OF INSTREAM CONCENTRATIONS

Due to the uncertainty associated with measured stormwater concentrations, with respect to both the large standard deviations and the timing of sampling and storm events, a stochastic approach was utilized to simulate runoff loadings and resulting acute SBC concentrations. The @Risk addin program to Microsoft Excel was used in conjunction with the runoff model and the water quality statistical analyses. @Risk is a stochastic modeling tool that incorporates and quantifies the uncertainty of specified input parameters by using Monte Carlo simulations to run a given model for a large number of iterations while randomly sampling input probability distributions for each stochastic parameter at each iteration. The resulting output variables are presented in the form of CDFs of expected values.

The SBC model was set up as a stream mixing spreadsheet where 24-hour stormwater hydrographs are combined with assumed upstream and groundwater flow and concentration conditions to calculate expected downstream concentrations. The model assumes instant and complete mixing in the stream. For each simulated storm event (2, 5, and 10 year 24-hour storms) runoff hydrographs were determined for each sub-basin externally using the runoff model. These hydrographs were then used as the flow inputs to the mixing model. The PDFs for stormwater and groundwater concentrations at each sub-basin were input as stochastic variables to the instream model. Correlation coefficients between each of the sub-basin PDFs were also input.

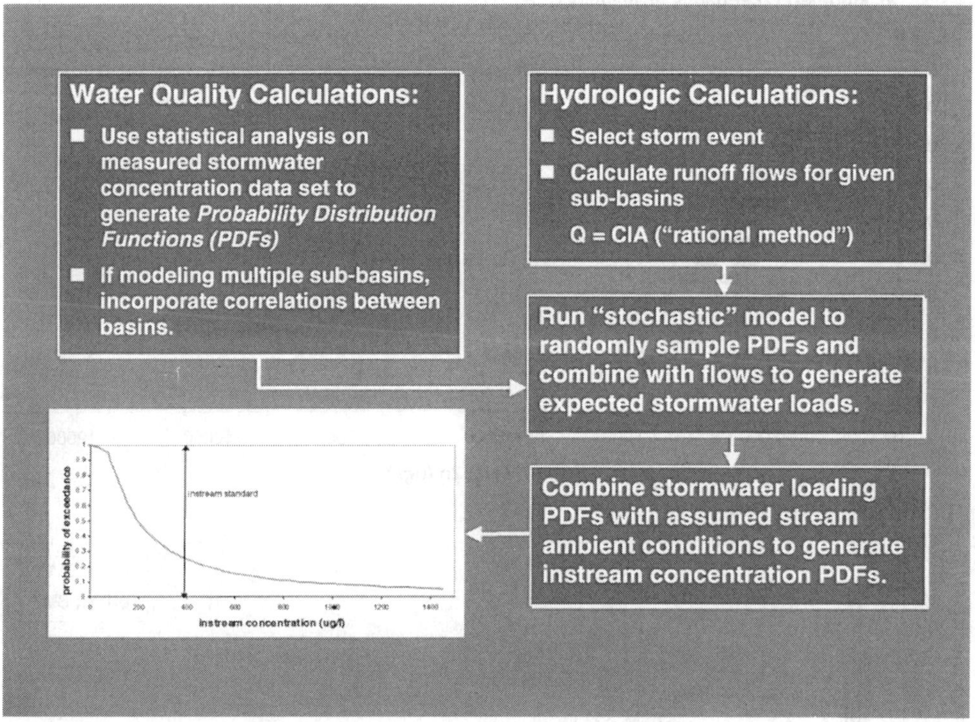

Figure 3. Stochastic modeling overview.

During a given simulation, the model, at each timestep, randomly sampled the sub-basin PDFs (incorporating the appropriate correlations) to generate stormwater/groundwater concentrations. These concentrations were combined with the runoff flows and mixed with ambient stream conditions to generate instream concentrations throughout the reach of interest. This process was repeated thousands of times within a given simulation. The model output was in the form of CDFs that predict 24-hour average instream concentration exceedance probabilities at selected points along SBC. Figure 3 summarizes the modeling process.

Separate simulations were performed for each of the contaminants of concern, for each of the three storm events (2, 5, and 10 year), and for each of the two site characteristic scenarios (pre and post BMP conditions).

6 RESULTS AND DISCUSSION

6.1 *Predicted Instream CDFs*

Figure 4 shows an example CDF generated by the stochastic model, in this case for instream total Zn concentrations at the furthest point downstream in the SBC reach of interest. Included on Figure 4 are CDFs for pre and post BMP model results under each of the three simulated storm events. As an example of how to interpret the CDFs, an exceedance probability of 0.4 represents a 40% chance of exceeding the corresponding concentration as a 24-hour average for the given storm event. Histogram distributions of measured total Zn concentrations are provided for comparison with the modeled CDFs.

The results shown in Figure 4 indicate that significant improvements in predicted total Zn concentrations have likely occurred following BMP implementation, e.g., about a 35% reduction in the 20% exceedance concentration. The results also indicate that differences in total Zn concentrations due to varying magnitudes of storm events are very small. Furthermore, while

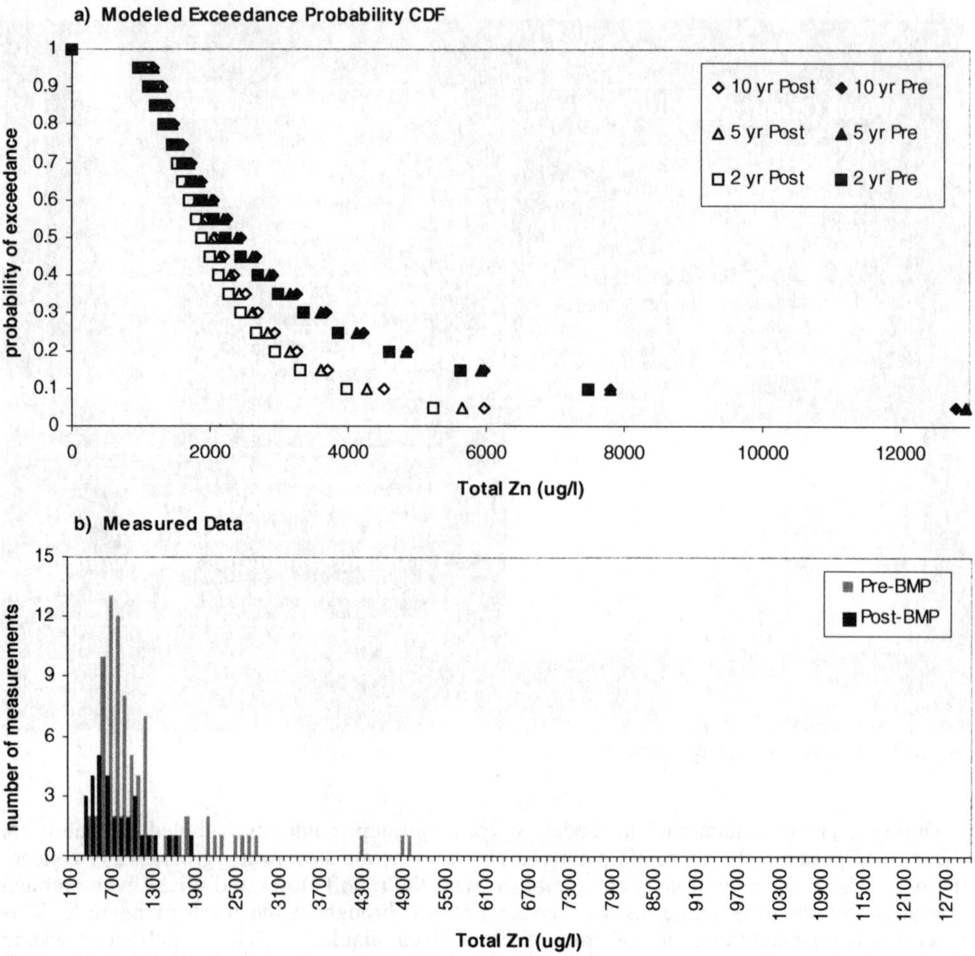

Figure 4. Modeled and measured results for Silver Bow Creek.

measured total Zn concentrations fall within the CDF ranges predicted by the model, they tend to cluster near the lower concentration end, indicating the influence of sample collection during ambient rather than storm event periods. Similar results were obtained for dissolved Zn and total and dissolved Cu.

6.2 *Model Sensitivity Analyses*

To further evaluate model uncertainty, three key non-stochastic model parameters were investigated in terms of model sensitivity: groundwater inflow, Blacktail Creek flow (upstream steady flow), and the correlation coefficient matrix. The uncertainties associated with these model inputs were not captured in the stochastic approach. For each of these parameters, the model input values were varied over a reasonable range and the resulting instream CDFs were compared to the original modeled results. Groundwater inflows, which were included as steady flows in the instream mixing model at both the Missoula Gulch outfall and as a portion of the Metro Storm Drain contribution, and the Blacktail Creek upstream flows were varied ± 100% during the sensitivity analysis. The correlation coefficient matrix was varied from 0.0 to 1.0. Total copper at the most downstream location on SBC was predicted for the 10-year storm, post-BMP scenario.

The results of the sensitivity analyses indicate very little model sensitivity to groundwater flow (Figure 5) and a slightly higher sensitivity to the correlation coefficient matrix (Figure 6), with exceedance concentration differences reaching as high as +19% but generally below ±15%. The model was most sensitive to Blacktail Creek flow variations (Figure 7), with modeled concentration differences of approximately ± 20% throughout most of the distribution. The sensi-

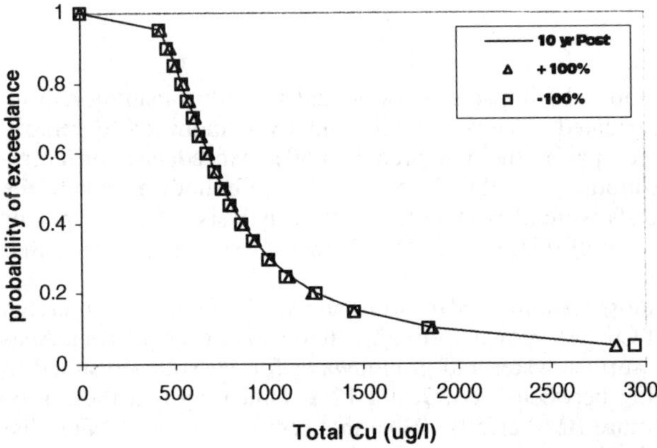

Figure 5. Groundwater flow sensitivity for 10-year, 24-hour storm.

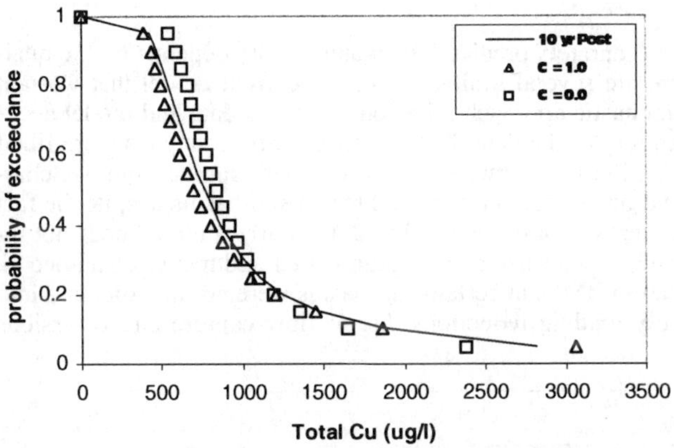

Figure 6. Correlation Coefficient sensitivity for 10-year, 24-hour storm.

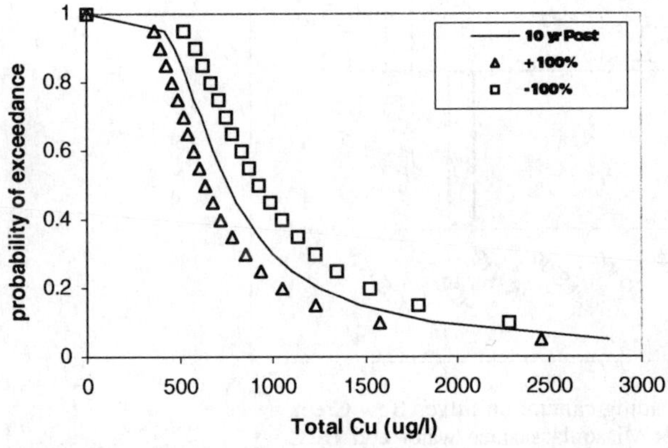

Figure 7. Blacktail Creek flow sensitivity for 10-year, 24-hour storm.

tivity to Blacktail Creek is expected as the Blacktail Creek headwater flow is a primary dilution factor in the model. The flow value used in the stochastic modeling represents a conservative assumption of baseflow conditions in Blacktail Creek. Increased upstream flow in Blacktail Creek due to storm conditions was not incorporated into the scenarios modeled here but could easily be altered for future uses of the model.

6.3 *Future BMP Implementation*

To help guide future BMP implementation, the impacts of individual sub-basin loading removals on instream concentrations were evaluated. The post-BMP model was modified by removing loads one sub-basin at a time and comparing the new predicted 50% exceedance concentration with the original model concentration for the 2-year storm. Groundwater and the wastewater treatment plant effluent loads were also included in this analysis. These types of simulations only allow for the comparison of relative effects of *isolated* loading removal rather than of any combined loading removal.

Concentrations at the downstream-most station on SBC were analyzed. Figure 8 represents an example of the results for dissolved Cu. As shown, individual removals of the Warren, Anaconda, and Missoula sub-basins (both surface water and groundwater for Missoula) resulted in percent dissolved Cu reductions ranging between about 7 and 12%, indicating that these sub-basins would be good areas to focus future BMP efforts. Note that removal of the WWTP discharge results in a negative percent reduction (or increase) because it represents a dilution component for dissolved Cu.

6.4 *Model Limitations*

The ability of the stochastic model to accurately predict SBC water quality depends on the quality of the data input variables. There are several limitations of the current model that warrant further characterization. First, the stochastic approach relies on measured data and model accuracy is limited by the quality and quantity of the data. For this case study, data gaps were filled by assuming similar concentration distributions between sub-basins with similar land-use characteristics. Second, a constant correlation coefficient was used for all sub-basins despite the fact that the value most likely varies between sub-basin pairs. Third, the current model does not include a component for storm-induced re-suspension of contaminated sediments contained in SBC. Finally, the impacts of reclamation BMPs in certain sub-basins were not incorporated into the current model. In particular, only loading reductions due to flow capture and diversions

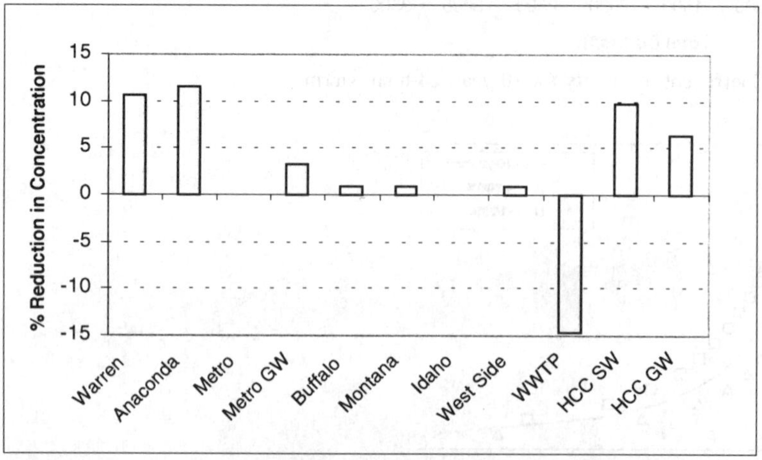

Figure 8. Effects of isolated sub-basin loading capture on Silver Bow Creek dissolved Cu concentration. HCC SW is Missoula surface water and HCC GW is Missoula groundwater.

were incorporated in the current model, whereas BMPs that may have reduced storm water concentrations were not. With adequate post-BMP concentration data, reclamation BMPs could easily be included.

7 SUMMARY AND CONCLUSIONS

The stochastic approach to modeling stormwater and receiving stream concentrations at the Butte hillside proved useful for characterizing the uncertainty associated with stormwater quality data. This approach enabled prediction of SBC water quality resulting from various storm events, evaluation of critical data needs, and characterization of the impacts of BMPs implemented at the site. Most importantly, the model provides a tool for guiding subsequent data collection, so that future BMP activities can be focused to provide maximum benefit. This approach is considered extendable to similar mining waste sites where stormwater runoff is impacting stream water quality.

REFERENCES

CDM 1990. *Phase 2 Priority Soils Operable Unit Runoff Modeling, Butte Area NPL Site, Butte, Montana*, Prepared for the U.S. Environmental Protection Agency by CDM Federal Programs Corporation. September 21, 1990.

CDM 2000. *Draft Technical Memorandum, Regulatory Considerations for Storm Water Management at the Silver Bow Creek/Butte Area NPL Site, Butte Priority Soils Operable Unit, Butte, Montana.* Prepared for the U.S. Environmental Protection Agency by CDM Federal Programs Corporation. March 20, 2000.

DEQ 1998. *Montana Water Quality Bureau Circular WQB-7; Montana Numeric Water Quality Standards.* Prepared by the Montana Department of Environmental Quality Planning, Prevention, and Assistance Division – Standards and Economic Analysis Section. November, 1998.

ESA 1998. *Butte Priority Soils RI/FS and Lower Area One Expedited Response Action, Phase II Monitoring Work Plan and Sampling and Analysis Plan, Silver Bow Creek/Butte Area NPL Site.* Prepared for ARCO Environmental Remediation, LLC by ESA Consultants, Inc. July, 1998.

ESA 1999. *Final Butte Priority Soils RI/FS and Lower Area One Expedited Response Action, Phase II Monitoring Quarterly Report: October 1 through December 31, 1998.* Prepared for the BPSOU PRP Group by ESA Consultants, Inc. February 26, 1999.

NOAA 1988. *Climatological Data Annual Summary, Montana.* Volume 91, Number 13. National Oceanic and Atmospheric Administration.

Remediation and reclamation

Tailings and Mine Waste '01 © 2001 Balkema, Rotterdam, ISBN 90 5809 182 1

Evaluation of long-term seepage impacts from a uranium tailings facility

T.G. Michel
Hydro-Engineering, L.L.C., Casper, Wyo., USA
T.W. Hardgrove
Pathfinder Mines Corporation, Casper, Wyo., USA

ABSTRACT: The development of a ground water Corrective Action Program (CAP) is examined and the performance of the CAP is evaluated. Restorative progress in a local Surficial aquifer system is quantified as a component in predicting future seepage impacts. The remaining quantity of the seepage source in the tailings is changing as a result of dewatering and ongoing seepage and infiltration, and future dewatering is considered in predictions. A numerical ground-water flow and transport model is used to predict long-term (several decades) impacts of the ground water discharge to a perennial stream which intercepts the seepage impacted ground water. Various operational scenarios for the CAP and dewatering are considered in predicting future seepage impacts. Continued operation of the CAP and dewatering does affect the future seepage impacts, but each incremental extension of the operation of the CAP and dewatering has less benefit in reducing the eventual seepage impacts on the surface water.

1 INTRODUCTION

The Pathfinder Mines Corporation (PMC) Shirley Basin Mine site is located in southeast Wyoming approximately 35 miles south of Casper. Uranium milling began at this site in 1971 and continued through 1992 when the last ore was processed. The mill utilized a conventional acid leaching process, and the tailings were hydraulically placed in two constructed tailings impoundments. The tailings impoundments were constructed across the Mine Creek drainage, which is tributary to Spring Creek (see Fig. 1). Seepage from the tailings has reached shallow sands which are hereafter described as the Surficial aquifer. The Surficial aquifer is hydraulically connected to Mine Creek and Spring Creek, and consequently, the seepage impacted water discharges to the remaining section of Mine Creek and to Spring Creek northeast of the tailings.

Seepage from the tailings has resulted in elevated concentrations of hazardous and non-hazardous constituents in the Surficial aquifer, and migration of those constituents in the general Mine Creek area was identified prior to 1984, when a Corrective Action Program (CAP) was undertaken. The CAP has evolved over time and currently consists of collection wells, injection wells, and recharge lines to both contain and retrieve seepage impacted ground water. Dewatering of the tailings was initiated in 1993, and currently consists of approximately 46 operating dewatering wells.

2 GROUND-WATER RESTORATION

The primary seepage impacts are occurring in the Mine Creek area largely because of historic deposition patterns that have resulted in greater permeability in the vicinity of the Mine Creek channel. Restoration efforts have centered on the Mine Creek area. The seepage-impacted ground water eventually discharges to Spring Creek as does unaffected Surficial aquifer water.

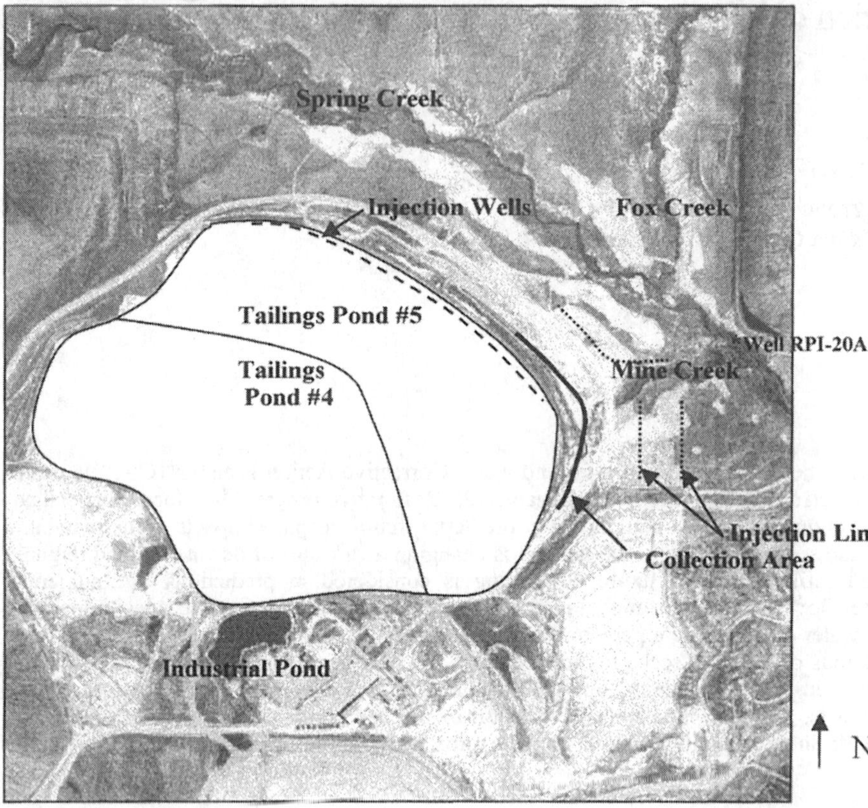

Figure 1. PMC Shirley Basin tailings site. Approximate Scale: 1" = 1600'

2.1 *Corrective Active Program*

The corrective action program consists of complimentary collection and injection to retrieve seepage-impacted water and to affect containment by reversing the gradient back towards the tailings area. The first collection wells were made operational in late 1984, and the first recharge line for the injection of fresh water was installed in 1986. The current system consists of 36 collection wells, three separate recharge lines with individual distribution systems, and 22 fresh water injection wells. In addition to the Surficial aquifer collection/injection system, there are currently 46 operating tailings dewatering wells. The corrective active system has been operated in essentially its current configuration for approximately four years.

2.2 *Seepage Impacts*

Prior to initiation of the corrective action program, a significant plume of contamination was identified in the Mine Creek area extending both north and south of the channel. Figure 2 presents the TDS and uranium concentration for Surficial aquifer well RPI-20A, which is located near the confluence of Spring Creek and Mine Creek and was one of the more severely impacted wells early in the restoration process. The combination of collection and injection has restored concentrations of these constituents to near background levels. This level of restoration is typical of downgradient wells in the Mine Creek area. The corrective action system has successfully restored the seepage impacted area outside of the tailings with the exception of the capture zone in the immediate area of the collection wells and small local areas.

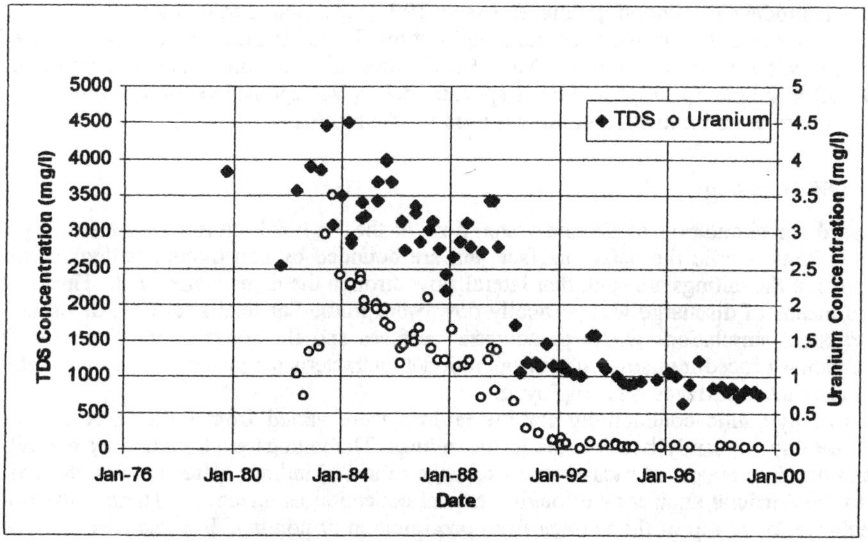

Figure 2. Historic TDS and uranium concentration in well RPI-20A.

2.3 *Ground-Water Discharge to Surface Water*

Spring Creek serves as the receiver and hydraulic boundary for the Surficial aquifer on the east side of the tailings area. As the boundary for the seepage impacted area of the Surficial aquifer, the ground-water impacts are consolidated to Spring Creek through discharge to the surface water. There is no measurable discharge of seepage-impacted ground water to other surface waters surrounding the tailings area. Base flows in Spring Creek and tributaries were measured in 1982 (Hydro-Engineering, 1982) and 1999. The measurements were taken in the fall when the flow is typically smallest. The identification of a base flow in Spring Creek is important because as the rate of surface flow decreases, the constituent concentrations in the composite of surface flow and ground-water discharge increase.

Two tributaries combine with Spring Creek in the reach adjacent to the tailings. Flow from Fox Creek, a perennial stream extending to the north and east from Spring Creek enters Spring Creek upstream of the confluence with Mine Creek. The flow in Fox Creek was measured and is typically 15% to 25% of the base flow in Spring Creek. Currently, the discharge from Mine Creek to Spring Creek is supported by the fresh-water recharge. The series of flow measurements indicates that there is approximately 24 gpm gain in Spring Creek through the measured reach when the Fox Creek flow is subtracted. The majority of this flow is believed to be ground-water discharge to Mine Creek and Spring Creek from the west side of Spring Creek.

The smallest 1999 flow measurements for Spring Creek show nearly a 50% increase over the 1982 measurements and appears to be representative of a late season base flow for a relatively wet recharge cycle. It is likely that base flow is affected by the quantity and nature of precipitation for one or more years prior to the actual measurement. The 1982 measurements came at the end of a 3-year drought cycle, while precipitation in the two years prior to the 1999 measurements was above normal. For the purposes of transport modeling, the base flow in Spring Creek was assumed to be 290 gpm, or approximately the sum of the upstream base flow and the Fox Creek flow from 1982.

3 TRANSPORT MODELING

The MODFLOW model (McDonald and Harbaugh, 1988) was used to model ground-water flow in the Surficial aquifer at the Shirley Basin site. The MODFLOW model is a multi-layer finite difference model that allows pseudo-three-dimensional flow modeling with distributed properties. The MT3D model (S.S. Papodopulous & Assoc. 1992) was used to model transport of constituents in ground-water flow in the Surficial aquifer at the Shirley Basin site. The MT3D used the cell by cell flow terms produced by the MODFLOW model. The same features incorporated in the MODFLOW are incorporated in the transport modeling. An updated version of the MT3D model that incorporates an implicit solution was used for the modeling.

3.1 *MODFLOW Modeling*

The model used for simulation of ground-water flow in the Surficial aquifer was a two layer model. The tailings overlie the native surface and are bounded by constructed tailings dams. The water levels in the tailings are such that lateral flow through the dams is minimal. Thus, the only modeled means of discharge was vertically downward to the Surficial aquifer. All simulations were transient simulations where piezometric surfaces and flux rates could change with time. A calibration procedure using current operational characteristics to model long-term stability of the piezometric surfaces was employed.

The values of hydraulic conductivity for the tailings were varied from 0.024 feet/day for slimes to 18 feet/day for sandy beach areas in the tailings. The tailings are bounded by the tailings dams and the only means of measurable exchange of stored tailings water is through vertical leakage to the Surficial aquifer or through artificial collection or injection. There is the potential for recharge to the top of the tailings from precipitation or ponding, but this was offset in the modeling by use of a net tailings dewatering rate rather than a gross dewatering rate. The extension of low-permeability interim cover and the eventual construction of the infiltration barrier will nearly eliminate infiltration into the tailings. The water levels in the tailings are too low for appreciable horizontal drainage through the dams or losses through evaporation. The vertical conductance was estimated to be greater near the original Mine Creek channel and to be smaller in the northern portion of Pond #5 where borehole lithology has indicated the presence of a thick clay layer.

A combination of drain cells, river cells, and well cells was used to model the exchange of water with both the tailings and Surficial aquifers. River cells were used to model water exchange between the Surficial aquifer and Spring Creek. Drain cells were used to model water exchange between the Surficial aquifer and Mine Creek. Well cells were used to model all collection, injection and recharge line operation in both the Surficial and tailings aquifers. The elevation of each river cell was derived from surface topography and a total of 143 river cells were used. The depth of water in Spring Creek was assumed to be one foot. Spring Creek is a receiving stream in the modeled reach.

Drain cells were used to model the discharge to Mine Creek. Drain cells were used rather than river cells because the model output summary tabulations of discharge to the drain cells are separate. This allowed distinction between ground water discharging to Spring Creek and ground water discharging to Mine Creek. Collection wells have been used to intercept seepage-impacted Surficial water for over 16 years and current rates of well extraction were used in the modeling.

Recharge lines are used to create hydraulic barriers to further seepage in the Surficial aquifer and to reverse gradients back to collection wells. Although the recharge lines are constructed as buried perforated pipelines, the lines were modeled as adjacent wells. This allowed precise control of the flux into the drain lines which is not possible with a drain cell approach. Injection wells are used to create a hydraulic barrier in the Surficial aquifer on the crest of the No. 5 dam.

Dewatering wells are the primary means of removing water from the tailings. Dewatering rates were reduced from measured well yields to reflect some offset of dewatering effect by ongoing recharge. The dewatering rates were also reduced with time in the various simulations to reflect the reduction in well yields as the water level in the tailings is lowered.

Recharge to the Surficial aquifer is occurring in areas outside of the immediate tailings area due to precipitation and ponding of runoff or other fresh waters. In the modeling, the active areas of recharge were the area south of the tailings, and the area between the tailings and Spring Creek. No recharge was used in areas where the Surficial aquifer is not present or active, or on the northeast side of Spring Creek. The recharge rate is increased in the industrial pond area to reflect the current ponding and the eventual use of the area as a surge pond. The recharge is reduced slightly in the Spring Creek area to reflect the usage of water by vegetation.

3.2 *MT3D Modeling*

The tailings is the source of elevated concentrations for a variety of constituents and the modeling was focused on the transport of uranium. Seepage from the tailings carries these constituents to the Surficial aquifer, where they mix with resident water. There are natural measurable concentrations of uranium and these were incorporated in the modeling.

With the extremely heterogeneous hydraulic conductivity in the Surficial aquifer and the resulting non-uniform flow field, the dispersivity of the Surficial aquifer does have some impact on the transport process. However this impact is moderated by the long duration of the source term. The seepage from the tailings will approach a steady-state rate lasting for decades. This has the effect of dramatically reducing the influence of longitudinal dispersivity, and to a lesser extent, transverse and vertical dispersivity. The longitudinal dispersivity used in the modeling was 10 feet for both the tailings and Surficial aquifers with a ratio of transverse or vertical dispersivity to longitudinal dispersivity of 0.2 for both aquifers. The initial concentration for constituents in the tailings was assumed to be uniform, so the dispersion process has virtually no impact. The diffusion coefficient was set at zero because diffusion is trivial in comparison to convection driven processes.

The adsorption process was not considered in the general transport modeling to add a measure of conservatism. Adsorption/desorption processes in the absence of decay processes act to slow down constituent migration and to spread a "pulse" of the constituent. Decay processes or other permanent removal processes were not considered in the modeling.

The initial concentrations in the tailings were set to reflect the mobile concentrations as evidenced by the measured concentrations in the Surficial aquifer. Geochemical processes described in Hydro-Engineering L.L.C. (2000) limit the mobile concentration of uranium and other constituents. The baseline concentrations in the Surficial aquifer were set at the average background concentrations except directly beneath or directly adjacent to the tailings. In these areas, the concentrations were increased in direct proportion to proximity with the original Mine Creek channel to reflect the continuous seepage impacts for the Surficial aquifer directly beneath the tailings. Recharge and injection to the Surficial aquifer were at the average background concentration or activity.

3.3 *Modeling Approach*

The seepage from the tailings to the Surficial aquifer is ongoing, and it is not feasible to completely evacuate the water from the tailings. With the installation of the infiltration barrier in the near future, the recharge to the tailings will likely be reduced to a fraction of a gpm. The seepage rate will gradually decline as the remaining volume in the tailings is reduced, and over a period of decades, the rate will approach the recharge rate to the tailings. At this eventual seepage rate, the seepage impacts will likely be undetectable.

The available means of reducing long-term seepage impacts include continued operation of the corrective action system and continued dewatering of the tailings. The injection rate to the Surficial aquifer has exceeded the collection rate from the Surficial aquifer for many years. This has created a mound or "reservoir" of fresh water within the aquifer that will start to dissipate as soon as the excess injection ends. This "reservoir" of fresh water is located between the tailings and Spring Creek and will delay and attenuate the seepage impacts at Spring Creek when the ground-water restoration program is terminated. At present, the collection/injection system has produced containment of the seepage and restoration of the bulk of Surficial aquifer between the tailings and Spring Creek. There is some minor progress in restoration and gradual shrinkage of the remaining contaminated zone immediately around the collection wells. However, the remaining area to be restored is very small and final ground-water restoration of the area around the collection wells would require a fairly dramatic change in the collection/injection system configuration. The Surficial collection/injection system is functioning primarily as a post-restoration seepage containment system.

The tailings dewatering program reduces the volume of water that will seep to the Surficial aquifer, and by reducing the head and saturated "footprint" of the tailings, will also reduce peak seepage rates following termination of dewatering. As the water level in the tailings declines, it

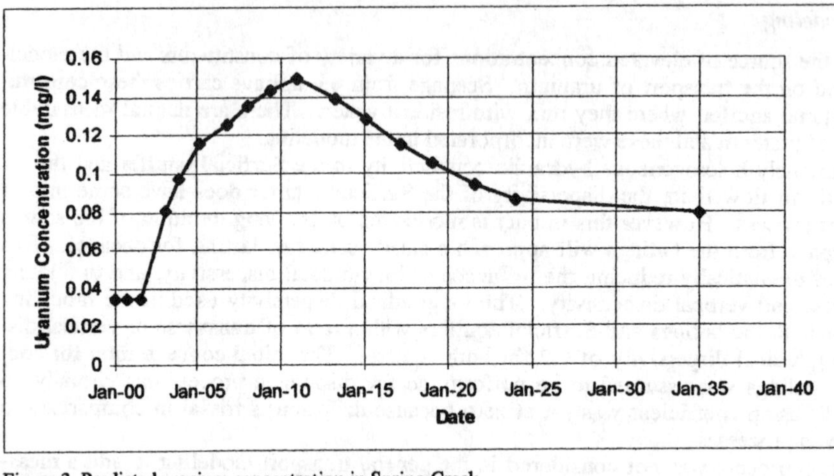

Figure 3. Predicted uranium concentration in Spring Creek.

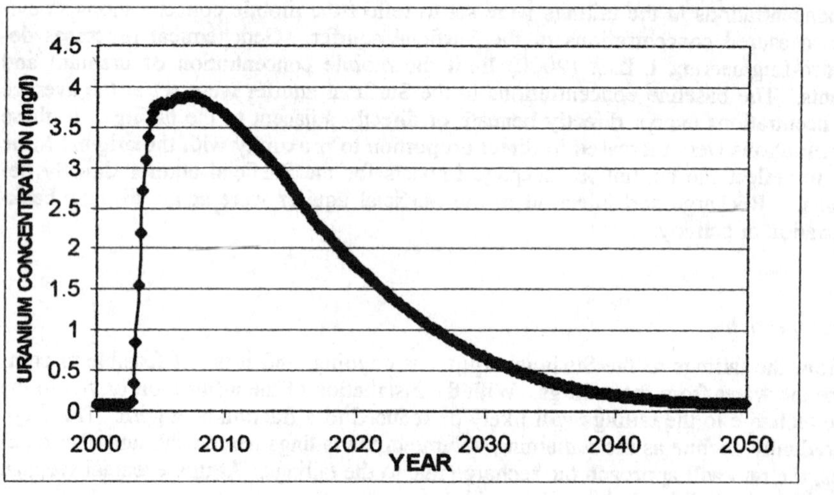

Figure 4. Predicted uranium concentration in well RPI-20A.

becomes more difficult to extract water through wells, and eventually, dewatering of the tailings becomes impractical.

3.4 *Modeling Scenarios*

Three scenarios were considered in the modeling of long-term seepage impacts as indicated by modeling of uranium transport to Spring Creek. The first scenario included immediate termination of the CAP and dewatering efforts. The second scenario included continued operation of both the CAP and the dewatering through mid 2001. This scenario is projected to remove approximately 30% of the drainable water in the tailings while maintaining seepage containment. The third scenario included continued operation of the CAP through mid 2003 and continued dewatering through late 2004. In this scenario, injection components of the CAP are phased out earlier than 2003 because they become counterproductive to dewatering efforts. Approximately 55% of the remaining drainable water in the tailings is projected to be removed in the third scenario. In comparing the three scenarios, it quickly becomes obvious that continued CAP operation and continued dewatering will reduce long-term seepage impacts. However, the incremental benefit for continued operation of these systems diminishes rapidly with time. This is illustrated by a comparison of projected dewatering success between scenarios #2 and #3. With an extension of the dewatering program from 1½ year to 5 years, the projected volume of water extracted is less than doubled.

3.5 *Modeling Results*

The transport of uranium to Spring Creek was modeled for scenario #2. The retardation factor used in the model was one, giving no simulated retardation. Based on comparisons between observed chloride movement rates and uranium movement rates, it is apparent that there is some retardation of uranium transport. However, the retardation does appear to be inversely proportional to permeability of the aquifer with the least retardation in the immediate Mine Creek area. This is consistent with the expectation that a larger fraction of silts and clays in the aquifer would offer more adsorptive surface and reduce the permeability at the same time. The net effect of retardation in a situation where the source is slow leakage from a relatively large source volume, is a slowing of the movement rate and a modest reduction in peak concentrations at a given observation point. Hence, the use of no retardation in the modeling adds a measure of conservatism in the transport approach.

A significant portion of the ground water entering Spring Creek has been impacted by seepage from the tailings. This water combines with the base flow in Spring Creek to produce composite surface water concentrations that are much lower than the typical ground-water concentrations. The MODFLOW model predictions of ground-water discharge to the river and drains cells representing Spring Creek and Mine Creek were extracted for each of 14 stress periods over a 50-year simulation period. The average concentrations for a series of drain and river cells was then multiplied by the discharge to each type of cell to give a constituent "load" to Spring Creek. The composite concentration of the ground-water discharge and the 290 gpm base flow in Spring Creek was then the concentration at a point where Spring Creek exits the impacted area. Mine Creek was represented by 20 drain cells, and the concentration of every other drain cell was averaged at regular intervals through the simulation period to represent the Mine Creek discharge. This quantity was then multiplied by the total drain discharge rate for each stress period. Since the drain cells are located roughly in a line parallel to the gradient from the tailings, the contaminant front reaches the cells sequentially. The maximum concentration in the Mine Creek discharge represented by the drain cells occurs long after the fresh-water mound from injection has decayed, and the contaminant has reached the last drain cell at the confluence of Mine Creek and Spring Creek. A similar approach was used in determining the average concentration of ground water entering the river cells in Spring Creek over time. Twelve indicator river cells along Spring Creek were used in this averaging. The indicator cells were selected to provide a distribution that would reflect the varying rates of discharge due to the larger conductivity near the Mine Creek confluence. When the base flow in Spring Creek and the ground-water discharge to river cells and drain cells are combined at the end and middle of each stress period, a concentration curve for the Spring Creek flow is developed. The maximum predicted concentration of uranium in Spring Creek is 0.15 mg/l. Figure 3 presents the projected uranium concentration for Spring Creek at a location downstream of the tailings under the assumption of a late season base flow of 290 gpm in Spring Creek. Figure 4 presents projected uranium concentration in well RPI-20A.

3.6 *Modeling Prediction Summary*

The transport of uranium in the Surficial aquifer to eventually discharge to Spring Creek was modeled for scenario #2. There was approximately a three-year delay in the predicted appearance of elevated uranium concentrations at well RPI-20A, which is less than 100 feet from the Spring Creek channel. This corresponds with the appearance of elevated concentrations in Spring Creek. This delay is both a result of continued seepage containment until mid-2001 and the fact that the ground water has been restored to near background conditions in the zone between the tailings and Spring Creek. The predicted maximum uranium concentration at well RPI-20A is consistent with historic data prior to initiation of the CAP, but Figure 4 exhibits the decay in uranium concentration at this well as the seepage rate from tailings declines. This particular well is in the heart of the historic seepage plume and is also located in the higher permeability zone where the seepage velocity is greater. Thus, well RPI-20A exhibits both a large magnitude and rapid response to the cessation of the CAP. Other locations at a similar distance from the tailings will show a lagged and attenuated response in comparison to well RPI-20A.

The predicted uranium concentration in Spring Creek shown in Figure 3 illustrates the combination of un-impacted base flow in Spring Creek and variously impacted Surficial aquifer discharge to Spring along the reach adjacent to the tailings. There is great heterogeneity in the degree of seepage impacts to the Surficial aquifer, as well as the transport velocity. This results in an "averaging" of ground-water discharge to the creek, and a smoothing of the composite concentration curve.

The continued CAP and dewatering system operation until mid-2001 for scenario #2 produces a significant reduction in long-term seepage impacts. The simple removal of significant portion of the water in the tailings through dewatering causes a dramatic reduction in peak post-dewatering seepage rates. A scaling of the seepage rates for MODFLOW simulations of scenario #1, where both the CAP and dewatering programs were terminated, produces concentrations that are roughly 150% of those discussed in this section. In contrast, scenario #3 with its greater than three-fold increase in operation time for the CAP only produces a reduction in the peak uranium concentration in Spring Creek from 0.15 mg/l to 0.08 mg/l.

4 SUMMARY AND CONCLUSIONS

The CAP has proven to be very effective in containing seepage and in restoring the quality of Surficial aquifer ground water in seepage impacted areas. At present, off-site seepage impacts are nominal. However, there will be some water left in the tailings after reclamation that will inevitably migrate to the Surficial aquifer. Modeling predictions indicate that significant reductions in the magnitude of long-term seepage impacts to Spring Creek can be achieved with continuation of the current CAP and dewatering programs for a period of one to two years. Extension of the operation of these programs becomes progressively less productive as dewatering rates decline. When this is coupled with an expected dramatic increase in the unit cost of dewatering extraction with declining well yields, there comes a point when continuation of a very costly and inefficient dewatering system is no longer justified by very modest reductions in seepage impacts.

REFERENCES

Hydro-Engineering, L.L.C., 2000. Hydrologic Monitoring for Shirley Basin's Tailings Seepage Control Plan. *Consulting Report for Pathfinder Mines Corporation, Shirley Basin Mine.*

McDonald, M.G. and A.W. Harbaugh, 1988. *A Modular Three-Dimensional Finite-Difference Ground-Water Flow Model (MODFLOW), Book 6, Modeling Techniques*, United States Geological Survey.

S.S. Papadopulos and Associates, Inc., 1992. *MT3D, A Modular Three-Dimensional Transport Model, Version 1.5*. Bethesda, Maryland, 20814.

Tailings and Mine Waste '01 © 2001 Balkema, Rotterdam, ISBN 90 5809 182 1

Reclamation progress at the Summitville Mine Superfund Site, Rio Grande County, Colorado

A.M.Campbell
Colorado Department of Public Health and Environment, Denver, Colo., USA
M.J.Gobla
US Bureau of Reclamation, Denver, Colo., USA

ABSTRACT: Since the government took over this challenging high-altitude site in 1992, numerous techniques for treating, controlling and preventing acid mine drainage have been researched and applied. While water treatment was an immediate concern for the site, recent reclamation activities have been the major focus of the mine closure. Combinations of conventional, innovative and experimental methods are being applied to mitigate acid drainage from the site. Initial results indicate that the prevention and passive treatment techniques are improving water quality. Storm-water runoff from the Cropsy Valley area comprising 1/4 of the site is being released directly to the environment without the need for active water treatment. Other reclaimed areas are also showing significant gains in water quality improvement.

1 INTRODUCTION

The Summitville Mine Superfund Site is an open-pit gold mine that was abandoned in December of 1992 when the operating company, Summitville Consolidated Mining Company Incorporated, and its parent, Galactic Minerals International, declared bankruptcy. At the request of the Colorado Department of Public Health and Environment (CDPHE), the Environmental Protection Agency (EPA) initiated emergency response actions to prevent the release of cyanide solutions from the heap-leach pad. The EPA engaged the U.S. Bureau of Reclamation (BOR) to provide engineering design, contract administration, and construction management services for the project. The site was added to the National Priorities List in June of 1994. Since 1995, Colorado State University has conducted research to optimize procedures for site revegetation. Transition from EPA management to State-lead management was recently initiated by the CDPHE.

Approximately 55 % of the total expenditures at Summitville have been for water treatment. Reclamation activities have been directed at eliminating the generation of acid rock drainage (ARD) and have reduced water treatment costs from the initial \$1.0 million per month to approximately \$2.0 million per year. Site reclamation includes the detoxification and capping of the heap-leach pad, removal of mine wastes from waterways, plugging underground adits, reduction of steep waste banks, backfilling mine pits and depressions, construction of permanent drainage channels, limestone neutralization of disturbed surfaces, and revegetation using a native seed mix. Innovative techniques include placing limestone riprap in steep drainage channels as a form of passive treatment and neutralization of permanent haul road surfaces using agricultural limestone. Experimental technologies for controlling ARD are also being investigated.

Summitville Mine Superfund Site

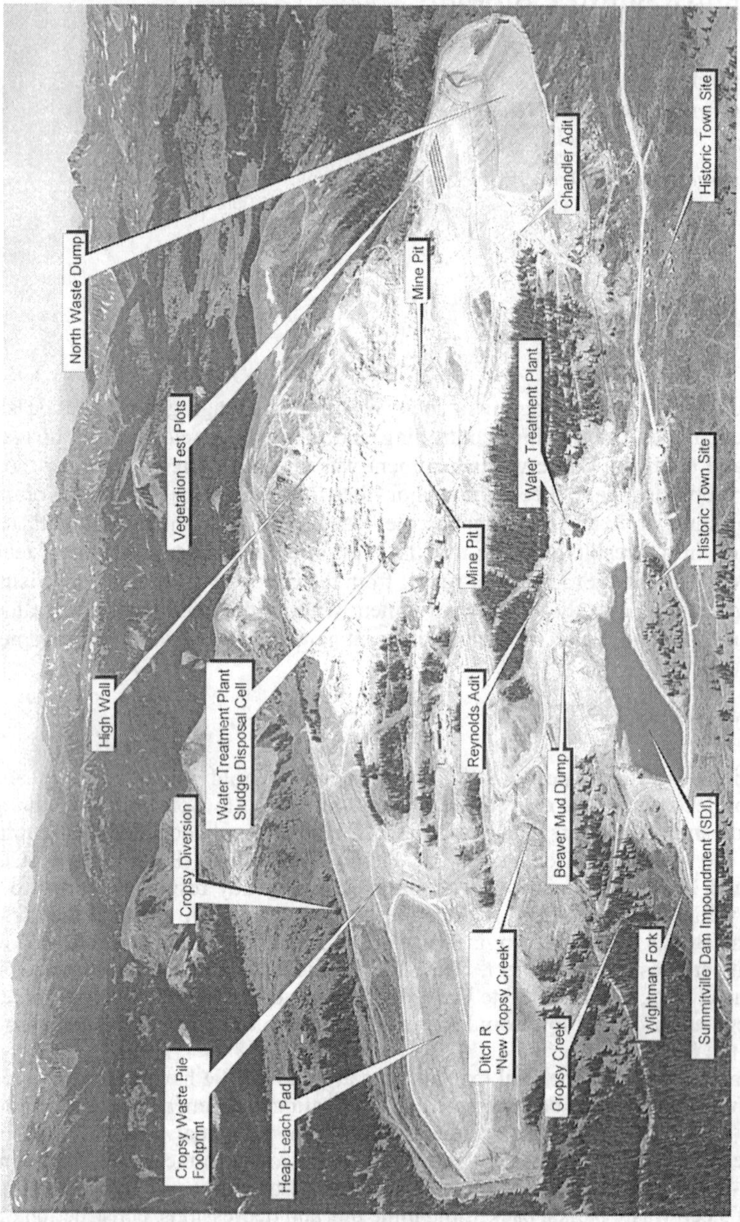

FIGURE 1

Angus Campbell, OU4 Site-Wide Reclamation Project Manager
Hazardous Materials and Waste Management Division

Colorado Department of Public Health and Environment

1.1 *Site conditions*

Summitville is a remote high-altitude site where frequent storms cause "white out" conditions in the spring and fall. The summer season brings intense lightning, heavy downpours, and mud. Poor weather often shuts down reclamation activities at the site as workers seek shelter, and heavy equipment loses traction in the mire. The occurrence of acid generating soils and rocks is widespread. In addition to the waste dumps and ore stockpiles, almost every disturbed surface, including haul roads, drainage ditches, and highwalls, are sources for the generation of ARD.

1.2 *Location and Topography*

The 700-acre site is located in the San Juan Mountains approximately two miles east of the Continental Divide in southwestern Rio Grande County, Colorado. The disturbed areas of the site, see Figure 1., are on the north side of South Mountain situated between 11,200 feet and 12,200 feet in elevation. The crest of South Mountain reaches an elevation of 12,555 feet. Flat ground is mainly confined to the historic town site and wetlands found below the site. Except for the backfilled mine pits, the disturbed areas are in steep terrain with 33% or greater slopes.

1.3 *Climate*

The site is characterized by harsh winter weather with cool and wet summers. Freezing temperatures are normally experienced from October through May and also occur infrequently during the summer season. The site receives an average of 55 inches of annual precipitation; most of this is in the form of snow (389 inches/year). Located on a north-facing slope surrounded by mountain peaks, direct sunlight on a majority of the site is limited from mid-morning to mid-afternoon during the short summer growing season.

1.4 *Geology*

The site lies on the northern margin of the Platoro caldera complex within the San Juan volcanic field. The mine is situated in the Tertiary age, Summitville caldera on the north slope of the South Mountain vent-dome complex (Lipman, 1974). The Tertiary rock units are quartz latite, quartz latite porphyry, and andesite (Steven and Ratte, 1960).

Following emplacement of the South Mountain quartz latite volcanic dome, the mineral deposit developed over the course of at least six different events (Gray, et al., 1993). The events are 1) early acid-sulfate alteration, 2) copper sulfide and gold mineralization, 3) widespread hydrothermal brecciation, 4) deposition of base-metal-sulfide bearing barite veins, 5) emplacement of small-scale kaolinite matrix breccias, and 6) supergene oxidation and secondary sulfide enrichment.

The principal rock type at the site is quartz latite that exhibits varying stages of hydrothermal alteration from propylitic alteration at the margin of the deposit to an intensely altered, vuggy silica zone at the innermost core of the mineralized structures. The andesite is most prevalent in the vicinity of the historic town site area below the mine. The andesite, a hard black rock, is not considered a source of ARD unless it has been intruded by the younger latite.

2 SOURCE OF ACID ROCK DRAINAGE

Pyrite-bearing quartz latite is the source rock for the pervasive ARD problem at Summitville. Practically every disturbed surface within the caldera generates ARD laden with copper and other heavy metals. Although local concentrations can be very high, most of the quartz latite rock contains from 1 to 2.0% sulfur in the form of iron pyrite (FeS_2) and alunite ($KAl_3(SO_4)_2(OH)_6$). Small amounts of covellite (Cu_2S), enargite (Cu_3As_4), and native sulfur (S) also occur, however, pyrite is the principal mineral species that forms ARD at Summitville. The pyrite is fine-grained. In addition to pyrite and quartz, the latite contains relict crystals of feldspar that have been altered to a mixture of clays and alunite. Hard when blasted from the outcrop, the latite slakes upon exposure to air and water. The weathered latite is rapidly transformed into an ARD producing mud.

Acid-base accounting has been performed upon samples of the mine waste material and the topsoil. Because acid and sulfate rich waters altered the latite, there is almost no buffering capacity present in the mine waste or in the topsoil.

3 STATUS OF RECLAMATION ACTIVITIES

Previous cleanup activities conducted under EPA leadership include water collection and treatment, chemical drum removal, heap-leach pad detoxification, waste stabilization, grading, topsoil replacement, seeding of approximately 100 acres of land, and salvage operations. Reclamation activities conducted under the leadership of the CDPHE consist of Operable Unit 4, which is known as Site-Wide Reclamation. Site-Wide Reclamation was initiated in May of 1999.

3.1 *Heap-Leach Pad*

The pad has been pumped down and rinsed with water to reduce the cyanide levels. The gold processing facility at the mine was converted to a cyanide destruction plant that used hydrogen peroxide as the detoxifying agent to destroy the cyanide compounds in the rinsate. The water was then further treated to reduce the heavy metals content. The heap-leach pad has been capped with a geocomposite-clay liner to prevent saturation of the pile and to minimize ARD generation from the sulfide minerals contained in the pile. Dike 1, an embankment across Cropsy Creek that supports the heap-leach pad structure, has been rehabilitated. A toe drain has been installed, and additional fill has reduced the embankment slope, thereby enhancing the geotechnical stability of the structure.

3.2 *Cropsy Waste Pile*

In order to reduce the cost of water treatment at the site, response actions were taken to reduce or eliminate ARD generated from several sources (Ketellapper, et al., 1995). The 6-million cubic yard Cropsy Waste Pile was removed from Cropsy Creek where it had covered a number of active seeps and springs that were forming ARD. The waste was placed in the mine pits, and the disturbed surfaces in the Crospy Basin have been reclaimed.

3.3 *Mine Pits*

The bottom of the North Pit and South Pit were covered with a compacted-clay drainage barrier and then covered with a neutralizing layer of lime-kiln dust. Mine wastes from the Cropsy Waste Pile and the historic mill tailings from other areas of the site were backfilled into the pits. A geocomposite-clay liner was placed on top of the South Pit backfill to minimize infiltration into the fill. The mine waste backfills have been found to have a permeability of 5×10^{-5} cm/sec. Infiltration to the underground workings has been greatly reduced as a result of the pit backfill operations.

3.4 *Summitville Dam Impoundment*

In 1994 and 1995 an existing tailings pond was modified and enlarged into a water storage dam. Renamed the Summitville Dam Impoundment, the facility is used to collect and store ARD from the site. By collecting the various sources of ARD from the site in the impoundment, the water is mixed to a uniform chemistry, see Table 1., allowing a single treatment plant to be utilized instead of the multiple plants that were formerly in use. The water is pumped from the impoundment to the "Metals Removal Plant" where conventional precipitation methods are used. Hydrated lime and a polymer are used in the plant, which has a capacity to treat 1,050 gallons per minute of ARD. A recirculating sludge circuit has increased sludge density, which has reduced water treatment sludge volumes and disposal costs.

Table 1. Typical analysis of water removed from Summitville Dam Impoundment for treatment.

Date	pH	Conductivity us	Temperature Degree C	Fe ppm	Mn Ppm	Cu ppm	Zn ppm
9/15/97	2.9	2,150	13.4			31.0	
9/15/98	3.1	1,471	12			32.0	
9/27/99	2.94	2,500	9.9	233	26.5	35.4	12.0
8/15/00	3.14			125	34.6	40.0	15.33

3.5 *Adit Plugs*

Excavation of the mine pits encountered old underground mine workings and lowered the ground surface to` within a few hundred feet of the level of the historic Reynolds adit. The ARD issuing from the Reynolds adit increased in quantity while the water quality declined. In 1993 and 1994, hydraulic plugs were placed in the Reynolds and Chandler adits. This helped re-establish the original ground water levels in South Mountain and reduced the quantity of ARD requiring collection and treatment. Negative aspects of the plugs are the reactivation of natural acid seeps and springs issuing from the lower portions of the site, leakage around the plugs, and the maintenance requirements. A hydraulic adit plug is in effect a dam; routine monitoring and maintenance must prevent failure. Annual inspections are performed to observe seepage and record water pressure acting on the plug.

3.6 *Operable Unit 4 - Site-Wide Reclamation*

The goals for Operable Unit 4 are directed at improving water quality. These goals are accomplished by removing, reducing, stabilizing and/or containing non-point sources of acid rock drainage to prevent further releases to the site; eliminate or minimize non-point source impacts to aquatic receptors in Wightman Fork, the Alamosa River, and Terrace Reservoir; eliminate or minimize potential water treatment for non-point source contaminated waters; and enhance and/or improve on-site biota habitat. Past actions at the site followed a dual track of water collection and treatment on one track, and prevention of ARD on the other track. Operable Unit 4 is focused principally on prevention and control techniques. Grading, water diversion, construction of permanent drainage ditches, soil neutralization with the addition of organic amendment, and revegetation were identified as major activities needed to reclaim the site (Gobla, 1999). To minimize project impacts, the reclamation program has been designed to limit additional disturbance to vegetated areas, protect cultural resources, control safety hazards, and minimize dust and sediment generation. OU-4 (1999-2001) will utilize the following materials to reclaim 300 acres of the site: 80,000 cubic yards erosion protection rock and 70,000 square yards of geotextile and geomembrane to establish five miles of permanent drainage ditches; 45,000 tons of crushed limestone soil amendment, 12,000 tons of organic compost soil amendment, 45 tons of fertilizer, and three tons of native seed to revegetate disturbed areas.

4 APPLICATION OF ARD MITIGATION TECHNIQUES

Mitigation methods at Summitville include a combination of conventional, innovative, and experimental methods. Conventional methods include diversion, waste stabilization, capping, and revegetation. Innovative techniques include the use of limestone riprap in steep drainage channels and liming of permanent haul roads. Experimental prevention and treatment methods are being evaluated to help solve Summitville's two most difficult ARD problems, runoff from the sloughing mine highwall, and acid seepage from waste banks and the underground mine workings.

4.1 *Diversion Ditches*

Diversion ditches are a cost effective method to keep clean water away from ARD source materials. At Summitville the Cropsy Diversion was installed by the mining company to carry the headwaters of Cropsy Creek around the heap-leach pad. This ditch has good water quality, however, its steep eroded banks will be reclaimed under OU-4 to reduce sediment generation.

The 550 Ditch was also installed by the mining company to divert runoff from a reclaimed hillside away from the heap-leach pad. The 550 Ditch had poor water quality because it has been constructed from mine waste and it allows some infiltration of water into the heap-leach pad. The ditch has been reclaimed by excavating to reduce a steep waste bank, neutralized with limestone, and revegetated. Flows from small hillside seeps and springs have been diverted away from the 550 Ditch into Cropsy Drain to minimize water infiltration into the heap-leach pad.

Cropsy Drain is a diversion ditch installed in 1998 by the government to minimize water ponding against and infiltrating into the heap-leach pad. The drain excavation includes installation of a 54" diameter high-density polyethylene pipe to ensure that slope failures and avalanches will not block the drainage path.

In the year 2000, two diversion ditches were installed in the Exploration Benches area lying high on the site above the mine pits. One ditch diverts clean surface water runoff from the top of South Mountain away from the mine pits, while the second ditch diverts the runoff from the reclaimed Exploration Benches area. The two ditches redirect over 20 acres of runoff away from the collection and treatment system.

4.2 *Waste Removal from Riparian Areas*

Wastes are being removed from riparian areas and wetlands. In addition to the six-million cubic yards of mine wastes removed from Cropsy Creek, approximately 20,000 cubic yards of waste has been removed from below the Chandler Adit and along the Wightman Fork to reestablish a small wetland area. A portion of Cropsy Creek below the heap-leach pad also has been excavated to remove latite mine wastes from the stream bed.

4.3 *Stabilization of Eroding Waste Banks*

Nearly one-million cubic yards of mine wastes have been removed from steep, unstable road slopes and pond and ditch embankments. The slopes were cut back to a 3 horizontal to 1 vertical slope (3H:1V) prior to revegetation to eliminate erosion problems and ensure that the topsoil layer will not slide off the slopes. The mine waste from the slope excavations has been used to fill depressions by establishing a minimum 5% slope across the mine pit backfill to eliminate standing water. The excavated waste is also being used to create 3H:1V fill embankments to cover exposed rock highwalls. The 700-foot highwall that cuts into South Mountain will not be reclaimed.

4.4 *Liners and Caps to Reduce Infiltration*

Liners and caps are being employed to reduce the infiltration of surface waters. Compacted bentonite clay liners were placed in the mine pits to minimize seepage out of the backfill. Geosynthetic clay liners were placed as caps over the South Pit and the heap-leach pad to minimize surface water infiltration. A 4-foot thick soil cover has been placed over the clay caps for frost protection, and then the surface has been topsoiled and revegetated.

There are several small acid seeps at the toe area of the North Waste Dump. The seeps are fed in part by water percolating through the dump, and in part by springs and seeps present in the dump foundation area. A geocomposite liner consisting of a PVC membrane bonded to a geotextile has been installed in the drainage ditches on the North Waste Dump. The drainage ditches are rather flat, therefore it was decided that once snowmelt and surface stormwater runoff had reached a drainage ditch, all of the water should be carried off of the dump in order to minimize water percolation into the dump. The majority of the dump surface consists of 3H:1V slopes that will be amended, topsoiled, and seeded to form a vegetative cover.

4.5 *Vegetative Covers*

The prescription for establishing a vegetative cover at Summitville (Redente and Richard, 1998) consists of amending the upper 1.0 ft. layer of disturbed ground surface with limestone and organic material, replacing 0.5 ft. of topsoil, applying fertilizers, tackifier, seed, and covering and crimping mulch.

The top 12 inches of disturbed ground (mine waste) surface is amended with 130 tons per acre of ground limestone and 40 tons per acre of organic compost. Limestone amendment raises the pH of the disturbed ground from a range of 2.0 to 3.5 to a range of 5.5 to 7.0.

The topsoil has a pH range of 3.5 to 5.0. From 10 to 40 tons per acre of agricultural limestone is added to the 6- inch layer of topsoil for neutralization prior to seeding. The topsoil is also fertilized with 150 lbs./acre of nitrogen and 200 lbs./acre of phosphorus, seeded, tacified, and mulched. Long slopes are not mulched, instead straw blankets are unrolled and staked down to control erosion prior to vegetation establishment.

The native seed mix includes tufted hairgrass, alpine timothy, alpine bluegrass, spike trisctum, slender wheatgrass, mountain brome, and western yarrow. Mountain brome is added as a cover crop because of its quick germination, however, it is not native to the site like the other plants. All of the seed species are acid tolerant. Approximately equal numbers of seed from each species are applied at a rate of 300 total seeds per square foot.

4.6 *Innovative Methods*

Approximately 2,000 cubic yards of limestone riprap has been placed in reclaimed ditches to provide partial passive treatment. Limestone was placed in the bottom section of the ditch along steep ditch segments and below cascades and waterfalls where water velocities exceed 10 ft./second. In some ditches water velocities are as rapid as 25 feet per second. Based upon success in eastern coal fields (Ziemkiewicz, et al., 1997), high water velocities are expected to scour off iron and aluminum hydroxide rinds to allow continued dissolution of the limestone. Limestone sand and gravel has also been applied to the ditches as chinking to fill the void spaces between the andesite riprap.

The mine road surfaces are also a significant source of ARD at Summitville. Roads that will be used on an infrequent basis are being revegetated. Major site access roads that must remain will be neutralized to minimize the generation of acidic runoff from the road surface. The roads will be ripped, amended with 130 tons per acre of ground limestone, and compacted.

4.7 *Experimental Methods*

Experimental investigations conducted for the project include prevention, passive treatment, and active water treatment methods. Bactericide was tested on North Waste Dump as a soil amendment to prohibit ARD formation and enable plant growth. This is a temporary prevention measure and is no longer under consideration for use at the site. Conventional soil amendment with agricultural limestone is effective at Summitville.

The Cropsy Drain highwall, the 700-ft. mine highwall, and a few other unstable rock slopes at the site are too steep and dangerous to be reclaimed by the conventional vegetative cap approach. These slopes will continue to generate surface ARD unless treated in some manner. Conceptually, a spray-on chemical could be safely applied and may reduce or eliminate ARD from this source. The following treatments have been considered:

Do Nothing - Collect and treat runoff from the mine highwall, the current plan.
Polymer Sealant - Temporary and too expensive to apply to a large area.
Hydrated Lime Slurry - Reasonable cost, unknown longevity.
Phosphate - Reasonable longevity, unknown cost.

A commercial phosphate microencapsulation treatment has been tested in the laboratory as a means of preventing ARD formation. Column leach investigations on samples from Summitville show that the phosphate treatment raises the pH of the mine waste to around 6.5, and it has stayed at this level for at least six months.

Passive water treatment experiments under consideration include a "successive alkalinity producing system" this treatment utilizes an iron settling pond followed by an anoxic organic substrate on top of limestone followed by a settling pond. The passive treatment experimental systems were installed in the year 2000 to treat water seeping past the Reynolds Adit plug. In the Cropsy Valley a bog treatment scheme designed to produce siderite is being tested. The methodology utilizes raw manure, limestone and hay with the seep waters percolating down through the substrate.

Active water treatment experiments conducted at Summitville include "zeolite cation exchange," direct lime addition and Microfiltration. None of the methods justify replacement of the conventional lime precipitation plant at the site at this time, however; treatability studies involving these technologies are being conducted.

The zeolite rock, although effective in initial treatment, must be regenerated on a periodic basis by acid stripping. Eventually the rock will require replacement after several cycles of treatment.

The direct lime addition uses an hydraulicly driven feed system to feed quick lime directly in to a reaction trench which in turns flow into a settling tank. The technology is a product produced by "Aquafix", and was installed in 2000.

Microfiltration removes metals by filtering on the molecular level minimizing the formation of sludge. Experiments have shown the effectiveness of this method. One concern is the high pumping pressures required to force water through the ceramic filters and the freeze thaw effects on the filters during winter shutdown periods.

5 PRELIMINARY RESULTS

The recent surface water monitoring results are encouraging. The data indicate that reclamation activities implemented to date at the Summitville Mine are having positive effects on water quality leaving the site while at the same time consolidating the contaminated source water on site needed for treatment.

As shown in Table 1, the copper concentration in the SDI is increasing. The increase could be attributed to the realization of one of the design objectives, of diverting runoff from reclaimed areas away from the SDI. The result of this is a concentration of major source loads from the site, specifically adit seepage and groundwater seeps, into the SDI.

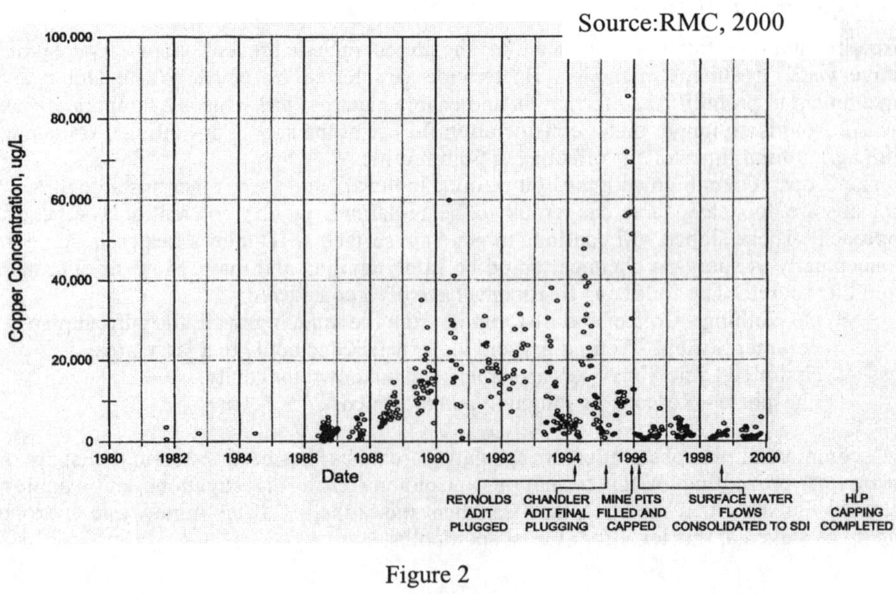

Figure 2

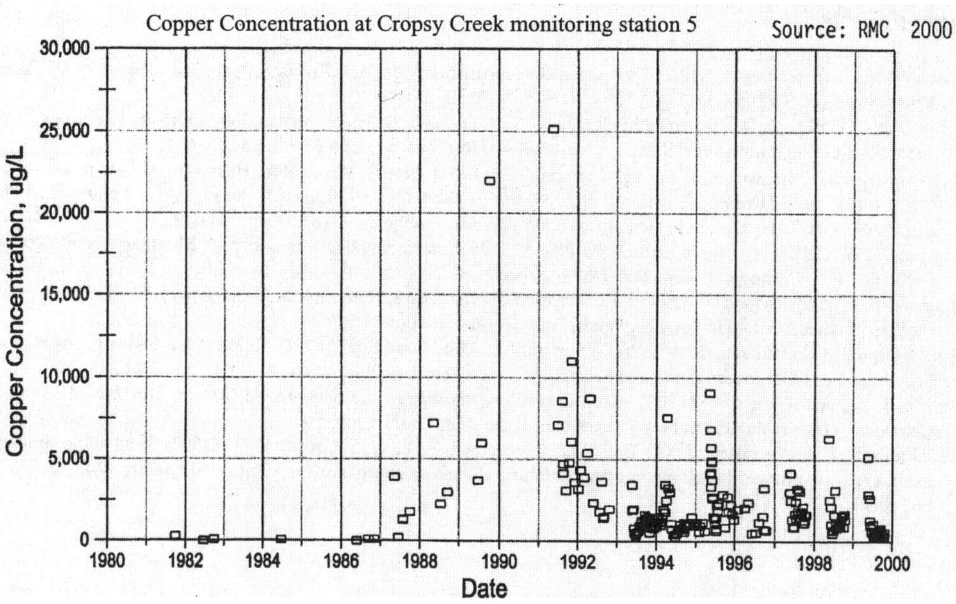

Figure 3.

Additionally, trends in down stream surface water monitoring show a marked decrease in metal concentration with some increase in pH. Recent copper concentrations measured in the Wightman Fork at the site boundary have decreased to levels observed during SCMCI's first year of operation (RMC, 2000) see Figure 2.

This trend showing improvement in water quality is also reflected in the Cropsy Valley. The upper Cropsy Valley and heap-leach-pad were reseeded in 1998, and the reclamation of the lower portion of the Cropsy Valley was completed in 1999. This trend in decreasing copper concentration at the lower end of Cropsy Creek is reflected in Figure 3.

6 CONCLUSIONS

The geology, wet climate, steep slopes, presence of surface and underground mine workings, wetlands, and cultural resources combine to form a complex site that is difficult and costly to reclaim. Multiple techniques are necessary to mitigate ARD at Summitville. Water treatment requirements are declining as the reclamation work is completed and vegetation is reestablished. ARD prevention is the key. While conventional methods such as regrading, limestone neutralization, and revegetation are successfully mitigating ARD at Summitville, two difficult problems remain:

1. Acid stormwater runoff from the mine highwall.
2. Acid seepage from the underground mine workings and from the french drain under the heap-leach pad.

Long-term active water treatment for these remaining ARD sources is always a solution, but not a desirable one unless treatment costs can be greatly reduced. Experimental methods may be a key to prevention or providing low-cost passive treatment, which is why promising techniques are being pursued. Once all reclamation activities are completed, annual monitoring and maintenance will be conducted at the site for a period of up to ten years. The monitoring will evaluate the reclamation and identify the needs for repair of erosion features, reseeding, and other maintenance.

REFERENCES

Gobla, M. J. (1999) Summitville Mine site-wide reclamation design. *Tailings and Mine Waste '99*, Balkema, Rotterdam, pp. 797-806.

Gray, J. E., Coolbaugh, M. F., and Plumlee, G. S. (1993) Geologic framework and environmental geology of the Summitville acid-sulfate mineral deposit. U.S. Geological Survey, Open File Report 93-677, 1993.

Ketellapper, V. L., Cressman, J. E., and Carmody, C. (1995) Cropsy Waste Pile, Beaver Mud Dump, Cleveland Cliffs and Mine Pits response action. In *Proceedings Summitville Forum '95*, Posey, H. H., Pendleton, J. A., and Van Zyl, D. (eds), Colorado Geological Survey Special Publication No. 38, pp. 121-126.

Lipman, P. W. (1974) Geologic map of the Platoro Caldera area, southeastern San Juan Mountains, southwestern Colorado. U.S. Geologic Survey, Map I-828, 1 sheet.

Redente, E. F., and Richard, C (1998) Reclamation alternatives at Summitville Mine Superfund Site: greenhouse and field studies. Colorado State University, Fort Collins, Colorado, 56 p.

Rocky Mountain Consultants (RMC) (2000) Summitville Mine Superfund Site, Draft Remedial Investigation Report, Site-Wide Remedial Investigation and Feasibility Study, February 2000.

Steven, T. A., and Ratte, J. C. (1960) *Geology and ore deposits of the Summitville District, San Juan Mountains, Colorado.* U.S. Geological Survey Professional Paper 343, Washington, DC.

Ziemkiewicz, P. F., Skousen, J. G., Brant, D. L., Sterner, P. L., and Lovett, R. J. (1997) Acid mine drainage treatment with armored limestone in open limestone channels. Journal of Environmental Quality, Vol. 26, p. 560-569.

Tailings and Mine Waste '01 © 2001 Balkema, Rotterdam, ISBN 90 5809 182 1

In situ treatment of groundwater impacted by acid mine drainage using permeable reactive materials

D.J.A.Smyth, D.W.Blowes & A.M.Hulshof
Department of Earth Sciences, University of Waterloo, Ont., Canada
S.G.Benner
Geological and Environmental Sciences, Stanford University, USA

ABSTRACT: Groundwater impacted by acid-mine drainage is typically characterized by elevated concentrations of sulfate, ferrous iron and other dissolved metals. An approach to the control and remediation of acid mine drainage is the passive interception and in situ treatment of groundwater using permeable reactive barriers (PRBs). The reactive materials, which incorporate various forms of organic carbon, promote microbially mediated sulfate reduction, the generation of hydrogen sulfide, and the subsequent precipitation of stable iron and other metal sulfide minerals. Full and demonstration scale field applications and supporting laboratory columns have shown that several thousands of mg/L sulfate, more than 1,000 mg/L iron, and several tens of mg/L of other metals can be removed from plume or tailings groundwater. Organic materials have been applied in PRBs to remediate contaminated groundwater at a mine site and an industrial property, and as reactive amendments mixed directly with tailings at an active mine.

1. INTRODUCTION:

A pervasive issue for environmental management in the mining industry involves the generation and discharge of acidic effluent from mine-waste facilities. Acid-mine drainage (AMD) arises from the oxidation of residual sulfide minerals in mines and mine-wastes. The oxidation of an iron sulfide mineral such as pyrite in the presence of oxygen and water can be expressed as:

$$2FeS_{2(s)} + 7O_2 + 2H_2O \rightarrow 4SO_4^{2-} + 2Fe^{2+} + 4H^+ \qquad (1)$$

The resultant effluent is acidic and contains elevated concentrations of sulfate, reduced iron (Fe(II)) and dissolved trace metals. Projections of the AMD-generating potential for many mine-waste sites extend from decades to centuries.

A typical mine-tailings impoundment in a humid climatic setting is shown in Figure 1. Sulfide oxidation occurs in the vadose zone established within the tailings. AMD-impacted water is transported through the subsurface with groundwater flow from the mine-tailings impoundment and is discharged to a receiving surface-water body. Although buffering of the impacted water by mineral reactions may increase the pH to between 4 and 7 along the length of the groundwater flow system, the oxidation of Fe(II) to Fe(III) occurs as a consequence of the exposure of AMD-impacted groundwater to the atmosphere or surface water. This reaction generates acidic conditions and results in the precipitation of ferric oxyhydroxides. The reaction can be represented by:

$$4Fe^{2+} + O_2 + 10\,H_2O \rightarrow 4Fe(OH)_{3(s)} + 8H^+ \qquad (2)$$

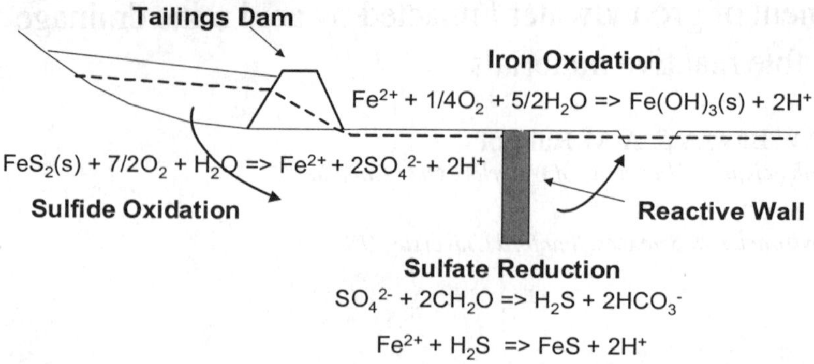

Figure 1. Schematic diagram showing generation, impacts and remediation of AMD from tailings.

The discharge can have adverse impacts on surface water ecosystems by lowering the pH and enhancing the mobility of trace metals in surface water.

The conventional approach to the environmental control and management has typically involved the collection of the AMD effluent. The effluent is treated by lime neutralization and settling. Within the past decade, other remedial techniques have been developed, but these have proven to be difficult to implement at the scale necessary to achieve success. For example, the performance of low permeability covers to prevent the introduction of oxygen to mine wastes may be difficult to predict and maintain (Bennett and Ritchie, 1991). Constructed wetlands may have limitations in achieving treatment to objective levels for discharge from full-scale mine operations (Eger et al. 1994; Sobolewski et al. 1995). In recent years, the use of permeable reactive barriers (PRBs) to intercept and treat AMD–impacted groundwater has been proposed (Blowes and Ptacek, 1994; Blowes et al., 1994, 2000; Benner et al., 1997; Waybrant et al., 1998). As indicated in Figure 1, the purpose of the PRB system is to remove metals from solution and decrease the acid-generating potential of the groundwater prior to its discharge to surface water.

A PRB contains permeable materials and is constructed in a manner to facilitate the passive interception of contaminated groundwater. Treatment of the contaminants of concern occurs within or down-gradient of the PRB system by a combination of physical, chemical or biological processes. In the remediation of AMD-impacted groundwater, the purpose of a PRB system is to provide a source of carbon to enhance bacterially mediated sulfate reduction and promote the precipitation of low solubility metal sulfides. Sulfate reduction and the consumption of simple organic carbon produces hydrogen sulfide and alkalinity. The reaction can be represented by:

$$SO_4^{2-} + 2CH_2O \rightarrow H_2S + 2HCO_3^- \qquad (3)$$

The production of dissolved sulfide in the presence of metals can result in the precipitation of low solubility metal sulfides:

$$Me^{2+} + H_2S \rightarrow MeS + 2H^+ \qquad (4)$$

Me^{2+} denotes a metal such as Fe, Cd, Ni, Co, Cu, Zn, As or Zn. If the hydrogeological setting is maintained, and the PRB remains below the water table, the sparingly soluble metal sulfide precipitates are expected to be stable for the very long term.

The University of Waterloo has participated in three field-scale applications of organic-carbon based PRB technology to remediate groundwater containing elevated concentrations of sulfate and iron or other metals. The systems have been operating for between one and five years, and each provides strong evidence of effective treatment of contaminated groundwater. The field-scale applications suggest that PRB systems can form an important component of environmental control of AMD and similar problems. The reactive materials offer the advantage of being widely available, low cost and, on occasion, being waste products of municipal or industrial activity.

2. PRB AT AN INACTIVE MINE-TAILINGS SITE

At the Nickel Rim Mine near Sudbury, Ontario, a plume of groundwater affected by sulfide oxidation has migrated from inactive pyrrhotite-rich tailings through an aquifer and discharged to a lake approximately 160 m from the tailings impoundment (Bain et al. 1995). The Nickel Rim tailings area is approximately 10 ha, and approximately 25% of the total discharge from the tailings area contributes to this plume. The aquifer is fine-grained alluvial sand and is bounded by bedrock. The plume contains elevated concentrations of sulfate (1,000-5,000 mg/L) and iron (200-2,500 mg/L) and pH ranges from 4 to 6. Some surface seepage develops at the toe of the tailings dam, so acidic surface water is present over portions of the aquifer throughout the year. This surface water either recharges the aquifer between the tailings dam and lake, or discharges directly to the lake.

Benner et al. (1997, 1999) installed a PRB system to treat AMD-impacted groundwater in the shallow aquifer in 1995. The PRB extends across the bedrock valley, and is 20 m long, 3.6 m deep and 4 m thick in the direction of groundwater flow. The reactive materials consist of a mixture of municipal compost (20% by volume), leaf mulch (20%), wood chips (9%), limestone (1%) and gravel (50%). Zones of sand approximately 1 m thick bound both the up- and down-gradient faces of the PRB. The PRB system was installed in a backhoe-excavated trench and, after installation, was covered by a clay cap to minimize direct infiltration of surface water into the reactive materials. The installation and material cost of the PRB was $35,000 Cdn ($25,000 US). The horizontal component of groundwater velocity in the vicinity of the PRB is approximately 15 m/a, with local velocities of up to 30 m/a.

After five years of performance, it is clear that water quality has been improved significantly within and down-gradient of the PRB. Within the first year, alkalinity increased by as much as 2,500 mg/L (expressed as mg/L $CaCO_3$), sulfate decreased by as much as 3,000 mg/L, iron decreased by as much as 1,300 mg/L and nickel decreased by as much as 30 mg/L within the PRB. Cross-sectional distributions along the centerline in the direction of groundwater flow are shown for sulfate, iron, alkalinity and nickel one year after the installation of the PRB (Figure 2). The alkalinity increase was initiated in the up-gradient sand zone. It was associated with an increase in pH to approximately 6 and was likely a consequence of calcite dissolution. Most of the observed increases of alkalinity occurred in the reactive organic carbon material. The groundwater was transformed from net acid producing to net acid consuming through removal of dissolved ferrous iron and the addition of alkalinity as the water moved through the PRB. The water leaving the PRB has the capacity to consume acidity in the down-gradient portion of the aquifer and in the receiving lake.

The decrease of sulfate within the reactive organic carbon is attributed to microbially mediated sulfate reduction. Evidence for sulfate reduction was provided by the presence of sulfate-reducing bacteria, the presence of dissolved sulfide (17 mg/L) and isotopic enrichment of ^{34}S in remnant sulfate (Benner et al. 1999). The identification of iron mono-sulfide solids including mackinawite on cores from within the PRB was achieved using scanning electron microscopy with energy-dispersive x-ray analysis (SEM-EDX) (Herbert et al. 2000). Benner et al. (2000a) suggest the most active sulfate-reduction occurred during the three-month period immediately following installation of the PRB and that reactivity gradually decreased with time thereafter.

The removal of iron from solution within the PRB is attributed to the formation of iron-sulfide minerals. The total mass of solid-phase sulfide precipitates accumulating in the barrier is consistent with the calculated mass of S and Fe removal (Benner et al. 2000b). These results

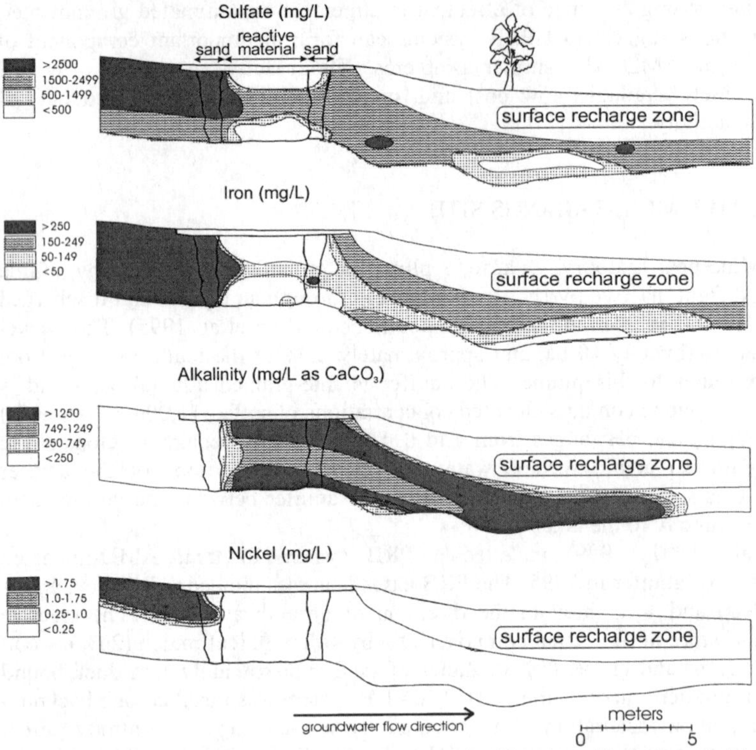

Figure 2. Cross-sections showing the concentrations of sulfate, iron, alkalinity and nickel in groundwater in June 1996 in the vicinity of the PRB at the Nickel Rim Mine near Sudbury, Ontario (from Benner et al. 1997).

are further confirmed by geochemical-speciation modelling using MINTEQA2 (Allison et al. 1990), with a modified database (Ball and Nordstrom 1991), which indicated the groundwater was saturated with respect to mackinawite and amorphous iron sulfide (Benner et al. 1999).

Monitoring of the performance of the PRB system has indicated spatial and temporal variations in the extent of remediation. Heterogeneity of hydraulic conductivity, loss of reactivity with time, and fluctuations of temperature all may contribute to variations in treatment. As shown in Figure 2, there is a zone of less removal of sulfate and iron and less production of alkalinity in the middle portion of the centerline profiles parallel to the direction of groundwater flow. Benner et al. (1997; 1999) attributed this pattern to the presence of higher hydraulic conductivity materials and higher rates of groundwater flow through this central zone of the PRB. The hydraulic conductivity in this central zone is believed to be a factor of approximately three higher in comparison to other zones of the barrier. The average residence time of water within the central zone of the PRB was approximately 60 days, whereas it was as much as 165 days above and below this zone. With lower residence time, less complete sulfate and iron removal occurs. The fraction of sulfate and iron removed in the zones of slower flow was more than twice that in the zone of faster flow in the central part of the PRB (Benner et al. 2000a). The preferential flow of water through the central zone may also preferentially flush labile organic carbon from this part of the PRB (Benner et al. 2000b).

Variations in temperature also appeared to influence the rates of treatment within the PRB. Temperature variations at ground surface range from less than 0°C in the winter to in excess of 25°C in the summer months. Temperatures within the shallow PRB fluctuate with air-temperature variations, and are also influenced by the temperature of surface water that infiltrates to the groundwater up-gradient of the barrier. Benner et al. (2000b) found higher levels of sulfate and iron removal in the late summer months relative to those in the post-winter months over four years of monitoring, and attributed this trend to increased activity of sulfate-reducing bacteria when groundwater temperatures were warmest.

Performance monitoring also indicated the reactivity of the PRB has decreased with time (Benner et al. 2000b). Using data for the first three years of operation in proximity to the central zone of higher flow, vertically averaged concentrations of sulfate and iron entering the PRB were approximately 27 mmol L^{-1} (2,600 mg/L) and 10 mmol L^{-1} (560 mg/L) respectively. The average concentrations down-gradient of the PRB increased from less than 17 mmol L^{-1} (1,630 mg/L) to more than 23 mmol L^{-1} (2,200 mg/L) for sulfate and from less than 1 mmol L^{-1} (56 mg/L) to approximately 6 mmol L^{-1} (335 mg/L) for iron.

Processes that may lead to decreasing reaction rates in PRBs include consumption of reactive material, depletion of the reactive component of a treatment media, or decreases in reactive surface area arising from precipitation of secondary minerals on reactive materials, clogging or preferential flow. Benner et al. (1999) suggested that the PRB theoretically contained sufficient organic carbon to sustain sulfate-reduction and removal of contaminants for more than 100 years. Only a portion of the organic carbon, however, is likely readily available for sulfate-reducing bacteria. Benner et al. (2000b) indicated that the decreasing reactivity of the PRB with time might have been a consequence of the depletion of the reactive fraction of the organic carbon from the zone of higher groundwater flow in the central portion of the barrier. Increasing the residence time of the contaminated groundwater within the PRB could improve treatment. This could be achieved by increasing the thickness of the PRB or by construction designs that facilitate the uniform distribution of groundwater flow through the PRB.

3. SULFATE-REDUCTION PRB AT AN INDUSTRIAL SITE

Sulfide-rich ore material was introduced to shallow soils as a consequence of the historical storage of ore concentrate at a former industrial property in Vancouver, British Columbia. Most of the iron sulfide minerals associated with the ores, which originated at distant mine sites, had been removed in the concentration process and were not shipped to the industrial site. The oxidation of the sulfide minerals such as chalcopyrite and sphalerite has introduced elevated concentrations of sulfate and metals including copper, zinc, cadmium, nickel and lead to the shallow groundwater flow system. The shallow aquifer consists of fill and alluvial sand and gravel, which contain cobbles and boulders. The site borders on a tidally influenced channel (McGregor et al. 1999; 2000).

In 1997 a demonstration-scale PRB was installed to evaluate its potential for the removal of dissolved metals from groundwater. The test wall was constructed using slurry-trenching techniques. The slurry used to support the trench was a biodegradable guar-gum liquid. The reactive materials, which consisted of a mixture of leaf compost (approximately 15% by volume), limestone (approximately 1%) and pea gravel (approximately 84%), were placed through the slurry using a clamshell bucket so that the trench could be filled from the base upwards. The approximate dimensions of the PRB were 10 m in length, 6 m in depth and 2.5 m in thickness in the direction of groundwater flow. The guar gum is soluble in water and is biodegradable, so it was likely that the slurry disappeared from the PRB within a matter of weeks. Furthermore, the guar gum is a potential energy source for sulfate-reducing bacteria, and its presence may have promoted the development of active sulfate reduction in the PRB within days of installation. McGregor et al. (2000) estimated the residence time of water within the PRB to be three days based on hydraulic conductivity values calculated from single-well tests and on hydraulic head levels when low-tide conditions existed in the receiving channel.

Monitoring over a three-year period indicated excellent attenuation of the metals of concern. The predominant removal mechanism has been sulfate reduction and the subsequent precipitation of metal sulfides. In the zones where influent concentrations of sulfate were of the order of 1,500 mg/L, several hundred of milligrams per liter were removed by sulfate reduction within the PRB. Evidence for sulfate reduction included the development of low Eh conditions,

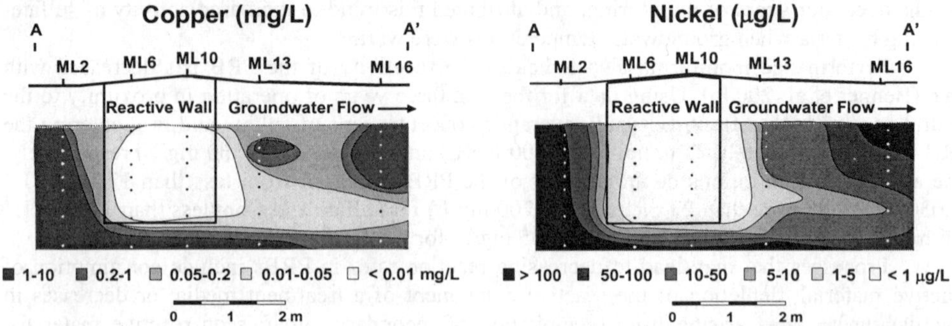

Figure 3. Cross-sections showing the concentrations of copper and nickel in groundwater in the vicinity of the PRB at an industrial site in Vancouver. The data were collected 21 months after installation of the PRB (from McGregor et al. 2000).

the generation of dissolved sulfide concentrations of as much as 1.7 mg/L and the production of alkalinity within the PRB. Geochemical modelling using MINTEQA2 (Allison et al. 1990) suggested that conditions within the PRB favored the precipitation of iron and copper sulfide minerals. Average contaminant concentrations decreased from 0.0159 to 0.0001 mg/L for cadmium, from 4.510 to 0.0077 mg/L for copper, from 0.118 to 0.0065 mg/L for nickel, from 2.396 to 0.082 mg/L for zinc and from 0.0038 to 0.0019 mg/L for lead between monitoring wells immediately up- and down-gradient of the PRB almost two years after its installation (McGregor et al. 2000). The distribution of copper and nickel in groundwater in the vicinity of the PRB is shown in Figure 3.

Column tests using the same reactive materials and site groundwater were performed in the laboratory subsequent to the installation of the trial PRB. The reactive mixture in one column was amended by the addition of zero-valent iron (10% by volume). The tests demonstrated that the addition of reactive iron can enhance the rate of sulfate reduction and the removal of metals relative to that achieved by the standard compost-limestone-gravel mixture. The results of the laboratory testing imply that a residence time for the site groundwater of approximately five days within a full-scale PRB is required to achieve removal of metals to the low target levels. This estimate incorporates allowances for lower reactivity of the reactive materials at the lower temperatures in the field relative to those of the laboratory, and decreasing reactivity of materials with time.

4. REACTIVE LAYERS IN MINE TAILINGS

PRB technology has potential applications directly in mine-tailings impoundments. This approach is particularly attractive because it can be initiated in conjunction with the disposal of tailings, and has the potential of immobilizing AMD contaminants within the tailings and preventing off-site discharge. Blowes and Ptacek (1991) described the concept of incorporating organic carbon below the water table with mine tailings to promote the treatment of AMD within the tailings. Sulfide oxidation occurs within the tailings in the zone above the water table, consistent with Figure 1. The products of sulfide oxidation (sulfate, iron and other metals) are transported from the vadose zone to the groundwater zone. The organic carbon in the groundwater zone is intended to promote favorable conditions for sulfate reduction and the formation of metal sulfide minerals. If hydraulic conditions within the tailings are maintained, the metal sulfide minerals will remain sparingly soluble and stable. The removal of sulfate, iron and other metals from solution within the tailings would improve seepage-water quality and decrease the negative effects of AMD.

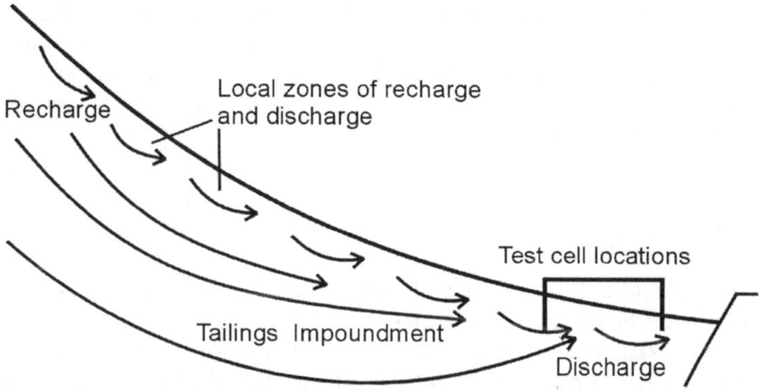

Figure 4. Cross-sectional diagram showing general pattern of groundwater flow within the tailings at the Kidd Creek Metallurgical Site. The test treatment cells were installed in close proximity to the perimeter drain.

Concentrator-tailings at the Kidd Creek Metallurgical Site near Timmins, Ontario are deposited from a series of central spigot points as a thickened tailings-slurry containing 62 weight per cent solids. The tailings form a conical pile of approximately 15 m in depth and several thousand meters in diameter within the 1,200-hectare impoundment. Sulfide minerals including pyrite, pyrrhotite, chalcopyrite and sphalerite comprise as much as 25% of the tailings. Natrojarosite, a by-product of the zinc-refining process, has been co-disposed with the tailings since 1985, and releases sulfate, iron, sodium and trace metals to the tailings pore water in addition the products of sulfide oxidation in the central part of the tailings cone (Al et al. 1994). Groundwater flow within the tailings occurs downward in the central area of the cone and horizontally towards the perimeter of the impoundment (Figure 4). In the general vicinity of the perimeter drain, there is an upward component to groundwater flow, and the water table occurs within several tens of centimeters of the surface of the tailings. Surface water, which discharges at the perimeter of the tailings cone is collected and treated prior to release.

In September 1999 three treatment cells were installed within tens of meters of the perimeter drain. The cells were constructed of cold-rolled steel sheet piles to a depth of approximately 2 m. The cells were approximately 3 meters square. In two of the cells, different organic-rich waste materials from the forestry industry were incorporated in the upper meter of tailings (approximately 15% by volume) by physical mixing with a backhoe bucket. In the third cell, the upper meter of tailings was physically disturbed with the backhoe bucket, but no materials were added to the tailings. Each cell was instrumented with two sets of multi-level stand-pipe piezometers at depths from 0.5 to 2 m below ground surface. Similar instrumentation was installed at two adjacent locations in undisturbed tailings.

Monitoring of the tailings pore water by piezometer sampling and the collection and squeezing of soil cores in and adjacent to the cells has been conducted on four occasions. The shallow pore water is characterized by very high concentrations of sulfate (approximately 25,000 mg/L), iron (6,000 to 8,000 mg/L) and zinc (500 to 800 mg/L). Initial evidence suggests that sulfate reduction is occurring in one of test cells containing organic forestry wastes. The respective concentrations of sulfate and iron are of the order of 2,000 mg/L lower in the pore water within one meter of ground surface in this treatment cell in comparison to those in the tailings cell and the undisturbed tailings. Similarly, the concentrations of zinc in the same depth interval are several hundreds of milligrams per liter lower within the active treatment cell. Treatment has not been as pronounced in the second test cell. Although complete removal of sulfate, iron, zinc and other metals has not been achieved, the preliminary results indicate that significant contaminant removal has occurred in at least one of the treatment cells. This performance needs to be confirmed over a period of years. Furthermore, the mechanism for emplacement of the reactive materials on a tailings-basin scale needs to be developed, but the field trial has indicated that contaminant removal can be achieved. At on-going or inactive tailings facilities such as the Kidd Creek Metallurgical Site, in situ removal of contaminants by reactive organic mixtures in the tailings could decrease demands on and costs for conventional water-treatment. If reactive materials were incorporated with the mill wastes from the start of operations at new mining operations, the benefits and cost savings for environmental management of tailings are potentially greater.

5. SUMMARY AND CONCLUSIONS

Demonstration and full-scale applications have indicated that the use of PRB technology has the potential to provide significant advantages and benefits to the environmental control of AMD and similar groundwater. Reactive materials, including mixtures of organic compost, limestone or gravel, have been shown to enhance the activity of sulfate-reducing bacteria. The transformation of sulfate to sulfide is accompanied by the precipitation of iron and other metal sulfides, which are sparingly soluble and stable in many hydrogeological environments. A product of the microbial activity is the generation of alkalinity, and this may be increased further by interaction between the contaminated groundwater and limestone. Through the removal of iron and other metals and the increase in alkalinity, the quality of effluent is significantly superior to that of groundwater entering a PRB. The effluent water contains low concentrations of metals. Furthermore, the water can be transformed from net acid-producing to net acid-consuming within and down-gradient of the PRB.

The field experience to date has shown that the reactivity of organic materials is variable, and that rate of treatment is directly dependent on the rates of sulfate reduction. Preliminary laboratory experiments suggest that the rates of reactivity and the extent of removal of metals can be enhanced through the addition of zero-valent iron to the reactive materials. The applications have shown that residence time of contaminated groundwater within the PRB has an important influence on the extent of treatment achieved. The residence time of contaminated groundwater within sulfate-reduction PRBs must typically be a minimum of several days. The residence time of water and contaminants within the PRB is directly proportional to barrier thickness. Uniform hydraulic-conductivity characteristics are also important to prevent preferential flow, and shorter residence times, in isolated zones within the PRB.

The longevity of PRBs for AMD treatment requires further evaluation. The prototype barrier at the Nickel Rim mine was installed in 1995, where peak influent concentrations of ferrous iron were very high (>2,000 mg/L). Although the rates of treatment are decreasing with time, the PRB continues to remove significant concentrations of dissolved iron from solution, and decrease the acid-producing potential of the groundwater. Based on the field experience, it appears that the construction of PRBs to last at least three to five years is achievable, and that longer life spans may be feasible. Prolonged treatment of copper, zinc, nickel and lead to levels that meet regulatory guidelines at the Vancouver site suggest the PRB technology may be well suited to treating persistent low flows of low concentrations of dissolved metals.

The use of reactive materials to remove AMD contaminants from groundwater within tailings impoundments has shown some promising results. The preliminary work has been conducted on an existing active tailings basin, and the removal of significant concentrations of iron and other metals has been achieved. Issues related to the placement of the materials at the tailings-basin scale at active or inactive tailings and the longevity of performance require further investigation. The technology does, however, have the potential to reduce the flux of AMD-impacted groundwater from tailings basins, thereby decreasing the need for active water treatment and the negative effects on receiving surface water.

The application of the PRB technology may have significant cost advantages over conventional collection and treatment systems for AMD-impacted groundwater. Installation of the full-scale PRB at Nickel Rim was completed for an approximate cost of $25,000 US. The Nickel Rim tailings area is approximately 10 ha, and approximately 25% of the total discharge from the tailings area is treated by the PRB system, without the requirement for additional infrastructure or maintenance expenditures.

ACKNOWLEDGMENTS

Funding for the research and projects was provided in the form of grants and contracts to the University of Waterloo by Falconbridge Limited, the Natural Sciences and Engineering Research Council of Canada, Environment Canada, Conor PacificEnvironmental Technologies Inc, the Province of Ontario (Premier's Research Excellence Award to D. Blowes) and the Ontario Centre for Research in Earth And Space Technology. The authors appreciate the assistance of Chris Hanton-Fong.

REFERENCES

Al, T.A., D.W. Blowes, J.L. Jambor, and J.D. Scott 1994. The geochemistry of mine-waste pore water affected by the combined disposal of natrojarosite and base-metal sulphide tailings at Kidd Creek, Timmins, Ontario. *Canadian Geotechnical Journal*, 31:502-512.

Allison, J.D., D.S. Brown and K.J. Novo-Gradac 1990. *MINTEQA2/PRODEFA2, a geochemical assessment model for environmental systems: User's manual.*. Athens GA: U. S. Environmental Protection Agency.

Bain, J.G., D.W. Blowes and W.D. Robertson 1995. The hydrogeochemistry of a sand aquifer affected by drainage from the Nickel Rim mine tailings impoundment. *Sudbury '95, Mining and the Environment*: 715-723. Ottawa: CANMET.

Ball, J.W., and D.K. Nordstrom 1991. *User's manual for WATEQ4F, with revised thermodynamic data base.* U.S. Geological Survey, OFR 91-83.

Benner, S.G., D.W. Blowes and C.J. Ptacek 1997. A full-scale porous reactive wall for prevention of acid mine drainage. *Ground Water Monitoring and Remediation*, XVII: No. 4, 99-107.

Benner, S.G., D.W. Blowes, W.D. Gould, R.B. Herbert Jr., and C.J. Ptacek 1999. Geochemistry of a reactive barrier for metals and acid mine drainage. *Environmental Science and Technology* 33:2793-2799.

Benner, S.G., D.W. Blowes, C.J. Ptacek and U. Mayer 2000a. A permeable reactive wall for treatment of mine drainage: Long-term results. In: *ICARD 2000, Proceedings from the Fifth International Conference on Acid Rock Drainage (Society of Economic Geologists), Denver, Colorado, May 21-24*: 1221-1225. Littleton: Society for Mining Metallurgy and Exploration.

Benner, S.G., D.W. Blowes, C.J. Ptacek and U. Mayer 2000b. Rates of sulfate reduction and metal sulfide precipitation in a permeable reactive barrier. *Applied Geochemistry* In revision.

Bennett, J.W. and A.I.M. Ritchie, 1991. Measurements of the transport of oxygen into two rehabilitated waste rock dumps. *Proceedings of the Second International Conference on the Abatement of Acidic Drainage (MEND), Montreal, September 16-18*: Ottawa: CANMET.

Blowes, D.W. and C.J. Ptacek, 1991. Treatment of mine tailings. U.K. patent 2,219,617 filed June 9, 1988, issued August 19, 1992, U.S. patent 4,990,031 filed June 7, 1989, issued February 5, 1991, Canada patent 1,327,027, filed June 7, 1989, issued February 15, 1994.

Blowes, D.W. and C.J. Ptacek, 1994. System for treating contaminated groundwater. U.S. Patent 5,362,394 filed March 3, 1992, issued Nov. 8, 1994, Canada patent 2,062,204, filed March 3, 1992, issued July 7, 1998, Europe Patent 92103559.8, filed March 2, 1992, allowed, January 2000.

Blowes, D.W., C.J. Ptacek, E.O. Frind, W.D. Robertson and J.W. Molson, 1994. Acid-neutralization reactions in inactive mine tailings impoundments and their effect on the transport of dissolved metals. *Proceedings of the Third International Conference on the Abatement of Acidic Drainage, Pittsburgh, PA, April 24-29*. 1 429-438. U.S. Department of the Interior, Bureau of Mines SP 06A-94

Blowes, D.W., C.J. Ptacek, S.G. Benner, C.W.T. McRae and R.W. Puls 2000. Treatment of dissolved metals and nutrients using permeable reactive barriers. *Journal of Contaminant Hydrology:* 45, 123-137.

Eger, P., J.R. Wagner, Z. Kassa and G.D. Melchert, 1994. Metal removal in wetland treatment systems. *Proceedings of the Third International Conference on the Abatement of Acidic Drainage, Pittsburgh, PA, April 24-29*. 1 80-88. U.S. Department of the Interior, Bureau of Mines SP 06A-94.

Herbert Jr., R.B., Benner, S.G., and Blowes, D.W., 2000. Solid phase iron-sulfur geochemistry of a reactive barrier for treatment of acid mine drainage. *Applied Geochemistry:* 15, 1331-1343.

McGregor, R., D. Blowes, R. Ludwig, E. Pringle, M. Choi, R. Booth, M. Duchene and M. Pomeroy, 1999. Remediation of heavy metal-contaminated groundwater using a porous reactive wall in an urban setting. In: *Proceedings of the Sudbury'99 Mining and the Environment II Conference*, September 13-17, 1999, Sudbury, Ontario, 645-653.

McGregor, R.G., Blowes, D.W., Ludwig, R. and Choi, M., 2000. The use of an *in situ* porous reactive wall to remediate a heavy metal plume. In: *ICARD 2000, Proceedings from the Fifth International Conference on Acid Rock Drainage, (Society of Economic Geologists), Denver, Colorado, May 21-24*: 1227-1232. Littleton: Society for Mining Metallurgy and Exploration.

Sobelewski, A., L. Gormely and R.U. Kistritz, 1995. Copper removal from mine drainage by an experimental wetland at Bell Copper Mine, Smithers, B.C. *Sudbury '95, Mining and the Environment*: 683-692. Ottawa: CANMET.

Waybrant, K.R., Blowes, D.W., and Ptacek, C.J., 1998. Prevention of acid mine drainage using *in situ* porous reactive walls. Selection of reactive mixtures. *Environmental Science and Technology* 32: 1972-1979.

Tailings and Mine Waste '01 © 2001 Balkema, Rotterdam, ISBN 90 5809 182 1

A chemical and ecotoxicological assessment of the impact of marine tailings disposal

M.C. Blanchette, T.P. Hynes & Y.T.J. Kwong
CANMET, Natural Resources Canada, Ottawa, Ont., Canada
M.R. Anderson, G. Veinott, J.F. Payne & C. Stirling
Fisheries and Oceans Canada, St. John's, Nfld, Canada
P.J. Sylvester
Memorial University of Newfoundland, St. John's, Nfld, Canada

ABSTRACT: Marine tailings disposal (MTD) promises to be an efficient means of arresting sulfide oxidation and preventing the occurrence of acid drainage. Like disposal of reactive mine wastes under a fresh water cover, metal leaching from the submerged tailings and the subsequent biological impact under saline conditions remain major concerns. To assess the general applicability of MTD to coastal mines, two historic Cu mines in the Notre Dame Bay area of Newfoundland, Canada were selected for detailed study.

A three year, multi-disciplinary research program was initiated in the summer of 1999 by Natural Resources Canada (CANMET), Fisheries and Oceans Canada (DFO), and the Memorial University of Newfoundland. Detailed site assessments, including near-shore multi-beam acoustic surveys and field sampling, were carried out at both Tilt Cove and Little Bay in the first year. Selected results related to chemical characterization of the mine waste, leach testing and metals mobility under fresh and saline conditions, impact zone delineation through isotope and trace element ratios as well as ecotoxicological assessments to define and delineate risk associated with acute and chronic toxicity are presented.

1 INTRODUCTION

Mining is a major component of the Canadian economy and approximately 1.9 billion tonnes of reactive tailings now exist at various mine sites across the country (Feasby and Jones, 1994). During the last decade it has been shown that storage under water has been highly successful in preventing further oxidation, and where site conditions permit, this has become the preferred disposal option using non-productive natural lakes or man-made facilities. In coastal regions where deep, restricted submarine basins occur, mine wastes may potentially be disposed with little ecological impact (Poling and Ellis, 1993, Ellis *et al*, 1995). The chemistry and reaction kinetics of tailings stored under fresh water are now well known (Pederson and Losher, 1988, MEND 1990a&b, 1992), significantly less is known of storage under marine conditions.

1.1 *Project objectives*

The present work is a multi-stakeholder project developed by a steering committee composed of representatives from four federal departments, two provincial departments of Newfoundland, and Memorial University and is funded in part through the joint Environment Canada/Health Canada Toxic Substances Research Initiative. The project represents a multi-disciplinary effort to examine the long-term environmental and ecotoxicological consequences of historic marine tailings discharges off the coast of Newfoundland.

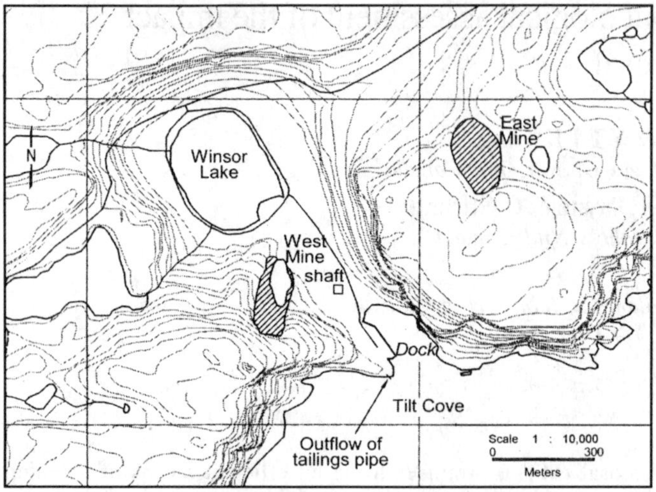

Figure 1: Schematic of Tilt Cove mine and surrounding area

1.2 *Site history and description*

Two abandoned mine sites in the Notre Dame Bay area of Newfoundland, both of which hosted massive sulfide mineralization, were selected for detailed assessment. The production history and geological and physical attributes of these sites have been compiled from available documentation (Dimmell, 1999), a brief overview of which is presented below. Both sites were considered to provide ideal conditions for evaluating the biological impact of complex mixtures of metals without the confounding effects of other major sources of industrial or urban contamination.

The Tilt Cove Mine is located on the Baie Verte peninsula of northeastern Newfoundland on the north side of Notre Dame Bay. Mineralization at Tilt Cove was discovered in 1857 and was first worked between 1864 and 1918. No tailings were produced in this early phase of mining, however at least one shipment of ore is reported to have floundered off Tilt Cove. The property was reactivated in 1957 and operated until 1967 at 2,000 tons/day, producing ~468,000 short tons of copper concentrate from approximately 7.4M tonnes of ore grading approximately 1.5% copper. In this production phase, tailings were slurried by pipeline into deep water (300 feet) in a high energy environment just to the southwest of the harbour. The total tonnage of tailings deposited into the marine environment is unknown but is at least 5-6 million tons, with estimated particle size of 95% less than 200 mesh and 70-80% less than 325 mesh.

The Little Bay Mine is located on the southwest side of Notre Dame Bay on the Springdale peninsula of north central Newfoundland and offers a low energy environment study site. The mine was first worked from 1878 to 1893 and again in 1918. The mine was reactivated in 1961 and operated until 1969 with production totaling ~2,572,000 tonnes of 0.8-2% copper. A total of ~1.8 million tonnes of tailings were generated during this phase of mining and placed in an impoundment area formed by damming off a shallow mixed marine/freshwater bay. An 8-10 m high dam was placed across the mouth of Shoal Bay. The dam breached in 1974 (Dimmell, 1999) and an estimated 30-50% of the tailings now form a beached area below the dam into Little Bay Arm.

1.3 *Project scope*

The 3 year project, initiated in 1999, consists of three major elements: chemical assessment; impact zone delineation; and, ecotoxicology. This integrated approach is providing the opportunity to evaluate the movement and fate of metals released from the tailings into the

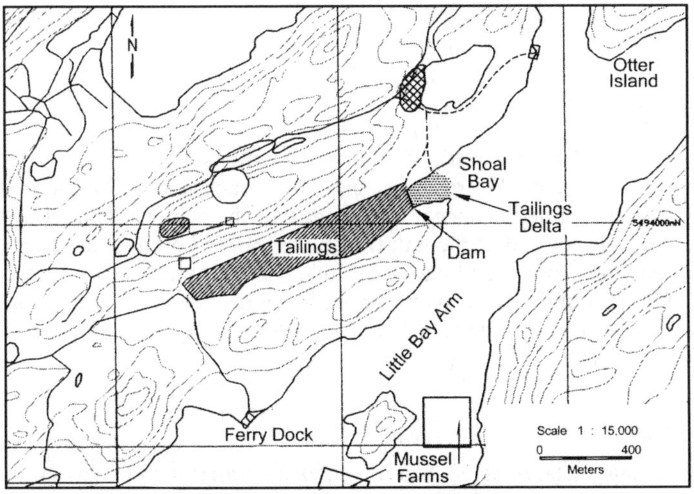

Figure 2. Schematic of Little Bay mine and surrounding area

marine environment and to assess the toxicological consequences for marine fish and shellfish at two historic MTD sites. Selected results from the first year of this study are presented.

2 CHEMICAL ASSESSMENT

Through analysis of the waste materials collected from Tilt Cove and Little Bay and their leaching characteristics the potential and actual availability of metals and oxidation of mine wastes under saline versus fresh water conditions is being assessed by CANMET.

2.1 *Reconnaissance Survey*

Reconnaissance surveys of the two study sites were conducted in 1999, and both water and solid samples collected. Both the Tilt Cove and Little Bay sites host massive sulfide and related deposits enriched in copper as chalcopyrite (see Tables 1 and 2). Little Bay tailings have a total sulphur content of ~1.5-2.0% and a net neutralization potential (NNP) of approximately -45 kg $CaCO_3$/tonne. The Tilt Cove tails contain highly reactive fine-grained pyrite. The total sulphur content is between 18-20% and the NNP is -534 kg $CaCO_3$/tonne. Selected metal concentrations in the two tailings are shown in Table 1.

At Little Bay visible evidence of surface oxidation (i.e. intensive iron staining) is most prominent at the most inland area of the tailings deposit where influx of surface runoff is evident. Seepage from the toe of the tailings deposit at this extremity is acidic (pH as low as 3.4). Progressing towards the dam breach and Little Bay Arm, water in the stream incising the tailings deposit exhibits increasing pH, Ec, as well as chloride and sulphate contents, suggesting increasing influence of saline water from the Arm.

At Tilt Cove iron staining is observed on the mineralized rock faces and waste rock. Bulk tailings deposited just inland of the cove gave a paste pH of 2.7 for the overlying oxidized material and 4.1 at depth (~0.6 m). The pH of the discharge from the closed shaft ranged from

Table 1. Bulk chemical composition of Little Bay and Tilt Cove Tailings

	Al	As	Ca	Cu	Fe	Mg	Mn	Na	Ni	S_{tot}	Zn
Tilt Cove	5.6	0.03	1.3	0.4	30.1	3.2	0.05	0.07	0.02	18.7	0.39
Little Bay	6.1	<0.01	1.9	0.04	13.8	3.7	0.13	0.65	0.008	1.73	0.01

3.4 to 5.9 depending on the season of sampling, while that of the pit water in the west mine varied from 7.1 to 7.5.

2.2 *Metal Mass Balance at Little Bay*

The quantity of tailings that have washed into Little Bay has been calculated in an effort to estimate the metal loadings to the area since the impoundment dam breached in 1974. During the 1999 survey, a reference line was defined connecting the centre of the dam breach and the head of the erosion channel at the western end of the tailings deposit. Based on measurements made along 13 transects of the erosion channel and taking the side channels into consideration, it is estimated that 25% (236,000 m^3) of the tails have escaped the containment dam. Taking into consideration the preferential movement of the fine fraction into the Bay, the total Cu and Zn loadings to Little Bay Arm are estimated at 25 and 21 tonnes, respectively.

2.3 *Batch leaching and sequential extraction*

To determine the extent of immediate metal release when impounded tailings at either site enter a different environment, tailings collected on land from both sites were subjected to an 18-hour regulatory leach test according to the US-EPA 1311 method. The method was modified such that local fresh lake water and saline water collected from Tilt Cove were used as the leach media. The test results for Cu, Ni, Pb, and Zn suggest that a saline water leach of these materials has significantly enhanced the release of Pb, to a lesser extent Ni and Zn, but not Cu.

In addition to batch leach testing, a total of 28 samples were analyzed by sequential extraction at the Geological Survey of Canada. Using a procedure modified after Hall et al. (1996), the analysis determined the partitioning of 29 elements among seven components in the samples: 1) exchangeable ions; 2) carbonate-bound; 3) manganese oxide; 4) amorphous iron oxyhydroxide; 5) crystalline iron oxides; 6) organics and sulfides; and, 7) residual (mainly silicates). Each successive leaching step reflects a greater difficulty of metal release and hence reduced bioavailability. Based on the results it is evident that marine sediments at both sites are depleted in sulphides relative to tailings on land. Both preferential washing of fines through the dam breach and/or dilution with natural marine sediments may account for this depletion. Once entered into the marine environment, co-precipitation with carbonates instead of crystalline iron oxides becomes an important mechanism in the attenuation of Cu transport and perhaps also Ni, making them less available to fish but not necessarily to bottom feeders like mussels.

2.4 *Galvanic interaction and long term lysimeter testing*

Galvanic interaction of sulphide minerals may lead to preferential metal leaching under a water cover despite reduced oxygen availability. Eight polished thin sections were prepared from mineralized specimens containing mixed sulphides from the two study sites and used to investigate metal leaching due to galvanic interaction. Conducted for CANMET at the National Hydrology Research Centre, the method described by Lawrence et al. (1997) was carried out using well oxygenated saline as well as fresh water as the circulating leach media but did not incorporate microbial inoculation. To assess the behavior of galena and sphalerite under these conditions, two additional sections from the Keno Hill mining district in Yukon Territory were included.

Petrographic examination of the polished thin sections at the end of the nine-week experiment shows clear evidence of preferential weathering (tarnishing or dissolution) of minerals with a lower electrode potential (e.g., pyrrhotite, galena, and chalcopyrite) in contacting sulphide pairs. The extent of weathering in many cases appears to be more intensive in the saline versus fresh water medium, likely because the former is a stronger electrolyte to enhance the electron transfer process. In addition, data clearly show that Pb leaching is enhanced under saline versus fresh water cover. Ni leaching also appears to be enhanced under a saline water cover while the data for Cu and Zn are inconclusive. Further investigation under more reducing conditions (e.g.

deep water or presence of decaying organics) are warranted but are not within the scope of the current project

Long term lysimeter testing is currently underway at CANMET to elucidate temporal changes and trends in metal influx to and/or efflux from tailings disposed under water. Conducted in four aquarium-type diffusion lysimeters, 0.25m thick layers of tailings from the two sites rest under oxygenated and well mixed, 0.25m deep fresh and saline water covers. Preliminary surface water quality results indicate a higher reactivity of the Tilt Cove tailings under both fresh and saline water covers relative to the Little Bay tailings. While it is too early to assess detailed differences in metal fluxes under saline versus fresh water environments, the available alkalinity in the saline waters has delayed acid generation and moderated the surface oxidation rate. These tests will be completed by March of 2002.

3 IMPACT ZONE DELINEATION

The movement and fate of metals from the tailings, the level of acute and chronic toxicity of ionic forms and the potential for accumulation in local biota are being assessed through the following activities: a Hydroacoustic survey to delineate physical contours of the submerged tailings; delineation of tailings dispersal using Pb isotope and trace element ratios; and sampling and subsequent chemical analyses to assess metal accumulation in biota.

3.1 *Hydroacoustic Survey*

The Canadian Hydrographic Service (CHS) of DFO conducted an EM 3000 multibeam acoustic survey at Little Bay and Tilt Cove in October of 1999, with over 8 km^2 of seafloor isonified in depths ranging from 1-100 m and with a resolution of ~ 8.0 cm. These data sets were used to generate detailed bathymetric
and backscatter imagery and to do detailed comparisons of historical and new multibeam data sets (Stirling and Roy, 2000).

Figure 3 is a bathymetric image of the Little Bay Arm and clearly depicts the morphology of the seafloor at the site. The most striking aspect is the trench-like feature emanating from the entrance to Shoal Arm and extending downslope into the east basin area of Little Bay Arm channel. A comparison of the backscatter results (not shown) with the depth contours indicate that the deposition of sediments occurs mainly in basin areas, at depths of 30 m or greater.

Single beam bathymetric data was collected in Little Bay Arm in 1960, prior to the dam breach. A detailed, three-dimensional comparison of historical and new multibeam data sets has demonstrated an excellent correlation outside the effected area of the east basin and along the

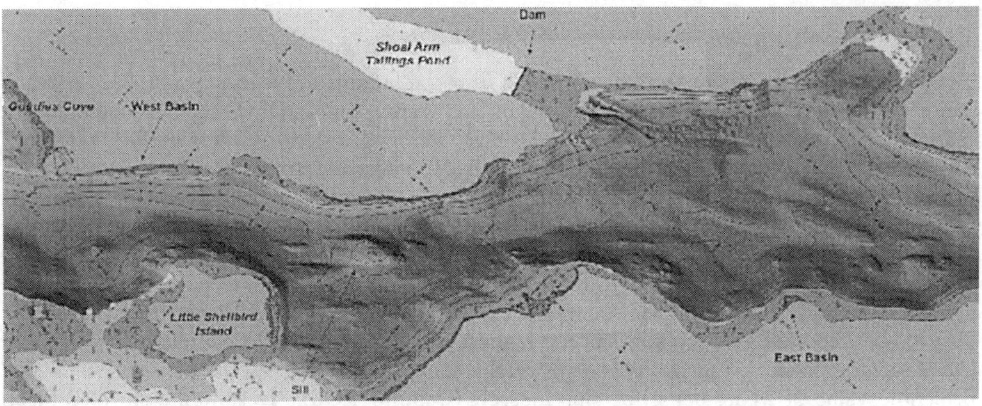

Figure 3. Bathymetric image of Little Bay Arm

Figure 4: Bathymetric image of Tilt Cove mine

lateral areas of Little Bay Arm. However the comparison indicates that as much as 4-5 m of tailings have been deposited at the entrance to Shoal Arm. Using the historical and multibeam data sets, two separate digital elevation models were created, the difference of which is a 'picture' of suspected mine tailings deposition with a resulting volume calculated to be approximately 500,000 m^3. The preferential washing of fine tailings and their subsequent deposition in the Little Bay Arm with less compaction (relative to the on-land tailings) probably accounts for the difference in tailings erosion volume as estimated in the reconnaissance survey. Despite exposure to full wave action of the sea, immediately south and to the west of the entrance to Tilt Cove Harbour (Shaw et al, 1999), the detailed bathymetric image suggested an area of deposition of coarse mine tailings material. The backscatter images (not shown) of the site support this. However, until samples are obtained and analyzed, the conclusion that this material is mine tailings cannot be verified. The comparison of current and historic data sets also shows good correlation outside the zone of impact. Differences in Tilt Cove Harbour, sometimes exceeding 5 m, can be attributed to landfill and wharf construction. As to whether or not the landfill consists of mine tailings cannot be assessed without sampling. Further offshore the possibility exists of finer sediments being deposited in the depressions between bedrock high spots.

3.2 *Metal Accumulation in Biota*

To determine the movement and fate of metals from the tailings at Little Bay and Tilt Cove and the potential for accumulation of the metals in nearby biota, DFO collected samples of sediment, mollusks and seaweed from the study sites and a control site (Smith's Harbour). Samples were collected from the tailings discharge sites and from locations progressively further from the discharge sites to approximately 2 km.

Concentrations of Cu, V, Mn, Co, and Fe were significantly higher in the soft tissue of blue mussels (*Mytilus edulis*) and soft shelled clams (*Mya arenaria*) in the vicinity of Little Bay compared to the control site. Significant species effects were also detected with the blue mussels generally having higher concentrations of all metals. The Cu concentrations in the blue mussel reported here are some of the highest levels found in wild populations with mean Cu concentration of 45 ug per g of dry weight being double the World Mussel Watch 85th percentile value of 21 ug per g for blue mussels (Cantillo 1998). Despite the elevated metal values in the soft tissue, both wild blue mussels and soft shelled clams are plentiful in the study site and commercial mussel farms operating ~ 1 km south from the Shoal Bay outlet in Little Bay Arm.

Seaweed species *Fucus anceps* and *Ascophyllum nodosum* from the Little Bay area displayed similar patterns of enrichment. Concentrations of Cu and Co were elevated in both species while concentrations of V, Cr, Fe, Ni and Se were elevated in *Fucus* sp. only. Again, both species were plentiful in the study site at Little Bay.

Sediments appear to play a key role in controlling the metal concentrations in biota. At Little Bay, as with the biota, the sediment is elevated in Cu, Mn, Co, and Fe. Further, sediment at the control site shows a higher Pb concentration, again reflected in the higher Pb levels in the control site biota.

3.3 *Tracing the Fate of Metals*

Measurements of the chemical compositions of collected marine sediments were made at Memorial University by ICP-MS and XRF and compared to compositions of on-land tailings. In total six on-land and 17 marine sites, extending from north of the mussel farm to Otter Island, were analyzed. In addition, one marine sediment from Smith's Harbour was also taken as a control. All marine samples were retrieved from the surface of the sea floor. To make direct chemical comparisons of on-land and marine samples, all samples were sieved to a common grain size (<107 microns) prior to analysis.

On-land and marine tailings have similar contents of Cu, Fe, Ni, V, Mn, and Co, values high compared to unaffected estuarine sediments including that from Smith's Harbour. Levels of Zn and Cd are however typical of unaffected sites and, correspondingly, are not enriched in local mussels and clams as measured by DFO.

Most metals are remarkably homogeneous within the marine tailings depositional apron in Little Bay Arm. There is little evidence for systematic geographic variations in the contents of most metals. The total variation in Cu is <25% (except for one site), compared to a factor of 1.5 variation on-land. This suggests that the tailings were well mixed during or after marine deposition. As the Bay is a sheltered estuarine environment, mixing during marine deposition is more probable.

Concentrations of Pb, Sn and Sb do not show the same geographic homogeneity, rather increasing in abundance towards the opposing shore from Shoal Arm to values much greater than those found in the on-land tailings pile. An exotic (non-tailings) source of these metals is indicated, with lead isotopic compositions suggesting contamination by a radiogenic source. Analysis of sediments have been started to evaluate the possibility of contamination from antifouling paints released from the hulls of ships navigating waters in Little Bay Arm.

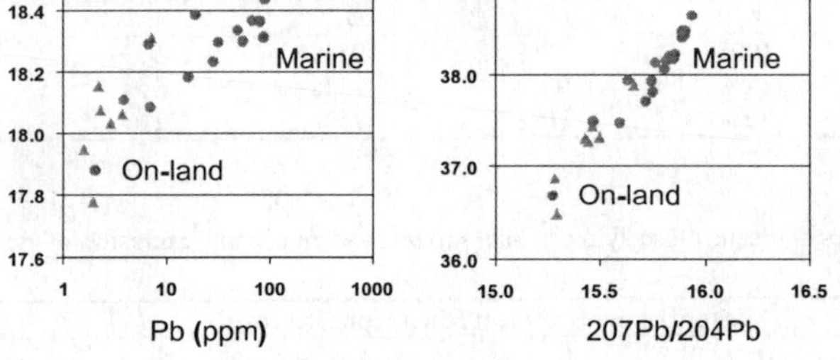

Figure 5. Lead Isotopic compositions of on-land and marine tailings

4 ECOTOXICOLOGICAL ASSESSMENT

The delineation of risks associated with chronic as well as acute toxicity at the study sites is being assessed by DFO through the activities below and preliminary findings on both microscale toxicity and biochemical toxicity indicate the potential for both physical and chemical toxicity at the Little Bay site. The activities are microscale toxicity testing, detailed histopathological studies on flounder and mussels, effects of metal exposure to the immune system of flounder and mussel, and biochemical toxicology and DNA damage studies in flounder and bivalves.

4.1 *Microscale toxicity testing*

A 10-day marine amphipod bioassay was carried out with *Rhepoxynium abronius* on tailings collected from the intertidal area in Little Bay. A key aspect of the assay was to account for grain size whereby physical and chemical mediated toxicity could be "proportioned". Tailings were sieved to obtain the <75 micron fraction. The reference fraction, also <75 microns, was from an estuarine sediment. Locally obtained sediment (DFO control) was homogenized and samples were spiked with tailings and reference fractions to a weight of 20% each. Test conditions for the amphipod bioassay followed Biological Test Method: Reference Method for Determining Acute Lethality of Sediment to Marine or Estuarine Amphipods (Report EPS 1/RM/35 December, 1998).

The slight difference in toxicity between the sediments spiked with tailings versus reference fines was not statistically significant. There was however, a distinct difference in toxicity between the DFO control sediments and those spiked with either tailings or reference fines, indicating the importance of physical toxicity (Table 2). The remaining amphipods exposed to the tailings and reference fines, unlike the control, were moribund at the end of the experiment (Table 3). Tailings often contain a high percentage of fines indicating the potential for physical toxicity alone to alter infaunal sediment communities to a considerable degree.

4.2 *Biochemical Toxicity*

It was established in exploratory studies with cell membranes that tailings from Little Bay had some potential for producing redox activity. Since clams are commonly found in the tailings in Little Bay, preliminary studies were carried out for evidence of an enhanced level of cellular lipid peroxidation in the population relative to the reference site (Smith's Harbour).

Approximately 20 clams were collected from each site and 9000 g fractions of hepatopancreatic and gill tissues prepared. Lipid peroxidation was assessed through the production of malondialdehyde, which reacts with thiobarbituric forming a colored complex

Table 2. Amphipod survival

Test Replicates	% Surviving in sediment (each replicate, n=20)				
	A	B	C	D	E
DFO- Control	100%	100%	95%	95%	95%
DFO- Reference	75%	85%	95%	95%	75%
Tailings	85%	80%	70%	75%	80%

Table 3. Amphipod morbidity – Percentages indicate those lying at the surface and moribund exclusive of dead animals

Test Replicates	% Moribund in sediment (each replicate, n=20)				
	A	B	C	D	E
DFO- Control	0%	0%	0%	0%	0%
DFO- Reference	100%	100%	100%	100%	100%
Tailings	100%	100%	100%	100%	100%

after Uchiyama and Mihara (1978). Intersex variability was determined not to be statistically significant and both sexes from each site were combined for analysis. It is of interest that lipid peroxidation was distinctly enhanced in both the hepatopancreatic and gill tissues of clams from the tailings in comparison to the reference site. Similar studies on mussel hepatopancreatic and gill tissues have shown no enhanced level of oxidative stress.

5 CONCLUSIONS

The main objective of the initial stage of this project has been to gain a better appreciation of the physical and chemical characteristics of the two historic MTD study sites and to collect appropriate samples for various laboratory testing. Preliminary results indicate the following:

1. Whereas submergence of tailings in a marine environment may limit the bioavailability of copper, the mobilization of lead may be enhanced.
2. Sulfide oxidation and metal leaching can still occur to varied extents under a water cover, for example through galvanic interaction, as long as the water column remains oxygenated.
3. Metal concentrations in biota analyzed appear to be controlled by the local sediment composition.
4. An exotic source of lead is apparently present at the Little Bay site.
5. There is potential for both physical and chemical toxicity associated with the Little Bay tailings.
6. Tailings with very low redox activity have potential for producing oxidative damage to biological systems under chronic exposure conditions.

The focus has now shifted to more detailed sample analyses, interpretation, and the integration of the physical, chemical, biological and toxicological components of the study, to be completed in the spring of 2002.

REFERENCES

Dimmell, P.M 1999. Report on research on the history of mining operations and tailings disposal at Little Bay and Tilt Cove, Newfoundland. Sponsored by Canadian Department of Fisheries and Oceans.

Ellis, D.V., G.W. Poling, & R.L Baer 1995. Submarine tailings disposal (STD) for mines: an introduction. *Marine Georesources and Geotechnology,* 13:3-18.

Feasby, G. and Jones, R.K. 1994. Report of Results of a Workshop on Mine Reclamation, Toronto, Ontario, March 10-11, 1994.

Hall, G.E.M., G. Gauthier, J-C. Pelchat, P. Pelchat, & J.E. Vaive 1996. Application of a sequential extraction scheme to ten geological certified reference materials for the determination of 20 elements. *Journal of Analytical Atomic Spectrometry,* **11**: 787-796.

Lawrence, J.R., Y.T.J. Kwong, & G.D.W. Swerhone 1997. Colonization and weathering of natural sulfide mineral assemblages by *Thiobacillus ferrooxidans.* Can. J. Microbiol. **43**: 178-188.

MEND 1990a. *Geochemical Assessment of Subaqueous Tailings Disposal in Buttle Lake, British Columbia.* MEND Report 2.11.1a, March 1990, CANMET, Ottawa.

MEND 1990b. *Geochemical Assessment of Subaqueous Tailings Disposal in Anderson Lake, Snow Lake Area, Manitoba,* MEND Report 2.11.1b, September 1990, CANMET, Ottawa.

MEND 1992. *Chemical Diagenesis of Submerged Mine Tailings in Benson Lake and Natural Sediments in Keogh Lake, Vancouver Island, British Columbia,* MEND Report 2.11.1c, June 1992, CANMET, Ottawa.

Pedersen, T.F. & A.J. Losher 1988. Diagenetic processes in aquatic mine tailing deposits. In *Chemistry and Biology of Solid Waste,* eds. W. Salomons and U. Forstner, pp.238-258, Berlin: Springer-Verlag.

Poling, G.W. & D.V. Ellis, eds 1993. *Case studies of submarine tailings disposal: Volume I – North American examples.* U.S. Bureau of Mines Open-Files Report 89-93.

Shaw, J.F., D.L. Forbes, & K.A. Edwardson 1999. Geological Survey of Canada Bulletin 532, Surficial Sediments and Placer Gold on the Inner Shelf and Coast of Northeast Newfoundland.

Stirling, C. & A. Roy 2000. EM 3000 Multibeam investigation of marine mine tailings at Little Bay Arm and Tilt Cove Notre Dame Bay, Newfoundland. Natural Resources Canada.

Uchiyama, M. & M. Mihara 1978. Determination of malondialdehyde precursor in tissues by thiobarbituric acid test. Anal. Biochem. 271-278.

Tailings and Mine Waste '01 © 2001 Balkema, Rotterdam, ISBN 90 5809 182 1

Conceptual regrading using three-dimensional GIS to evaluate mine reclamation

David L. Jordan
Daniel B. Stephens and Associates Incorporated, Albuquerque, N.Mex., USA

ABSTRACT: Regulators required evaluation of regrading of existing stockpiles at several mine sites in the Southwest to evaluate the viability of using traditional reclamation methods (soil cover and revegetation) at the site. Custom three-dimensional geographic information system (GIS) software was developed to perform conceptual regrading of all of the existing stockpiles, to estimate the volume of material that would have to be moved, and to estimate how much the stockpile regrading would increase the footprint of each stockpile. Initially, a terrain model of present-day topography was built in the GIS using existing AutoCad data provided by the mines. Each stockpile was then conceptually regraded to a less steep slope, and the resulting topography was generated. Results indicated that the volume of material that would have to be moved rendered the project technically impractical. Other calculations made using the GIS included calculations to look at potential increases in mass loading due to the larger footprint of the regraded stockpiles and calculated thickness isopachs of each stockpile. These results were fed into a spreadsheet model that calculated potential mass loading to the underlying aquifer.

1 INTRODUCTION

New software developed using the Avenue programming language for ArcView GIS, and utilizing the 3D Analyst extension, has been applied at several open-pit copper mines in the Southwest to evaluate stockpile regrading scenarios. Each of the mine sites is faced with evaluating reclamation of waste-rock stockpiles from both a technical practicability and cost standpoint. The existing stockpiles have sideslopes in the range of 1.5:1 or steeper and can be hundreds of feet high. State regulators were interested in evaluating the possibility of regrading the existing stockpiles to a 4:1 slope in an attempt to use traditional reclamation methods such as soil cover with revegetation. The GIS was used to develop a terrain model from existing electronic topography data available from AutoCad drawings. The terrain model was then manipulated to create the topography that would result from regrading each stockpile. The methodology was able to balance cut and fill volumes such that cut and fill were optimized to be approximately equal. That is, material was removed from the upper portion of each stockpile and redistributed along the lower portion of the stockpile to decrease the sideslope from 1.5:1 to 4:1.

Conceptual regrading of stockpiles allows assessment of proposed reclamation schemes by evaluating issues such as volume of material moved and the new footprint of the regraded feature. Estimates of cut and fill volumes allow consideration of cost issues for regrading, while evaluation of revised stockpile footprints allows assessment of the effect of regrading on features near the present perimeter of the existing stockpile. The conceptual regrading is especially useful in evaluating the potential increase in the footprint of a stockpile after regrading. In many cases, features proximal to a stockpile — such as buildings, roads, surface-water features, and tanks — may be inundated if the stockpile were to be regraded.

The methods discussed here are presented primarily as they apply to waste-rock stockpiles at hard-rock mines in the Southwest. However, these same methodologies can be applied to other types of mines, such as coal mines, as well as to construction and earth-moving problems in general.

2 METHODOLOGY

Figure 1 presents a two-dimensional representation of the regrading process. A cross section of an existing hypothetical stockpile with a slope of 1.5:1 is shown. In order to balance cut and fill

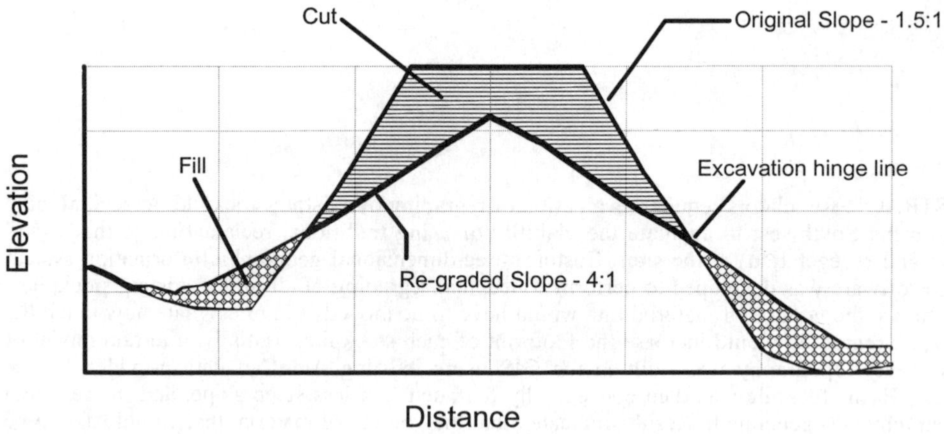

Figure 1. Cross-sectional view showing original and regraded topography of a stockpile

Figure 2. Raw topography data used to build a TIN

Figure 3. Three-dimensional visualization of TIN built from topography data shown in Figure 2

volumes, an iterative process is used to select what will be called the excavation hinge line. The excavation hinge line is defined as the line above which material is removed from the stockpile to create the new slope (in this case, 4:1), and below which material is added to create the new slope. The elevation of the excavation hinge line can be constant, or vary across the stockpile.

In this example, we are assuming a constant-elevation hinge line. For a stockpile with a relatively regular geometry, such as a truncated cone, a constant-elevation hinge line works well. For stockpiles with more complicated geometry, such as a stockpile built against a hillside, typically a line of varying elevation must be used to define the excavation hinge line. A trial-and-error approach must be used to define the excavation hinge line such that the cut volume balances with the fill volume. The cut and fill volumes should be balanced so that if the regrading were actually completed in the field the least amount of material would be moved, thus reducing earthmoving costs to a minimum.

The first step in the regrading process is to build an accurate terrain model of present-day topography for a particular stockpile. This can typically be done using existing electronic data, such as AutoCad files, or other data that are usually already available for most mine sites. The only condition is that the contour lines in the electronic file must have elevations assigned to them. This is typically the case for topography data used for mine-planning purposes. ArcView GIS, with the addition of the 3D Analyst extension, uses a construct called a triangulated irregular network (TIN) to represent three-dimensional surfaces. The TIN represents the surface using contiguous, non-overlapping triangular faces, with a height value for each triangle node, and preserves all observed data points. Figures 2 and 3 show examples of raw topography and the resulting TIN, respectively.

Once a TIN of topography has been constructed, the conceptual regrading process can begin. As discussed above, the first step in the regrading process is to select the excavation hinge line that allows cut and fill volumes to balance. This is an iterative, trial-and-error process by which an analyst must select a starting excavation hinge line, then calculate the cut and fill volumes associated with that particular hinge line. The cut volume is calculated by removing material to reach the desired slope above line hinge line (Fig. 1), and the fill volume is calculated by filling in below the line to reach the desired slope (Fig. 1).

Once a TIN of topography has been constructed, the conceptual regrading process can begin. As discussed above, the first step in the regrading process is to select the excavation hinge line that allows cut and fill volumes to balance. This is an iterative, trial-and-error process by which an analyst must select a starting excavation hinge line, then calculate the cut and fill volumes associated with that particular hinge line. The cut volume is calculated by removing material to reach the desired slope above line hinge line (Fig. 1), and the fill volume is calculated by filling in below the line to reach the desired slope (Fig. 1).

The approach for manipulating the TIN to develop the regraded surface is conceptually straightforward. Each of these steps is performed in an automated fashion, but they are discussed here by breaking the process down into its component steps. The process of removing material above the excavation hinge line (the cut calculation) will be discussed first. The fill process will be discussed second.

The excavation hinge line must be defined as a closed polygon of a known elevation. The hinge line can either have a constant elevation (such as a contour around the perimeter of a stockpile), or the elevation can vary along its perimeter. Once the hinge line is defined, a series of new contour lines that will define the regraded surface are developed by stepping inward and upward from the hinge line at a user-defined step interval.

For example, for a 4:1 slope, if the hinge line is a closed polygon at an elevation of 5100 feet, and the user has defined the step interval as 100 feet, then a new polygon is created at an elevation of 5200 feet that is inside the hinge line and is 400 feet from it at all points along its perimeter (Fig. 4). This type of polygon is also called a buffer. A surface defined by these two polygons will have a 4:1 slope. This process is repeated at successive intervals up the height of the existing stockpile. If the slope of a stockpile is decreased, the elevation of the resulting regraded stockpile will always be lower than the original stockpile.

Once a series of polygons and their elevations has been created, a TIN is automatically generated from the polygons (the regraded surface). A cut/fill calculation is then performed by subtracting the regraded surface from the existing surface, and determining the amount of material that must be removed to regrade to the desired slope. The cut volume can then be compared with the fill volume to check for mass balance. If the final location of the excavation hinge line has been determined, the existing surface inside the hinge line boundary is replaced by the regraded surface. This is performed by clipping out a portion of the existing topography TIN and replacing it with the corresponding clipped portion of the proposed topography TIN.

The fill calculation proceeds in a fashion similar to the cut calculation. Starting at the excavation hinge line, a series of polygons is generated by stepping out and down from the hinge line. To use the example given above, if the hinge line is a closed polygon at 5100 feet (Fig. 4), to create a 4:1 slope, a buffer is created that is at an elevation of 5000 feet and is at all points 400 feet outside the perimeter of the hinge line. This continues in a stepwise fashion until the lowest elevation on the existing surface is reached.

The existing surface must then be replaced by the regraded surface in those areas where it is appropriate. This is accomplished in two steps. First, a cut/fill calculation is performed between the existing surface and the proposed (regraded) surface. The cut/fill calculation will define an area of fill where the proposed surface is above the existing surface. That is, a standard

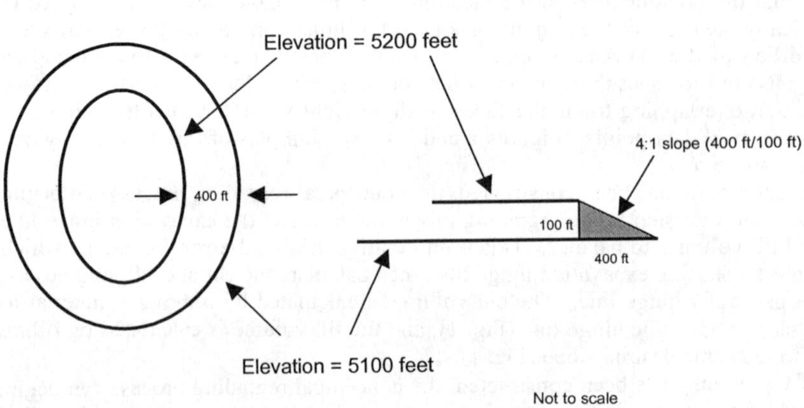

Figure 4. Schematic diagram of the process for creating constant-slope feature

cut/fill calculation will show that where a proposed surface is above an existing surface, material must be added (fill) to the existing surface to bring it up to the level of the proposed surface. The outer boundary of the fill area defines a polygon that indicates the area of the existing topography that will be replaced by the proposed topography.

In the second step, the portion of the existing topography that is contained within the fill area is removed and replaced by the proposed topography. This is done by clipping an area of the existing topography TIN and inserting the corresponding clipped portion of the proposed topography TIN. The two TINs are then combined to create a single TIN, as described above.

For the fill calculation, the user may mask off certain areas that are not allowed to be filled. For example, if a stockpile were adjacent to an open-pit mine, one could mask off the mine excavation area so that it would not be filled during the conceptual regrading process.

3 EXAMPLE REGRADING

Figures 5 and 6 present existing and regraded topography, respectively, for a hypothetical stockpile. The existing slope of the stockpile is 1.5:1, or about 30 degrees. Figure 5 shows the placement of the excavation hinge line. As discussed above, the location of this line is determined by balancing cut and fill volumes to the most accurate extent possible. Figure 6 presents the results of the conceptual regrading of the stockpile to a 4:1 slope, or about 14 degrees. The figure also presents the locations of the footprints of both the original stockpile and the regraded stockpile. Note that the area of the footprint of the stockpile has increased substantially due to material being removed from the top of the stockpile and added to the bottom perimeter in order to reconfigure the stockpile slope to the desired 4:1 slope.

In addition to estimating cut and fill volumes, the ability to evaluate the revised footprint of the regraded stockpile is a powerful tool in evaluating the technical practicability of a potential regrading operation. For example, knowing that a group of buildings or a stream channel would become inundated during a particular regrading scenario is an important factor for planning purposes.

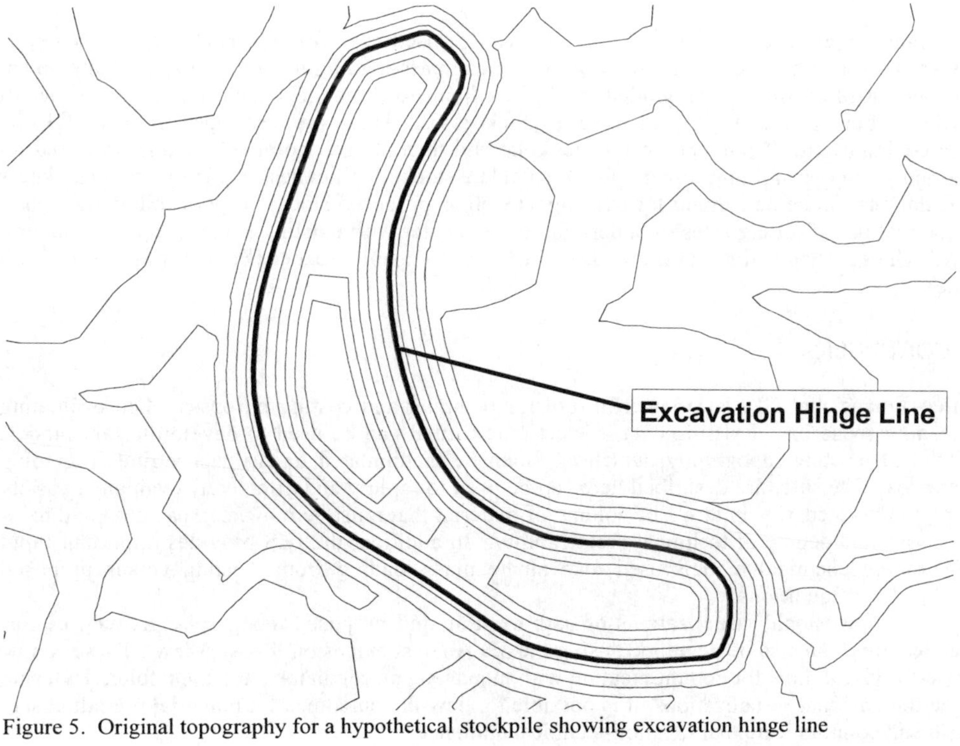

Figure 5. Original topography for a hypothetical stockpile showing excavation hinge line

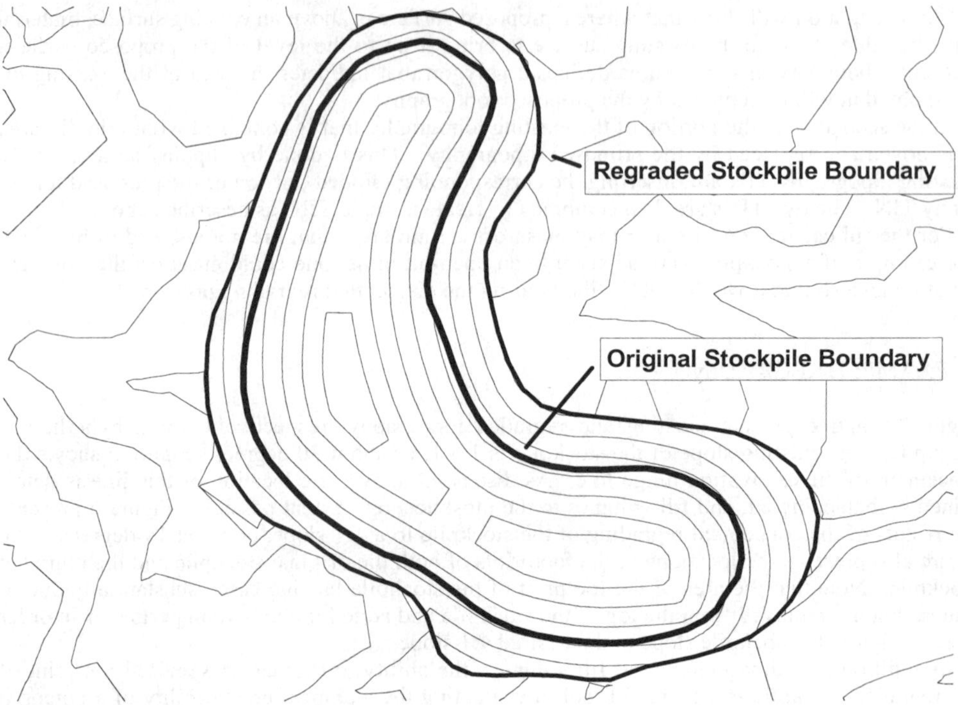

Figure 6. Regraded stockpile showing topography, original footprint, and revised footprint

4 DEVELOPING INPUT FOR A GEOCHEMICAL MASS-LOADING MODEL

A terrain model of the proposed topography can be used to develop isopach maps of stockpile thickness, which are then input into a geochemical mass-loading model used to evaluate potential mass loading from the regraded stockpile. Input to the geochemical model is a series of thickness bins, or a histogram of material thicknesses. These are developed from the GIS by discretizing a grid of points onto the stockpile and then at each point subtracting proposed topography from pre-mining topography to calculate stockpile thickness at each point. Thickness calculations can be performed for existing as well as proposed stockpiles, thus allowing a comparison of mass loading rates to underlying groundwater with existing versus proposed topography. The geochemical modeling is described in a companion paper (Earley et al., in this volume).

5 CONCLUSIONS

Three-dimensional GIS is a powerful tool for planning and costing purposes when evaluating regrading scenarios. Existing data in electronic format can be used to develop terrain models (TINs) of existing topography, which can then be manipulated to evaluate various regrading scenarios. The software described here can be used to rapidly and effectively evaluate a variety of regrading scenarios in terms of volume of material that must be moved, excavation and hauling costs, and degree of technical practicability. In addition, the GIS provides important input data for geochemical models used to evaluate mass loading from existing versus proposed stockpile configurations.

Three-dimensional visualizations of both existing and proposed topography, as well as surface features, can also be developed using the 3D Analyst extension for ArcView. These can be powerful visual tools for communicating with management, regulators, and the public. By using three-dimensional visualizations, it is possible to view the outcome of a potential regrading scenario and communicate that to a non-technical audience.

Tailings and Mine Waste '01 © 2001 Balkema, Rotterdam, ISBN 90 5809 182 1

Plant growth on aggregated materials generated from oil sands operations

X. Li
Alberta Research Council, Vegreville, Alb., Canada
Y. Feng
University of Alberta, Edmonton, Alb., Canada
M. Fung
Syncrude Canada Limited, Alb., Canada

ABSTRACT: Re-establishment of self-sustaining ecosystems is a major challenge in the reclamation of disturbed land from oil sands mining operations. We developed an aggregation technique that was able to create soil-like materials for plant growth using by-products from Alberta oil sands operation process. A research site on Mildred Lake of the Syncrude Canada Ltd., Alberta, Canada was established to assess plant performance on the soil-like material created with this technique. The site was established using composite tailings (CT) as a sub-material. Four different materials were placed on the top of CT (20 cm). Three of the materials are the 1:1 mixture of mature fine tailing (MFT) with CT, amended by 5%, 10% and 15% of peat moss, respectively. These mixtures were produced through a soil aggregation technology developed by Alberta Research Council Inc., Canada. The land reclamation material was used as control. Jack pine (*Pinus banksiana,* Lamb.), white spruce (*Picea glauca* Moench Voss), trembling aspen (*Populus tremuloides* Michx.), hybrid poplar (*Northwest*), dogwood (*Cornus stolonifera* Michx.) and American green alder *(Alnus crispa* [Ait.] Pursh) were planted in 1997. Soil electrical conductivity and soluble salts were monitored along with other physical, chemical and biological properties. Trembling aspen and hybrid poplar out-performed other species. The growth rate for all of species increased with increasing amount of peat moss amendment added in the materials. A significant downward salt movement was observed between the fall of 1998 and the spring of 1999. This may have contributed to the better plant performance during 1999. It is possible to use the oil sand processing by-products to reclaim the disturbed land, however, the challenge is to control soluble salts in rooting zone.

1 INTRODUCTION

Surface mining and water-based extraction of oil sands in northeastern Alberta have generated large volumes of tailings sand and fine tails. Tailings slurry from the extraction plants contains significant quantities of water, residual bitumen, and coarse and fine particles, all of which are pumped into large settling ponds. The coarse particles quickly settles to form the dykes and beaches, which is known as "tailings sand"(TS). The fine particles, silts and clays, and residual bitumen settle in the pond to form an accreting layer, which consolidates very slowly. When consolidation reaches 30% solids by weight in a period of 1 to 2 years, the resulting deposit is known as "mature fine tails"(MFT). Currently some 350 million cubic metres of fine tails have been produced and are being stored in large tailings ponds. It is possible that mature fine tails will exceed a billion cubic metres in the ponds by the year 2025. To accelerate the consolidation of fine tails, the gypsum (calcium sulfate) was added into the fines-enriched tailing slurry to induce a very rapid release of water. The consolidated dense materials, now with a solids content of approximately 80% by weight, are termed "composite tails"(CT). The CT consists of 18 to 20% of fine particles, mainly silt-size particles, and 80-82% sand.

The traditional method to reclaim these materials is capping with topsoil. With a large disturbed area during oil-sand operation in Northern Alberta, Canada, the available soil has become limited. A research site has been established in 1997 to explore possibilities of using oil-sand processing tailings to create soil-like materials that can be used as plant growth medium. An aggregation technique was successfully developed under field conditions to produce aggregates using the structure-less tailings materials amended with peat moss (Li and Fung, 1997, 1998). It was hypothesized that the mechanical force during the process of artificial aggregation acted as the means of binding soil particles together to form aggregates, accomplishing a function of physical entanglement usually performed by roots and fungal hyphae in the natural aggregate formation process. As soon as the aggregates had been created and plant communities established, the soil ecosystem is expected to progress towards forming more stable aggregates with the help of roots, hyphae, and other biological binding agents.

The objective of this long-term study is to assess the capability of the aggregated tailings materials to support a self-sustaining ecosystem. In this paper the survival rates of tree and shrub species, growth of plant seedlings and its relationship with soluble salts and their movement in soil profiles during the first two growing seasons were describe. It was aimed to develop a set of parameters can be used to describe the development of aggregated materials to be able to support a self-sustaining ecosystem.

2 MATERIALS AND METHODS

2.1 Experimental Design

A research site, located on Syncrude Site at Mildred Lake, was established in 1997. The site was established on composite tailings (CT), i.e. composite tailings as a sub-material. It consisted of the following treatments as the topsoil material[1].

1. 1:1 MFT:CT mixture with 5% peat moss
2. 1:1 MFT:CT mixture with 10% peat moss
3. 1:1 MFT:CT mixture with 15% peat moss
4. Land reclamation material (control).

An artificial aggregation process was used to create soil-like materials using these oil sands processing residues. The procedure was detailed in Li and Fung (1997). A randomized block design was used. Each treatment was replicated four times. A total of 16 plots were set up at the site.

2.2 Plant species

After the aggregation process and fertilization, plant seedlings, were planted in August 1997. The plant species planted were

Tree and shrubs:

1. jack pine, (*Pinus banksiana,* Lamb.)
2. white spruce, (*Picea glauca* Moench Voss)
3. trembling aspen, (*Populus tremuloides* Michx.)
4. hybrid poplar, (*Northwest*)
5. dogwood, (*Cornus stolonifera* Michx.) and

---

[1] The following abbreviations appeared in the text, tables and figures are: RM=land reclamation materials; 5%=1:1 ratio of MFT:CT with 5% peat moss; 10%=1:1 ratio of MFT:CT with 10% peat moss; 15%=1:1 ratio of MFT:CT with 15% peat moss.

Figure 1. Overview of the site in 1997 (above) after planting and in fall of 1999 (below).

6. American green alder *(Alnus crispa* [Ait.] Pursh).

Grass and legume:

1. highlander (*Elymus trachycaulus subsp. Trachycaulus*)
2. birdsfoot trefoil (*Lotus Corniculatus*)
3. alfalfa (*Medicago sativa*)
4. mixed native grasses

Within each plot, 25 seedlings from each tree or shrub species were planted. These seedlings were planted in 15 rows. Each row was divided into two sections, resulting in a total of 30 sections. Five groups, each with five seedlings from each tree or shrub species, were planted in randomly assigned sections. The seedlings were planted at 1 x 1 m spacing. The grass and legume seeds were sown between the rows of tree/shrub seedlings to provide ground coverage.

2.3 Survival Rate and Growth of Trees and Shrubs

Plant survival rate was calculated by the numbers of alive tree divided by the total trees were planted.

Plant height and diameter of trees and shrubs were measured at least twice a year since 1997. Five plants for each species per plot were measured. Data were averaged among 20

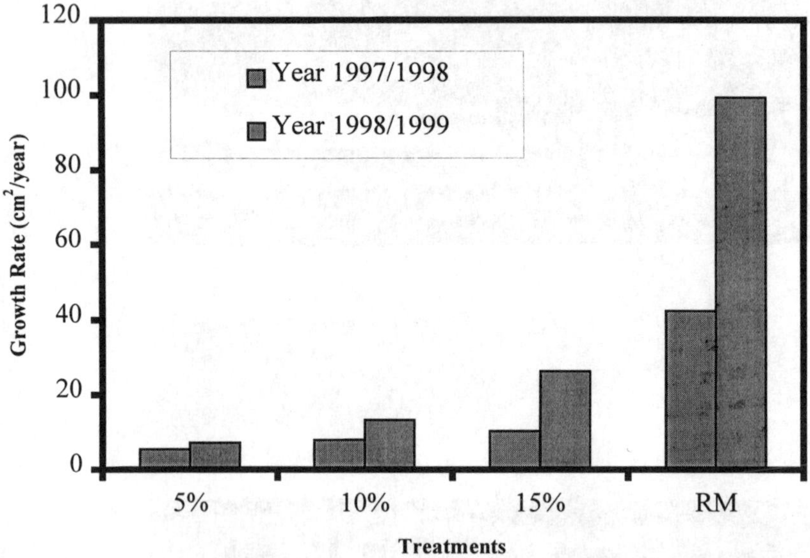

Figure 2. The growth rate of the trembling aspen grown on the indicated treatments. The values were calculated by using the equation [1]. The error bar represents standard error from 20 measurements.

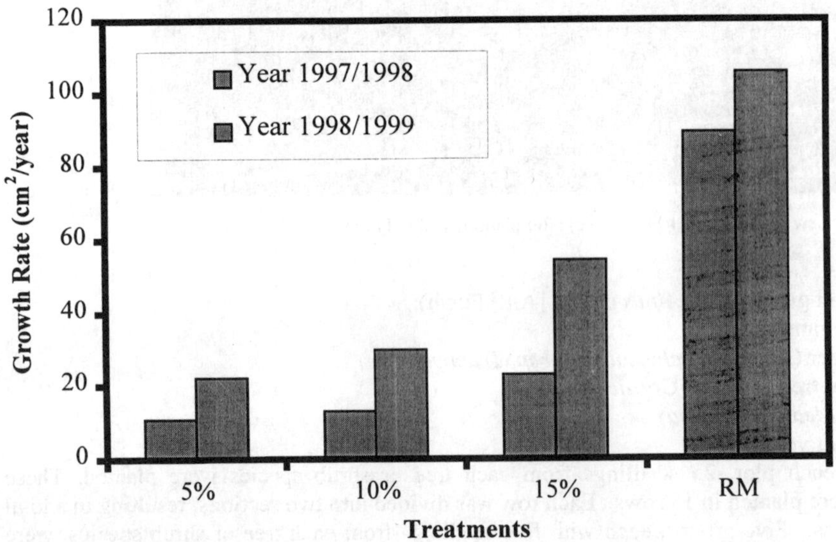

Figure 3. The growth rate of the hybrid poplar grown on the indicated treatments. The values were calculated by using the equation [1]. The error bar represents standard error from 20 measurements.

measurements for each species and each treatment. The results are presented as growth rate. It is defined as

$$GR(\Delta HDcm^2 / year) = HD_{currentyear} - HD_{previousyear} \quad [1]$$

where HD was defined as

$$HD\ (cm^2) = Height\ (cm) \times Diameter\ (cm) \quad [2]$$

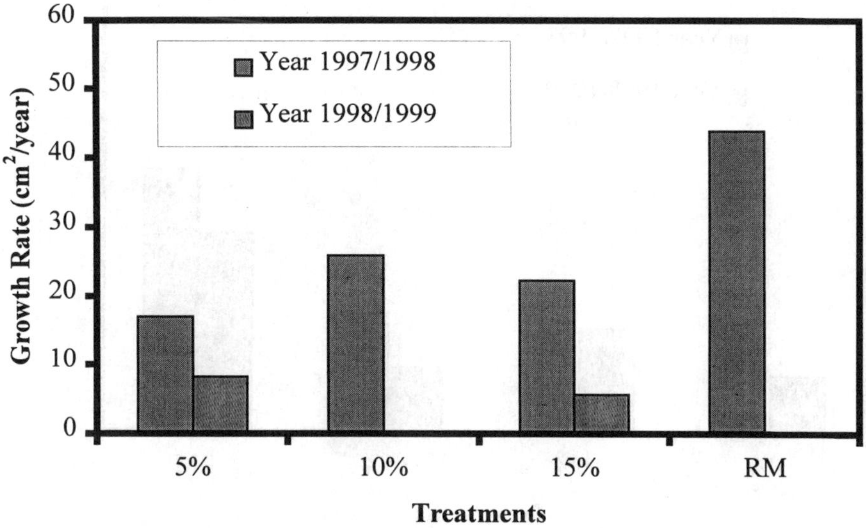

Figure 4. The growth rate of the dogwoods grown on the indicated treatments. The values were calculated by using the equation [1]. The error bar represents standard error from 20 measurements.

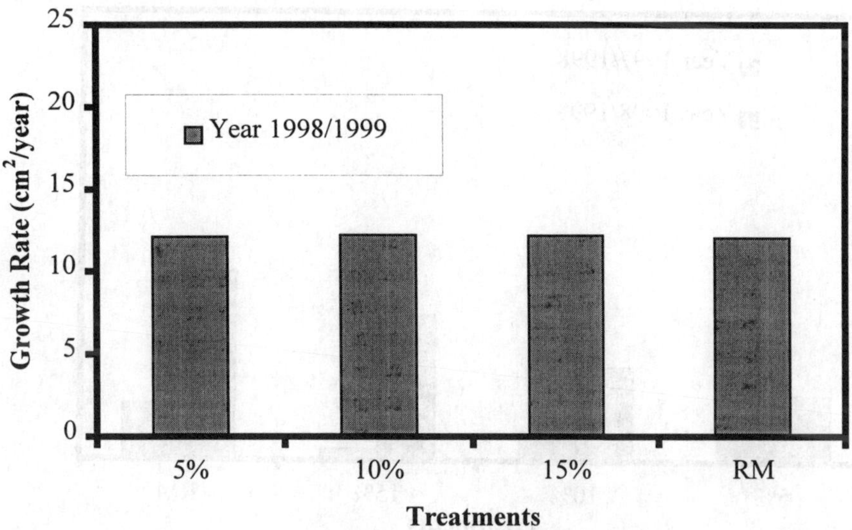

Figure 5. The growth rate of the green alder grown on the indicated treatments. The values were calculated by using the equation [1]. The error bar represents standard error from 20 measurements.

2.4 Sample Analysis

Soil samples were taken with a 2.5 cm diameter soil sampler to 120 cm depth. Two soil cores were taken for each plot. Soil samples were analyzed for soluble salts, including Na^+, $Cl^-$and electrical conductivity from saturated paste. The analytical procedures described in Li and Fung (1997) were followed.

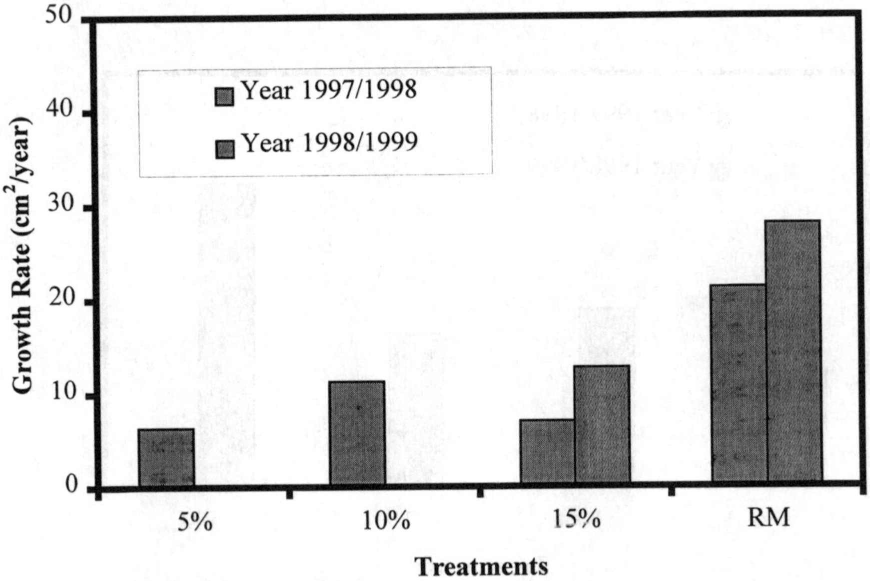

Figure 6. The growth rate of the jack pine grown on the indicated treatments. The values were calculated by using the equation [1]. The error bar represents standard error from 20 measurements.

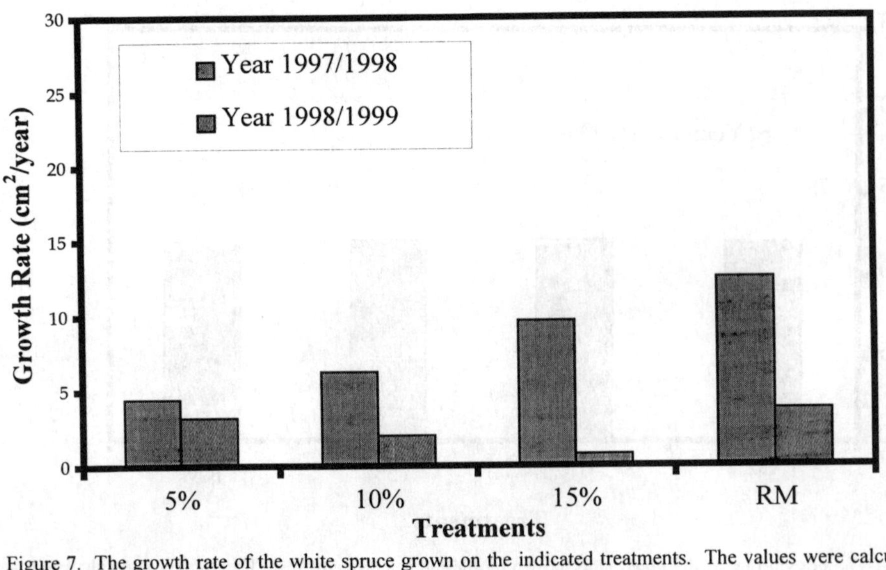

Figure 7. The growth rate of the white spruce grown on the indicated treatments. The values were calculated by using the equation [1]. The error bar represents standard error from 20 measurements.

3 RESULTS AND DISCUSSION

3.1 Growth of Trees and Shrubs

Overviews of the site (Figures 1 and 2) in 1997 and 1999 indicated that the trees, shrubs and vegetation are generally growing well. Most of trees and shrubs were survived after two years (Table 1). Trembling aspen, hybrid poplar and dogwoods had about 5 to 10 mortality in the treatment with 5% peat moss. While the majority of jack pine did not survive after two years with the treatments of 5% and 10% peat moss and the white spruce had a significant mortality in the treatment with 5% peat moss.

Table 1. Survival rate of tree and shrub species after two growth seasons. Assessment was made in 1999 (the data for alder was after one growing season). The different letters after number indicated statistically different at 95% confidence level.

Species	Survival rate (%)			
	5%	10%	15%	Rm
Aspen	90 (a)	99 (b)	100 (b)	100 (b)
Poplar	95 (a)	98 (b)	100 (b)	100 (b)
Dogwoods	100 (b)	100 (b)	100 (b)	100 (b)
Alder	100 (b)	100 (b)	100 (b)	100 (b)
Jack pine	2 (a)	5 (a)	7 (a)	95 (b)
White spruce	70 (a)	90 (b)	95 (c)	95 (c)

Plants grown on control treatment, land reclamation materials, performed significantly better than those grown on the rest of the treatments for most species (Figures 2 to 7). This indicated that the aggregated materials might contain unfavorable substances for plant growth; some of these substances may be toxic enough to kill the jack pine. Or it may be the lack of some critical substances needed to support fast plant growth. The growth of plants were generally better with increasing amount of peat moss amendments in the materials, for most of the species. This illustrates that the organic amendments are extremely important for plant growth. Although jack pines showed significant growth in 1998 for all treatments, the plants grown on CT treated with 10% and 5% peat moss amendment did not survive in 1999. However, the plants that did survive, on the treatment with 15% peat moss amendment, showed significant growth in 1999. Although dogwoods had a good growth in the year 1988 most of plants did not grow in 1999. This is unexpected since dogwoods was considered as one of salt tolerant species. It requires further observations. White spruce had a similar growth pattern. The alder had a very good growth in the year 1999. But it requires more observations before any conclusion can be drawn.

The much superior growth performance of the trembling aspen and hybrid poplar indicated that these two species are more favorable than the other species (Figures 2 and 3). Plant performances improved generally with increasing amount of peat moss amendments in the tailings materials. Although plants grown on the control plots (land reclamation material) out performed those grown on amended tailings materials, the differences are getting smaller with time. Differences between plants grown on control and on amended tailings materials in 1999 were less than that in 1998, especially for poplar.

3.2 Salt Movement in Soil Profiles

The EC and Na^{-} concentration profile for control and the treatment with 5% peat moss amendment, as measured in the fall of 1998 and spring of 1999, are shown in Figures 8 and 9. These results illustrate that EC is generally higher in the treatment with 5% peat moss than in the control. This may contribute to the high mortality rate for jack pine since it is a salt sensitive species. However, the values measured at the spring of 1999 was lower that that in the fall of 1998, for all depths in the profile. This indicates that there was a significant downward movement of soluble salts during this period. The single most important contributing factor may have been the infiltration of spring snowmelt. The improved growth rate for tree and shrub species might be the results of lower salt concentration in the profile.

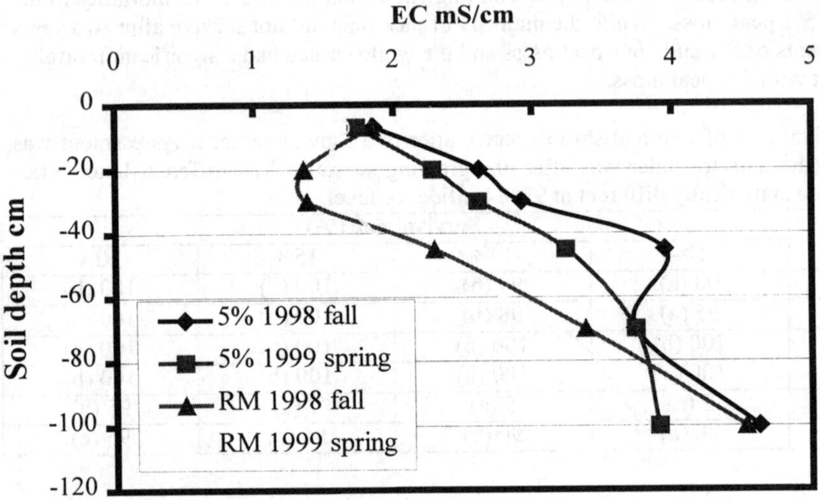

Figure 8. Concentration of EC from saturated paste as a function of soil depth for different treatments. The values were average of four replicates.

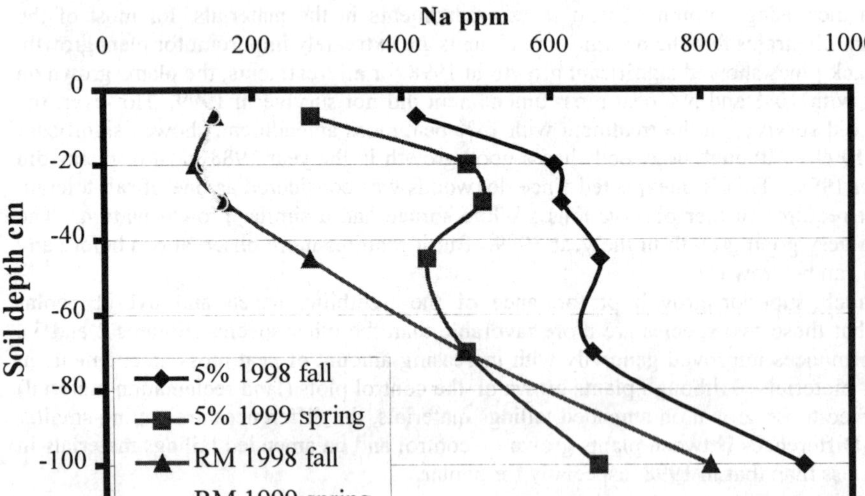

Figure 9. Concentration of Na^{+1} from saturated paste as a function of soil depth for different treatments. The values were average of four replicates.

4 CONCLUSIONS

Trembling aspen and hybrid poplar out performed other species. Growth rate of both aspen and poplar increased in 1999 as compared to 1998. The difference between control and treatment with 15% peat moss decreased with time. Jack pine and white spruce struggled for survival. Although dogwoods survived in the first two years growth rates were very low. Overall the growth rate increased with increasing amount of peat moss amendment in the surface materials. A significant downward salt movement was observed between the fall of 1998 and the spring of 1999. Salt level and organic matter content are two major important factors affecting plant growth. The first one retarded plant growth; even slowly killed salt sensitive species while the organic matter content improved soil cation exchange capacity and biological activities (data not show) and hence improved plant growth.

ACKNOWLEDGMENT

Authors wish to thank Syncrude Canada Ltd and Alberta Research Council, formally Alberta Environmental Center, for their financial support. The special thanks to Jim Storey, Dayna McTnyer and Trish Rattray for their technical support in this study.

REFERENCES

Li, X. and M.Y. Fung, 1996. Creating a soil-like profile for plant growth using tailings sand and fine tails. Proceedings of the Petroleum Society 47th Annual Technical Meeting. p104-114. Calgary Convention Center, June 10-12, 1996. Alberta.

Li, X. and M. Y. P. Fung, 1997. Land reclamation using oil sands processing tailings Phase I pilot project. Report to Syncrude Canada Ltd. 31p.

Li, X. and M. Y. P. Fung, 1998. Creating a soil-like material for plant growth using tailings and and fine tails. Journal of Petroleum Technology V.37:44-47.

Li, X., Y. Feng and M. Y. P. Fung, 1999. Land reclamation using oil sand processing tailings Phase III. Capability of the aggregated materials: A survival of trees. Report to Syncrude Canada Ltd. P30.

Passioura, J.B., 1991. Soil structure and plant growth. Australian J. Soil Research. 29:717-728.

Quirk, J.P and Murray, R.S., 1991. Towards a model for soil structure behavior. Australian J. Soil Research. 29: 829-868.

Reprocessing, utilization, and treatment

Tailings and Mine Waste '01 © 2001 Balkema, Rotterdam, ISBN 90 5809 182 1

Tailings, mine waste and mined land as future resources

David L. Kuck
Geological and Mining Consultant, Oracle, Ariz., USA

ABSTRACT: Exploitable mineral deposits are rare and precious things. Major mines have values exceeding downtown Manhattan real estate. Mined lands, their waste dumps and tailings are future exploitable resources. Recent history has many cases of retreatment of tailings, slag dumps, reopening of old surface and underground mines and mining of waste dumps. These have become exploitable as economic, technological and political changes have occured through time. Political changes such as inclusion in parks, monuments and wilderness areas have destroyed mineral resources. Since mined lands, tailings and waste dumps may become future resources, they should be treated with care.

1 INTRODUCTION

Everything starts with mining. Your car was mined. The road you drive your car on was mined. The average house in the United States weighs 125 tons, eighty six percent of which was mined. The airplane you fly in was mined. The airport runway was mined. Fertilizer for agriculture is mined. The dishes and salt on the table was mined. Industry, and the jobs of workers in the United States are dependant on mining for their raw materials.

Many large mines exceed the value of downtown Manhatten real estate. The mineral concentrations are rare and special features, not easily available, except where they are found. Since almost everything begins with mining, our economy is absolutely dependant on mining for it's prosperity. The cost of materials mined determine the cost of products produced by industry. Every time we restrict or interfere with mining, we increase the cost of things we buy.

Every mine is different. The climate, topography, vegetation, culture, rock chemistry, size and nature of workings is different. The effects of these factors on the mine operation and ultimate reclamation are different. The reclamation should not be such that it makes future mining difficult or impossible, and thus in effect destroys future mining of mineral resources, and raise the cost of the products we all buy.

We must be careful what we do to old mines, waste dumps and tailings deposits. These will in the future, often produce income. Thus we must leave these special materials in a condition where they can be utilized in the future.

2 TAILINGS

Mine tailings have been retreated as economic and technological advances have made their retreatment profitable. Mineral processing plants at mines produce tailings deposits, some of which are huge, as at the Magma underground block caving mine at San Manuel, Arizona, which was closed in 1998, and other large low grade mines. These tailings are future resources. As mineral technology has developed through time, old tailings have turned into new resources. Often they are valuable for minerals other than the ones the rock was originally mined for. Tailings have a special property that rock with the same mineral content does not possess. Tailings have had the investment of mining, crushing and grinding, which are the largest costs of extracting minerals from rock. They are finely divided particles which may require partial regrinding, but the major costs of crushing and grinding (>$1.00/ton) has already been paid. Since they are on the surface, re-mining can be done either hydraulically or with simple loaders at a relatively low cost as opposed to the original cost of >$1:00/ton. Even if regrinding is necessary for the already finely ground rock, it is only a small fraction of the cost for fresh rock in place.

At Miami, Arizona, the Miami Mine's 1912 to 1926 tailings were placed in a high tailings dam which overlooked the town. In 1926 tailings thickeners were placed on top of the tailings, and the tailings from 1926 to 1960, when the mine was closed, were hydraulically transported six miles to the Solitude tailings dam. Since 1988, the old pre-1926 tailings are being mined hydraulically, reprocessed, transported by pipeline 3 miles and placed in the old Copper Cities Mine open pit. This placement in the open pit may be a bad move, for it may preclude reopening the Copper Cities open pit mine at some future time. It took about seventy years for the pre-1926 tailings to again become an economic resource. Thus a low grade valuable mineral can be recovered rapidly at a relatively low cost. Tailings have been resources in the past and will again be future resources.

3 MINE WASTE

Waste dumps from old large surface mines can again produce values, often without being remined, by hydro-metallurgical techniques. With changing economic conditions, it might even be possible to remine the old waste dumps, and crush and grind the waste for hydro-metallurgical treatment. Old waste dumps can become future ore deposits.

Waste rock from the Atlantic City iron mine in Wyoming (operated from 1962 to 1983) has been used as an aggregate/rip-rap source by local contractors and the Wyoming Department of Transportation, and as a source of sand for highways in winter. Rock boulders were stacked as snow fences rather than vertical slat snow fences on the tailings (elevation 2470 m (8100 ft.)) to control wind blown dust and snow.[6]

Open pit copper mine waste is producing copper around the world by bio-oxidation[7] and acid leaching of the waste dumps. Production by leaching and solvent extraction and electrowining (SX/EW) usually extends long after active mining has stopped.

Phelps Dodge's Morenci Mine, the largest mine in Arizona, is segregating and sometimes crushing mine waste, and stacking the waste rock for bio-leaching to improve copper recovery by waste dump leaching and recovery by SX-EW. Recently it was reported that Phelps Dodge had shut down it's sulfide flotation mills at Morenci, and is producing from dump and heap bio-leaching and SX-EW.[1] In 1998, Morenci moved over 1,000,000 tons of ore and waste per day.

4 MINED LAND

In mineral exploration, the first place we look for new mineral deposits is old abandoned mines and in proximity to operating mines. Withdrawal of old mine sites as historic sites or for other reasons, such as parks, wilderness areas or national monuments, robs the American people of possible future mineral wealth. This wealth that is taken from us supplies the raw materials for all other industry, which along with mining provide the jobs that feed our children.

Some of the largest mines have been in continuous operation for over a hundred years. However, many deposits have been operated for long periods with long periods of closure and abandonment. Again, changing economic conditions and advances in technology have allowed these deposits to be reopened. In 1900, copper mines had to have at least 6% copper ore to operate. In the 1950s copper ore had to have more than 0.4% to 0.5% copper to be ore. In 1998, the Morenci Mine considered everything over 0.1% (2 lb/ton) copper to be ore. Again, changes in technology and methods have made what used to be waste into ore. Proposals to backfill abandoned large open pit copper mines could preclude reopening these mines in the future. This amounts to criminal destruction of the nations future mineral resources.

In South Dakota, was the backfilling of the Richmond Hill Mine with the Spruce Gulch waste repository sulfide waste rock a good move or is it the destruction of mineral resources that could be mined in the future? It certainly is likely to preclude reopening of the mine in the near future. Was the enormous expense of back filling the pit justified?[5]

The Copper Cities open pit mine near Miami, Arizona was the site of the Girard Mining Co. gold and silver mine in 1896 and reincorporated the same year as the Lost Gulch Mining Co. In 1909 the Lost Gulch United Mines was organized to operate the property, which was succeeded by the Louis d'Or Gold Mining Co. in 1912, and a little gold-silver-lead ore was produced. In 1928 the Pinto Valley Co. forclosed on the defunct Louis d'Or Co. and organized the Porphyry Reserve Mining Co. and 350,000 lb. Cu was produced from stream gravels cemented by copper carbonates and silicates in Tinhorn Wash. In 1942 the Copper Cities Mining Co., a subsidiary of Miami Copper Co. purchased these claims at a sheriff's sale. Stripping started in 1951 and mill production started in 1954, upon depletion of the Castle Dome open pit mine west of Miami, Arizona in 1953.[3] Castle Dome was reopened by Magma Copper in 1973 as the Pinto Valley open pit mine, which was again closed in 1998 by B.H.P.[4] The filling of the Copper Cities Mine with old Miami Mine tailings might prevent Copper Cities from being reopened sometime in the future.

Copper is still being produced by leaching and SX-EWof the caved ground over the old pre-1960 Miami Copper Mine and the waste dumps of the Castle Dome - Pinto Valley Mine.

Developments in bio-oxidation of sulfides is making marginal sulfide mineral deposits and waste dumps of precious metals, copper, nickel, cobalt, zinc and lead attractive exploration targets. Heap and dump leaching with bio-oxidation is making these waste and marginal mineral resources competitive in spite of the present depressed metal prices.[7] High pressure oxidation and leaching is making mixed and refractory mineral deposits into ore.[7]

Huge areas of the western United States and Alaska are withdrawn from mineral entry for mining or petroleum exploitation, even though they are known to contain mineralization, some of which was mined in the past. This robs the American people of both the income from possible development of these mineral resources, and the income from jobs in other industries these mineral resources support.

5 SEWAGE

In the early 1900s Cincinnati, Ohio started generating electricity by burning sewer gas given off by the city's sewer system septic tanks. How many towns and cities today take advantage of this resource? Phoenix, AZ sewage is used for cooling water for the Palo Verde Nuclear power plant. Treated Arizona sewage effluent is often used for watering golf courses.

Sewage sludge has been used as fertilizer in agriculture for many years for contained trace minerals.

Is discharging sewage into the nearest body of water such as a stream, lake or the ocean, only contamination, or is it also a waste of energy, mineral and water resources?

6 MUNICIPAL GARBAGE LANDFILLS

Many municipalities are now placing plastic covers over old abandoned landfills to trap methane given off by decomposition of the hydrocarbons in the garbage. This natural gas is then marketed or utilized for electric power generation.

There are many other metals and mineral materials in the landfills. Once the organic phase has been depleted by bio-decomposition of the organic component of garbage to produce natural gas, the residue may be processed in the future for contained metals and other raw materials. Our land fills will go from problems and liabilities to valuable resourses in the not to distant future.

7 RADIOACTIVE WASTE

It is being proposed to place hazardous radioactive waste deep underground where we will not have access to the waste. This is a bad idea, for this places the waste where we can't reprocess it at some time in the future, when more advanced technology can utilize some component of the waste. The very fact of the radioactivity of elements in the waste, makes elements in the waste special. Who are we at this point in the infancy of our knowledge of radioactivity, to say that these elements will be of no value or that the material can't be made safe and usefull, in the future? Why should we contaminate large blocks of ground with radioactive material that might start to migrate through groundwater as the radioactivity deteriorates the containment materials? Other countries reprocess their radioactive waste, but we refuse to out of fear. Fear is a poor substitute for understanding in dealing with any problem. Let us work to gain understanding of how to reprocess these wastes and to utilize their special properties. Bad decisions now can come to haunt us in the future.[8]

8 CONCLUSIONS

The mining industry is fleeing the United States because of the unfriendly economic and political climate, two things essential for development of a mining industry. The United States used to be friendly to the mining industry, which supplied abundant reliable raw materials to the developing manufacturing industries. Manufacturing in the United States takes place in the heartland, an average of one thousand miles from seaboard. Raw materials from foreign countries must be shipped in, unloaded from ships, transported one thousand miles to the factory, then the product transported back to seaboard in order to compete with foreign manufacturers in Europe and the far east with better access to seaboard. Any manufacturer in Europe is less than three hundred miles from seaboard. How long can United States manufacturing industries compete with foreign competitors who have closer access to seaboard?

All basic commodities lose real value with time! Copper at its lowest value of \$0.06/lb. in 1930 was worth more in real value than it is at \$0.80/lb. In 1930 you could hire a man to work for \$1.00/day, if he could find the work. Today, minimum wage is \$40.00/day. This trend in the value of basic commodities will continue as technnology continues to lower the cost of producing basic commodities.

Our prosperity is absolutely dependant upon our domestic mining industry to supply the raw materials for everything we produce and consume.

REFERENCES

1. "Phelps Dodge Regroups", TMIM, Engineering & Mining Journal, Aug. 2000, page ww 13.

2 "Bio Leaching", Shuey, Scott A., Technical Editor, Engineering & Mining Journal, May 1999, p 16BB-NA.

3 Titley, Spencer R., Carol L. Hicks "Geology of the Coper Cities Mine" University of Arizona Press 1966.

4 "BHP curtails sulfide operations at Pinto Valley"; Mining Engineering April 1998, page 17.

5 "SouthDakota gold mining Regulation and Environmental history"; Thomas V. Durkin, Mining Engineering April 1998, page 27.

6 "Tailings basin reclamation at Wyoming's Atlantic City iron mine"; by James J. Gusek, Mining Engineering April 1998, page 33.
7 "Advances in bio-oxidation of sulfides" Mining Opportunity Bulletin, Summer 2000, page 5.
8 Sailor,William C., David Bodansky, Chaim Braun, Steve Fetter, Bob van der Zwaan "A Nuclear solution to Climate Change?"; Science Vol. 288, page 1177, 19 May 2000.

Tailings and Mine Waste '01 © 2001 Balkema, Rotterdam, ISBN 90 5809 182 1

Waste disposal of dimension stones in the Lagoa Santa EPA

A.Curi, F.A.R.Carmo, J.L.Morais, J.G.Lima & R.C.Reis
Departamento de Engenharia de Minas, Universidade Federal de Ouro Prêto, Brazil

ABSTRACT: In order to preserve a limestone site internationally renewed by their archaeological content including caverns and evidences of pre-historic ages the Brazilian government has created the Lagoa Santa Environment Protection Area –EPA-. This place is located in the proximity of the Belo Horizonte international airport. In the past two decades have increased in these area the exploitation of dimension stones to use from the residential construction to exportation. This exploitations have been done without a mine planning promoting several environment impacts. On the other hand, this is an important activity for the region in terms of employment. The objective of this paper is to analyze these activities identifying the main impacts and to propose measures considering the mitigation of the impacts including the adequate disposal of the waste generated by the limestone quarries.

1 INTRODUCTION

The Lagoa Santa Environmental Protection Area - EPA - was created by the Brazilian Federal Government on January 25, 1990, through the statute 98.881, published in the Act 6.902 of April 27, 1981 and Act 6.938 of August 31, 1981 and in the CONAMA (Conselho Nacional do Meio Ambiente) Resolution number 10 of December 14, 1988.

Lagoa Santa EPA is located in the metropolitan area of Belo Horizonte in the State of Minas Gerais, in the eastern margin of the Das Velhas River.

The mining of Lagoa Santa dimension stone for use as construction material is concentrated, mainly, inside of the limits Protection Zone of the Cultural Patrimony–PZCP, constituting one of the conflicts of use of this zone.

Although the dimension stone mining of limestone is only a temporary occupier of the land surface, it causes serious environmental impact associated with overburden, hauling roads, unvegetated surfaces, coarse rejects, fine-grained tailings, topsoil and stockpiles. The quality of the air, water and land is affected. Dust adversely affects air quality. The physical quality and quantity of surface water and groundwater can be affected if reclamation is not practiced. Watercourses may be disturbed and flow rates altered. Erosion may be excessive, and the surface and ground water may become mineralized. The topography, drainage, vegetation, appearance, and surface texture of the mine site may all be seriously impacted. The gradients of sidecasts spoil may make the topography unsuitable for the planned post-mining land uses (Williams, D.J. et al, 1997). The impacts of mining also include the destruction of the landscape, degradation of the visual environment, destruction of agricultural and forest lands and degradation of recreational lands.

The mining of Lagoa Santa's dimension stone is developed out of the technical, environmental parameters and in disagreement with the legislation. For the economic use of that ore considering the sustained development, are necessary changes in the productive process.

In this article are commented measures to improve the environmental quality in this area of environmental protection through the use of methodologies adapted to the closure of small mines. EPA can be implanted in lands of public domain or of private domain, previously inhabited. These areas of environmental protection are destined to maintain the well being of the urban populations and to conserve or to improve the local ecological conditions. In those areas, of private or public domain, the government organizations will establish norms, limiting or prohibiting some activities with objectives of protecting local interest for its specific environmental characteristics (Act 6.902/81). The CONAMA Resolution number 10 (1988) establishes the economic and ecological exploration norms, not allowing activities of earth movement, mining and excavations that cause damages to the environment and represent danger to people or to the biosphere, making restrictions to the those activities in areas close to caves, waterfalls and natural monuments. All those activities will depend on special licensing.

The mining is considered an activity that degrades the environment, in agreement with the CONAMA Resolution number 1 of January 23, 1986, what implies certain restrictions, and even prohibition, for its exercise in certain areas defined as untouchable as, for example, conservation areas and indigenous areas, where it can only be exercised with authorization of the National Congress. This characteristic of degradation of the mineral activity doesn't imply that its exercise is prohibited in certain Units of Conservation. However, the previous authorization of the responsible government environmental organizations is demanded by the administration of the Unit of Conservation (Act 7805/89).

The environmental fragility of that EPA, for being a limestone's area, is enough reason to call the attention of everybody involved directly or indirectly, in the sense of trying to divide responsibility of its rational use, being still considered the inherent important aspects to its biological, cultural nature, landscape, as well as its social and economic characteristics (Barreto, M. L. & Sirotheu, G. J.).

In this context, is inserted, as conflict of use of the area, the mining and the treatment of Lagoa Santa dimension stone for industrial use. These mining activities of small size should receive mechanisms of environmental control on its disordered expansion and technical and operational attendance with relationship to its handling which is the main objective of that work.

2 THE AREA OF ENVIRONMENTAL PROTECTION OF LAGOA SANTA

2.1 *Location of Lagoa Santa EPA*

Lagoa Santa EPA is located in the metropolitan area of Belo Horizonte of the State of Minas Gerais, in the eastern margin of the Das Velhas river (Fig. 1). It could be reached by the highway MG-424, in the direction Belo Horizonte–Pedro Leopoldo–, or by the highway MG-010, in the direction Belo Horizonte– Confins International Airport.

This EPA, with an area of 35.600 hectares incorporates the municipal districts of Lagoa Santa, Pedro Leopoldo, Matozinhos, Funilândia and Confins (Cabral, J.A et al, 1998).

The mining of Lagoa Santa dimension stone for use as construction material is concentrated, mainly, inside of the limits of the Cultural Patrimony Protection Zone – CPPZ-, constituting one zonal use conflicts.

2.2 *The main characteristics and objectives of the CPPZ*

2.2.1 *Functions*:

To protect and to promote the landscape and the regional culture, represented by the Sumidouro arqueo-paleontological environmental System patrimony, for the Lapinha archaeological patrimony and Fidalgo historical patrimony;

To protect the humid ecosystem and the remaining species in the environment, especially the fauna associated to the Sumidouro lake.

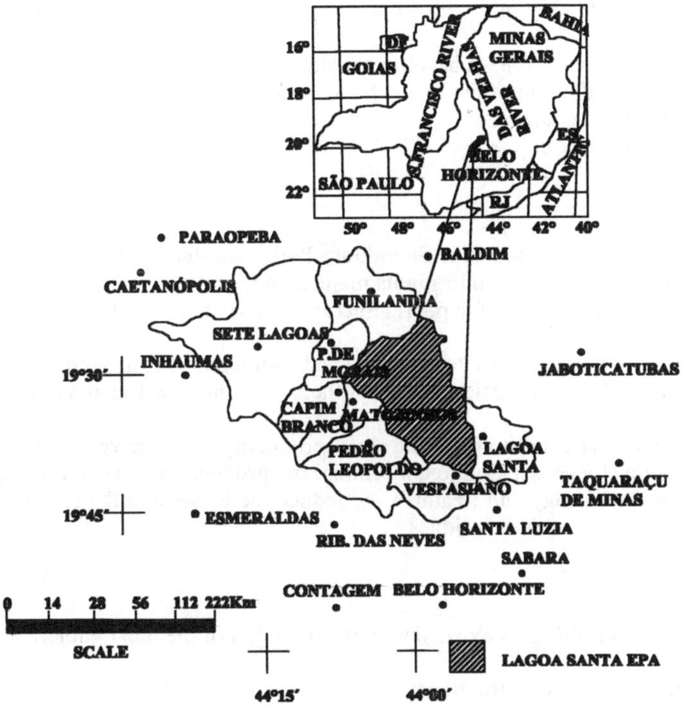

Fig1. Location Map of Lagoa Santa EPA
(Source: CPRM - IBAMA)

2.2.2 *Elements to analyze*

Sumidouro lake, especially the features of the calcareous landscapes;

Rosário chapel;

Sumidouro lake, an important humid system of great relevance for the survival and the maintenance of the diversity of the regional fauna, including species that are threatened of extinction;

Lapinha cave group, of great tourist, cultural, archaeological, paleontological and speleologic importance;

Fernão Dias House: A historical and cultural reference, where the EPA reference center is located;

Fidalgo and Quintas of Sumidouro Urban environmental context;

Group of native vegetation (denominated " cerrados ") that is located to the north of EPA and in all the proximities of the Sumidouro lake ;

The Jenipapo water course, a calcareous area with mountainous, and low resistance lands subject to the erosion processes and mass sliding;

Non calcareous areas more vulnerable to the erosion processes;

Expressive species diversity the humid EPA environment, represented, especially, for the Sumidouro lake, in the interface with the "cerrado'' nuclei.

2.2.3 *Conflicts of Use*

Activities of sands extraction in the Das Velhas river, causing significant impacts to the fluvial system and the surrounding vegetation;

Inadequate disposition of urban solid residues and of the residues of the mining;

Extraction of Lagoa Santa dimension stone, with impacts on areas of great environmental fragility;

Impact to the speleologic patrimony, to the archaeological and cultural patrimony and to the landscape, due to the activities of production of the Lagoa Santa dimension stone;

Blockage and pollution of the water sources;

Guidelines of Environmental Protection

2.2.4 *Incentives*

To reactivate the mechanisms for the creation of the Sumidouro Park, with the objective to protect the patrimony and to integrate local communities in its maintenance and operation;

To establish a managing plan for execution of support structures for visitation and to increase the value of the patrimony;

To establish technical and financial conditions to elaborate and to implant the managing plan for the urban nuclei of Fidalgo and Lapinha, seeking to adapt them with the new functions foreseen for the area;

To establish technical and financial conditions for the small producing cooperative organization of the Lagoa Santa dimension stones, seeking to concentrate the production areas in mining districts, to guide the extraction technology and treatment, to reduce the losses and the environmental impacts of the stone's cutting machines residues.

2.2.5 *Main Restrictions*

To discipline the use of the soil, controlling its expansion in the direction of the most vulnerable areas;

To restrict the use of the soil for urban constructions;

To establish occupation patterns that guarantee the maintenance of low population densities;

To control the agricultural activities that causes pollution in the Sumidouro lake;

To control the mining activities, especially the Lagoa Santa dimension stone extraction and the calcareous exploitation.

2.2.6 *Norms and Guidelines of Use*

2.2.6.1 *Allowed uses*

Reforestation with native species, seeking to the recovery of the vegetation and the renewal of the fauna, mainly in the proximity of the areas of natural vegetation;

Scientific research;

Agricultural, wild activities or activities of the creation of animals in areas with a decline less than 45% that use handling techniques that are compatible with the natural processes of the ecosystems;

Ecological tourism, that uses techniques with low impact on the environment to be preserved;

Subsistence fishing;

Recreational ranches, with minimum fraction of 5.000 m2, being 20% of the area destined to the vegetation with species of the native flora;

Urban equipment and visitation infrastructure destined to organize the tourist and cultural activities of the area.

2.2.6.2 *Tolerated uses*

Extraction and irregular treatment of the Lagoa Santa dimension stone, conditioned to its regulation and the participation of the producers in a program of environmental extension, destined to the reduction of the impacts of the activity, technological improvement and residues reduction;

Agricultural and wild activities and creation of animals already existent and conditioned to the reduction of disconformities such as: use of areas with decline of more than 45% and with handling practices that cause the degradation and pollution of the soil and the underground waters, hindered the expansion of the cultivations already existent.

Activities of mineral extraction already existent and regularly approved by the Organizations of Environmental Control, with appropriate treatment systems and disposition of liquid residues and of solid waste, which promote the environmental recovery of the degraded areas, hindered the expansion of the mines already existent.

Urban establishments already installed in inadequate areas, since endowed with systems of garbage collection, disposition and treatment of sanitary residues, adapted to the demands of the calcareous environment;

Industries already existent, since licensed by the competent organizations of environmental control and with appropriate treatment systems and disposition of liquid effluents and solid waste, hindered the expansion of the industrial areas.

2.2.6.3 *Forbidden uses*

Activities of mineral extraction that cause any risks to the environmental, archaeological, paleontological and speleologic patrimony;

Intensive creation of animals;

Agriculture with intensive handling and with use of herbicides and fertilizers;

Division of the soil in small lots, with urban purpose or for recreational small farms;

To install and to operate new industries;

Use of areas for disposition and treatment of sanitary effluents, domestic or industrial solid residues, under any conditions;

Disposition of chemical effluents, including fertilizers.

3 THE DISTRICT OF FIDALGO

The district of Fidalgo represents most of Pedro Leopoldo municipal territory and has the prevalence of activities destined for the creation of animals. However, the activity that characterizes the district is the presence of small industries for production of the Lagoa Santa dimension stones , in the margins of the Lake of Sumidouro, linked to the irregular extractions, that absorb practically more than 90% of the local population. The Fidalgo district is well known for the quality of the material extracted there. Fidalgo is located about 16 km from Pedro Leopoldo. The mines are located approximately 3 km from the town of Fidalgo.

The current population of Fidalgo, which is the area of direct influence of the mines, is approximately 4.000 inhabitants. As spoken before, about 90% of the active population work in the production of dimension stones for civil construction. The remaining population works in the creation of animals, mainly cattle. This percentage of the local population that works in the production of dimension stones includes children starting from 12 years of age. The average workers' salary is US$ 150,00 a month.

The area is considered rural, being largely destined to recreation and tourism. Thus, are commons in the Sumidouro district the ranches and small farms in the lake margins? The cultivation of agricultural products is only made for subsistence, because, according to previous citation, the labor used in Fidalgo is, in its majority, working in the main economic activity of the area, the small mining.

4 THE MINING AND THE WASTE DISPOSAL

Bench cutting is the quarrying method utilized in which primary irregular blocks are first isolated from the hillside and then cut into slices which are toppled to the floor and finally divided into dimensional stones commercial size. Cutting technologies chiefly consist of explosive splitting and wedge shearing. The monthly production of dimensional stones of the 50 quarries of the area ranges from 5000 to 10,000 m^2. with a quarry yield (recovery of commercial dimension stones) range between 40 and 80 %. The Labour force generally consists of 5 workers for each quarry, although in the largest quarry more than 20 men are employed, 5 of them in the sawing and polishing plant. Investments range between US$ 5000,00 and US$ 100,000,00. A great amount of waste is produced due to the selection of blocks. Overburden, defective material and scraps from squaring and shaping operations are dumped nearby.

The area hosting the quarries is characterized by the presence of gently steeping slopes covered by "cerrados" trees. Fauna chiefly consists of various reptile species, small rodents and birds. Limestone quarrying gives the major contribution to the economy of the Fidalgo neighboring zone.

In the area of Fidalgo, substantial increase in mining waste dumps due to irregular dimension stones exploitation has stretched the self-sustaining capacity of the ecosystem to the point where its delicate environmental equilibrium is threatened. Excavation of limestone's deposits involves removal of overlaying soil and rocks debris as overburden. This overburden modifies the natural land topography, affects the drainage systems, prevents natural succession of plant growth resulting in acute problems of soil erosion and environmental pollution (Singh et al. 1998). The mining operations in the area of Fidalgo causes harvesting of forest, destruction of habitat and the accumulation of huge quantities of overburden dumps in and around the mining areas. Surface mining not only destroys the existing natural vegetation but also causes the loss of soil both in pedological and biologic sense. Flora, fauna, hydrological relations and soil biological systems are drastically disturbed (singh et al. 1994). Management of overburden dumps to accelerate the rehabilitation process for ecological equilibrium or return such areas to an acceptable post mining land use is a challenging problem. Mine site waste disposal and rehabilitation operations comprise the actions of disposing of the waste products, preparing mined land for future land uses after the ore has been removed, and returning the disturbed land to a future use. They also include those actions that stabilize mined land and improve the land environment. The purpose of waste disposal and rehabilitation operations are to reduce the environmental impacts of mining and to meet the satisfactory requirements (Williams, D.J. et al, 1997).

A disposition project of waste in a pile should be elaborated for each small mine in activity. The Brazilian norms NBR 13029 of ABNT (Associação Brasileira de Normas Técnicas) should be followed, seeking to assist safety conditions, hygiene, operation, economy, abandonment and mitigation of the impacts to the environment, according to the legal patterns (Brandi,I.V.,1994).

The waste piles should be disposed, if possible, inside of the mine pit exploited, or close to the areas already degraded. It should be avoided to dispose the material in valleys, mainly, with larger inclination than 18 degrees; natural surfaces drainages, water courses; areas of permanent preservation; unstable areas, flooded lands; areas with exuberant native vegetation; areas with fertile soils. As the external and internal geometry of the pile, the following limits and cares should be observed: maximum height of banks of 10 meters; minimum width of berms of 6 meters; maximum height of the pile of 200 meters; existence of maintenance accesses; to reduce the angle among banks, for values less than the natural angle of rest of the waste; berms with longitudinal and traverse minimum gradient of 1% and 5%, respectively; drainage in the pile.

Other parameters in the conception of the project are: - size classification of the materials to be disposed of to take advantage of the maximum resistance and drainage characteristics of each material; adaptation of the formation and classification of the pile with the stages of waste removal; execution of the pile in an ascending way; protection of the slopes with vegetation; removal and deposit of the organic soil of the pile for future use; interns, superficial and outlying drainage system; system of retention of sediments originating from erosion; maintenance system and control. Dimension stone production reject is other great problem that can be solved, for example, with its use in a cement industry, in the construction of houses or roads, in general. The best option should be verified for its use.

In any case it doesn't have an economic use this material should be transported and arranged in piles controlled with a contention system of fines (Dike).

5 REHABILITATION OF THE DEGRADED AREAS

Typically the mining activity lasts over just a few years and the miners "walk away" leaving existent or future problems for others to deal with. Operation at a mine cease when the quarry can no longer be mined at a profit. The exact timing of this event is uncertain due to uncertainties about reserves, costs and commodity prices. As a general rule, in a few years closure does happen, and this presents the mining with a new challenge. Mine sites are, by their nature potential sources of very long term environment impact, for example inadequate disposition of wastes. The sites themselves can sometimes be restored to something like their former appearance, although for many open pit mines this is not (yet) seen as a viable or desirable option. For a small mining company closing down a mine, the preferred solution is to be able to "walk away" leaving existent or future problems for others to deal with(Sinding 1998). Unfortunately, the "walk-away'' option is still feasible in this part of Brazil but as time passes and hitherto hidden problems become apparent, undertaking the preventive efforts that are and will increasingly be required for this approach may become prohibitively costly. An alternative is to require (and enforce) closure plans that specify long term site monitoring to ensure that no unexpected impacts occur. In essence, this is the equivalent to more or less indefinite ecological accounting for each former mining site. Reporting of site performance may be a further element in the management of such sites (Sinding 1998). This form of long term commitment, however, carries with the beginning of other problems.

A project of rehabilitation of degraded areas should be elaborated by the mines, following the norms NBR 13030 of ABNT, therefore, should be made the adaptation of the landscape to ensure the harmonization of the mined areas with its proximities, with the purpose to minimize the visual impact and to do the topographical adaptation objectifying the future use of the area.

It is necessary to make the rehabilitation (group of procedures through which is returned of the productive function and the natural processes of the area, seeking adaptation of future use) or the recovery (group of procedures through which is made the recovery of the area degraded for the establishment of the original function of the ecosystem) or the restoration (group of procedures through which is made the replacement of the exact ecological conditions of the area degraded by the mining, in agreement with the established planning) to define a future use (use foreseen for a certain area, considering its aptitudes, intentional use and fragility of the physical and biological environment).

6 CONCLUSIONS

Fidalgo quarrying district incorporates a number of quarries sharing common features (alike type of material, similar exploitation methods, competition in to the same market), whose concentration is higher than 1 unit per km^2 on some areas, although its limits cannot be defined with a general rule.

It is obvious that the overall impact on the environment produced by the district as a whole is not merely the sum of the impacts of individual quarries considered as isolated units but must be calculated taking in to account that (Ciccu *et al.* 1998):

-only some the quarries and related infrastructures are visible from a point of view moving across the land;

- the exposed area of the faces and dumps decreases with the distance from the observer;

- nuisances like noise, dust and vibration are gradually attenuated away from the sites;

- flora and fauna are affected only within a limited area around each quarry so that no superimposition of effects takes place if the distance between them is sufficiently large;

-social, economic and cultural aspects can benefit from the quarrying activity over a broad area far exceeding the limits of the district itself.

Although the dimension stone mining is only a temporary occupier of the land surface, it causes dramatic and serious environmental impacts associated with overburden, haul roads, unvegetated surfaces, coarse rejects, fine-grained tailings, topsoil and stockpiles. The quality of the air, water and land is affected. Dust adversely affects air quality. The physical quality and quantity of surface water and groundwater can be affected if reclamation is not practiced. Watercourses may be disturbed and flow rates altered. Erosion may be excessive, and the surface and ground water may become mineralized. The topography, drainage, vegetation, appearance, and surface texture of the mine site may all be seriously impacted. The gradients of sidecasts spoil may make the topography unsuitable for the planned post-mining land uses (Williams, D.J. et al, 1997). The impacts of mining also include the destruction of the landscape, degradation of the visual environment, destruction of agricultural and forest lands and degradation of recreational lands.

The mining of Lagoa Santa's dimension stone is developed out of the technical, environmental parameters and in disagreement with the legislation. For the economic use of those quarries, inside of the approaches of sustained development, are necessary changes in the productive process, as:

Qualified professional technical support (mining engineers and geologists), seeking a better planning of the mining operations is necessary. This technical support can be made possible by the Small Producing Cooperative organization of the Lagoa Santa dimension stones, assisting the needs of all the associated quarries;

To assure measures to mitigate the impact caused to the environment and for the controlled disposition of the sterile generated by the production of the dimension stone;

Study of economic use of the reject of the production. This study can be developed in partnership with Institutions of superior level of the mining area, objectifying to find use for it;

Technical study with the purpose of decreasing the loss of the dimension stone of Lagoa Santa during the production process;

Use of research methods, such as, seismic refraction to identify the fractures in the mines reducing the material loss during the mining exploitation.

Finally, the establishment and the growth of the plants should be observed with the purpose of return of the native fauna. The fertility of the soil, the pH and the salinity should be controlled.

REFERENCES

Barreto, M. L. & Sirotheu, G. J.1992. Mineração em Áreas Ambientalmente Sensíveis – Um Estudo Comparativo. Conselho Nacional de Desenvolvimento Científico e Tecnológico. 18 pp.

Brandi,I.V 1994. Estudo da efetividade dos planos de recuperação de área degradas de atividades de explotação de minérios de ferro na região do quadrilátero ferrífero –MG –Dissertação de Mestrado. USP-EESC. 114 pp.

Cabral, J.A et al., 1998 Zoneamento Ambiental da APA Carste de Lagoa Santa. IBAMA, CPRM, Brasília, Brasil

Ciccu, R, Mocci, G. & Imolesi, E. 1998. A rational approach to the assessment of environmental issues in stone quarrying districts. *Proceedings of the Environment Issues Waste Management in Energy and Mineral Production* (ed. A. A. Balkema) Rotterdam, pp. 87-93.

Norma NBR 13028 da ABNT (Associação Brasileira de Normas Técnicas), Coletânea de Normas de Mineração e Meio Ambiente , Companhia Vale do Rio Doce, Rio de Janeiro, 1993.

Norma NBR 13029 da ABNT (Associação Brasileira de Normas Técnicas), Coletânea de Normas de Mineração e Meio Ambiente , Companhia Vale do Rio Doce, Rio de Janeiro, 1993.

Williams, D.J; Wu,Y; Morris, P.H.1993. Systems analysis of engineered mine site rehabilitation, *Proceedings of The Fouth Intern. Conf. on Tailings and Mine Waste*, Fort Collins, Colorado, Rotterdam: A.A. Balkema.

Sinding, K 1998 Environment impact assessment and management in the mining industry. *Proceedings of the Environment Issues Waste Management in Energy and Mineral Production* (ed. A. A. Balkema) Rotterdam, pp. 81-86.

Singh, R.S., Tewary, B.K. & Dhar, B.B. 1994. Effect of surface mining on plant biomass and productivity in a part of Dhanbad coalfield areas. *Second National Seminar on Minerals and Ecology* (ed. Banerjee, S.P.), Oxford & IBH Pub., New Delhi, pp. 103-109.

Singh, P.S., Chaulya, S.K. & Singh, S.K. 1998. Eco-friendly technology for waste dumps management in India. *Proceedings of the Environment Issues Waste Management in Energy and Mineral Production* (ed. A. A. Balkema) Rotterdam, pp. 463-468.

Tailings and Mine Waste '01 © 2001 Balkema, Rotterdam, ISBN 90 5809 182 1

The effect of microstructural evolution on the physical properties of paste backfill

T. Belem, B. Bussière & M. Benzaazoua
Unité de Recherche et de Service en Technologie Minérale, Université du Québec en Abitibi-Témiscamingue, Rouyn-Noranda, Qué., Canada

ABSTRACT: One tailings sample from hard rock mine was used for this study. The paste backfill mixture was made with 5 wt% binder (which is made of a ratio of 80:20 blast furnace slag to ordinary Portland cement). The paste backfill cylinders were then cured in a fogroom for periods of 0, 1, 2, 5, 7, 14 and 28 days. The performed tests were SEM investigations, permeability, suction and MIP tests. The results demonstrate that paste permeability decreases rapidly from 0 to 7 days of curing time before stabilizing between 7 to 28 days. The suction increases gradually from 0 to 7 days of curing time before reaching a plateau after 14 days. The paste backfill free mercury porosity decreases gradually and becomes low from 28 days of curing time. SEM observations showed that permeability, suction and porosity results are related to increasing precipitation of hydrate minerals in the paste backfill.

1 INTRODUCTION

The use of paste backfill is currently practiced in many modern mines throughout the world, particularly in Canada and the United States because of increasingly stringent environmental regulations (Weaver & Luka 1970, Barsotti 1978, Viles & Davis 1989, Petrolito *et al.* 1998). Paste backfill is a hardened composite material made from a mixture of mine tailings, hydraulic binders (such as ordinary Portland cement, fly ash, blast furnace slag or a combination of these) and water. The two main benefits of paste backfill are lower operating costs and a reduction in the amounts of waste material sent to the tailings facility for disposal. This reduction of the amount of waste sent to the tailings facility decreases the environmental impact and decreases future capital expenditures related to the tailings facility (Barsotti 1978, Landriault & Tenbergen 1995, Hassani & Archibald 1998). Because the role of hydraulic binders is to generate mechanical strength, one of the challenges is the use of cemented paste backfill as secondary pillars, to ensure the stability of the underground excavations (Thomas *et al.* 1979, Mitchell & Wong 1982, Mitchell 1989, Stone 1993, Amaratunga & Hein 1997, Ouellet *et al.* 1998, Petrolito *et al.* 1998). A number of studies which have been published in the last 10 years have allowed a better understanding of the properties of paste backfill (Ouellet *et al.* 1998, Benzaazoua *et al.* 1999a & b, Bernier *et al.* 1999). These previous works have indicated the influence of many of the parameters on paste backfill which affect short-term and long-term strength acquisition. Some of these parameters are tailings mineralogy and particle size distribution, moisture content, binder type and proportion, the degree of saturation and finally, chemical weathering (e.g. Benzaazoua *et al.* 1999a & b, Benzaazoua & Belem 1999, Hassani & Hossein 2000).

The physical behaviour of a paste backfill is complex because of the continuous evolution of the paste from its mixture in the backfill plant to storage and hardening in the stope. One can consider that paste fill is similar to a soft clay at the early stage; after 28 days curing time, similar to a compacted silt and, after 91 days of curing time it can be harder than a consolidated soil and approachs the properties of a soft rock. This change corresponds to the evolution of the backfill microstructure because of binder hydration which will alter the water flow characteristics (hydraulic conductivity and percolation rate) and mechanical strength of the paste.

The purpose of this paper is to relate the microstructure evolution of the paste backfill to its physical characteristics. To reach this goal, different tests have been performed on paste backfill samples. The backfill sample studied is a mixture containing 95% tailings and 5% binder. The tests performed are permeability tests, suction tests and mercury intrusion porosimetry after curing periods of 0, 1, 2, 5, 7, 14 and 28 days. The backfill microstructure has also been studied with a scanning electron microscope (SEM).

2 EXPERIMENTAL PROCEDURES

2.1 *Samples preparation*

One mill tailings from an underground hard rock mine which contains approximately 16% sulphur was sampled for the cemented paste backfill preparation. This corresponds to approximately 30% sulfides essentially pyrite and a small amount of pyrrhotite; the other significant minerals are silicates and aluminosilicates. For this study, the same backfill mixture ~~of~~ as the one used at the mine was chosen. The hydraulic binder used consists of 80% blast furnace slag (BFS) and 20% ordinary Portland cement (OPC). The final mixture contains 95% tailings and 5% binder. The process water from the mine was used as the mixing water for sample preparation. The final mixture contains approximately 78% solids (water-to-solids ratio of 0.28) and the water-to-cement ratio (w/c) was 8.

The resulting mixtures were poured into plastic cylinders (10 cm diameter and 20 cm long) which gives a height to diameter ratio of 2. The cylinders were then sealed and cured in a fogroom maintained at approximately 70% humidity (similar to underground mine conditions) for periods of 1, 2, 5, 7, 14 and 28 days. After curing, resulting specimens were used for permeability, suction, mercury intrusion porosimetry and uniaxial compression tests.

2.2 *Permeability tests*

Permeability is the velocity of fluid flow through a porous medium. The permeability of the cemented paste backfill samples were evaluated with rigid-wall variable head permeameter tests. For this study, the paste backfill was directly cast in the permeameter cells and the device was assembled and conditioned in a controlled moisture room. The permeameters have an internal diameter of 8 cm, and the sample height is limited to 15 cm. The walls are made of translucent Plexiglas (cf. Aubertin *et al.* 1996). The permeability tests were performed under falling head conditions at different curing times, following the ASTM method. By applying the Darcy's law, the coefficient of permeability or hydraulic conductivity, k (cm/s), is defined by the equation :

$$k = 2.3 \frac{a}{A} \bullet \frac{L}{t} \bullet log \frac{h_1}{h_2} \qquad (1)$$

where a (cm^2) is the cross-sectional area of the tube; A (cm^2) is the cross-sectional area of cemented paste backfill sample normal to direction of flow; h_1 (cm) and h_2 (cm) are the static water heads, and L (cm) is the sample height.

2.3 *Suction tests*

When a soil is unsaturated, the water is subjected to a negative pressure relative to atmospheric pressure. This negative pressure, called suction, is caused by the surface tension which exists between air and water in contact with the soil matrix (matrix suction) as well as the affinity between water and solids (osmotic suction). The soil-water characteristic curves, also called water retention curves, are mainly used to represent the behavior of unsaturated soil medium (e.g. Bear 1972, Kovàcs 1981, Fredlund and Rahardjo 1993). These water retention curves represent the relation between the volumetric water content (θ) and suction (ψ). The main characteristic of the water retention curve is ψ_a, the pressure for air entry, also called the air entry value (AEV), which corresponds to the beginning of desaturation. In this study, the evolution of the ψ_a of the studied cemented paste backfill with respect to curing time, was investigated.

The ASTM D 3152 standard defines the test procedure used to obtain the water retention curves. The paste backfill sample was placed on a porous ceramic plate (which retains water until a pressure of 5 bars) in a pressurized cell commonly called a Tempe Cell. Then, the sample is subjected to a positive air pressure and the outflow water pressure is measured. By increasing the pressure and measuring the volume of evacuated water with each pressure change, one can establish the water retention curve, which is representative of the volumetric water content (or degree of saturation) against the suction. In this study, the ψ_a value was evaluated using the method proposed by Aubertin *et al.* (1998) which corresponds to the value of ψ_a at 90% degree of saturation.

2.4 *Mercury Intrusion Porosimetry (MIP)*

The theory of a mercury porosimeter is based on the principle that a non-reactive and non-wetting liquid will not penetrate pores until sufficient pressure is applied to force the entrance of the liquid into the pore spaces. The relationship between the applied pressure and the pore size into which the mercury will intrude is given by the Laplace's law or the Washburn equation:

$$D = -\frac{4\gamma \cos\varphi}{P} \quad \text{or} \quad r_p = -\frac{2\gamma \cos\varphi}{P} \tag{2}$$

where D or r_p is the opening pore diameter or radius, P is the applied pressure, γ is the surface tension of mercury (480 dyne/cm) and φ is the contact angle between mercury and the pore wall (usually near 140°).

The pore diameter was calculated from applied pressure assuming cylindrical pores, to give both cumulative and incremental mercury intrusion porosity versus pore diameter or radius data for mesopores (0.002 μm < D < 0.05 μm) and macropores (D > 0.05 μm). The mercury intrusion porosity (*MP*) was evaluated using Micromeretics® AutoPore III 9420 mercury porosimeter. The MIP data allows one to determine total porosity (MP_{tot}), free access porosity (MP_{free}) and trapped porosity (MP_{trap}) of the cemented paste backfill. The calculation of the two former types of porosity requires two cycles of mercury intrusion. The backfill samples which have undergone the first cycle of mercury intrusion (MP_{tot}) are stored in a desiccator for 24 hours to drain the trapped mercury before undergoing the second cycle of mercury intrusion (MP_{free}). The difference between MP_{tot} and MP_{free} corresponds to MP_{trap}. When a pore family observed after the first mercury intrusion was not observed after the second intrusion, it is because the intruded mercury was entirely trapped by that pore family.

2.5 *Scanning electron microscope observations*

The scanning electron microscope (SEM), a Hitachi®3500-N, was used to see the microstructure and the texture of the studied backfill samples. The SEM investigation was performed with several different magnifications. The back-scatter electron (BSE) imaging mode was chosen because it produces images with the chemical contrast superimposed on the relief contrast. The operating conditions are the following ones: 20 kV of voltage, a DC from 150 to 200 μA, a lowered pressure of 25 Pa to avoid the harmful electrical charge effects on image quality and a working distance of 15 μm. The electronic images give information on (i) the particle size distribution and grain morphology of the composite, (ii) the texture of the cement phase, (iii) the morphology of the hydrated cement phases and, (iv) the porosity of the composite.

3 RESULTS

3.1 *Evolution of cemented paste backfill hydraulic conductivity*

Figure 1 shows the evolution of hydraulic conductivity, *k*, of the studied paste backfill with curing time. The overall trend of this curve is a rapid decrease of the hydraulic conductivity of the paste backfill with curing time. This evolution suggests that the hydraulic conductivity of the paste backfill decreases rapidly during the first 7 days due to the formation of the cement hydrates which gradually fill in the initial porosity. After 7 days, any change in hydraulic conductivity (near 10^{-8} cm/s) is insignificant.

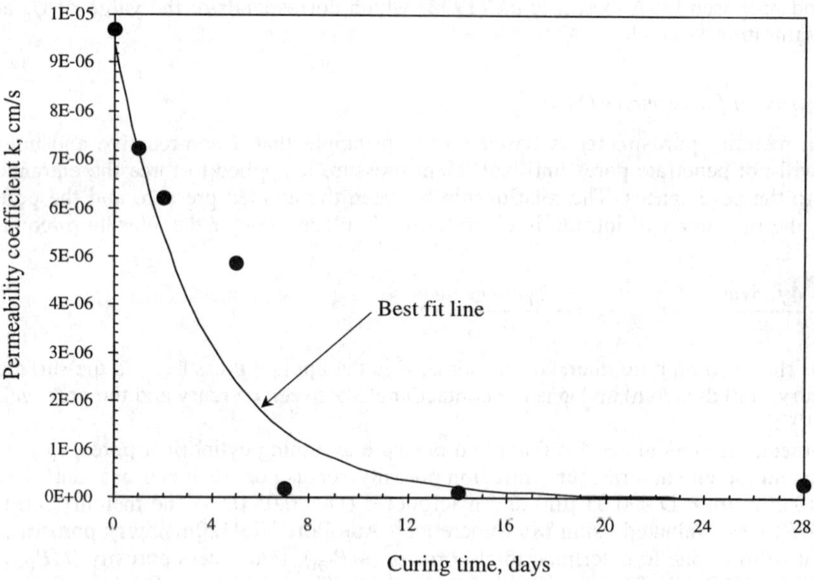

Figure 1. Evolution of the paste backfill hydraulic conductivity with curing time

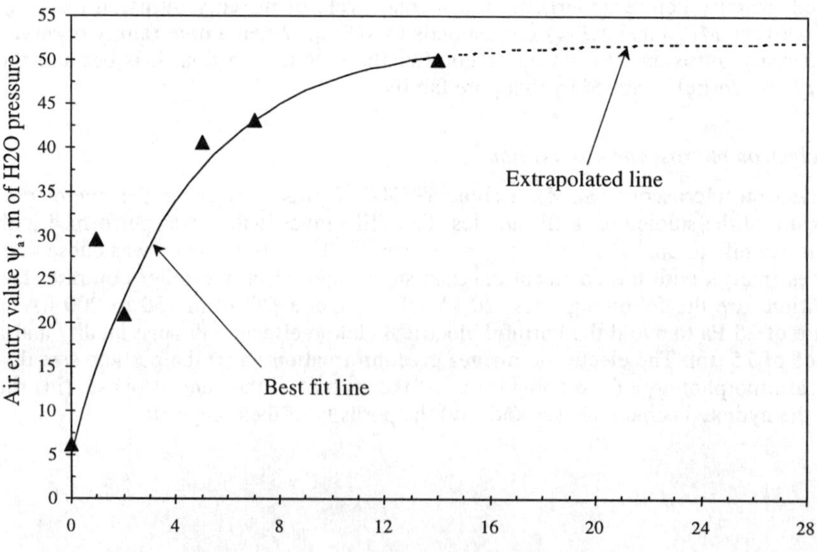

Figure 2. Evolution of the paste backfill water retention capacity through the air entry value

3.2 *Suction test results*

The main results of the suction tests performed on the paste backfill samples at different curing times are presented on Figure 2. It should be noted that the suction test was not performed for the curing time of 28 days because the applied pressure exceeded the maximum capacity of the porous ceramic plate. This figure represents the evolution of the air entry value ψ_a with curing time. Just after molding the fresh paste backfill, ψ_a is at approximately 6.5 m of water pressure. After less than two days of curing time, the air entry value ψ_a reaches values of between 20 and 30 m of water pressure. Beyond this date, ψ_a continues to increase, but less rapidly with increasing curing time. After 14 days of curing time, ψ_a reaches a value of approximately 50 m of water pressure which seems to be the maximum value obtainable for suction for this paste backfill sample.

3.3 *MIP results*

Figure 3 shows the MIP curves of the studied cemented paste backfill after 28 days of curing time. It should be recalled that the first MIP curve allows one to obtain the MP_{tot} value, while the second MIP curve allows one to obtain the MP_{free} value (Fig. 3a). The difference between the two porosity values corresponds to the trapped mercury porosity, MP_{trap}. The incremental porosity curves (Fig. 3b) allow one to look at the MIP results in term of modal distribution of the pore families compared to their median diameter (D_m). On Figure 3b, one can notice that there are two pore families within the paste backfill. Among these two pore families, the second pore family is the more significant and the more representative of this paste backfill sample (peak of the histogram). The first pore family consists largely of macropores having median diameter ranging from 123 and 125 μm while the second family consists of mesopores having median diameter ranging from 0.2 and 1.2 μm.

Figure 4 allows one to compare the pore distribution network of the paste backfill with that of the tailings used for its preparation. The difference between the curves of these two types of material accounts for the difference between their textures and microstructure. From this, one notices the direct influence of binder hydration on the evolution of the spatial geometry of the pore network which is much finer in the paste backfill sample than in the tailings sample. From Figure 4, the MP_{tot} value is 42% for the tailings and 41% for the paste backfill; the MP_{free} value is 40% for the tailings and 8% for the paste backfill and, the MP_{trap} value is 2% for the tailings and 33% for the paste backfill. One can notice that the cemented paste backfill free porosity is much lower than the porosity of the tailings.

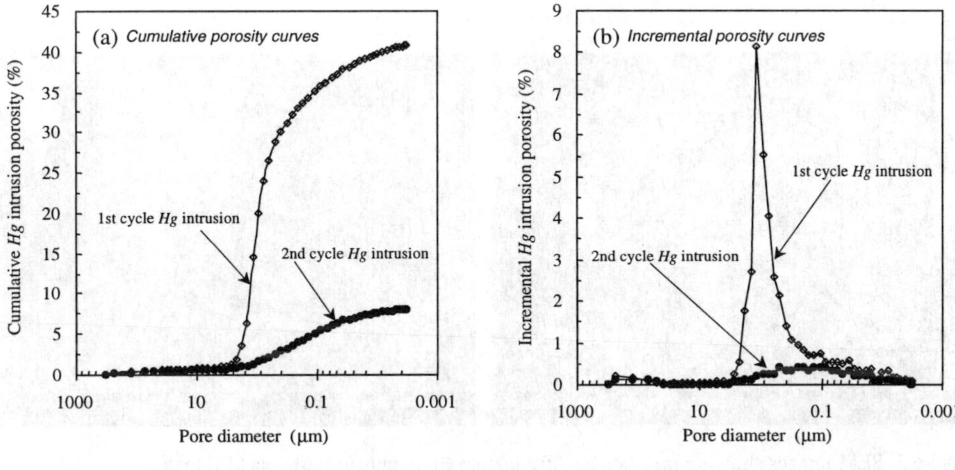

Figure 3. Typical MIP curves of the cemented paste backfill after 28 days curing time

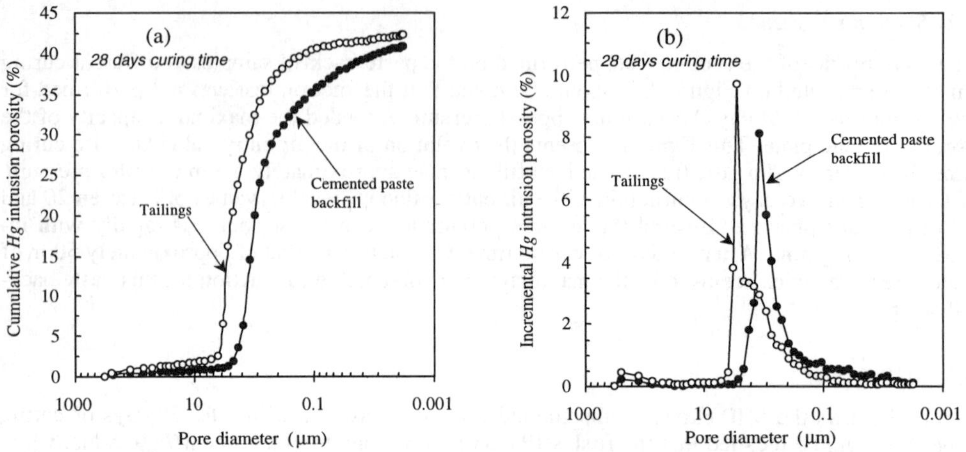

Figure 4. MIP curves of the tailings likened to the MIP curves of the cemented paste backfill

3.4 *SEM investigations*

The SEM observations of the studied paste backfill samples showed an even grain dispersion pattern, filled intergranular spaces, a dense matrix (Fig. 5), an absence of cracks, the presence of macropores and the presence of secondary precipitation of $AlMgSiFeCaSO_4$ phases. The voids between the tailings coarse particles are filled with ultra-fine particles and a cemented matrix (Fig. 5b). The paste texture has a direct effect on the hydraulic properties of the cemented paste backfill. Figure 5a shows the paste backfill matrix to be well cemented.

The SEM observations also showed the significant role of secondary precipitation in the reduction of free porosity during the course of curing time (Fig. 6). Secondary precipitation representing all the phases occurring after the hardening of backfill due to the formation of the hydrated primary phases (Fig. 6a). This contributes to the cohesion of the backfill when it fills intergranular spaces, as well as mesopores and macropores (Fig. 6b). However, these precipitates become harmful when they are expansive (example of sulfatic attack due to formation of the ettringite and the expanding secondary gypsum). For the studied samples, these secondary precipitates were especially beneficial in the acquisition of mechanical strength by thickening the matrix of the backfill and precipitating in macropores.

Figure 5. SEM images showing (a) paste backfill texture (b) details of paste backfill matrix

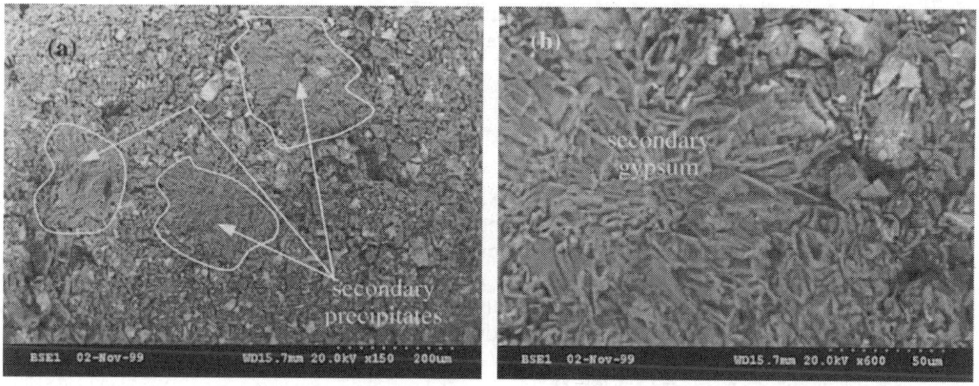

Figure 6. SEM images showing (a) macropres filled with secondary gypsum (b) details of a secondary gypsum within the studied paste backfill after 28 days curing time

4 DISCUSSION

The results of the permeability tests, suction tests and SEM observations agree on the evolution of the paste backfill microstructure during the curing time. This microstructure evolution results in a reduction of the paste backfill water filtration capacity and an increase in its water retention capacity. One can note that as the microstructure evolves, the hydraulic conductivity decreases and the water retention characteristics increase. The high water retention capacity of the backfill will increase its resistance to sulphate attack or chemical weathering (Lawrence 1992) as well as reduce its self-heating potential (exothermic effect due to the extreme oxidation of sulphurous minerals present in the backfill). Moreover, a high degree of saturation decreases the ability of the backfill to acquire strong mechanical strengths (very slow hydration) and the backfill can become more susceptible to liquefaction induced by blasting or mobile equipment vibrations

It should be noted that even if the total porosity of the paste backfill is very close to that of the tailings, these two materials are clearly distinguished by their fluid trapping capacity. A detailed and thorough study on the use of MIP results (Remy 1993) indicated that free porosity is directly related to permeability. The MIP tests carried out after 28 days curing time clearly show that the paste backfill had a low free porosity and a strong capacity for fluid trapping which is different from the properties of the tailings. This means that the cemented paste backfill becomes less permeable after 28 days of curing time. Indeed, with a MP_{trap} value of 33% the cemented paste backfill can be considered as trapping the all of intruded mercury. This phenomenon is possibly caused by the irregular pore geometry in the pore network due to the formation of hydrates in the binder. SEM observations confirm that this geometry is irregular and could indeed support the fluid trapping phenomenon. The SEM imagery also demonstrates that after 28 days curing time the paste backfill matrix becomes very dense and less permeable. This agrees with the permeability test results shown in Figure 1.

Figure 1 shows that in terms of permeability, the paste backfill microstructure evolves very rapidly in the first week and remains stable afterwards. A general regression equation describing the evolution of the studied paste backfill permeability according to the curing time $k(t)$, has been found; the exponential equation is

$$k(t) = k_0 \bullet exp(-\alpha \cdot t) \tag{3}$$

where, k_0 (cm/s) is the initial permeability of the paste backfill or the fresh paste backfill hydraulic conductivity immediately after its mixture (without any hydration activity), α is backfill constant and can be considered as a proportionality factor depending on the binder type, its proportion and the physical properties of the tailings, t is curing time. From the regression, k_0 = 9.4097e-6 cm/s and α = 0.2741 with a coefficient of correlation R = 0.96. Assuming that the hydraulic properties of fresh paste backfill and tailings are similar, the following equation proposed by Aubertin *et al.* (1996) can be used to evaluate the initial permeability:

$$k_0 = c_0 \cdot \frac{\gamma_w}{\mu} \cdot D_{10}^2 \cdot C_u^{1/3} \cdot \frac{e^{3+m}}{(1+e)} \qquad (4)$$

where, c_0 is tailings parameter, γ_w is the unit weight of water (9.79 kN/m^3 at room temperature), μ is the viscosity of water ($\approx$ 9.8 x 10-6 N.s/cm2 at 20°C), D_{10} (μm) is particle diameter at which 10% of the tailings passes, C_u is the coefficient of uniformity, *e* is the voids ratio and *m* is tailings parameter. In their experiments on homogenized tailings, Aubertin *et al.* (1996) found that the average value of *m* is equal to 2.16. By combining equations 3 and 4 the following general equation can be proposed for predicting the cemented paste backfill hydraulic conductivity with respect to the curing time, *k*(t), by the following equation:

$$k(t) = c(t) \cdot \frac{\gamma_w}{\mu} \cdot D_{10}^2 \cdot C_u^{1/3} \cdot \frac{e^{3+m}}{(1+e)} \cdot \exp(-\alpha \cdot t) \qquad (5)$$

where *c*(t) is backfill hydraulic constant depending on binder type, its proportion, tailings properties and curing time. From experimental results of the studied paste backfill, *c*(t) is given by the following empirical equation:

$$c(t) = c_0 \cdot (0.9537 \cdot \exp(-0.2727 \cdot t)) \quad \text{with} \quad R = 0.97 \qquad (6)$$

where c_0 is tailings or fresh paste backfill hydraulic constant.

The suction test results (Fig. 2) showed that the microstructure of the studied paste backfill increased its water retention capacity during the curing period. This increase is progressive and seems to reach a plateau after 14 days of curing time. This evolution is directly related to the mode of binder hydration and its velocity. A regression equation describing the evolution of the studied paste backfill air entry value with respect to curing time, ψ_a(t), has been found

$$\psi_a(t) = \psi_{max} \cdot (a_1 - \exp(-a_2 \cdot t)) \quad \text{and} \quad \psi_{max} = a_0 \cdot \psi_0 \qquad (7)$$

where ψ_{max} is the maximum air entry value that the backfill can have after 28 days curing time, a_0 is a proportionality constant, ψ_0 is fresh paste backfill or tailings air entry value, a_1 is a constant depending on binder type and its proportion and water-to-cement ratio (w/c), a_2 is a constant depending on tailings physical properties, *t* is curing time. From the regression, ψ_0 = 6.2 m of water pressure, a_0 = 7.6, ψ_{max} = 47.1 m of water pressure, a_1 = 1.1 and a_2 = 0.235 with a coefficient of correlation R = 0.99.

5 CONCLUSION AND PROSPECTS

A Canadian underground hard rock mine tailings was sampled in order to produce a cemented paste backfill with the same composition as used at this mine. The resulting mixture contains 95% tailings and 5% binder which consists of 80% blast furnace slag and 20% ordinary Portland cement. Several types of tests were performed on the paste backfill samples to observe either directly or indirectly the evolution of the texture and the microstructure of this material during the course of curing. These tests were MIP, suction and permeability tests and finally, the SEM observations.

The results of this exploratory study led to the following conclusions: (i) the paste backfill permeability decreases very rapidly in the very short term (i.e. between 0 and 7 days of curing time) before stabilizing after 7 days of curing time, (ii) the paste backfill suction increases gradually before reaching a possible plateau after 14 days of curing time, (iii) the mercury free porosity of this paste backfill is very low after 28 days curing time and, (iv) the SEM observations allow us to better understand that the decrease in the paste backfill permeability and porosity is directly related to the phenomena of minerals precipitation. From the obtained experimental results, regression models were found to predict the hydraulic conductivity, *k*, and the suction, ψ_a, of the studied paste backfill.

A better understanding of the evolution of pastefill properties in practice is important for different reasons: (i) the strength of pastefill in the filled stope is related to the water flow which is a function of the pastefill properties, (ii) the self-heating potential of a sulphidic paste will depend on the water content of the pastefill in the stope which is again a function of the material properties and, (iii) the use of surface pastefill as waste management technique for sulphidic tailings will be interesting if the material has a high water retention characteristics and a low hydraulic conductivity. In short, the results of this study allowed us to better understand the evolution of the matrix of the paste backfill from the fresh paste to its hardening and opens up several interesting prospects for the continuation of this work. The next step will be to increase the variety of the binder types used to refine the proposed regression models.

ACKNOWLEDGMENTS

This research was supported by the Fond de l'Université du Quebec en Abitibi-Témiscamingue (FUQAT). The authors gratefully acknowledge their support. The authors would also like to thank our mining partner, Aur-Novicourt-Teck (Louvicourt Mine) for their collaboration in the completion of this work. Also, thanks to Hugues Bordeleau, technician at URSTM, for performing the actual suction and permeability tests.

REFERENCES

Amaratunga, L.M. & G.G. Hein 1997. Development of a high strength total tailings paste fill using fine sulphide mill tailings. *Proc. 29th Ann. Conf. Can. Min. Process*: 293–305. CIM-CMP.

Aubertin, M., J.F. Ricard & R.P. Chapuis 1998. A predictive model for the water retention curve: application to tailings from hard rock mines. *Can. Geotech. J.* 35:55–69.

Aubertin, M., B. Bussière & R.P. Chapuis 1996. Hydraulic conductivity of homogenized tailings from hard rock mines. *Can. Geotech. J.* 33:470–482.

Barsotti, C. 1978. The evolution of fill mining at the Ontario Division of Inco Metals. *Proc. 12th Can. Rock Mech. Symposium, Mining with Backfill.* CIM special volume 19:37–41.

Bear, J. 1972. Dynamics of fluids in porous media. Dover Publications Inc., New York.

Benzaazoua, M. & T. Belem 2000. Optimization of sulfide-rich paste backfill mixtures for increasing long-term strength and stability. In M. Sanchez, S. Castro & F. Vergara (eds), Proc. 5th Conf. Clean Techn. for Mining Industry: Vol. II:947-957. Santiago: Chile.

Benzaazoua, M., T. Belem & D. Jolette 1999a. Investigation de la stabilité chimique et de son impact sur la qualité des remblais miniers cimentés. *Rapport d'activité de recherche IRSST.* 152p.

Benzaazoua, M., J. Ouellet, S. Servant, P. Newman & R. Verburg 1999b. Cementitious backfill with high sulfur content: Physical, chemical and mineralogical characterization. Cement and Concrete Research: 29:719–725.

Bernier, R.L., M.G. Li & A. Moerman 1999. Effects of tailings and binder geochemistry on the physical strength of paste backfill. In N. Goldsack, P. Belzile, Yearwood and G. Hall (eds), Sudburry'99 – Mining and the environment II: Vol. 3:1113–1122.

Fredlund, D.G. & H. Rahardjo 1993. *Soil Mechanics for Unsaturated Soils.* John Wiley & Sons, inc., New York.

Hassani, F. & M. Hossein 2000. Strength development in underground high-sulphide paste backfill operation. *Mining Millennium 2000, Metro Toronto Convention Centre, 5-10 March 2000*: CD-ROM paper No. W–7.

Hassani, F. & J.H. Archibald 1998. *Mine backfill.* CD-ROM.

Landriault, D.A., & R. Tenbergen, R. 1995. The present state of paste fill in Canadian underground mining. *Proceedings of the 97th Annual Meeting of the C.I.M. Rock Mechanics and Strata Control Session, Halifax, Nova Scotia, 14-18 May 1995.*

Lawrence, C.D. 1992. The influence of binder type on sulfate resistance. *Cement and Concrete Research* 22:1047–1058.

Mitchell, R.J. 1989. Stability of cemented tailings mine backfill. In Balasubramaniam *et al.* (eds), Proceedings of Computer and physical modelling in geotechnical engineering: 501–507. Rotterdam:Balkema.

Mitchell, R.J. & B.C. Wong 1982. Behavior of cemented tailings sands. *Can. Geotech. J.* 19:289–295.

Ouellet, J., M. Benzaazoua & S. Servant 1998. Mechanical, mineralogical and chemical characterisation of paste backfill. Proc. 4th Int. Conf. Tailings and Mine Waste: 139–146 Rotterdam: Balkema.

Petrolito, J., R.M. Anderson & S.P. Pigdon 1998. The strength of backfills stabilised with calcined gypsum. In M. Bloss – Austr. Inst. Min. Metal. – (eds), Minefill'98 – Proc. 6th Int. Symp. Mining with Backfill.: 83–85. Brisbane: Australia.
Remy J.M. 1993. Influence de la structure du milieux poreux carbonaté sur le transfert d'eau et les changements de phases eau-glace. Application à la durabilité des roches calcaires de Lorraine. *INPL Doctoral Thesis*, Nancy, France.
Stone D.M.R. 1993. The optimization of mix designs for cemented rockfill. *Minefill'93 – Proc. of 5th Int. Symp. Mining with Backfill* : 249–253. Johannesbourg:SAIMM.
Thomas, E.G. 1966. The important properties of hydraulic fill, with particular reference to mechanized cut and fill mining operations. *Proc. Australian Inst. Min. Metall.* 220:1–20.
Thomas, E.G., J.H. Nantel & K.R. Notley 1979. Fill technology in underground metalliferous mines. *International Academic Services Ltd.* Kingston:Canada.
Viles, R.F., & R.T.H. Davis 1989. New material technologies applied in mining with backfill. In F. Hassani *et al.* (eds), *Innovation in Mining Backfill Technology:* 95–101. Rotterdam: Balkema.
Weaver, W.S., & Luka, R. 1970. Laboratory studies of cement-stabilized mine tailings. *Canadian Mining and Metallurgical Bulletin* 64(701):988–1001.

New technologies and approaches

Tailings and Mine Waste '01 © 2001 Balkema, Rotterdam, ISBN 90 5809 182 1

Neural network analysis for acid drainage in waste rock pile

M.A.B.da Silva & M.de S.S.Almeida
COPPE/UFRJ, Rio de Janeiro, Brazil

ABSTRACT: Throughout the world, acid drainage is widely recognized as one of the most costly and sensitive environmental issues facing the mining industry. To address this problem, environmental managers have devised an array of techniques for predicting and minimizing acid drainage from mine sites. This paper describes how neural networks can be used as a modeling technique of acid mine process in mining. The specific objective is to develop a system able to predict the acid drainage generation in a specific waste pile. This system must use only data obtained from the monitoring data. The developed model shows the advantage of neural nets with respect to conventional techniques leading rapidly to useful results with reasonable precision. The authors believe that this paper may encourage other new applications of neural networks in industrial process modeling, specially in mineral technology.

1 INTRODUCTION

Acid drainage from waste rock pile containing sulfide minerals is a serious environmental problem. The accurate prediction of the potential magnitude and duration of the contaminant loading is of concern to both mine operators and regulatory agencies. Mathematical modeling is a recommended tool for evaluating the extend of the problem and for providing direction for assessing management and decommissioning strategies (Scharer et al. 1994).

A computer model based on neural networks has been developed to simulate the generation of acid drainage in a waste rock pile from an uranium mine. Data from monitoring programs developed by the operator and controlling organ were collected and statistically analyzed. The set of data consists of the rainfall measurements and concentrations of the ion SO_4^{2-}. A neural network based on the set of data was obtained.

1.1 *General description of uranium mining environmental impacts*

Uranium mining facilities are generally classified as underground or open pit operations. The mining and milling of ores to produce uranium required to fuel the nuclear power reactors produce large quantities of wastes which must be safely managed. The mining of ores containing on an average about 0.1 to 0.2% of uranium leaves large amounts of waste rock residues which need planned disposal. If current practices of using uranium are considered, the volume of mining and milling wastes may vary between 40,000 and 60,000 m^3 per nuclear gigawatt.

Sources of pollutants to the environment include acid drainage from mine and waste rock pile, and tailings, and airborne radionuclides, the primary one being ^{222}Rn and its short-lived daughters. Besides the radionuclides from the uranium series, non-radioactive heavy metals may also be present in the ore. Some ores may contain small quantities of Mn, Cu, Se, Mo, V, Ni, As and Zn. Rarely do metals such as these exist in uranium ores in economically attractive concentrations. However, these elements will also be present in the effluent and may need to be considered in the evaluation of environmental effects.

Brazil's first Uranium Mining and Milling Facility (UMMF) is located in the mineral region of the Poços de Caldas Plateau. Since the beginning of its operations in 1982, several technical works and scientific studies have come out dealing with: radiological impact assessments, the transport of radionuclides in surface and groundwater, and the transference of radionuclides to plants from the soil (Fernandes et al. 1996).

However, no attempt has been made to integrate the findings of these studies into an effective plan of environmental management. In the present case, the environmental pollution of nearby and groundwater by UMMF results from waste rock piles, acid mine drainage and tailings effluent. A schematic location map of UMMF is shown in Figure 1.

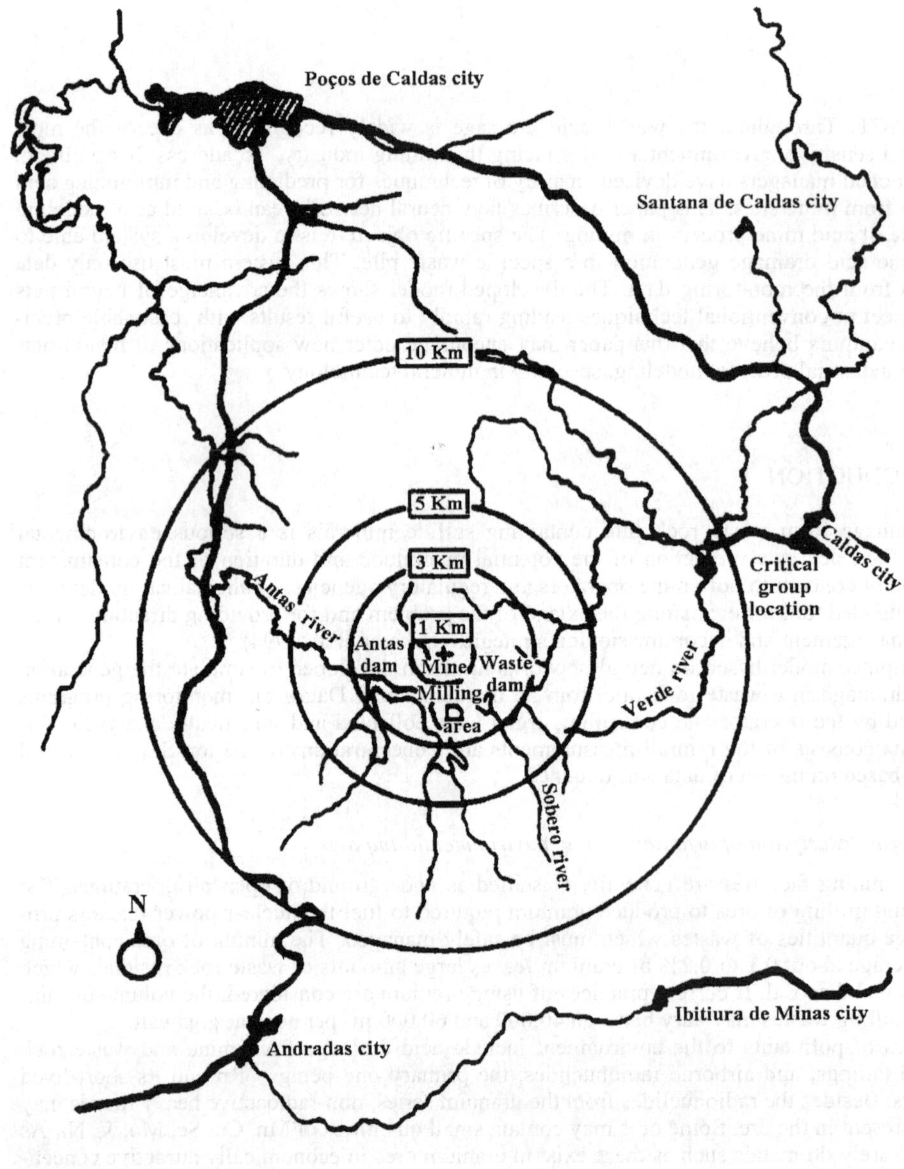

Figure 1. Location map.

In UMMF, uranium ore is mined in an open-pit operation. It was estimated that an amount of 94.5×10^6 tones of rock were removed due to mining operation. Only 2% of this amount was destined for physical and chemical processing. The rest is stocked in two main waste rock piles ("bota-foras", BF). One of these is the BF-4 which is the object of our interest. The BF-4 has 12.4×10^6 m^3, height of about 90 m, and occupying an area of about 0.57 km^2.

1.2 *Description of the general environment at UMMF*

The UMMF of Poços de Caldas is located in one of the most important alkaline intrusions in the world. The site can be best described as a round hollow, 35 km in diameter, covering an area of 1000 km^2. Drill cores from the underlying rock showed extensive rock alteration due, in part, to weathering, partially overprinted by hydrothermal activity. Sharp redox fronts are one of the most outstanding features of the mine area: the overlying (oxidized) rock is brownish-red due to iron oxyhydroxides, while the underlying (reduced) rock is gray.

The uranium is generally associated with zirconium ores, while the thorium is associated with manganese and iron oxides. Two hydrographic basins, the Antas and Verde, cross the UMMF region; the Antas basin drains about 70% of the plateau and the Verde basin 20%. The waters from the Antas River are used for crop irrigation and for cattle watering. The Verde River waters are also used for irrigation, but to a lesser degree since the field topography is not suitable for agricultural activities.

The annual average temperature is 19°C varying between 1°C and 36°C. The total maximum annual rainfall is 1700 mm, with more than 120 days of rain each year (Fernandes et al. 1994).

2 THE IMPACT OF ACID GENERATION IN URANIUM MINE WASTES

Experience around the world has led to the conclusion that sulfuric acid generation is one of the major concerns associated with the decommissioning of sulfide-containing tailings and waste rock areas. Acid-leached uranium tailings are often significant acid producers even though the sulfide mineral content in the ore may originally have been very low. Strong acid generation can occur because most or all of any natural mineral alkalinity has been removed in the uranium extraction process by a hot sulfuric acid leach.

The chemistry of acid generation can be summarized by the following reactions (Feasby et al. 1994):

$$2\,FeS_2 + 7\,O_2 + 2\,H_2O \rightarrow 2\,Fe^{2+} + 4\,SO_4^{2-} + 4\,H^+ \qquad (1)$$

$$2\,H^+ + 1/2\,O_2 + 2\,Fe^{2+} \rightarrow 2\,Fe^{3+} + H_2O \qquad (2)$$

Although these reactions can occur in a biologically sterile environment, chemolithotrophic bacteria, particularly *Thiobacillus ferrooxidans*, can accelerate the iron oxidation rate. Ferric iron in acidic solution can also oxidize metal sulfides (MS) as follows:

$$MS + 2\,Fe^{3+} \rightarrow M^{2+} + S^0 + 2\,Fe^{2+} \qquad (3)$$

Low pH (<4.0) seepage waters are often associated with mine wastes where sulfide oxidation is occurring, but frequently alkaline minerals such as calcite or dolomite present in the wastes can neutralize the acidity produced and near neutral conditions (pH ~ 7) can result until the alkaline minerals are consumed or rendered unreactive by surface coatings. Acidic drainage from mine wastes will solubilize many metals including aluminum, copper, zinc, nickel, manganese, and lead. For uranium-rich deposits, uranium and thorium are also dissolved. The metal contaminants mobilized by acid production are the most important concern in the management of pyritic uranium wastes. Waste rock piles are very porous to water and oxygen, essential ingredients for sulfide oxidation, because the piles are usually constructed on porous foundations, this "acid rock drainage" has been very difficult to prevent and control at sites around the world.

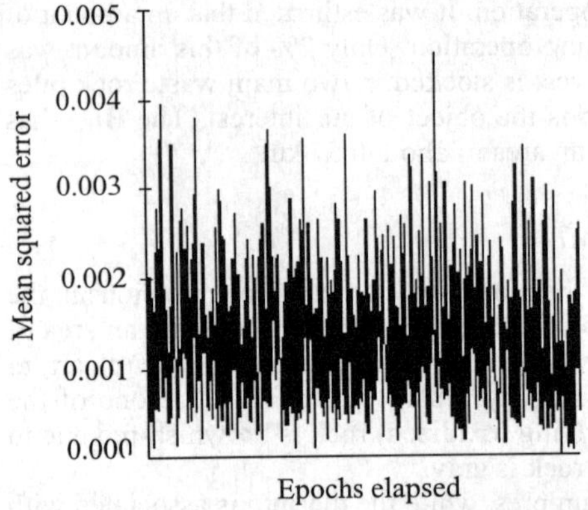

Figure 2. Error associated with the training set.

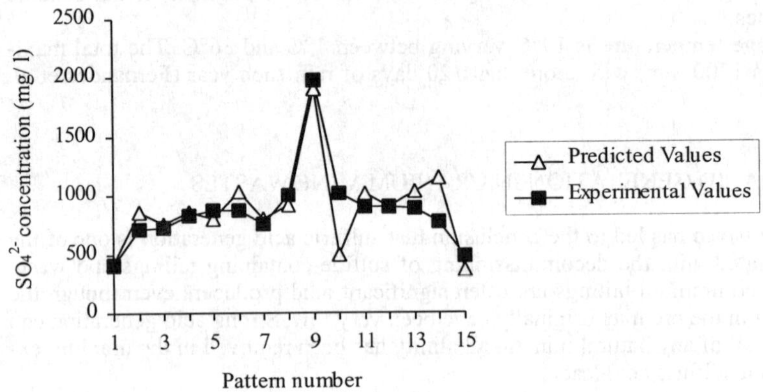

Figure 3. Prediction of the trained neural net.

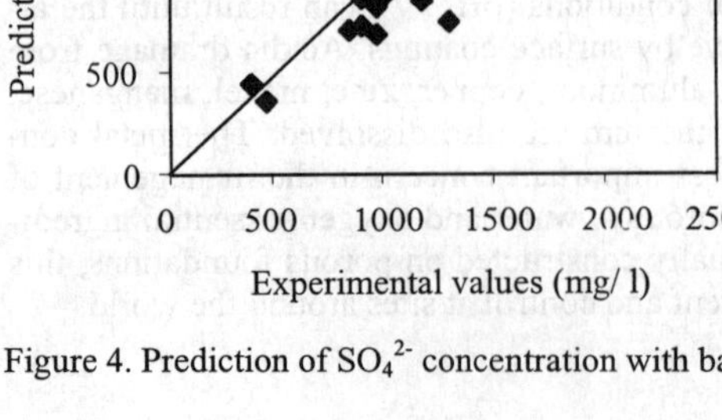

Figure 4. Prediction of SO_4^{2-} concentration with back propagation net.

3 NEURAL NETWORKS

Neural networks are essentially connection systems consisting of collections of simple, massively interconnected computational elements, and their characteristics stem from the collective behavior of these elements, also known as (artificial) neurons, neurodes, processing elements, processing nodes or processing units. Although many different structures have been described to date, back propagation neural nets have proved to be particularly useful for the modeling of complex or ill-defined process systems, where modeling from first principles is not technically or economically feasible (Aldrich et al. 1994).

In these types of networks (which were also used for the work described in this paper) processing elements are typically divided into disjoint subsets or layers, in which all the process units possess essentially similar computational characteristics. The layers comprising these systems are usually categorized as either input, hidden or output layers, to denote the way in which they interact with the information environment of the net. Process units in these layers are linked to other units in successive layers by means of weighted connections (adjustable numeric values). Training, which entails the adjustments of the weight matrix of the net, occurs by means of learning algorithms usually designed to minimize the mean square error between the desired and actual output of the net. During the learning process information is propagated back through the net in order to update its weight, thereby allowing the net to form an internal representation of the relationship between the inputs and the outputs presented to it (Haykin 1994).

4 ACID DRAINAGE MODELING

The acid drainage generation in a waste rock pile depends in a complex way on many parameters, such as, biological, chemical, and physical. In all, one parameter, rainfall, was used to simulate the generation of SO_4^{2-}, which is considered to give a extent of the acid drainage generation.

The main advantage of modeling techniques based on the use of neural nets, is that *a priori* assumptions with regard to the functional relationship between the parameters, rainfall (mm) and SO_4^{2-} (mg/l) are not required. The net learns this relationship instead, on the basis of examples of related rainfall-SO_4^{2-} vector pairs, examples, or pattern numbers. Once a model has been constructed, it is moreover possible to make use of general or special purpose neuralware to decrease computation time by several orders of magnitude, thus enabling, for example, the use of the model in on-line control systems.

A set of data consisting of a total of 66 vectors was obtained from the BF-4 site and randomized and, after statistical treatment, subdivided into a training set (15 vectors) and a test set (5 vectors). The training set consisted of exemplars presented to the neural nets during training (weight adjustment of the nets), while the test set was used to monitor the performance of the nets subsequent to training. This procedure is essential to verify the capability of the net to generalize the relationships between parameters correctly, instead of just learning to reproduce the data in the training set.

A back propagation net with an input layer, comprised of one processing element or artificial neuron, one hidden layer comprised of eight processing elements or artificial neurons, and an output layer comprised of one processing element or artificial neuron, was used to model the relationship between the rainfall and SO_4^{2-} concentration in the same period (month).

The input and output layers consisted of elements with linear translation function, while the hidden layer consisted of elements with hyperbolic tangent translation function. The use of sigmoidal transfer function in the hidden layer also yielded reasonable results, although not as good as those obtained with hyperbolic tangent transfer function. Gradient descent search strategies was used during the training. The mean (average) squared error is varying between 0.001 and 0.002 as shown in Figure 2.

In Figure 3 the output of the net is compared with the experimental data. Despite a relatively training set, the net was able to generalize the relationship between the input an output variables.

The output of the net after training is compared with the experimental data, as shown in a scatter graph in Figure 4. After training, the net was capable of predicting the training data with an accuracy characterized by a correlation coefficient (R^2) of 0.853.

5 CONCLUSIONS

The results obtained in this work point out that the neural network process model described could simulate the relationship between the rainfall and concentration of SO_4^{2-} (R^2= 0.853). The net also shows to be a powerful tool for modeling complex phenomena, especially when these phenomena are ill-defined and even when comparatively few data are available, such as in this case where acid drainage was modeled.

The authors believe that further research is needed to develop a useful model based on the neural network process for predicting with an accuracy acid drainage in waste rock pile at UMMF.

REFERENCES

Aldrich, C., Van Deventer, J. S. J. & Reuter, M. A. 1994. The application of neural nets in the metallurgical industry. *Minerals Engineering*. 7 (5-6): 793-809.

Feasby, D. G., Chambers, D. B., Scharer, J. M., Pettit, C. M., Dakers, R. G. & Goldsworthy, M. H. 1994. International perspective on the role of acid generation in selecting decommisioning techniques for uranium mining sites in Eastern Germany. *Third International Conference on Abatement of Acidic Drainage, Pittsburgh, 25-29 April 1994*: 67-76.

Fernandes, H. M., Veiga, L. H. S., Franklin, M. R., Prado, V. C. S. & Taddei, J. F. 1995. Environmental impact assessment of uranium mining and milling facilities: A study case at the Poços de Caldas uranium mining and milling site. Journal of Geochemical Exploration. 52: 161-173.

Fernandes, H. M., Franklin, M. R., Veiga, L. H. S., Freitas, P. & Gomiero, L. A. 1996. Management of uranium mill tailing: Geochemical processes and radiological risk assessment. J. Environ. Radioactivity. 30 (1): 69-95.

Haykin, S. 1994. *Neural networks: A comprehensive foundation*. New York: Macmillian College Publishing Company.

Scharer, J. M., Petit, C. M., Chambers, D. B. & Kwong, E. C. 1994. Mathematical simulation of a waste rock heap, *Third International Conference on Abatement of Acidic Drainage, Pittsburgh, 25-29 April 1994*: 30-39.

Tailings and Mine Waste '01 © 2001 Balkema, Rotterdam, ISBN 90 5809 182 1

Deep injection of warm tailings for thawing of frozen layers

P.G. Landine, N. Holl, D. Rezansoff & E. Yeates
Cameco Corporation, Saskatoon, Sask., Canada
H.K. Mittal
HK Mittal and Associates, Saskatoon, Sask., Canada

ABSTRACT: The Rabbit Lake In-pit tailings management facility (RLITMF) is located in northern Saskatchewan, Canada. Due to the cold climate at the mine site, tailings deposited during the winter months become frozen. Full consolidation of the tailings mass is required in order to minimize long term contaminant transport. The presence of frozen layers would prevent consolidation, leading to advective contaminant release. A full scale trial program of warm tailings injection via inclined injection wells was begun in 1999 and continued in 2000 for the purpose of thawing frozen tailings. A major portion of the 1999 summer tailings production was injected. The basic design and operational procedures for the 1999 injection system are described, along with physical observations of injection effects and the results of a 1999 borehole investigation program. The program has shown that deep tailings injection under gravity head is possible and that positive thawing benefits can be realized.

1 INTRODUCTION

The Cameco Rabbit Lake Operation is a uranium mine and mill complex located in northern Saskatchewan, Canada, approximately 700 km north of Saskatoon.

Tailings from the Rabbit Lake mill are deposited in the Rabbit Lake In-pit Tailings Management Facility (RLITMF), located within the mined-out Rabbit Lake pit. The principle of operation of the TMF is known as the "pervious surround" method, i.e. the tailings solids are surrounded by a pervious material (Gulf, 1981). This pervious material allows drainage of excess water contained in the tailings to an internal seepage collection system, it also allows water contained in the surrounding host rock to be collected, thereby maintaining a hydraulic gradient toward the facility. Upon final decommissioning and return to normal hydrogeological conditions, groundwater will flow preferentially through the pervious surround rather than the low permeability tailings.

Tailings deposition in the RLITMF began in 1985. The tailings have been discharged subaerially at the edge of the facility from two alternating locations throughout the year. In the winter period, deposited tailings become frozen in layers from 1 to 5 m in thickness. During the summer period, previously frozen tailings become buried under continuously accumulating new tailings.

Monitoring and borehole drilling investigations, Cameco (1995), and Cameco (1998) of the RLITMF have shown that layers of ice and frozen tailings have remained frozen for at least 13 years after deposition. Frozen tailings will not consolidate as required. These layers of ice and frozen tailings would also impede consolidation of the unfrozen tailings to some degree by preventing movement of water to the drain system.

Thermal modeling (Cameco 1994) of the deposit has indicated that complete thawing of the deposit by natural means will likely require several decades; this would either delay decommissioning significantly or necessitate special thawing measures at the end of tailings deposition (e.g. warm water injection). Accordingly, Cameco proposed a trial program to initiate thawing of these frozen layers by injection of warm fresh tailings below the frozen layers. The proposal submitted for regulatory approval on June 1, 1999 was for a long term test program, which could lead to a full scale injection program.

Benefits expected from the injection of tailings into the RLITMF include: the thawing of existing frozen tailings, minimization of tailings segregation across the tailings surface, enhanced consolidation of the tailings body, increased storage volume availability and faster and more economic decommissioning of the RLITMF.

2 TAILINGS INJECTION SYSTEM DESIGN

2.1 *General Description*

The 1999 injection system was composed of 4 major components: injection wells; injection pads; piping system; and an operation and monitoring system. Injection pressure for the system was supplied by a static head difference of 85 m between the mill and the tailings surface.

Three pairs of angled injection wells were installed at 3 injection pads located along the southwest to southeast perimeter of the pit, as shown in plan view on Figure 1. The injection wells were installed with the discharge point located beneath the lowermost frozen layer or about 60 m below the tailings surface. A pipeline system with appropriate valves and controls was installed to deliver tailings to the wellheads. The tailings flow was divided between two injection wells at any time.

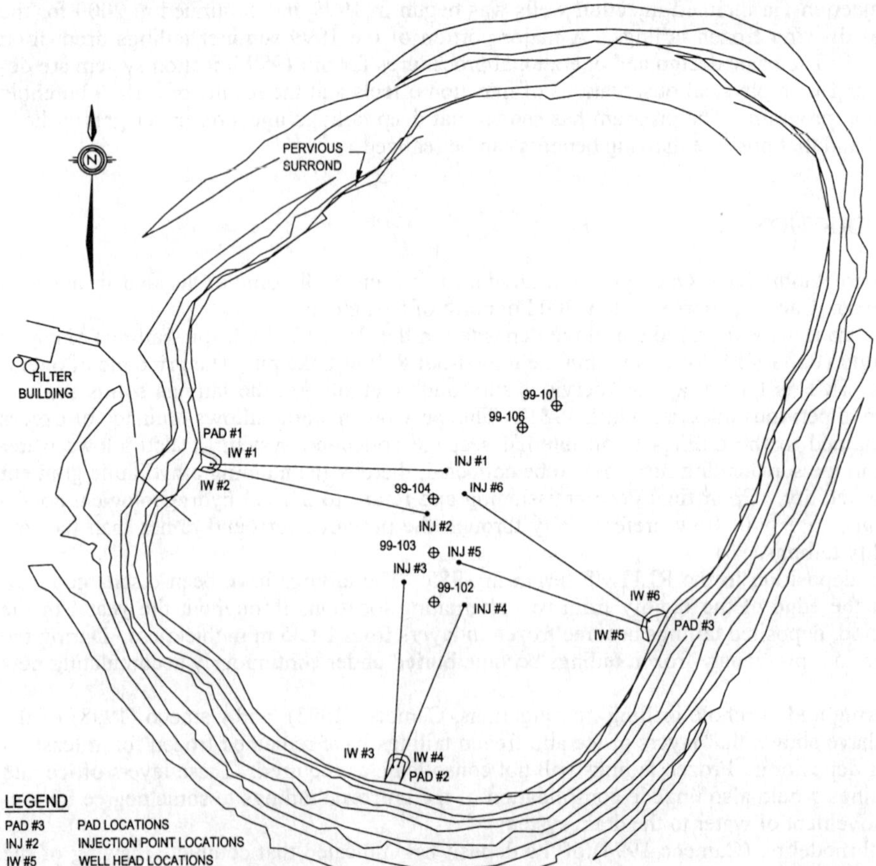

Figure 1. Plan view of Rabbit Lake In-pit Tailings Management Facility showing layout of tailings injection wells and locations of investigation boreholes.

2.2 *Injection Pad Design And Construction*

The positions of the injection pads were chosen to maximize borehole separation from pit walls while minimizing borehole length. Three injection pads were used in this trial program. Pad size was governed by the space required for a drill rig to install the wells. Injection pads were constructed by pushing sand onto the tailings surface until a stable base was achieved. The pads were then protected with geotextile and covered with additional sand.

2.3 *Injection Well Design*

A set of design criteria for the injection wells was established, based on the objectives of prevention of damage to the pervious surround system, achieving the required flow capacity, and injection below the lowest frozen layer. The primary criteria was that the endpoint of the injection wells must be located a minimum of 10 m from the nearest point of the pervious surround to ensure that injected tailings would not impact the pervious surround. Secondly, during drilling of the injection borehole, the drill stem was not allowed to pass through the filter sand portion of the pervious surround. Therefore, in the middle depth portions of the borehole a minimum separation distance of 5 m between the casing and the pervious surround was specified. Finally, the elevation of the injection points was set based on the elevation of the lowermost frozen layer found in the 1997 borehole investigation of the facility.

The as-built injection well locations are shown in plan view in Figure 1 and a typical cross-section view in Figure 2. Injection points were spaced about 50 to 70 m apart laterally to avoid interference from adjacent wells. The bottom elevation of the wells was about 60 m below the tailings surface.

The injection wells consisted of 100 mm (4 inch HWT) steel well casing installed using a diamond drill rig. Water, amended with drilling mud as required, was used as the drilling fluid. At critical times during drilling (just prior to making the closest pass to the pervious surround), the dip angle of the borehole was measured by acid-testing to ensure that the well maintained a safe distance from the pervious surround. After completion of each injection well, a final acid test was done to determine the horizontal and vertical position of the injection point.

2.4 *Piping System Design*

The tailings pipeline distribution system was extended from the existing south discharge line to six injection wellheads located at 3 drill pads, as shown schematically in Figure 1.

A four way splitter fitting was installed at the end of the main tailings line to divide flow be-

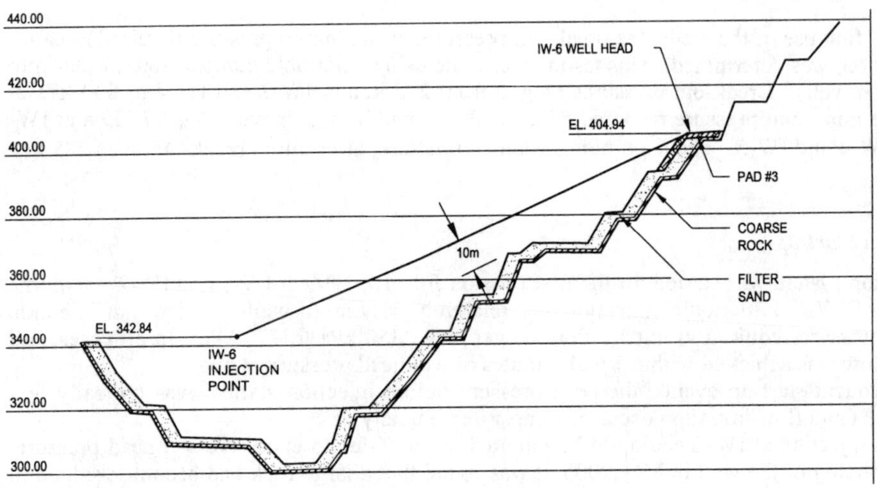

Figure 2. Cross-section view of a typical tailings injection well.

tween: the two branches of the injection pipe manifold, a 10" diameter sub-aerial bypass line, and a 4" diameter blow-out line equipped with a 145 psi rupture disc. The purpose of the rupture disc was to prevent tailings from backing-up into the mill, in the event of a plugged line and to protect the main line above this point from over-pressure conditions.

A system of Tees and Wyes was installed at each injection pad, to divide flow between two wells or to allow sub-aerial bypassing of tailings.

Pressure gauges were installed at each wellhead and at the main feeder line (four way splitter fitting).

Butterfly valves were positioned throughout the system to direct flows to the appropriate manifold sections, sub-aerial bypass lines or injection wells.

2.5 *Operational Procedures*

A set of operational procedures were designed to: ensure that a consistent and orderly process was followed during each injection cycle, prevent plugging of tailings pipelines, protect tailings lines and injection wells from freezing during non-operational periods, prevent spillage of tailings outside of the pit pervious surround and prevent damage to injection system components due to upset operation conditions.

At the beginning of each mill operating cycle, tailings were discharged by sub-aerial bypass until a steady flow was obtained. At this time, the main bypass valve was closed and tailings were directed along the manifold line to an injection pad. At the injection pad the sub-aerial bypass valve was closed, followed by opening of the injection wellhead valves one at a time.

2.6 *Monitoring Procedures*

Monitoring procedures were designed to measure pressures and flows during injection for documentation of operating characteristics, determine the location of fresh tailings breakthrough points at the surface of the existing tailings, develop techniques for detecting upset conditions, and gather operational data for system optimization and general record keeping.

System pressures were monitored at approximately 1 hour intervals by field technicians. The location of upwelling tailings was noted in field records and photographed periodically. The Total Suspended Solids (TSS) of water pumped from the pit seepage collection system was monitored twice daily to determine if tailings solids were entering the seepage collection system.

3 OPERATING SUMMARY AND DISCUSSION

Prior to the first use of the wells, the break-out pressure, or maximum pressure required to cause fluid injection, was determined. This testing was done using a portable pump to force water into the injection wells. Break-out pressures ranged from 275 kPa at IW-2 and IW-4 to 830 kPa at IW-3. The minimum pressure recorded after break-out had been achieved, was 175 kPa at IW-1, IW-2, IW-4 and IW-5. The maximum sustained pressure, after initial break-out, was 275 kPa at IW-6.

3.1 *System Start-up*

The injection system was started for the first time on July 10, 1999 at IW-1 and IW-2. With the exception of IW-6, initiation of injection was relatively easy at all wells. In the first injection cycle the pressure required to initiate flow ranged from 350 kPa to 770 kPa. In all cases, full injection flow was achieved within 1 to 4 minutes of wellhead pressurization.

In subsequent start-up events, the peak pressure before injection started, was typically less than 350 kPa and flow initiation occurred almost immediately.

Tailings injection at IW-6 could not be initiated in 1999 due to excessive wellhead pressure. During servicing of the well in May 2000, it was found that a large rock had become wedged in the casing shoe (during original installation), effectively plugging the pipe. The rock had been deposited in the tailings at an earlier time in conjunction with pond cleaning elsewhere on the site.

3.2 *Injection Pressures and Flow Rates*

Upon initiation of flow in the wells, wellhead pressures dropped to a typical range of 175 kPa to 245 kPa and remained relatively stable during any operating period. Occasional maximums of up to 420 kPa and minimums of 85 kPa occurred, due to variations in the tailings flow rate from the mill.

Flow rate measurements were made using a portable ultrasonic flowmeter. However, it was found that the readings were erratic and responded slowly to changes in system operation. In contrast, pressure responses were immediate, providing an excellent means of detecting any problems in the system.

For the purpose of determining injected volumes of slurry, accurate measurements of daily tailings production quantities were available from the mill operating records.

3.3 *Upwelling Tailings*

Tailings injection resulted in upward flow or "upwelling" of slurry/water to the tailings surface in various locations. The upwelling occurred in distinct areas of about 1 to 2 m in diameter.

The first indication of upwelling tailings occurred near Pad #2 about 3 hours after injection started at IW-3 and IW-4. This took the form of small boils (2 to 5 cm in diameter) above the centerline of the injection wells. Shortly after the flow was switched back to IW-1 and IW-2 (Pad #1) upwelling began to occur in standing water about 30 m offshore from Pad #3. Upwelling material initially consisted of a dark gray foamy substance; this was followed by reddish brown slurry consistent with the injected tailings colour.

Upwelling continued throughout the remainder of the program with only short interruptions. The primary upwelling began near Pad #3, but gradually moved to a position closer to the centre of the pit.

On a few occasions, upwelling would stop for a few hours. The tailings surface was observed to be swelling or doming on two of these occasions. When the pressure within the tailings had reached a critical level, tailings slurry initially shot into the air up to 1.5 m above the surface, then gradually declined to a steady "boil". In association with these events, large blocks (3m to 5m side lengths) of ice/frozen tailings came to the surface.

The TSS of water pumped from the pit seepage collection system remained low (normal) throughout the injection project.

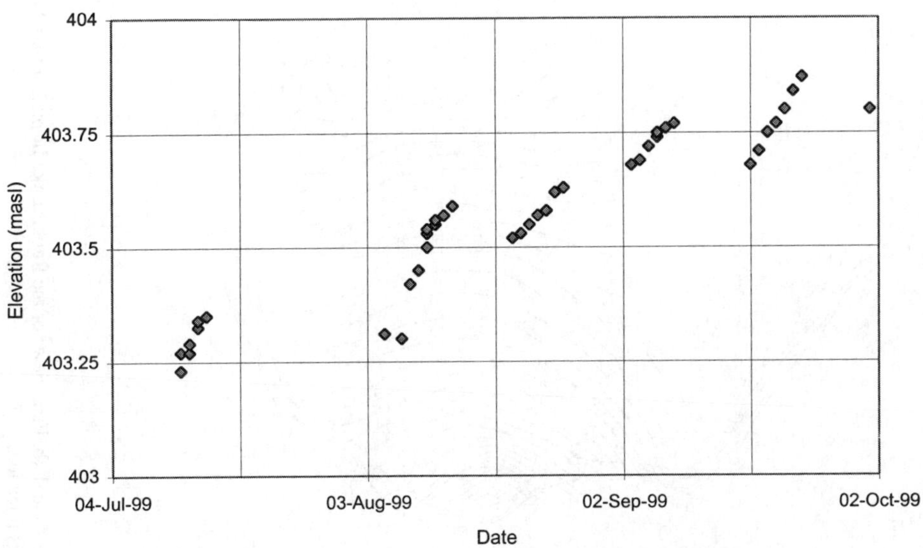

Figure 3. Water elevation on the tailings surface during the tailings injection program.

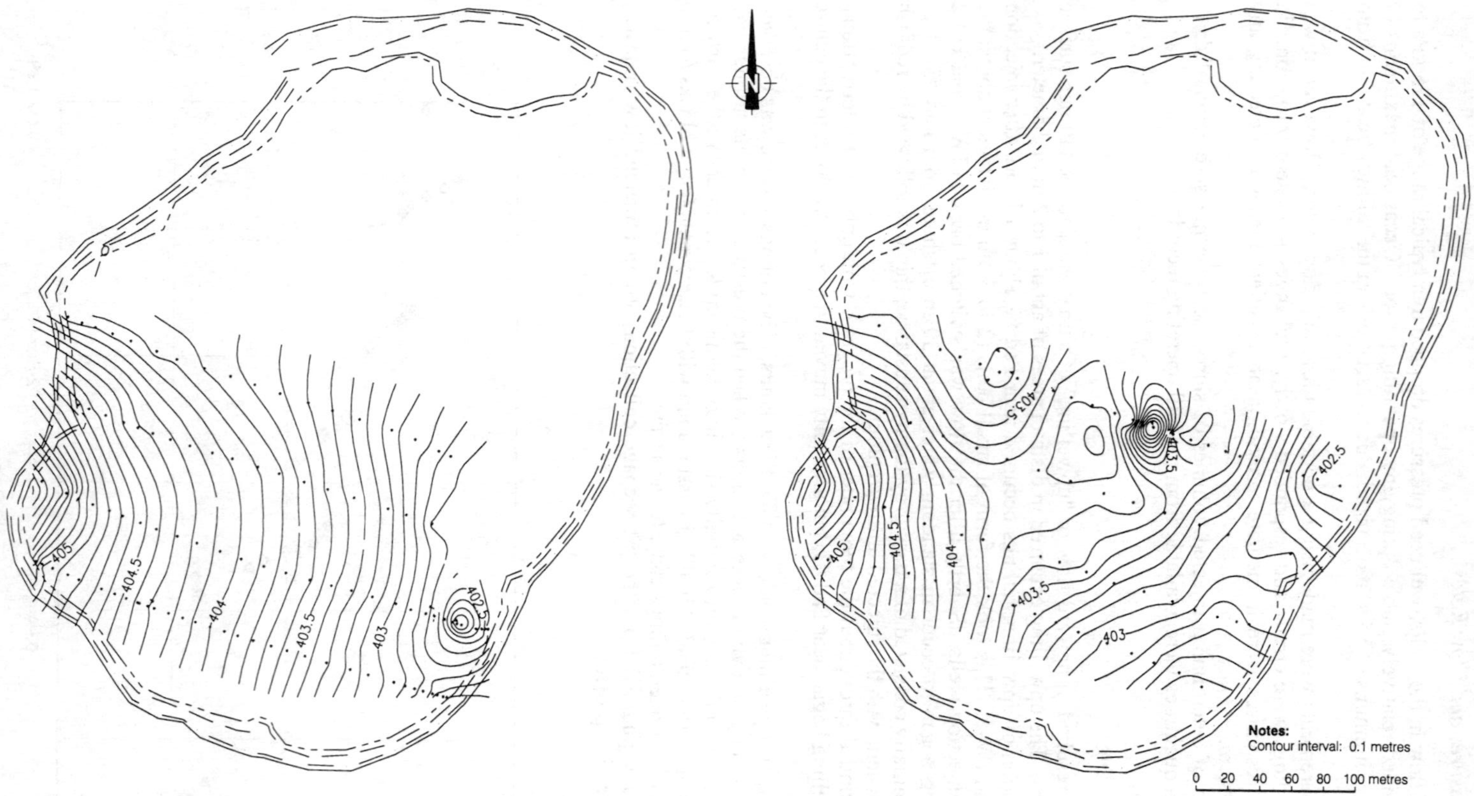

Figure 4a and 4b. Plan views of the Rabbit Lake In-pit Tailings Management Facility, showing tailings surface elevation contours, a) prior to tailings injection and b) after tailings injection.

3.4 *Surface Water*

The tailings surface is normally partially covered (25%) by water throughout the summer. During the injection project, the water covered area increased to about 85 to 90% of the tailings surface.

A staff gauge was placed near the edge of the surface pond to record water level changes. The water level elevation change recorded on this staff gauge is plotted versus time in Figure 3. The plot shows that the water level rose steadily during each operating week. The drop in water level during the non-operating weeks was due to on-going drainage through the pervious surround. A portion of this water level rise may be attributable to release of water from within the existing tailings caused by tailings injection. The volume of water released to the pond due to tailings thawing could not be estimated due to the continuous seepage of water into the surround, variable input from precipitation and the on-going consolidation of surficial tailings.

3.5 *End of Season*

The 1999 injection system was not designed for winter operation. Difficulties with frozen lines, gauges and valves occurred during the week of October 1 to October 7. Therefore, the trial injection was shut down on October 7,1999.

4 YEAR-END SUMMARY

4.1 *Injected Volume Of Tailings*

The total volume of tailings slurry injected during the 1999 season was approximately 104,500m^3, based on mill production records and well operating periods.

It has been found through various tailings investigations, that shortly after subaerial deposition, the void ratio of the tailings is typically about 4. The tailings volume, shortly after deposition, assuming a void ratio of 4 was 85,300 m^3. This volume of tailings, spread over the entire upper tailings surface, would have raised the tailings level by about 0.75 m. The increase would have been about 1.5 m if confined to the area directly affected by tailings injection.

4.2 *Tailings Surface Surveys*

The tailings surface in the vicinity of the injection program was surveyed prior to the start of the injection program on June 26, 1999. Contour data from the pre-injection survey are shown on Figure 4a. The contours show a deltaic landform with smooth contours, as would be expected due to sub-aerial discharge of tailings from a spigot.

A post injection survey was performed on October 24, 1999. Contour data from the post injection survey are shown on Figure 4b. This figure shows a remnant deltaic landform in the southwest part of the pit; however, the surface shape over the rest of the pit surface has changed significantly. Circular contours near the centre of the facility indicate the location of a steep cone shaped depression in the tailings surface. Upwelling tailings were observed in this area during the last week of injection.

In summary, the post-injection survey showed that the overall tailings surface elevation had changed very little, in comparison to the expected increase based on injected volume as discussed in Section 4.1. This may indicate that injection was causing some thawing and/or release of trapped water.

4.3 *Tracer Test*

Tracer testing was performed to determine how quickly tailings slurry was moving from the injection point to the tailings surface, and if possible, what percentage of the injected tailing was returning to the surface. Sodium bromide solution was added to the tailings stream immediately upstream of the wellheads.

The tracer tests did show that a portion of the slurry is returning to the surface about 1 hour after injection; however, they were unable to show what portion of the injected slurry was returning to the surface.

5 FOLLOW-UP INVESTIGATION

A borehole investigation was conducted in the injection area near the end of the 1999 injection season to determine the effects of tailings injection on tailings temperature.

A total of 5 test boreholes (RLP99-101 to RLP99-104 and RLP99-106) were drilled at locations shown on Figure 1. The depth of investigation ranged from 22.7 m to 71.6 m below the tailings surface. Field measurements of insitu temperature were made on tailings solids samples within 15 minutes of recovery from the borehole.

5.1 *Temperature Profiles*

Tailings temperatures for boreholes RLP99-104 and RLP-106 are plotted in Figure 5 along with temperature data for borehole 97-R2 drilled in the 1997 investigation program.

The general pattern seen in borehole 97-R2 has been repeated in RLP99-104 with temperatures varying from slightly below zero in frozen zones to a maximum of approximately 6 °C between the frozen layers. Below a depth of 25 m, the temperature profile is similar between 1997 and 1999, but the temperature of the non-frozen layers is generally higher in 1999. These data do suggest a general warming of non-frozen layers since 1997.

The tailings injection program dramatically impacted temperatures at borehole RLP99-106. This borehole was drilled at the centre of a boil (location of upwelling tailings) caused by tailings injection. The data clearly show that the tailings temperature has been raised to values greater than 6 °C. The temperature drops sharply at a depth of 21 m where a frozen layer was encountered.

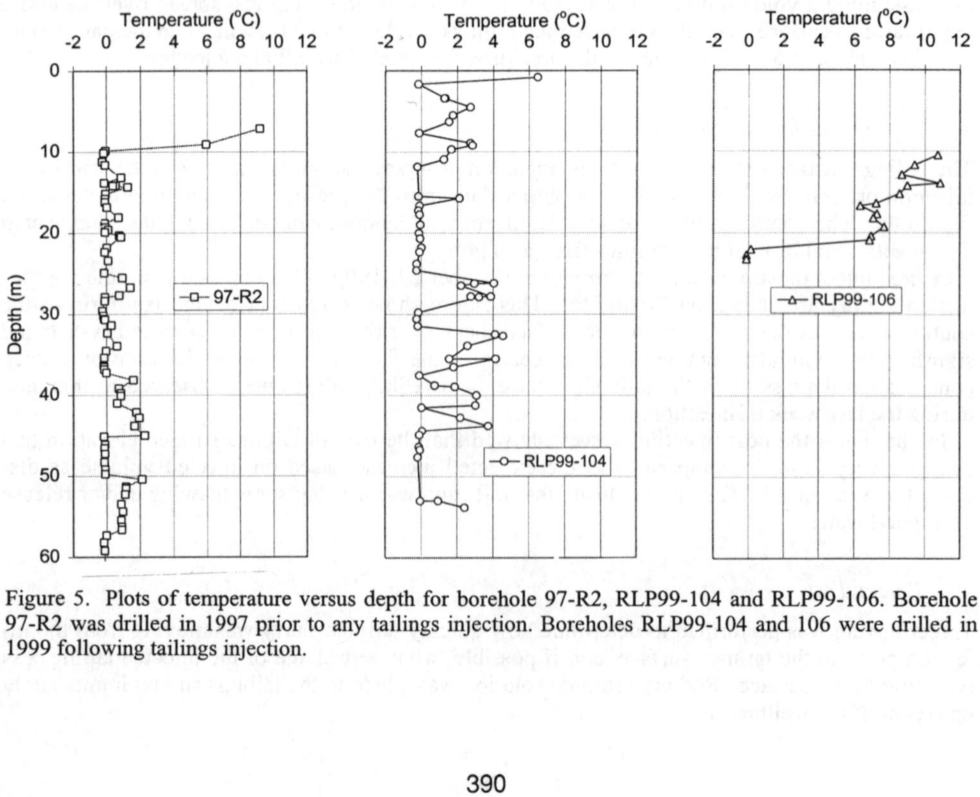

Figure 5. Plots of temperature versus depth for borehole 97-R2, RLP99-104 and RLP99-106. Borehole 97-R2 was drilled in 1997 prior to any tailings injection. Boreholes RLP99-104 and 106 were drilled in 1999 following tailings injection.

6 SUMMARY AND CONCLUSIONS

A number of observations may be drawn from the trial injection program in 1999.

The injection system was easy to start and required little attention during operation. At the start of each mill cycle it was simply a matter of diverting the tailings flow to the well heads and then opening the valve at the wells. Injection pressures were stable over periods of hours and days.

Injection start-up pressures tended to decrease with each operating cycle. This suggests that once a well has been "conditioned" by establishing flowpaths in the tailings body, the expansion of those flows paths on each cycle becomes easier.

The average injection pressure during continuous injection declined slightly over the duration of the program. This may indicate that once the formation has been fractured (opened), flowpaths remain open for extended periods.

Tracer testing showed that injected tailings arrived at the tailings surface within 1 hour of injection. However, these tests did not indicate what portion of the slurry returned to the surface or what percentage of previously deposited tailings was entrained with the flow.

Tailings injection did not have an observable effect on the pervious surround, as measured by the TSS of seepage water collected from the pit drainage system.

Survey data indicate that the tailings surface elevation declined in areas that were not subject to accumulation of upwelling tailings solids. This decline is partly attributable to normal ongoing consolidation, and partly to the injection project. Settlement caused by the injection project may be partly due to release of water trapped beneath frozen layers and/or due to melting of ice within frozen layers.

The greater than usual amount of water on the tailings surface conforms with the assertion that tailings injection caused the release of trapped water or caused melting of ice in frozen layers.

A 20 m deep zone at the surface in the vicinity of borehole RLP99-106 was fully thawed. This zone was subject to upwelling tailings during the latter stages of the injection program. The zone of upwelling migrated along a path about 130 m in length during the course of the program. It is reasonable to expect that a similar depth of thawed tailings developed over the full length of this path.

The temperature of non-frozen (summer deposition) layers was shown to have increased from the time of the 1997 investigation. These warmer zones will transfer heat to the adjacent frozen layers.

REFERENCES

GMCL (Gulf Minerals Canada Limited), 1981. Environmental Impact Statement Collins Bay B-zone Development. Second Addendum. June. Saskatoon

Cameco, 1994. Rabbit Lake Operation In-pit Tailings Disposal System Performance Report, 1985 to 1993. May 1994.

Cameco, 1995. 1994 Rabbit Lake In-pit Drill Programme and Instrumentation. Engineering and Projects Division. May, 1995

Cameco, 1998. 1997 Rabbit Lake In-pit Tailings Management Facility Drill Program. Engineering and Projects. November 1998.

Tailings and Mine Waste '01 © 2001 Balkema, Rotterdam, ISBN 90 5809 182 1

Application of cone penetrometer technology to phosphate mining exploration

W.L. Bratton
Applied Research Associates Incorporated, Richland, Wash., USA

J.T. Alex
IMC Phosphates, Mulberry, Fla., USA

ABSTRACT: A novel new approach was demonstrated for the site characterization and exploration of phosphate bearing materials at a site in Florida. The new approach consists of using Cone Penetrometer Technologies (CPT) to determine in-situ and real-time the amount of overburden, the quality (i.e. phosphate content) of the phosphate bearing materials, as well as the engineering load bearing capacity of the site. The exploration tools consisted of a cone penetrometer push rig and a standard CPT probe connected to a gamma spectroscopy module. The CPT sensors were used to define the soil stratigraphy, groundwater level and the strength of the soils. The gamma spectroscopy module consisted of a downhole sodium iodine crystal combined with an uphole spectroscopy system designed to detect gamma emitting particles. Although phosphate is not a natural gamma emitter, it is co-deposited with low levels of uranium that are sensed by the gamma tool. Initial correlations of the gamma response and the phosphate quality show a strong correlation. The gamma tool correctly identified the appropriate depth of mining in two demonstration penetrations that were conducted. Both gross gamma and spectral gamma measurements were obtained during the demonstration. The details of the test results are discussed.

1 INTRODUCTION

Cone Penetrometer Technology is a new innovative technology that is becoming more common in the geotechnical and environmental communities as an alternative to drilling techniques (Bratton & Timian 1995). Similar to drilling, CPT is used to gather information and create access to soils at various depths in the subsurface. As opposed to drilling where a borehole is essentially excavated, the CPT creates a borehole by pushing the soil aside making room for the instrumented probe. Because of this insertion technique, CPT techniques are limited to unconsolidated soils and not rock materials. Penetrations have been conducted as deep as 300 feet in mine tailings and typically achieve 100 to 200 feet in most soil regions.

The advantages of CPT technology are that penetrations are typically made faster than drilling and the soil disturbance is minimized. Approximately 200 to 400 feet of CPT penetrations are routinely accomplished in a single day. Typical CPT penetrations are 1.4 inches to 2 inches in diameter compared to 4 to 6 inch cores that are taken during most drilling operations. The most significant advantage of CPT for exploration work is that through the use of sensors detailed information is available in real-time, compared to sending soil samples to the laboratory for analysis that could take days to weeks to produce results. This permits the exploration program to be more adaptive in nature and cost efficient. Waste disposal costs are also minimized for CPT operations.

2 CONE PENETRAION EQUIPMENT

2.1 Probe insertion equipment.

Cone penetrometer tests can be conducted using a wide array of push rigs including customized track mounted rigs and drill rigs. For a truck system the penetrometer equipment is mounted inside a van body attached to a ten-wheel truck chassis with a diesel engine as shown in Figure 1. Ballast in the form of weights is added to the truck to achieve an overall push capability of 20 to 30 tons. Penetration force is supplied by a pair of large hydraulic cylinders bolted to the truck frame. All operations occur within the truck, making the presence of the truck inconspicuous at the site. Since the truck is fully enclosed, weather does not have an impact on the CPT operations. For operations in hazardous locations, the probes and equipment are automatically steam cleaned as they are retracted back into the truck.

2.2 Standard CPT Geotechnical Probe

The standard CPT geotechnical probe consists of sensors to measure tip stress, sleeve stress and pore pressure. From these three parameters, the soil type, soil strength (SPT blow count, friction angle, and undrained shear strength) and groundwater table elevation can be determined (Robertson & Campanella 1988). The probe has a conical tip and a friction sleeve that independently measures vertical resistance beneath the tip as well as frictional resistance on the side of the probe as a function of depth. A schematic view of ARA's penetrometer probe is shown in Figure 2. A pressure transducer in the cone is used to measure the pore water pressure as the probe is pushed into the ground (Piezo-CPT). Additional sensors that are occasionally added to the CPT probe consist of inclination and soil moisture/resistivity. Inclinometers are used to determine the number of degrees off vertical in both the X & Y directions. This information can be used to track the exact location of the probe at all times during the penetration. Finally, a CPT based soil moisture sensor can be added to the probe to determine the volumetric moisture content of the soils during the penetration.

Electronic data acquisition equipment for the cone penetrometer consists of signal conditioning and digitizing boards within the probe and a data acquisition computer with a graphics monitor within the CPT truck. Analog signals are amplified and filtered at 1 Hz before being digitized in the probe. Once a second, the data acquisition computer polls the probe for its data. The digitized data are then read into memory, displayed on a computer screen in the CPT truck and written to the internal hard disk for future processing. A set of data is normally recorded each second, for a minimum resolution of about one data point every 0.8 inch of cone advance. This represents a high density data and provides very detailed profile of the parameters being measured as a function of depth. Upon completion of the test, the penetration data are directly plotted in the CPT truck.

Figure 1. Photograph of CPT Truck.

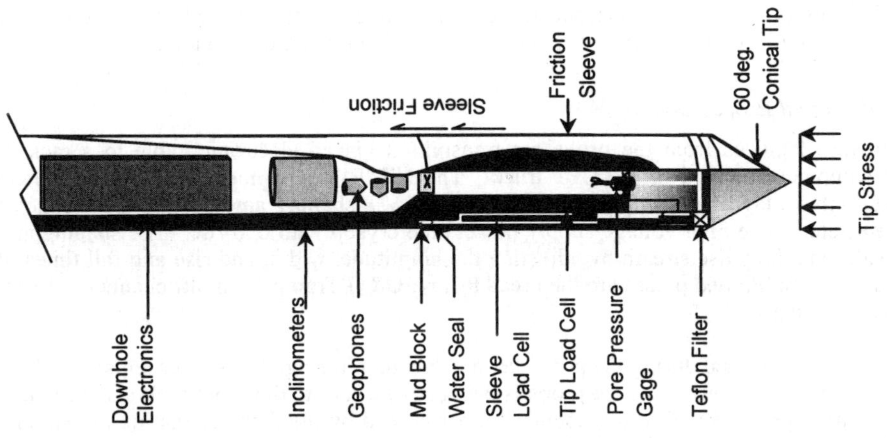

Figure 2. Schematic of Standard Geotechnical Cone Penetrometer Probe

3 SPECTRAL GAMMA CPT PROBE

ARA's gamma cone penetrometer module has the ability to detect gamma radiation as both a gross gamma reading during a penetration and as a spectral reading when the penetration is paused. Figure 3 shows the complete gamma system. The gross gamma reading is a measure of the total gamma radiation detected from all sources. For phosphate detection in the sate of Florida, the gamma signals are the result of low levels of co-mingled natural uranium and it's associated decay products (Glasser 1988). Other natural occurring radionuclides are also present but consist of the background. The spectral gamma measurement provides a means of determining the type of source emitting the radiation.

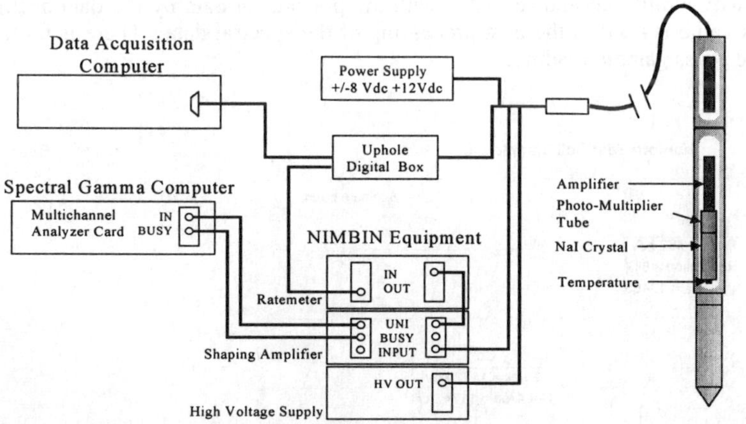

Figure 3. Schematic of the Gamma CPT system

3.1 Downhole Sensor

The gamma module mounts directly behind the standard geotechnical CPT probe. The gamma module houses a 0.75 inch diameter by 4.0 inch long NaI crystal. When gamma radiation of sufficient energy interacts with the scintillating crystal a photon is emitted. A photo-multiplier tube attached to the crystal then converts the photons to a voltage signal. The voltage signal is a stream of pulses proportional to the occurrences and energy of the emitted gamma ray. The pulse stream is amplified in the gamma module before being transmitted uphole.

3.2 Uphole Signal Processing

The stream of pulses from the probe are transmitted via an electronic cable to a rack of NIMBIN equipment located in the CPT truck. The NIMBIN equipment includes an EG&G Ortec 478 high voltage bias supply, an EG&G Ortec 671 shaping amplifier, and a Tennelec TC525 ratemeter. The high voltage supply powers the crystal with 630Vdc. The shaping amplifier conditions the pulse stream by adjusting the amplitude, width, and rise and fall times of the pulses. The conditioned pulses are then read by an EG&G Trump 2K multi-channel analyzer card for spectral data.

The ratemeter is used during the gross gamma data monitoring. It provides a display of the pulse rate in counts per second as the probe is being advanced into the ground. It also produces an analog output proportional to the count rate that is read by the data acquisition system and recorded every second and displayed with the CPT probe data.

For collecting a spectral gamma reading an EG&G Trump 2K multi-channel analyzer card is used. The multi-channel analyzer card receives shaped pulses from the shaping amplifier and records the energy of each incoming pulse. The results are plotted in a spectrum display as shown in Figure 4 and saved to disk. Figure 4 is a spectrum gathered with known calibration sources of Cs^{137} and Co^{60}. The Cs^{137} peak is shown at an energy of 662KeV and the Co^{60} is shown as twin peaks at 1173KeV and 1332KeV (Heath 1964). The large low-energy peak that is not fully plotted in Figure 4 is the Compton scattering of the gamma rays from the probe mandrel and crystal housing. A spectrum is gathered for a typical duration of 5 minutes while the probe is stationary in the ground. Specific radiation sources will produce peaks at specific energies on the spectrum, thus allowing quantifying of a radiation source.

The spectral response of the crystal is sensitive to temperature. The friction caused by the soil on the sides of the probe during penetration can greatly increase the temperature of the probe. Because of this, a temperature sensor is mounted on the bottom of the crystal. The temperature is continuously monitored and recorded with the piezo-cone data by the data acquisition computer. This value is used in the post processing of the spectral data. There is no temperature effect on the gross gamma reading.

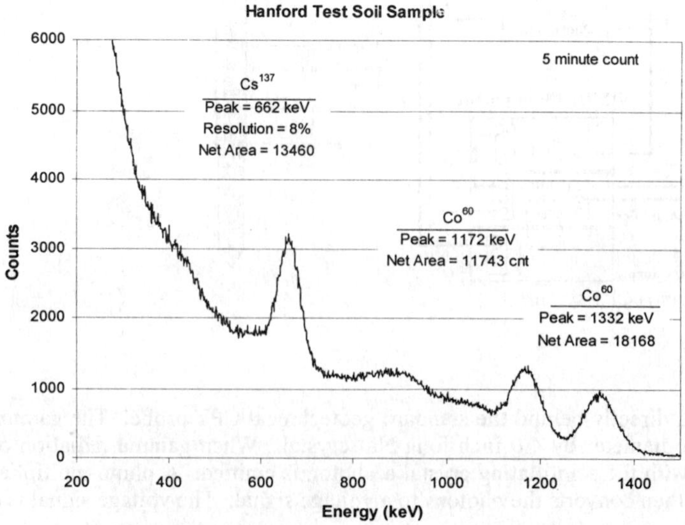

Figure 4. Gamma Spectrum showing Cs^{137} and Co^{60} peaks obtained using CPT gamma probe.

4 GAMMA PROBE RESULTS DUE TO PHOSPHATE

4.1 Laboratory Testing

To confirm the hypothesis that a gamma CPT module could be used to detect phosphate minerals in soils, a collection of 9 soil samples were sent from the IMC Phosphate mines in Florida, to Applied Research Associates, Inc for testing in the laboratory using the gamma CPT module. Each of the received samples were laid on the probe and the counted for approximately 10 minutes. Table 1 presents both the soil description, the gamma counts as well as the BPL (bone phosphate level) measurements provided by IMC Phosphates. Figure 5 plots the data in table and performs a linear regression on the data. The regression is quite strong indicating a correlation between the BPL values and the gamma count rate.

4.2 Field Testing

Based upon the success of the laboratory results, a two day demonstration of the gamma CPT was planned at the Plant of IMC Phosphates in Mulberry, Fl. The first day was spent conducting penetrations at locations in an area actively being mined. The first penetration was conducted in Area 2 and the second penetration in Area 5. The second day of the demonstration

Table 1 – Description of soil samples and gamma module test results.

Sample Name	Soil type Description	Chemical Results (bpl)	Ratemeter (CPS)
D 6-1	brown/gray sand	4.16	23
D 6-2	lt gray silty sand	30.97	40
D 6-3	med gray clay	20.93	30
D 17-1	brown sand	13.85	31
D 17-2	tan silty sand	16.76	31
D 17-3	gray silty sand	17.62	28
D 17-4	gray clay	18.87	28
D 17-5	b&w speckled sand	22.46	29
D 17-6	tan sand w/ speckles	50.56	43
D 17-7	green hard clay	9.24	24

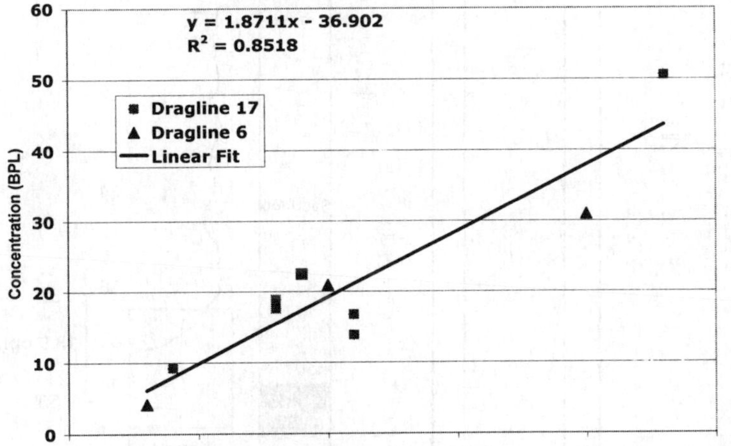

Figure 5. Comparison plot of the gamma count rates and the quantity of phosphate in the soil sample.

was conducted at the Fort Green site to gather data to determine if this site should be mined at a future date. The results from all of the penetrations are discussed in the following paragraphs.

The first penetration was conducted at Area 2 of the IMC Phosphates Plant near Mulberry, Florida and reached a total depth of 54.6 feet. The CPT profiles from this penetrations are presented in Figure 6. The upper 30 feet is a medium to coarse sand as shown by the tip stress values of 2000 psi. From 30 to 48 feet is a loose silty sand material with tip stress values averaging 1000 psi. The pore pressures track right along the hydrostatic line in this material. The depth to the groundwater is 33.6 feet. At a depth of 48 feet a clay unit begins and transitions to a clay silt at the bottom of the penetration. The gross gamma response at this location was below 150 counts per second (cps) until a depth of 39 feet and then increased rapidly, indicating the presence of phosphate materials. The gamma profile is clipped between 44 and 50 feet as the count rate exceeded the setting on the ratemeter. The ratemeter was adjusted on the next penetration to permit larger count rates. These measurements agree favorably with the depths that were actively being mined by the dragline as well as the economically mineable reserves (EMR) data provided by IMC Phosphates. The EMR log sheets indicate 38 feet of overburden at this site followed by material that is to be mined with an average phosphate level of 18.9 BPL between 38 and 50 feet.

The second penetration was conducted in Area 5 at the IRC-AGRICO plant and extended to a depth of 59.6 feet as shown in Figure 7. The soils at this location are much finer grained as indicated by the soil classification that was determined. There also appears to be two dense sand layers; one at 12 feet and another at 22 feet. Other than that, the soils are generally a clayey silt to clay material at this location. The gamma responses at this location were the highest recorded during the entire demonstration. The gamma is below 200 cps until a depth of 12 feet where is rises to 200 cps. The gamma count rate maintains this value until a depth of 25 feet where it increases to over 1000 cps. The count rate then stays over 500 cps for the majority of the remainder of the penetration. These results again compare favorably with the EMR data. The EMR data for the area indicates a phosphate matrix from 26.9 feet to 61.4 feet. Although the gamma tool indicated phosphate at a depth of 25 feet, this could be accounted for by dipping planes, as the core data and the CPT were not conducted at the same location. Additional testing that is co-located is needed to verify that the differences are due to dipping planes.

The final two penetrations were conducted at the Fort Green site in central Florida. Penetration 2462 reached a depth of 47.7 feet and penetration 2262 reached 53.8 feet. Both soil profiles are similar as expected since they were less than 400 yards apart and consist of mostly

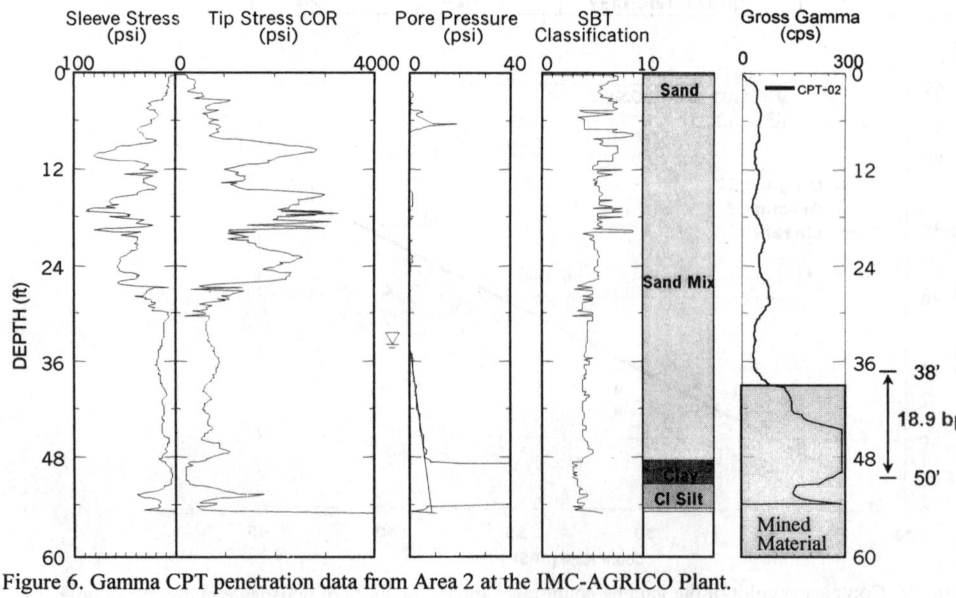

Figure 6. Gamma CPT penetration data from Area 2 at the IMC-AGRICO Plant.

sandy materials. Figure 8 presents the profile from penetration 2262. Both locations have a shallow water table at approximately 8 feet. A dense sand layer is present in both profiles at a depth of 6 feet. This material is very stiff. The gamma profiles from both locations also exhibit similar trends. Both profiles show a peak at a depth of 23 feet. Based upon the drilling log information location 2262 shows moderate to abundant phosphate levels from 44 feet down to 55 feet. The gamma log from this location shows counts above 200 cps from 45 feet to the bottom of the penetration as well as the three-foot zone at 23 feet.

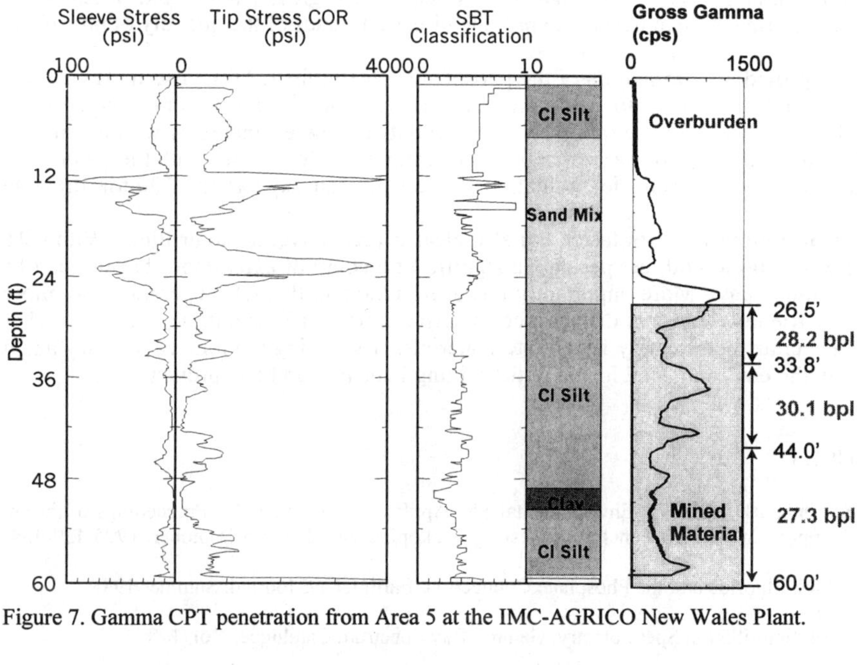

Figure 7. Gamma CPT penetration from Area 5 at the IMC-AGRICO New Wales Plant.

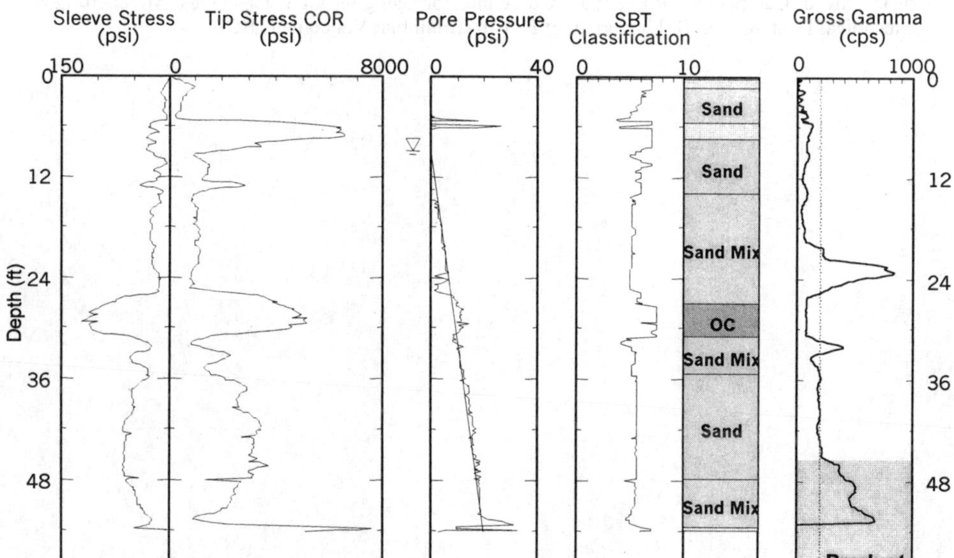

Figure 8. Gamma CPT penetration 2262 showing phosphate product as indicated by the gamma CPT tool.

5 SUMMARY

Based upon both the laboratory data and the field data that has been collected to date, the gamma cone penetrometer tool appears to be an effective tool for the identification of phosphate mineral in the Florida geologies. The gamma-CPT tool provides real-time results at less cost and with less disturbance than the conventional drilling programs. Currently the tool has only been used as a binary indicator of the presence of phosphate when the count rate exceeds 200 cps. Using the results from the EMR data this count rate is appropriate for sufficient phosphate to warrant mining. With additional data collection the gross gamma and possibly the spectral gamma information can be used to provide direct estimates of the quantity of phosphate present.

In addition to providing information of the phosphate content, the CPT approach also gathers the necessary geotechnical data to determine support conditions for the mining equipment. From the CPT data, standard penetration test blow counts can be estimated as well as the soil friction angle and the undrained shear strength. This information combined with the groundwater table information will permit determination if the soils can support the 100-ton dragline equipment.

Finally CPT operations are both faster and also cleaner that conventional drilling. With CPT production rates of 200 to 400 feet per day, a significant amount of exploratory borings can be completed in a single day. More importantly however is the fact that CPT operations are more environmentally friendly with less disturbance to native soils. CPT operations leave a smaller borehole and generate significantly less waste materials. By having the mining industry adapt CPT technology the environmental image of the mining industry can be improved.

6 REFERENCES

Bratton, J.L. & Timian, D.A., 1995. Environmental Site Applications of the CPT. Proceedings of the International Symposium on Cone Penetration Testing. Linkoping, Sweden, 4-5 Ocotober 1995:429-434.

Glasser, G.C., 1988. Fluoride and the Phosphate Connection. Earth Island Journal, Summer 1998.

Heath, R. L., 1964. Scintillation Spectrometry, Gamma-Ray Spectrum Catalogue, Vol. 1.

Robertson, P.K. & Campanella, R.G., 1988. Guidelines for Using the CPT, CPTU and Marchetti DMT for Geotechnical Design, Vol. II, University of British Columbia, Vancouver, BC.

Tailings and Mine Waste '01 © 2001 Balkema, Rotterdam, ISBN 90 5809 182 1

Hydrochemical and geotechnical properties of cemented uranium paste tailings

D.R.Jones & H.-Y.Li
Earth Water Life Sciences Pty Limited, Darwin, N.T., Australia

T.D.Waite & B.Fenton
School of Civil and Environmental Engineering, University of New South Wales, Sydney, N.S.W., Australia

ABSTRACT: The results of laboratory studies on the strength and hydraulic conductivity of cemented uranium tailings, and the ability of the tailings mass to retain major (Mg^{2+}, SO_4^{2-}), minor (Mn^{2+}), and trace (UO_2^{2+}) contaminants are reported. A solids content of the tailings of more than 70%, and greater than 4% addition of cement was required to achieve the UCS (28day) value of 750 kPa needed for primary stope backfill. The permeability of uncemented tailings is in the order of 10^{-7} m/s. The permeability decreases to 10^{-8} m/s with 4% cement addition, and to 10^{-9} to 10^{-10} m/s with 8% cement addition. Maximum removal of solutes occurred for 2% or higher additions of cement, with the concentrations of U in the porewater reduced to below 1 μg/L. The highest removals (on a mass basis) of solutes from the tailings water occurred for Mg and sulfate. Essentially all of the Mn present in the tailings porewater was converted to insoluble form at 2% addition of cement.

1 INTRODUCTION

1.1 *Background*

It has been proposed to use cemented paste tailings for the underground disposal of uranium ore tailings at the Jabiluka mine in northern Australia (approximately 250km east of Darwin), in the event that a processing plant is constructed on-site at the mine (Jabiluka Mill Alternative). Approximately half the tailings produced would be used as backfill in the mined-out stopes, with the residual being placed in underground tailings storage "silos" excavated from non-mineralised rock adjacent to the orebody. This option has received regulatory approval *via* the Environmental Impact Assessment process conducted according to the legislative requirements of the Australian Commonwealth and Northern Territory Governments (ERA 1998).

The Jabiluka mining lease lies within the boundaries of Kakadu National Park, a designated World Heritage Area, so particular care is being taken during the planning phases of the project to ensure that groundwater in the park adjacent to the mine will not be deleteriously affected by tailings disposal. The primary requirement is for effective containment of tailings over a 10,000 year time frame to minimise the long term movement of process fluids from the tailings. Preventing the leaching of solutes from the tailings mass by a combination of reduced permeability and decreased solubility of key solutes will provide the primary barrier to the escape of solutes from the tailings mass. Published literature indicates that the addition of Portland cement (OPC) to paste produced from mine tailings has the potential to increase strength, reduce permeability and immobilise heavy metals (Verburg 1997).

Pastes are defined as dense, viscous mixtures of tailings and water which, unlike slurries, do not segregate when allowed to rest (Cincilla et al. 1997). The most important distinguishing characteristic is the grain size distribution of the tailings solids. Empirical data indicate that to form a paste, tailings must contain at least 15% (by weight) of fine solids that pass a 20 μm filter. Tailings pastes tend to have low permeabilities because of the presence of the ultrafine (<20 μm) particles that fill voids between the larger particles.

In order to determine the likely benefits of cemented paste technology in the context of minimising leaching of solutes from the tailings mass at the proposed Jabiluka mine, the hydro-geochemical and geotechnical properties of tailings paste have to be understood. Laboratory investigations have therefore been conducted to investigate these properties for "surrogate" tailings currently being produced from the geologically similar Ranger orebody, located about 22km from Jabiluka.

1.2 *Characteristics of Ranger Mine Tailings*

Golder Associates made an assessment of the paste forming potential of the Jabiluka ore using material from a diamond drill core (Golder Associates 1997). The core was crushed and processed in a metallurgical test laboratory to form tailings that would represent those to be produced by the proposed mine. Over 30% of the tailings was found to be less than 20 μm in size, indicating that the tailings would be suitable for paste formation. The results of rheological tests indicated that the properties of trial pastes made from tailings and water (i.e. without addition of cement) were "excellent", with low yield stress and good water retention. Run-of-mill tailings produced from the operating Ranger uranium mine have an even greater percentage of particles less 20 μm in diameter. Laser particle size analysis has shown that the particle size distributions were centred around a median diameter of approximately 8μm based on the total volume occupied by that size fraction within the sample (Fenton and Waite 2000).

At the Ranger mine an acid leach process is used to extract the uranium from the ground ore. Sulfuric acid is added to adjust the pH to approximately 1.7 and pyrolusite (manganese dioxide) is added as the oxidant. The uranium is then extracted from the clarified pregnant liquor by solvent extraction. The acidic barren tailings stream contains high concentrations of sulfate (up to 50,000 mg/L from the added acid) and magnesium (up to 9,000 mg/L from dissolution of the gangue minerals), and moderate concentrations of iron, aluminium and silica (also from dissolution of gangue minerals) and manganese (from dissolution of the added pyrolusite). The tailings stream is neutralised to approximately pH 5 with lime prior to discharge to the tailings storage facility (open pit) at the Ranger mine. The neutralisation step results in the rapid precipitation of gel-like ferric and aluminium hydroxide phases that have a high affinity for trace solutes such as uranium and radium present in solution prior to neutralisation. It is anticipated that a similar process as described above will be used for Jabiluka ore.

The high concentrations of sulfate present in the tailings porewater have potential implications for the strength and integrity of a cemented paste tails matrix. Kumar and Rao (1994) report that the deleterious effects of sulfate in conventional cement applications peak at 3000-4000 mg/L sulfate after which the impact on setting time and strength is not as severe. Indeed, the loss in compressive strength due to the presence of sulfate does not appear to exceed 25% of that observed in the absence of sulfate. While such reductions in concrete strength might be of concern in the construction industry, the magnitude of such changes would not appear sufficient to seriously limit the use of cement binder for tailings paste. However, this does flag the need to examine the impact of tailings water chemistry on the characteristics of the cemented paste.

Selecting and optimising the plant components needed to dewater the tailings and produce and convey the paste is the first of two critical components needed to assess the cost and feasibility of the paste tails option. Investigations of these aspects are taking place. However, it is the second dimension of the use of cemented paste tails– namely the environmental performance of cemented paste tails with respect to minimising the loads of solutes that can be leached from the tailings mass – that is the focus of this paper. We report the outcomes of pilot studies using pastes produced from Ranger tailings, with particular attention being given to the effect of cement addition on the strength and hydraulic conductivity of the cemented tailings, and on the ability of the mass to retain trace contaminants and resist leaching of key solutes.

2 TEST METHODS

2.1 *Tailings Dewatering*

Two methods were used to produce the required solids content for the test work reported here. Final sample preparation of run-of-mill tailings to produce the required solids content was done at two locations (University of NSW School of Civil and Environmental Engineering, Sydney; and the Ranger Mine, Jaburu). The bulk tailings sample dispatched from the Ranger mine to UNSW was dewatered to 70% solids using a large pressure filtration apparatus. In contrast, the samples prepared at the mine were dewatered to 70% and 75% solids using a combination of gravity settling and accelerated weight surcharge consolidation with vertical expression of pore water occurring through a geofabric filter layer placed on top of the tailings material. The prefixes UNSW and RUM will be used to denote samples prepared and tested at the University of New South Wales, and the Ranger mine, respectively.

2.2 *Addition of Cement and Casting of Test Cylinders*

2.2.1 *University of NSW (UNSW)*

The required amount (4% and 10%) of ordinary Portland cement (OPC) was added to the paste and the material mixed in a cement mixer for 30 minutes at a moderate speed setting after which further mixing was performed by hand to break up any clumps which may have formed during the mixing process. The cement-paste mixture was then cast in the form of small test cylinders. The cylinders were produced using a compaction mould and 2 Kg compaction hammer. Each cast was produced in three layers with equal compaction within each layer from the hammer which was dropped from a height of 50 cm, 15 times at each compaction point. After compaction, cores were produced by pushing a 25mm stainless tube through the sample with a Shimadzu hydraulic press and the cores then extruded manually using a hydraulic ram. The cores which had dimensions of 25mm diameter and 100mm height were then cured, in a fog room which had a constant temperature of 23±2°C and a constant relative humidity of 100%, for the appropriate time interval before the required testing was carried out.

2.2.2 *Ranger Uranium Mine (RUM)*

Ordinary Portland cement (OPC) was added to the tailings at rates of 2, 4, 6 and 8% based on the initial weight of solids (70% and 75%) present in the dewatered tailings (Table 1). A third variable investigated was that of pore water content. One of the bulk samples of tailings was washed with potable tap water to remove the bulk of saline process water prior to final dewatering to 70% solids.

Table 1. Summary of RUM series of test samples

Tailings-Initial wt% solids	Pre-Treatment	pH	Pore Water	% Cement added	Aging Time days
70	none	5.5	process	0	28
70	none	5.5	process	4	28
70	none	5.5	process	6	7
70	none	5.5	process	6	28
70	none	5.5	process	8	28
75	none	5.5	process	4	28
75	none	5.5	process	2	28
70	none	5.5	process	2	28
70	tap water wash	7	tap water	2	28
70	tap water wash	7	tap water	4	28
70	tap water wash	7	tap water	4	7

The cement plus tailings samples were manually mixed for at least 10 min, since the mixtures were typically too stiff to be mixed properly in a cement mixer. The tailings-cement mixtures were cast in steel pipe moulds with an internal diameter of 100mm and height of either 290mm or 240mm. Heavy duty plastic plugs, machined to ensure a tight fit, were used to seal each end of the tube. The sealed samples were cured at ambient temperature (22-37°C) for approximately 28 before testing.

2.3 *Geotechnical Testing*

Measurements, using standard methods, of unconfined compressive strength (UCS), direct shear, and permeabilty (using a triaxial cell) were made in Sydney on the UNSW test samples. Permeability and UCS measurements were also made on the RUM series of samples. The permeability was determined by the falling head permeameter method described in Australian Standard AS 1289.6.7.2. The intended curing period for samples produced for permeability testing was 28 days. The actual tests were, however, performed between 28 and 38 days after the samples were produced due to the numbers of tests that had to be done and limits on the availability of equipment. The test results indicate that there was no significant change of permeability over this time period.

2.4 *Geochemical Characterisation of Cemented Tailings Pore Water*

Pore water was extracted from the UNSW test cores in a triaxial cell rig being used to measure the permeability of the cemented samples. A rubber membrane was placed over the sample and the sample then placed into a perspex cell. The cell was filled with water which did not penetrate the sample and a cell pressure of 150 kPa applied. This pressure drove the membrane onto the sample and prevented any flow of fluid between the boundary of the sample and the membrane. A pore pressure of 100 kPa was then applied at the bottom of the sample to negate gravity effects and distilled water was driven through the core and collected at the top into a bleed tube. The flow of distilled water through the sample was monitored gravimetrically. The permeate was collected into a HDPE container, acidified, and stored at 4°C for subsequent elemental analysis.

The technique of falling head permeametry was used to measure permeability and produce the pore water samples for chemical testing from the RUM series of test cylinders. The machined endcaps were removed from the steel tubes. Ceramic filter disks were then inserted. Samples were not submerged in water prior to the start of the test but directly connected to the reservoir of the falling head apparatus. Tap water was introduced from the bottom of the cement cylinder and the expressed pore water was collected from the top. The first aliquot of water collected was limited to about 50 mL in volume to ensure that it represented undiluted pore water. Several aliquots were collected sequentially during each test to check for breakthrough of the leach front. In many cases the expressed water was slightly turbid, requiring filtration through a 0.45 μm membrane prior to filtration and acidification of the filtrate for chemical analysis of metals. An unacidified portion was retained to measure pH and electrical conductivity.

A combination of Inductively Coupled Plasma Atomic Emission (ICPAES) and Mass Spectromtery (ICPMS) techniques were used to analyse the collected porewater samples for the required suite of elements.

3 RESULTS AND DISCUSSION

3.1 *Geotechnical Testwork*

The range of physical test data measured for the UNSW series of samples is summarised in Table 2. The UCS increases with increment of cement added and with curing time. The permeability of the 70% solids samples is in the order of 10^{-8} to 10^{-9} m/s and decreases with cement content. It also can be seen that there is no significant change of permeability with curing time.

Table 3 contains the density, moisture and UCS data for the RUM series of samples. The UCS values are very similar to the equivalent formulations in Table 2. However, the larger range of test variables spanned by the RUM series permits a more detailed assessment of effects of variables on cement characteristics. The starting pH has little effect on the UCS. However, it is apparent that the high salt content of the tailings process water resulted in substantially lower UCS values after 28 days than for the samples that had been pre-washed with tap water.

The plots of UCS values in Figure 1 indicate that in order to achieve the 750kPa needed for underground backfill, the initial solids content should be more than 70%, and more than 4% cement added. Based on the data presented here the optimum solids content lies between 72% and 75% and the cement content between 5% and 7%.

The results reported here for cement additions greater than 4% are consistent with published data on the effects of cement addition to tailings produced from a number of different minesites and mineral processing systems, with uniaxial compressive strengths ranging from 250 kPa to more than 2100 kPa having been documented for Portland cement additions of 4 to 16%, with curing times between 7 and 28 days (Bodi et al. 1996).

Cement additions as low as 1% have been found elsewhere to produce a significant increase in strength with 7 day unconfined compressive strength (UCS) results of approximately 150 and 220 kPa being reported for 1 and 3% Portland cement additions to tailings (Cincilla et al. 1997). However, strength increases of this order for low addition rates (less than 4%) were not observed for the current work. This may be the result of the negative effect on cement strength of the high magnesium sulfate levels present in the process water. The much higher UCS developed in the 2% and 4% additions of cement to tailings that had been pre-washed with tap water supports this contention (Table 3).

Table 4 shows that the permeability of cemented tailings decreases steadily with increasing cement addition. Without addition of cement, the permeability of tailings with dry density of about 1.3 t/m^3 is of the order of 10^{-7} m/s. With addition of 4% cement, the permeability is in the order of 10^{-8} m/s or lower. With addition of 8% cement, the permeability of cemented tailings is in the order of 10^{-9} to 10^{-10} m/s.

Table 2 Geotechnical characteristics of the UNSW samples

Solids content (%)	Cement content (%)	Curing time (days)	Unconfined Compressive Strength (UCS) (kPa) Dry	Wet	Shear Strength c (kPa)	ϕ (deg)	Permeability (m/s)
70	0				1.5	25	
70	4	7	470	430			1.24E-08
70	4	28	590	480			2.25E-08
70	10	7	920	800			7.27E-09
70	10	28	1200	1020			4.62E-09

Table 3. Density, Moisture and UCS data for RUM samples

Solids content (%)	Cement content (%)	Start pH [a]	Pore Water	Curing Time (days)	Density (t/m^3) Wet density	Dry density	Moisture content (%)	UCS (kPa)
70	2	5.5	Process	28	1.80	1.33	35.6	20
70	4	5.5	Process	28	1.79	1.29	38.4	100
70	6	5.5	Process	28	1.78	1.29	37.9	470
70	8	5.5	Process	28	1.79	1.30	37.2	1080
70	2	7.0	Process	28	1.85	1.36	36.3	10
70	4	7.0	Process	28	1.78	1.25	41.8	100
70	6	7.0	Process	29	1.77	1.29	36.7	500
70	2	7.0	Tap water	28	1.78	1.29	38.2	40
70	4	7.0	Tap water	28	1.78	1.29	38.3	720
75	2	5.5	Process	29	1.81	1.38	31.2	30
75	4	5.5	Process	28	1.89	1.45	30.0	640
75	2	5.5	Process	29	1.81	1.38	31.2	30

[a] pH of tailings prior to addition of cement. Run-of-mill tails, pH 5.5; pH 7 produced by adding additional lime to run-of-mill tailings.

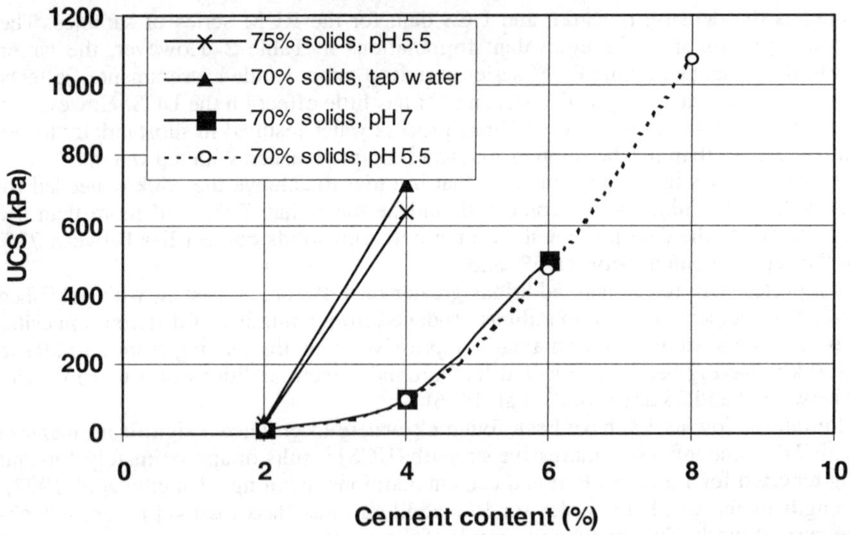

Figure 1. UCS as a function of added cement and percent solids content of the parent tailings

Table 4 Permeability data for RUM series of cemented tailings

Solids content (%)	Cement content (%)	pH	Pore Water	Curing period (days)	Dry density (t/m^3)	Moisture content (%)	Permeability (m/s)
70	0	5.5	Process	30		30.3	2.1E-07
70	6	5.5	Process	29	1.29	38.4	4.7E-09
70	8	5.5	Process	29	1.31	37.6	1.5E-10
70	8	5.5	Process	32	1.33	37.0	2E-09
70	2	7.0	Tap water	35	1.27	41.4	5.2E-08
70	4	7.0	Process	37	1.26	41.8	3.1E-08
70	4	7.0	Tap water	35	1.31	37.9	9.1E-10
70	6	7.0	Process	31	1.31	40.0	1.8E-09
75	4	5.5	Process	37	1.47	29.6	1.2E-09

3.2 *Geochemical Testwork*

The concentrations of important solutes in porewater samples extracted from the UNSW test cylinders are reported in Table 5. Comparison of the supernatant data for the cemented tails with the composition of the process fluid in the tailings prior to the addition of cement shows that there are orders of magnitude decreases in the concentrations of Mg, Mn, U and SO4 in the porewater samples extracted from the cemented tails.

A much more extensive range of variables was investigated for the RUM series of test cylinders. Cement additions of 2,4, 6 and 8% were used as well as two initial solids contents. The effect of initial tailings supernatant was specifically assessed by preparing cemented samples from which the original saline and metal rich process porewater had been washed out using tapwater, prior to the addition of cement. The porewater chemistry results for the samples subjected to falling head permeametry for the measurement of permeability are summarised in Table 6.

Maximal fixation of solutes occurs at cement additions of 2% or greater. Removal of salts by washing the tails with tap water prior to addition of cement had no material effect on the results, nor did pre-neutralisation (from pH 4.5 to 7) of the tailings.

The alkaline conditions produced by the addition of cement were expected to assist in "fixation" of a range of contaminants either through formation of insoluble oxide/hydroxide forms or through enhanced adsorption to solid phases (particularly the high surface area amorphous iron, manganese and silicon oxyhydroxide precipitates that will form as the initially acidic tailings

Table 5. Composition of Pore Water in Tailings Paste and Cemented Paste Tailings - UNSW series [a]

Sample	pH	Mg mg/L	Ca mg/L	Al mg/L	Fe µg/L	Mn mg/L	U µg/L	SO_4 mg/L
Tailings	4.42	477	478	2.45	197	2110	370	12952
Cement(%), age (d)								
4%, 7d	7.49	0.80	820	0.95	130	0.004	0.40	2031
10%, 7d	7.69	0.14	206	3.63	55	0.003	<0.1	1451
10%, 28d	10.64	0.63	279	1.87	65	1.09	1.7	523

[a] Pore water expressed by pressure extraction. Rowe cell used for tailings, pressurized triaxial cell used for cemented tails.

Table 6. Pore Water in Tailings Paste and Cemented Paste Tailings - RUM series (28 d cured)

%Solids	%Cement	Treatment	pH	EC µS/cm	Al µg/L	Ca mg/L	Mg mg/L	Mn µg/L	SO4 mg/L	U µg/L
70	0		5.5	15000	7.31	457	2010	796600	10926	3.29
70	2	Water Wash	9.6	1170	108	225	2.6	2.79	563	2.42
70	2		10.5	3300	120	495	0.7	0.73	1520	0.176
75	2		10.5	2030	176	408	0.8	1.18	1089	0.345
70	4	Preneutralised	11	3160	302	407	0.4	3.00	1604	0.030
70	4	Water Wash	11-12	--	447	318	<0.1	0.36	861	0.975
70	6	Preneutralised	11	--	126	433	1.2	5.13	1899	0.052
70	6		10.0	3640	1241	414	1.8	2.81	1554	0.061
70	8		10	-	260	71	1.3	0.27	293	0.029

stream is neutralised). However, prior to this testwork being conducted, concern had been expressed that the U could be more soluble in the cement matrix since the affinity of uranium for carbonate anions is sufficiently strong for U(VI) to be made soluble at pH values above about 8 (Waite et al. 1992, Waite et al. 1994). Several factors may counter the possible dissolution of U(VI) at high pH:

- The considerable amount of reactive silicate (initially dissolved at pH 1.7 during the U leach process) present in the tailings matrix may stabilise the uranyl ion either as insoluble uranyl silicates or in surface-bound U(VI)-silica ternary complexes;
- The formation of insoluble magnesium-uranium sulfates, magnesium-uranium carbonates, calcium-uranium carbonates or calcium-magnesium-uranium compounds may occur (Wanner and Forest 1992; Glasser 1997). Such solids have been identified as apparently controlling uranium solubility in high pH, high sulfate cement pore waters in northern Jordan (Linklater et al. 1996);
- Incorporation of U into lattice sites of secondary minerals such as ettringite (calcium aluminoferrite trisulfate, $Ca_6Al_2(SO_4)_4.24H_2O$). The ettringite structure can readily accommodate large cations, for example Pb^{2+} and UO_2^{2+}, and there is typically substitution of Ca by Fe^{3+}, Al^{3+} and Si^{4+}. This phase forms readily during the initial phase of hydration of portland cements and is stable over the pH range 10-14 (DeGroot et al. 1989, Atkins et al. 1991).

The data presented above indicate that the fixation mechanisms described above dominate the potential for remobilisation of U at elevated pH. Remobilisation of U will occur only if the pH of the cemented matrix decreases substantially. In the case of U the critical threshold appears to be pH 9. Provided that the pH of the cement matrix remains above this value there will be negligible re-dissolution of U.

Published data indicate that the solubility of U bound in prepared cement mineral phases decreases substantially with time (Atkins et al. 1991). For example, prolonged wet curing of cement systems has resulted in several orders of magnitude decrease in U solubility over 800 days of aging. Changes in the solubility of U through time will be assessed as part of the long term program of evaluation work planned for the cemented tailings cylinders produced for the work reported here.

4 CONCLUSIONS

The testwork with the surrogate Ranger tailings indicates that in order to achieve the UCS (28d) value of 750 kPa required for primary stope backfill the initial solids content of the tailings should be more than 70%. At this initial solids loading, greater than 4% cement would need to be added to produce the required strength. Based on the results from this initial testwork, the optimum solids content lies between 72% and 75% and the cement content between 5% and 7%.

The permeability of the cemented tailings decreased with increasing cement content. The permeability of uncemented tailings is in the order of 10^{-7} m/s. With addition of 4% cement, the permeability is in the order of 10^{-8} m/s or lower. The permeability decreased further to 10^{-9} to 10^{-10} m/s with addition of 8% cement.

Tailings strength is not so much of an issue for the material that will be disposed of in the purpose excavated underground silos owing to the dimensions of these chambers. However, both permeability and the concentrations of solutes in the tailings pore water are critical factors since these together will dictate the magnitude of the source term for seepage. Maximum removal of solutes was found to occur for 2% or higher additions of cement. The highest removals (on a mass basis) of solutes from the initial tailings pore water occurred for Mg and sulfate. Mn is also a very important ion with concentrations of up to 100 mg/L being present in process water. Essentially all of the Mn present in the tailings porewater was converted to insoluble form at 2% addition of cement. The concentration of U in the porewater was reduced to below 1 μg/L for cement additions of 2% or greater.

Groundwater modelling studies have been conducted on the potential impacts of underground disposal of tailings at Jabiluka (Kalf and Dudgeon 1999) assuming concentrations of solutes equal to those in run-of-mill neutralised tailings (see first line of data in Tables 4 and 5). The results of these calculations indicate no significant impact on ground and surface waters over a 1,000 year time frame. Addition of cement will decrease the concentration source term for U by orders of magnitude below that used in these model predictions, thus adding a further dimension of confidence to the long term integrity of the proposed method for tailings containment.

REFERENCES

Atkins, M., Glasser, F.P. & Moroni, L.P. 1991. The long term properties of cement and concretes *Proceedings Materials Research Society Symposium:* 212: 373-386.

Bodi, L., Hunt, G. & Lahnalampi, T. 1996. Development and shear strength parameters of paste backfill. *Proceedings of the Third International Conference on Tailings and Mine Waste '96*, A.A. Balkema Publishers, Rotterdam, pp. 169-178.

Cincilla, W.A., Landriault, D.A. & Verburg, R. 1997. Application of paste technology to surface disposal of mineral wastes. *Proceedings of the Fourth International Conference on Tailings and Mine Waste '97*: 343-355.A.A. Balkema Publishers, Rotterdam.

Connor, J.R. 1990. *Chemical Fixation and Solidification of Hazardous Wastes*. Van Nostrand - Reinhard, New York.

De Groot, G.J., Wijkstra, J., Hoede, D. & van der Sloot, H.A. 1989. Leaching characteristics of selected elements from coal fly ash as a function of the acidity of the contact solution and the liquid/solid ratio.

Stabilisation and Solidification of Hazardous and Radioactive Wastes, ASTM, Philadelphia, PA, pp. 170-183.
ERA 1998. *The Jabiluka Mill Alternative Public Environment Report*. Prepared by Kinhill Pty Ltd in association with ERA Environmental Services, June 1998.
Fenton, B.R. & Waite, T.D. 2000. *Geochemical, Geotechnical and Hydraulic Properties of Cemented Uranium Mine Tailings*, report No. 2000/3, Centre for Water and Waste Technology, School of Civil and Environmental Engineering, The University of New South Wales, Sydney, March 2000.
Glasser, F.P. 1997. Fundamental aspects of cement solidification and stabilisation. *J. Hazardous Materials.* 52:151-170.
Golder Associates 1997. *Laboratory Evaluation of Jabiluka Tailings for Suitability as a Paste Fill* , Report 961-9115 for North Ltd.
Kalf, F.R.P. & Dudgeon ,C.R. 1999. *Analysis of long-term groundwater dispersal of contaminants from proposed Jabiluka Mine tailings repositories*, Supervising Scientist Report 143, Supervising Scientist, Canberra.
Kumar, S. & Rao, C.V.S.K. 1994. Effect of sulfates on the setting time of cement and the strength of concrete. *Cement and Concrete Research.* 24:1237-1244.
Linklater, C.M., Albinsson, Y., Alexander, W.R., Casas, I., McKinley, I.G. & Sellin, P. 1996. A natural analogue of high-pH cement pore waters from the Maqarin area of northern Jordan: Comparison of predicted and observed trace-element chemistry of uranium and selenium. *J. Contaminant Hydrology.* 21:59-69.
Verburg, R.B.M. 1997. Environmental benefits associated with the use of paste for surface disposal of tailings. *Proceedings of the 50th Canadian Geotechnical Conference of the Canadian Geotechnical Society*: .484-491. Ottawa, Canada.
Waite, T.D., Payne, T.E., Davis J.A., & Sekine, K. 1992. Uranium Sorption. Volume 13 , *Alligator Rivers Analogue Project Final Report*, 106 pp.
Waite, T.D., Davis, J.A., Payne, T.E., Waychunas, G.A. & Xu, N. 1994. Uranium (VI) adsorption to ferrihydrite: Application of a surface complexation model. *Geochimica Cosmochimica Acta.* 58:5465-5478.
Wanner, H. & Forest, I. (Eds) 1992. *Chemical Thermodynamics of Uranium*. North Holland, Amsterdam.

Tailings and Mine Waste '01 © 2001 Balkema, Rotterdam, ISBN 90 5809 182 1

An alternative to conventional tailing management – "dry stack" filtered tailings

Michael P. Davies & Stephen Rice
AMEC Earth and Environmental, Vancouver, B.C., Canada

ABSTRACT: Development of large capacity vacuum and pressure belt filter technology presents the opportunity for storing tailings in a dewatered state, rather than as conventional slurry and/or in the "paste like" consistency associated with thickened tailings. Filtered tailings are dewatered to moisture contents that are no longer pumpable and need to be transported by conveyor or truck. Filtered tailings are placed, spread and compacted to form an unsaturated, dense and stable tailings stack (termed a "dry stack") requiring no dam for water or slurried tailings retention. This paper presents the basics of dry stack tailings management including design criteria and site selection considerations. Examples of several operations using dry stack technology are presented. Approximate operating costs for dry stack facilities are also included. Dry stack tailings are not a panacea for tailings management but present, under certain circumstances, an option to the tailings planner.

1 INTRODUCTION

Tailings management for the past several decades has largely involved the design, construction and stewardship of tailings impoundments. These impoundments are developed to store tailings slurry that typically arrives at the impoundment with solids contents of about 25% to 40%. The management of the traditional tailings impoundment is therefore a combination of maintenance of structural integrity and managing immense quantities of water.

The basic segregating slurry that has been used for conventional tailings management is only part of a continuum of products available to the modern tailings designer. Development of large capacity vacuum and pressure filter technology has presented the opportunity for storing tailings in a dewatered state, rather than as conventional slurry and/or in the "paste like" consistency associated with thickened tailings. Tailings are dewatered to moisture contents that are no longer pumpable. The filtered tailings are transported by conveyor or truck, and placed, spread and compacted to form an unsaturated, dense and stable tailings stack (a "dry stack") requiring no dam for retention. While the technology is currently (considerably) more expensive per tonne of tailings stored than conventional slurry systems, it has particular advantages in:

a) arid regions, where water conservation is an important issue
b) situations where economic recovery is enhanced by tailings filtration
c) where very high seismicity contraindicates some forms of conventional tailings impoundments
d) cold regions, where water handling is very difficult in winter

Moreover, "dry-stacks" have regulatory attraction, require a smaller footprint for tailings storage (much lower bulking factor), are easier to reclaim, and have much lower long-term liability in terms of structural integrity and potential environmental impact.

This paper will utilize the most common terminology in the industry. This includes:

• slurry tailings – the typically segregating mass of tailings that are in a fluidized state for transport by conventional distribution systems

• thickened tailings –partially dewatered but still a slurry that has a higher solids content by weight than the basic tailings slurry but is still pumpable. Chemical additives are often used to enhance slurry tailings thickening

• paste tailings – thickened tailings with some form of chemical additive (typically a hydrating agent such as Portland cement)

• wet cake tailings – a non-pumpable tailings material that is at, or near, saturation

• dry cake tailings – an unsaturated (e.g. not truly dry) tailings product that cannot be pumped.

The terms "dry stacked" or "dry stack" tailings have been adopted by many regulators and designers for filtered tailings. As long as the designers, owners and regulators understand that the tailings are not truly dry but have a moisture content several percent below saturation, there is nothing wrong with continuing the use of this terminology.

2 CONTINUUM OF TAILINGS

Figure 1 shows the continuum of water contents available for tailings management today and includes the standard industry nomenclature.

Filtered tailings are typically taken to be the dry cake material shown in Figure 1. This material has enough moisture to allow the majority of pore spaces to be water filled but not so much as to preclude optimal compaction of the material.

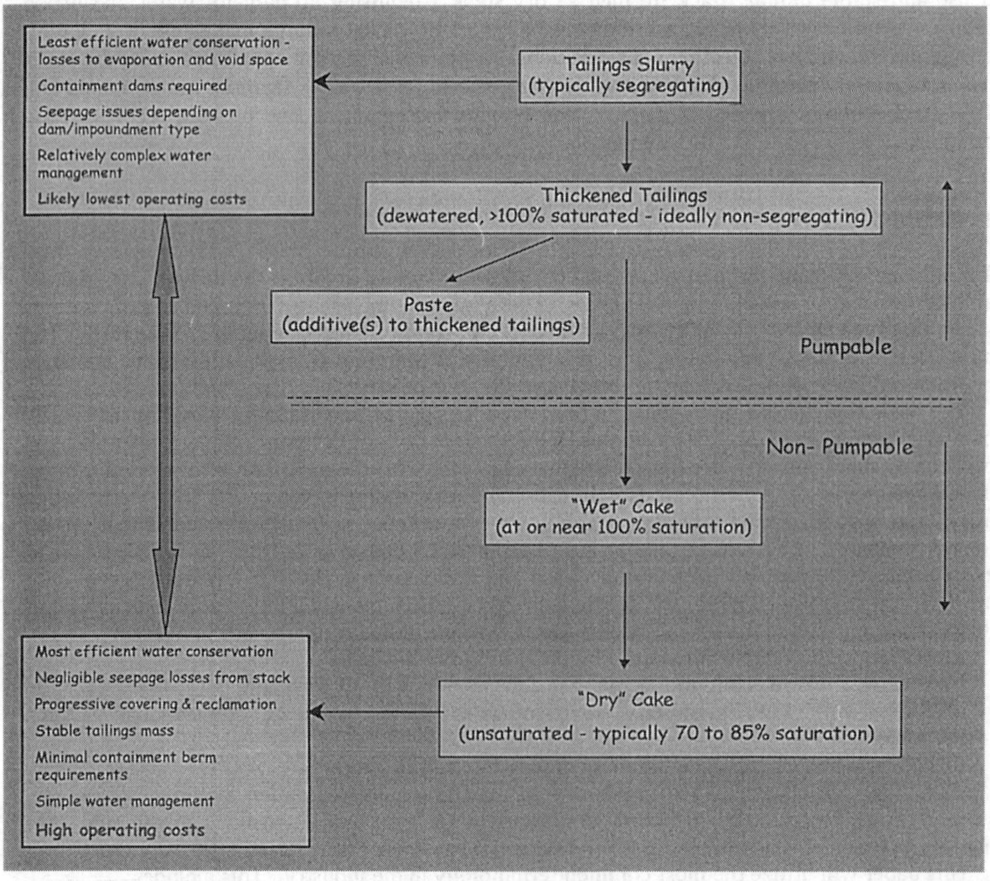

Figure 1. Tailings Continuum

Figure 2. Example of a Filter Plant

Filtering can take place using pressure or vacuum force. Drums, horizontally or vertically stacked plates and horizontal belts are the most common filtration plant configurations. Figure 2 shows a typical filter press. Pressure filtration can be carried out on a much wider spectrum of materials though vacuum belt filtration is probably the most logical for larger scale operations.

The nature of the tailings material is important when considering filtration. Not only is the gradation of the tailings important, but the mineralogy is as well. In particular, high percentages of <74 μm clay minerals (i.e.. not just clay-sized but also with clay mineralogy) tend to contraindicate effective filtration. Furthermore, substances such as residual bitumen (e.g. oil sands tailings) can create special difficulties for a filtration plant.

3 CONSIDERATIONS FOR FILTERED TAILINGS

As for any other form of tailings management, there are a number of issues that require careful consideration prior to selecting filtered tailings for a given project. Ultimately, these considerations are all about economics (capital, operating and closure liability) but require individual attention during the prefeasibility and feasibility stages of the project.

3.1 *General*

Whether filtered tailings are a candidate for a given project depends on the motivation to consider alternatives to a conventional slurried tailings impoundment. The motivation could include a more favourable, or timely, regulatory process or perhaps one of several technical issues presented by the site.

As noted in the introduction, filtered tailings could have application to meet technically challenging sites in:

a) arid regions
b) mines where dissolved metal recovery is enhanced by tailings filtration
c) high seismicity regions
d) cold regions
e) mine sites where space is limited as filtered tailings result in a lesser footprint than for slurried tailings.

In addition, site legacy issues are also part of the selection criteria as dry stacked tailings facilities are substantially easier to reclaim for mine closure, in most circumstances, when compared to conventional impoundments. Following is some elaboration on key issues to be considered.

3.2 *Water Management*

Where water is relatively scarce, either year round or seasonally due to extreme cold, sending immense quantities of water to quasi-permanent storage in the voids of a conventional impoundment can severely hamper project feasibility. By reclaiming the bulk of the process water in or near the mill, far more efficient recycle is achieved. Moreover, the amount of water "stored" in a dry stack facility will be typically >25% less than that in a conventional slurried impoundment even if 100% pond reclaim efficiency is achieved with the impoundment.

3.3 *Commodity Extraction*

Many mines, particularly those dealing with precious metals, can improve the bottom line by maximizing the amount of tailings water that can be reclaimed. Both the economic commodity (e.g. dissolved gold) and process chemicals (e.g. cyanide) can be recovered from the filtration water and one or more rinse cycles.

3.4 *Storage Availability*

Filtered tailings can be placed in a relatively dense state meaning that more solids per unit volume can be achieved. Furthermore, more aggressive use of available land (e.g. valley slopes) can be used with filtered tailings. As discussed in 3.5, lesser foundation conditions can also be considered in comparison to conventional impoundments.

3.5 *Geotechnical Issues*

The questionable manner in how some conventional impoundments are designed and/or operated provides support to considering the geotechnical advantages of filtered tailings. By objectively reviewing an instability database for conventional slurry tailings impoundments, over the past 30 years there have been approximately 2 to 5 "major" tailings dam failure incidents per year (Davies and Martin, 2000). There have been at least two events each year (1970-1999, inclusive). If one assumes a worldwide inventory of 3500 conventional tailings impoundments (a tenuous extrapolation at best), then 2 to 5 failures per year equates to an annual probability of between 1 in 700 to 1 in 1750. This rate of failure does not offer a favorable comparison with the 1 in 10,000 figure that appears representative for conventional water dams. The comparison is even more unfavorable if less "spectacular" tailings impoundment failures are considered. These impoundment failures, often equally economically damaging, are not just of older facilities constructed without formal designs, but include facilities designed and commissioned in the past 5 to 20 years - supposedly the "modern age" of tailings dam engineering.

The most common failure modes for slurry tailings impoundments are physical instability (including static and dynamic liquefaction) and water mismanagement issues (including lack of freeboard and seepage phenomena like piping). Filtered tailings placed in dry stacks are essentially immune to catastrophic geotechnical "failure" and can be readily designed to withstand static and seismic forces. A case can also be made for a reduction in the seismic design criteria based on failure consequence. This can significantly reduce operating costs. The unsaturated tailings mass is extremely resistant to saturation and seepage is governed by unsaturated hydraulic conductivities. Moreover, far less is required of foundation conditions as the unsaturated, largely dilatant tailings within a dry stack are not susceptible to static liquefaction or catastrophic breaching by an impounded pond should the foundation move creating substantive shear strains in the tailings mass.

3.6 *Reclamation/Closure Issues*

Dry stack facilities can be developed to consist of, or closely approximate, their desired closure configuration. Some form of assured surface runoff management plan is required. The tailings can be progressively reclaimed in many instances. In all cases, a closure cover material is required to resist runoff erosion, prevent dusting and to create an appropriate growth media for project reclamation.

The lack of a tailings pond, very low (if any) appreciable seepage from the unsaturated tailings mass and general high degree of structural integrity allows dry stacks to present the owner/operator with a comparably straight forward and predictable facility closure in comparison with most conventional impoundments.

3.7 *Environmental Stewardship*

Issues related to the environmental impacts from tailings dams were first seriously introduced in the 1970's in relation to uranium tailings. However, environmental issues related to mining have received attention for centuries. For example, public concerns about the effects of acid rock drainage (ARD) has existed for roughly 1,000 years in Norway. Today, environmental issues are growing in importance as attention has largely turned from mine economics and physical stability of tailings dam to their potential chemical effects and contaminant transport mechanisms. Recent physical failures such as Merriespruit, South Africa in 1994 and Omai, Guyana in 1995 and Los Frailes in Spain in 1998 illustrates this issue with most of the media reports highlighting the real or perceived environmental impacts of the failures.

Dry stacked tailings facilities have some tremendous potential environmental advantages over impounded slurried tailings largely because the catastrophic physical failures that define tailings management to non-supporters of the industry cannot occur. Moreover, leachate development is extremely limited due to the very low seepage rates possible. Oxidation processes are possible though the very slow rates for such, coupled with the limited seepage potential, limits or eliminates the concern of significant metallic drainage. Clearly, industry/regulatory standards of testing for potential operating and long-term impacts are essential. However, if the stack is operated to maintain its unsaturated character, any potential impacts should be predicted as acceptable except under unusual conditions.

Fugitive dusting, both during operation and upon closure, is a very real concern with dry stacks; particularly in arid environments. Progressive reclamation is the only effective method to address this concern.

3.8 *Regulatory Environment*

The regulatory environment worldwide is generally becoming less tolerant of one of humanities essential industries. Mining cannot exist without the creation of some form of tailings so the availability of a management strategy that is viewed (and correctly so) as both less invasive and less difficult to decommission as well as one that does not conjure up "massive" failure scenarios is a positive to the industry. As discussed elsewhere in this paper, the challenge is to get this regulatory friendly tailings management system to become cost-effective for those operations that would benefit (eg. in terms of the permitting process) from its consideration.

4 DESIGN CRITERIA

4.1 *General*

There are four main design criteria for filtered tailings:

1. filtering characteristics
2. geomechanical characteristics
3. tailings management
4. water management

In addition, the design must be compatible with an optimal closure condition (designing for closure). Implicit to the overall design criteria is project economics.

4.2 *Filtering Characteristics*

Determine the most cost-effective manner to obtain a dewatered product consistent with the other three design criteria (geomechanics, placement management and water management). Filter suppliers are both knowledgeable and helpful in this regard but some form of pilot test(s) is essential as every tailings product will exhibit its own unique filtering character. It is important to anticipate mineralogical and grind changes that could occur over the life of the project. The candidate filtering system(s) must be able to readily expand/contract with future changes at the mine with the least economical impact.

4.3 *Geomechanical Characteristics*

The strength, moisture retention and hydraulic conductivity characteristics of the tailings need to be established. The saturated tailings should be determined to "anchor" the results and tests as variable moisture contents are required to demonstrate the impact of the inevitable range of operating products. The other important geomechanical characteristic to determine is the moisture-density nature of the tailings. The unsaturated moisture-density relationship indicates in-situ density expectation as well as the sensitivity of the available degree of compaction for a given moisture content. From a compaction perspective, the filtered tailings should neither be too moist nor too dry. The optimal degree of saturation is usually between 60 and 80%.

4.4 *Tailings Management*

The design needs to be compatible with how the stack can be practically constructed using conventional haulage and placement equipment. Other than the capital and operating costs of the filtering process, the economics of dry stack management is the most important component of filtered tailings viability. Haul distance, placement strategy and compactive effort and additional works for closure and reclamation can make a larger incremental difference to the unit cost of a dry stack facility in comparison with a slurried impoundment.

The design should also clearly identify what contingency(s) will be in place if the filtering process experiences short-term disruptions. A temporary storage area or vessel is sound strategy. It is, however, the authors' experience that the filters should become part of the process plant under the management of the mill superintendent. The tailings processing then becomes integrated with the metal recovery functions and consequently down time is minimized because operation of the tailings system becomes critical to the overall mill performance.

4.5 *Water Management*

Surface water, particularly concentrated runoff, should not be permitted to be routed towards a dry stack. As important, the catchment and routing of precipitation (and any snow melt in colder climates) on the stack itself must be appropriately designed for. For the surface runoff within the overall catchment containing the dry stack, one (or more) of perimeter ditches, binds or under-stack flow through drains designed for an appropriate hydrological event(s) should be included in the design. For on stack water management, routing of flows to armoured channels and limiting slope lengths/gradients to keep erosion potential at a minimum are the best design criteria.

5 CONSTRUCTION AND OPERATIONAL CONSIDERATIONS

5.1 *General*

There are a number of construction and operational considerations that need to be accounted for in the design and planning of a dry stack. These considerations are very different from the construction and operational considerations normally associated with slurry tailings facilities. The main considerations are usually:

1. Site development

2. tailings transport and placement
3. water conservation and supply
4. reclamation and closure

In addition, there are often other considerations that need to be addressed on a site-specific basis for example co-disposal of waste rock in a combined mine waste management facility, storage of water treatment plant sludges etc.

5.2 *Site Development*

Site development for a dry stack normally consists of the construction of surface and groundwater control systems. There are normally two systems:

1. A collection and diversion system for non-contact water (i.e. natural surface water and groundwater from the surrounding catchment area that has not yet come into contact with the tailings). This system usually consists of ditches to divert surface runoff around the site and if necessary a groundwater cut-off and drainage system usually combined with surface water diversion. The cut-off system can range from simple ditches to sophisticated cut-off walls depending upon site conditions.
2. An interception and collection system for contact surface water, impacted groundwater, and seepage from the dry stack. This system usually consists of an underdrainage system of finger drains, toe drains, drainage blankets and french drains; collection sumps and ponds. Water collected in the ponds and sumps is usually used in process or pumped to a water treatment plant depending upon the site water balance. Liners for the facilities can also be components of the interception and collection system depending upon predicted impacts and regulatory requirements.

5.3 *Tailings Transport and Placement*

There are two methods in common use for transport of the filtered tailings to the tailings storage facility. These are conveyors or trucks and the equipment selection is a function of cost. Placement in the facility can be by a conveyor radial stacker system or trucks depending upon the application and the design criteria. Conveyor transport of tailings to the disposal site can be combined with placement by truck, so conveyor transport does not automatically result in placement by radial stacker.

The main issue associated with the placement of the filtered tailings by truck is usually trafficability. The filtered tailings are generally produced at or slightly above the optimum moisture content for compaction as determined in laboratory compaction tests (Proctor Tests). This means that a construction/operating plan is required to avoid trafficability problems. This is especially true in wetter environments since trafficability drops as moisture content rises and if the tailings surface is not managed effectively it can quickly become un-trafficable resulting in significant placement problems and increased operating costs. In addition, in high seismic areas there is often a design requirement to compact the tailings to a higher density in at least the perimeter "structural" component of the facility. This requirement increases the need for construction quality control. It is the authors' experience that the degree of compaction required for assured and efficient trafficability is often higher than the compaction required to achieve design densities.

Dry stack designs often incorporate placement zones for "summer/good weather" placement (dry, non-freezing conditions) and "winter/bad weather" placement (wet, or freezing conditions) with summer placement being focused on the structural zones. Again, this is especially true for facilities planned for wetter or colder climates were seasonal fluctuations are significant and predictable.

The key is to consider the environment and the design criteria and develop a flexible operating plan to achieve them.

5.4 *Water Conservation*

Often one of the main reasons to select dry stacked filtered tailings as a management option is the recovery of water for process water supply. This is particularly important in arid environments were water is an extremely valuable resource and the water supply is regulated (e.g. Northern Chile and Mexico). Filtering the tailings removes the most water from the tailings for recycle when compared with other tailings technologies as discussed earlier. This recovery of water has a cost benefit to the project, which offsets the capital and operating cost of the tailings system. It should be noted, that water surcharge storage needs to be factored in to the design of a filtered tailings system. Depending upon the application this can be a small water supply reservoir or tank.

5.5 *Reclamation and Closure*

One of the main advantages of dry stack tailings is the ease of progressive reclamation and closure of the facility. The facility can often be developed to start reclamation very early in the project life cycle. This can have many advantages in the control of fugitive dust, in the use of reclamation materials as they become available, and in the short and long term environmental impacts of the project. Progressive reclamation often includes the construction of at least temporary covers and re-vegetation of the tailings slopes and surface as part of the annual operating cycle.

6 ECONOMICS

6.1 *General*

It is hard to compare the economics of dry stack filtered tailings with other tailings options particularly conventional slurry tailings. This is mainly because of the difficulty of estimating the cost of closure and the potential costs associated with the long-term risk environmental liability associated with mine waste facilities. Therefore, the following discussion on economics is very subjective with a focus on perception.

6.2 *Capital Cost*

The capital costs are clearly a function of the size of the operation. Dry stack, filtered tailings currently appears to be limited to operations of 15000 tpd or less depending upon financial credits e.g. water recovery for use in process. Capital costs normally shift from the construction of engineered tailings containment structures to the dewatering (filter) plant. The capital costs may be further mitigated if the application is considered for small tonnage (less than say 4000 tpd) where the mine plan calls for paste backfill underground. Paste backfill requires a tailings processing plant with dewatering so incremental dewatering to produce filtered tailings make the economics more attractive. The capital cost appears to be much more attractive for operations under approximately 2000 tpd.

Other costs that should be factored into the equation are reduced costs associated with the smaller footprint, site development costs, and regulatory acceptance associated with dry stack tailings. These costs are often difficult to estimate accurately.

6.3 *Operating Cost*

The operating costs associated with the transport and placement of dry stack, filtered tailings are higher when compared with conventional slurry tailings, transported hydraulically and deposited in a tailings pond. The operating costs for a dry stack are difficult to summarize as every operation accounts for the costs differently. For example, if a mine uses a surface crew who do both tailings stack development as well as other duties, the cost/tonne will be much lower than a dedicated dry stack work force. Under the range of conditions for the presently operating dry stacks, the cost per tonne ranges from \$1 to \$10 but the average is more like \$1.50 to \$3. All costs are \$US and include filtering, transport, placement and compaction in the facility.

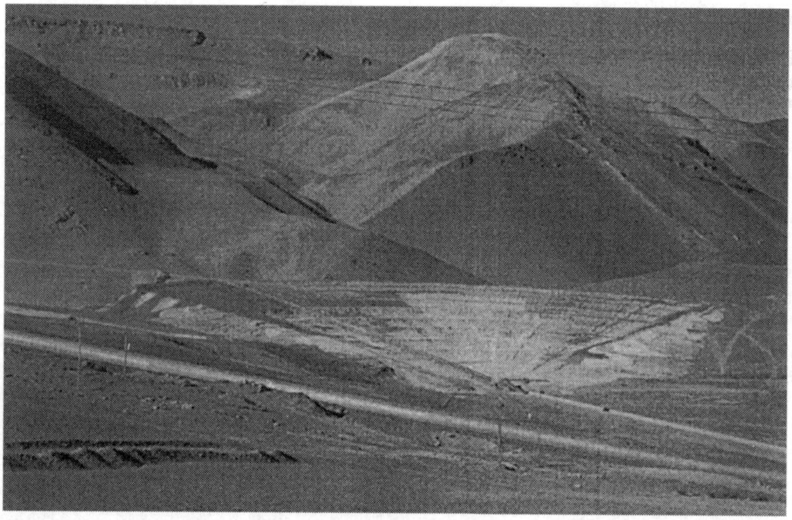

Figure 3. Dry-Stack Tailings Facility - Chile

6.4 *Reclamation and Closure*

Reclamation and closure costs are significantly reduced for dry stack tailings when compared with conventional tailings. This cost reduction is due to a reduced footprint and constructability. Other issues that need to be somehow factored into the "cost" of closure are the reduction in long-term risk and liability associated with dry stacks.

7 EXAMPLES

There are a growing number of dry stack facilities. At the same time, it would be fair to say that there is likely not any one of those operations who can point to an overall operating economic advantage to the practice. However, for at least three of the operating dry stack projects, the increased operating cost was sufficiently negated by other factors including regulatory issues and closure/liability costs.

The majority of the dry stacks are either in colder climates (e.g. Greens Creek, Alaska, Raglan, Quebec) or in arid environments (e.g. La Coipa, Chile). The La Coipa facility, developing at more than 15,000 tons/day, is one of the largest operating dry stacks. The La Coipa facility is located in a high seismic region with designed, and confirmed, structural integrity. Figure 3 shows the La Coipa facility a few years ago.

8 SUMMARY AND CONCLUSIONS

There are several candidate scenarios where dewatered tailings systems would be of advantage to the mining operation. However, dewatered tailings systems may have less application for larger operations for which tailings ponds must serve dual roles as water storage reservoirs, particularly where water balances must be managed to store annual snowmelt runoff to provide water for year round operation.

Filtered tailings, a form of dewatered tailings, are not a panacea for the mining industry for its management of tailings materials. Purely economic considerations rarely indicate a preference for dry stacked tailings facilities over conventional slurry impoundments. However, under a growing number of site and regulatory conditions, filtered tailings offer a real alternative for tailings management that is consistent with the expectations of the mining industry, its regulators and the public in general.

REFERENCES

Davies, M.P. and T.E. Martin (2000). Mine tailings dams: when things go wrong in proceedings of tailing dams 2000, Association of State Dam Safety Officials, U.S. Committee on Large Dams, March, Las Vegas, pp. 261-273.

Case histories

Tailings and Mine Waste '01 © 2001 Balkema, Rotterdam, ISBN 90 5809 182 1

Reclamation of the Big Springs subaerial tailings facility, Nevada

R.W.Reisinger, D.R.East, A.H.Gipson & J.E.Valera
Knight Piésold and Company, Denver, Colo., USA
J.A.Gorman
AngloGold (Jerritt Canyon) Corporaton, Elko, Nev., USA

ABSTRACT: The Big Springs tailings facility, located in northeastern Nevada, is presently owned by AngloGold (Jerritt Canyon) Corp. and was operated by its predecessors from 1988 through 1994. Initial construction of the 50-acre facility included a low-permeability soil liner with a full drainage blanket and zoned, earthfill embankments. Subsequent facility expansion was accomplished using a combination of upstream and downstream construction. Tailings were spigotted from approximately one-half of the perimeter dike using managed, subaerial deposition techniques. The design, construction, and operation of the subaerial facility resulted in a well drained, stable,and "full" facility with no surface water inventory at the time of reclamation in the summer of 2000.

The reclamation grading plan for the facility called for gently sloping the tailings facility into a mound-like structure using a combination of tailings and fill. The use of in-place tailings for sloping the facility resulted in using substantially less material to create the desired final graded facility than if the facility was graded entirely with random fill.

An important aspect of the grading plan included incorporating overbuild to compensate for anticipated settlement. To this extent, laboratory and field settlement tests were conducted in order to predict the long-term settlement within the graded facility.

Two basic cover alternatives were considered for the tailings facility. A vegetated soil evapotranspiration (ET) cover was selected based on the results of water-balance modeling.

1 LOCATION AND SITE HISTORY

The Big Springs Project was a gold mining and milling complex located in northern Elko County, Nevada, approximately 90 km (55 miles) north of Elko. The mill and subaerial tailings facility were constructed on alluvial sediments along the eastern flank of the Independence Mountains, typical to the Basin and Range Province. Tailings were produced at the mine from 1988 through 1994 at an average rate of approximately 900 tonnes per day (1,000 tpd). The footprint of the final tailings facility is approximately 20 hectares (50 acres). Since mill closure in 1994, the tailings facility has been used for site water management and for continued gold recovery. The facility was reclaimed during summer 2000.

2 FACILITY DESIGN AND CONSTRUCTION

The subaerial concept of tailings management, together with a combination of upstream and downstream construction, was employed into the design of the Big Springs tailings facility as described in a previous paper by Ulrich et al. (2000). Subaerial tailings deposition is described by Knight and Haile (1983) as a technique by which tailings are deposited in thin layers and allowed to settle, drain, and partially air dry prior to burial by subsequent layers. Through the incorporation of a blanket drain, increased densities can be realized with drainage and air drying serving to further increase the density.

The Big Springs tailings facility consisted of a four-sided embankment built on terrain that gently slopes toward the east and was designed as a zoned earthfill structure consisting of a low-permeability seal zone and a compacted structural zone. Longitudinal drains constructed along the embankment axis provided a means to intercept the phreatic surface. An aerial photo of the post-operational facility is shown in Figure 1.

A low-permeability soil liner was constructed within the basin of the impoundment, and a suitably-graded blanket drain was constructed upon the basin seal. Drainage of the tailings mass was also enhanced by a network of slotted pipes installed in a herringbone pattern within the blanket drain. Fluid collected by the underdrainage system was conveyed by HDPE pipe beneath the embankment to the external HDPE-lined reclaim water pond.

A supernatant water collection system was constructed in the topographically lower east embankment. This system consisted of sloping drains and pipework to collect and convey fluids by gravity to the reclaim water pond.

Tailings from the mill were pumped as slurry to a distribution system located along the inner slopes of the westerly embankments. Tailings were deposited in thin 100- to 150-mm (4- to 6-inch) layers in a controlled, rotational sequence to develop a large beach gently sloping toward the east. Rotating zones of deposition allowed the tailings to air dry before deposition of subsequent lifts. This method of deposition resulted in relatively coarse-grained tailings being deposited on the western portion of the facility and relatively fine-grained tailings being deposited on the eastern portion of the facility.

Expansion of the Big Springs tailings facility was planned to occur biannually with the easterly (pool-side) embankments being raised using downstream construction and the westerly (deposition-side) embankments being raised using the upstream technique with the drained and consolidated tailings serving as the foundation for the upstream raises. After construction in 1988, the facility was raised four times to accommodate additional requirements. Each expansion was sized based on mill production forecasts at the time of the design; thus, the design was quite flexible. Ultimately, the tailings facility was constructed in raises ranging from 1.2 to 4.0 m (4 to 13 feet) to a maximum height of approximately 20 m (70 feet).

3 RECLAMATION GRADING PLAN

This section discusses various grading alternatives considered for the facility, tests that evaluated the constructability of the selected grading alternative, and the overbuild design to offset anticipated settlement in areas of fill placement.

Figure 1. Aerial photo of the Big Springs tailings facility (looking to the southwest)

3.1 *Grading plan alternatives*

The overall design objective for reclaiming the Big Springs tailings facility consisted of minimizing infiltration into the tailings with a cost-effective design. Two general grading plans were considered that would form a gently sloping facility to encourage adequate drainage and avoid pooling of water in order to meet the overall reclamation objective. One option used fill from on-site borrow areas (*Fill Option*); the other option used tailings and embankment materials as a balanced cut/fill (*Tailings Option*).

The post-operational tailings surface generally sloped from approximately 0.6% to 0.8% from the western end of the facility to the seasonal supernatant pool at the eastern end of the facility. Local slopes within the facility on the western end of the facility were as low as 0.5%. In the area of the supernatant pool on the eastern end of the facility, the slope of the tailings surface approached zero.

The *Fill Option* called for adding fill from on-site borrow areas to create a 2% slope to the east for the top surface of the tailings facility. The desired grade could be obtained by sloping the tailings facility in a variety of ways. The option analyzed in this paper entailed sloping the facility at 2% to the east as shown in the schematic east-west section presented in Figure 2.

The *Tailings Option* called for generally grading the tailings surface to develop a 2% slope in both easterly and westerly directions with a crest located at approximately 275 m (900 feet) from the western end of the facility. To accomplish this, the relatively coarse-grained tailings from the western portion of the facility would be excavated and placed on the relatively fine-grained tailings on the eastern half of the facility. Fill also would consist of embankment materials that would be excavated to the elevation of the graded adjacent tailings surface. The maximum cut into the perimeter embankment would be approximately 4.6 m (15 feet) in the northwest corner of the tailings facility. A schematic east-west section of the *Tailings Option* is presented in Figure 2.

Both grading plans met the objective of promoting surface water runoff to minimize infiltration. However, the *Tailings Option* offered potentially significant cost savings due to the reduced amount of material moved. For the *Tailings Option*, an estimated 153,000 m^3 (200,000 cubic yards) of tailings and embankment material would be moved. For *the Fill Option*, the total amount of graded material would increase to an estimated 505,000 m^3 (660,000 cubic yards). Given the potentially significant cost advantage of the *Tailings Option* over the *Fill Option* due to differences in grading quantities, an evaluation of the constructability of the *Tailings Option* was pursued.

3.2 *Tailings Option constructability*

The *Tailings Option* required excavating tailings at depths reaching approximately 4.6 m (15feet) along the western portion of the tailings facility. The long-term data from vibrating wire pressure transducers installed in the western part of the tailings facility indicated that the phreatic surface within the tailings was relatively low (typically on the order of a few feet above the underdrain) and that downward flow of the process waters was taking place within the tailings (Ulrich et al., 2000). In addition, data from vibrating wire pressure transducers installed in the drainage blanket typically indicated that a relatively free-draining base existed.

To confirm the constructability of the grading plan under the *Tailings Option*, two tests were completed. The first test was performed in the fall of 1998 and involved excavating test pits in the western side of the facility to evaluate the physical characteristics of the tailings. This test

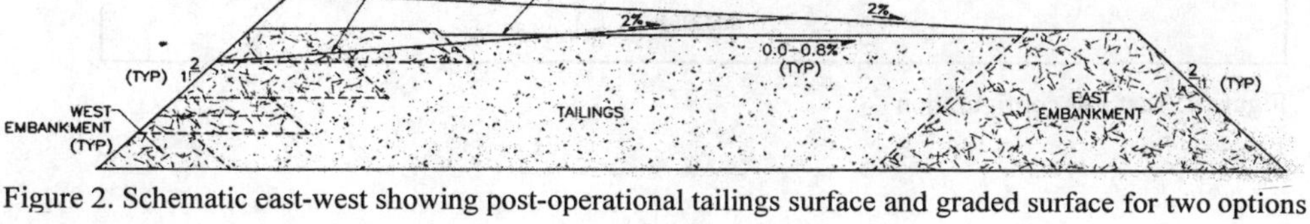

Figure 2. Schematic east-west showing post-operational tailings surface and graded surface for two options

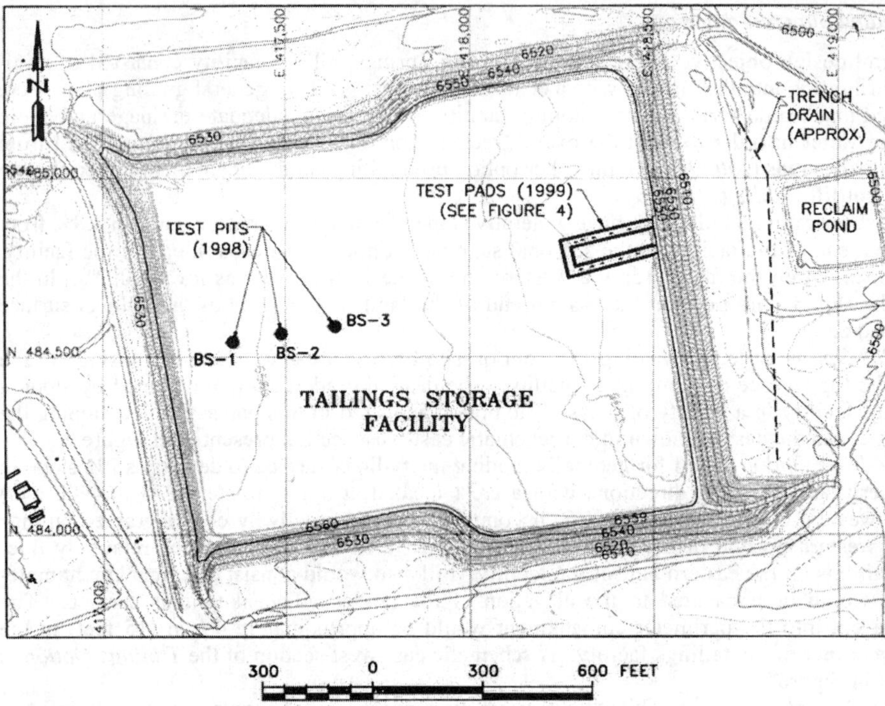

Figure 3. Location of 1998 test pits and 1999 test pits

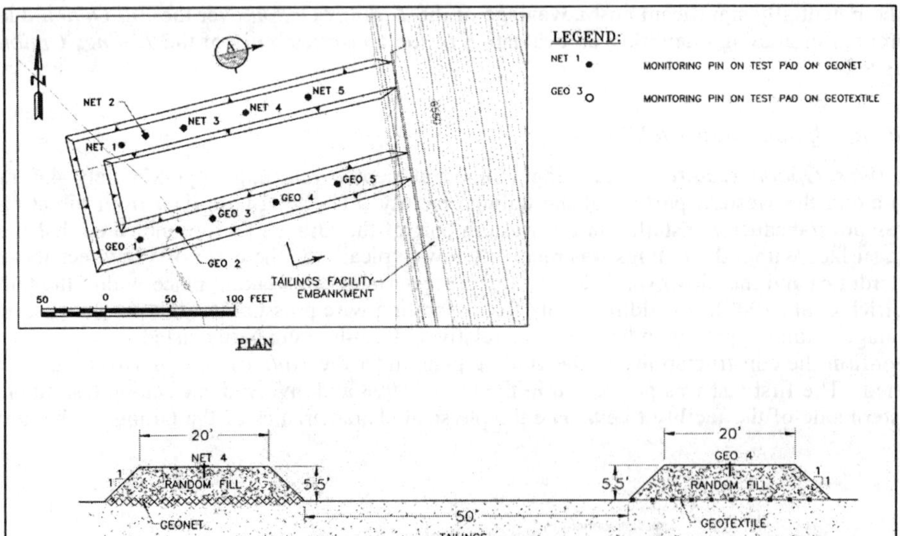

Figure 4. Test pad configuration

also evaluated equipment trafficability on the tailings. The second test, performed during the fall of 1999, involved the construction of two test pads along the eastern end of the tailings facility to evaluate construction procedures and to assess settling of the tailings after loading. This test was conducted three weeks after final evaporation of the seasonal supernatant pool.

3.2.1 *Fall 1998 test*

A preliminary test to evaluate the constructability of the *Tailings Option* was conducted in the fall of 1998. The test investigated the material characteristics of the tailings and equipment trafficability. Three test pits were excavated by a tracked backhoe on the western end of the tailings facility to a depth of approximately 4.6 m (15 feet). Figure 3 shows the location of the test pits. No water was present in any of the excavations. Samples collected from these test pits were classified as ML according to the Unified Soil Classification System (USCS), with a fines content ranging from 72 to 76 percent. Alternatively, these tailings could be classified as *silt with sand.* This is characteristic of subaerial deposition where relatively coarse-grained tailings were deposited on the western side of the facility.

The tracked backhoe used to excavate the test pits readily traveled across the tailings surface where the tailings were dry. However, in one area where runoff from the beach spray evaporation system had saturated the surface and shallow tailings, the trafficability was significantly reduced.

The results of the physical characterization and trafficability observations of the fall 1998 test suggested the constructability of the *Tailings Option* as long as the tailings surface was adequately dry.

3.2.2 *Fall 1999 test*

A second test was performed during the fall of 1999 to confirm the constructability of the *Tailings Option.* This test involved constructing two test pads on the eastern end of the tailings facility as shown in Figures 3 and 4. Each test pad measured approximately 6 m (20 feet) wide by 76 m (250 feet) long by 1.7 m (5.5 feet) high. Test pad fill materials consisted of tailings excavated from the southwestern corner of the tailings facility and random fill from an on-site borrow area. The two test pads were constructed approximately three weeks after the seasonal supernatant pool on the eastern end of the facility had completely evaporated.

Two construction procedures were evaluated in constructing the test pads. Under the first procedure, an excavator top-loaded tailings from the southwestern corner of the tailings facility onto scrapers. Scrapers hauled the tailings to the east embankment where they were placed onto the embankment. Tracked dozers pushed the tailings fill from the embankment onto the tailings surface. While this test began successfully, difficulties constructing the test pad were encountered after pushing the tailings in a single 1.7-m (5.5-foot)-thick lift approximately 8 m (25 feet) onto the tailings surface. These difficulties were attributed primarily to the use of excessively large equipment (a Caterpillar D-9 tracked dozer was used initially), repeated trafficking of the tailings fill by equipment, and not allowing the tailings fill to sufficiently dry prior to their placement.

The second construction procedure involved constructing the test pads using random fill hauled and placed by dump trucks (Figure 5). A tailings fill area between the test pads was constructed by dozers pushing tailings onto the tailings surface from one of the test pads in a single 1.7-m (5.5-foot)-thick lift. During the second construction procedure, the test pads were underlain by geosynthetics to enhance the stability of the tailings surface underlying the test pads. This construction method proceeded smoothly throughout the remaining construction of the test pads. No deformation of the tailings surface adjacent to the test pads was observed for the test pads with or without geosynthetics.

The information gained from the construction of the test pads led to the development of a conservative approach to implementing the *Tailings Option*. Under this approach, a complement of excavators and haul trucks would develop a series of access roads on the fill portion of the facility. These access roads would be constructed with random fill obtained from the embankment and on-site borrow areas. From these access roads, tailings material excavated from the western side of the facility would be hauled and in-filled by end-dumping from haul trucks.

Figure 5. End-dump truck loading random fill onto test pad

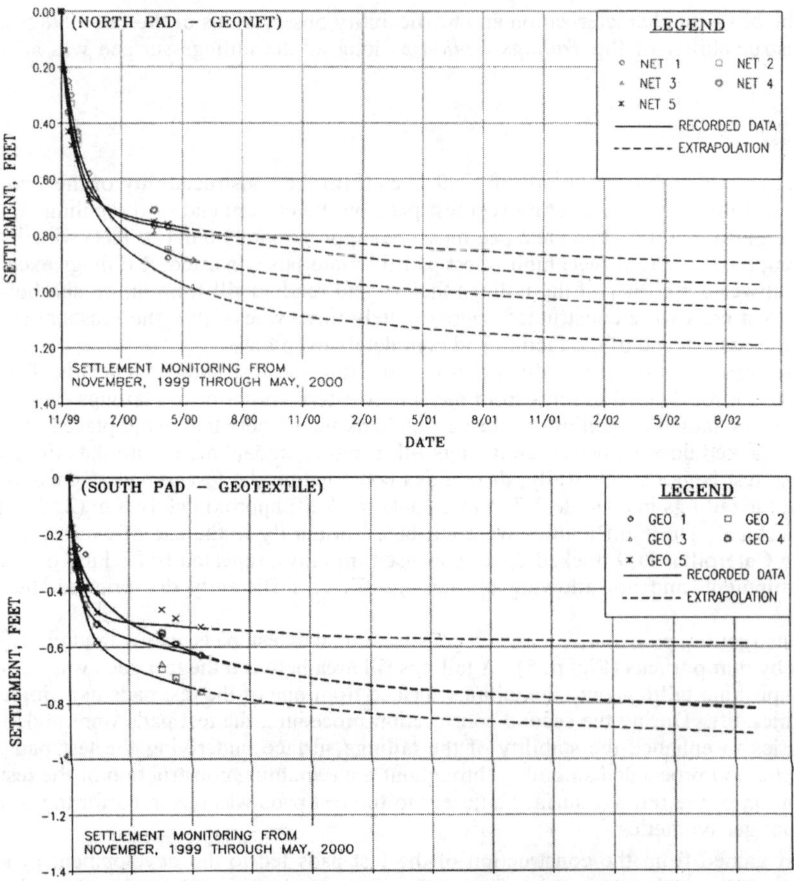

Figure 6. Results of settlement monitoring of test pads

The tailings would be graded with a fleet of low ground-pressure tracked dozers. Excavating, hauling, in-filling, and grading would continue until the necessary grading was attained.

A less conservative approach also was identified in which the *Tailings Option* was implemented using scrapers to excavate and haul tailings from the western side of the facility to the fill portions of the eastern side. Low ground-pressure tracked dozers would be used to redistribute and grade the hauled materials. The success of this approach would depend on having an adequately dry tailings surface prior to grading to maintain trafficability.

3.3 *Overbuild design*

The design of the final graded and covered facility required an estimate of the total long-term settlement due to the weight of the additional placed fill. This was necessary to ensure that the final slope of the surface of the fill corresponded as closely as possible to the 2% design value as shown on Figure 2.

Moisture content, dry density, degree of saturation, grain size analysis, Atterberg Limits, and consolidation tests were conducted on representative undisturbed samples of the near-surface tailings (depth of about 0.8 m (2.5 feet)) that were taken by pushing Shelby tubes near the eastern perimeter of the facility. The tailings classified as silt (ML) with Plasticity Index (PI) ranging from 1 to 11 and fines content from 93 to 100 percent. The degree of saturation ranged between 90 and 97 percent with a dry density of about 1.06 g/cm^3 (66.3 pounds per cubic foot) and a moisture content of about 54 percent.

A settlement analysis was conducted by first computing the applied vertical stresses induced in the underlying tailings by the test pad configuration shown on Figure 4. The settlement was computed at the center of the test pad since it corresponds to the maximum stress and thus the maximum settlement. The vertical stress at the center of the test pad was computed based on elastic theory using the procedure developed by Gray (1936). Settlements were computed for a 15-meter (50-foot)-thick tailings deposit according to the conventional one-dimensional consolidation theory described by Taylor (1948). The results of the laboratory consolidation tests were used in the settlement calculations assuming that the tailings were normally consolidated. The analysis was performed assuming that the total unit weight of the tailings at the surface increases by 0.4 percent per foot of depth. The total settlement of the 15-meter (50-foot)-thick tailings deposit was estimated to be about 0.40 m (1.3 feet) for the test pads.

A time rate of settlement analysis was also performed using the coefficient of consolidation estimated from the consolidation tests and assuming double drainage conditions. The time required for 90 percent consolidation was about four years.

In order to compare the estimated settlements with measured in-situ settlements, a field settlement monitoring program of the two constructed test pads was initiated. As shown on Figure 3, the two pads were constructed close to the eastern embankment where the seasonal supernatant pool was historically located and where the largest settlements would be expected to occur. The base of the North pad was underlain by geonet and that of the South pad by a geotextile fabric. Settlement monitoring pins consisting of steel rebar were driven at the center of the pads at the locations shown on Figure 4. A total of five monitoring points were established for each pad.

Settlements were monitored for a period of about seven months from November 8, 1999, through May 22, 2000. Plots of measured settlement versus time for each test pad are presented on Figure 6. The upper graph of Figure 6 corresponds to the North pad and the lower graph to the South pad. It should be noted that construction of the test pads was completed on November 4, 1999, but that the first settlement readings were not made until four days later on November 8, 1999. Readings at each of the ten monitoring points were made at time intervals ranging from four days to 18 days for the period from November 8, 1999 through December 12, 1999. Due to heavy snow and drifting conditions, no readings were made after this time for about three months until March 24, 2000. The next reading was made on April 14, 2000, and the final reading on May 22, 2000. After this time the test pads were decommissioned in order to reclaim the facility. It should be noted that the measured settlements include those of the fill itself; however this should be a minor component of the total measured settlement.

Our interpretation of the measured settlement data presented on Figure 6 is shown by a solid line that extends until May 22, 2000. The dashed line after this date is based on engineering judgment and extrapolation of the earlier data. The maximum settlement on May 22, 2000, ranges from about 0.24 to 0.30 m (0.8 to 1.0 feet) for the North pad and from about 0.15 to 0.24 m (0.5 feet to 0.8 feet) for the South pad. This is as expected since the North pad is underlain by geonet which would allow dissipation and drainage of excess pore water pressure from the surface of the tailings. This would not be the case for the South pad that is underlain by a geotextile. Monitoring points NET 5 and GEO 5 located closest to the embankment show the smallest settlement. This is due to the influence of the sloping upstream embankment.

The extrapolated settlement data shown by the dashed lines on Figure 6 indicate that the settlements would continue to occur but at a reduced rate. After about two years (August 2001), the majority of the settlement should have taken place. At this time the maximum settlement at NET 3 would be about 0.37 m (1.2 feet) and at GEO 2 and GEO 3 about 0.27 m (0.9 feet). This is in reasonable agreement with a maximum settlement of 0.40 m (1.3 feet) estimated from the settlement analysis. The time rate analysis indicated that it would take four years for 90 percent of the total settlement to occur instead of the two years indicated by the actual field data.

4 COVER DESIGN

4.1 *Cover alternatives*

The overall objective of the tailings facility cover was to minimize infiltration with a cost-effective design. For the graded facility, two basic cover alternatives were considered based on the site's high-altitude desert climate and available on-site cover materials. The *Evapotranspiration (ET) Cover* consisted of a 915-mm (36-inch) plant growth medium layer with an established vegetation community developed after seeding. The *Low-Permeability Cover* consisted of a 305-mm (12-inch) low-permeability material layer overlain by a 460-m to 610-m (18-to 24-inch) plant growth medium layer with an established vegetation community developed after seeding. In order to assess the effectiveness of the cover design to minimize infiltration, water balance modeling was undertaken. A discussion and general results of the water balance modeling are provided below.

4.2 *Water balance modeling*

Water balance modeling of the two alternative cover designs to evaluate the runoff, evapotranspiration, and infiltration values was conducted utilizing the Hydrologic Evaluation of Landfill Performance (HELP) model (Schroeder, et al., 1994). The water balance model runs used site-specific data collected from geotechnical investigations conducted during the construction of the tailings facility, data collected from investigations undertaken as part of the closure investigation, and typical values for the climate and vegetative conditions at the Big Springs site.

Daily precipitation and air temperature data over 15 years from the nearby Wildhorse Reservoir climatological station were used in the HELP model. Daily solar radiation data were sto-

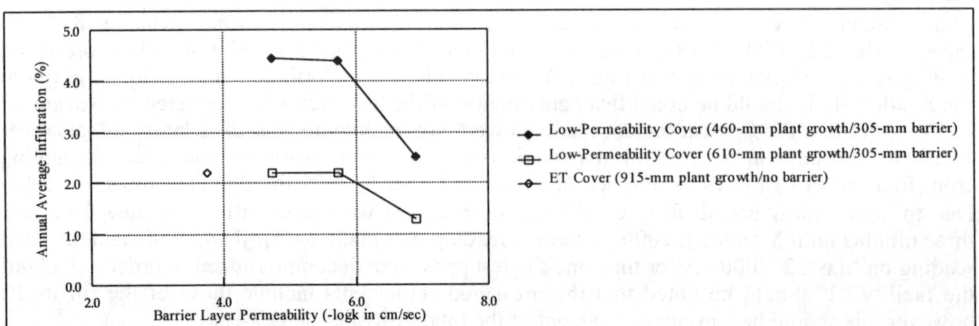

Figure 7. Infiltration rates for various covers predicted by the Help model

chastically generated by the HELP model from precipitation data from the Wildhorse Reservoir climatological station. Quarterly relative humidity and average annual wind speed from the Elko airport climatological station were used. The HELP model was run for a synthetic period of 100 years so that the initial conditions would not influence the model outcome.

Figure 7 presents select results of the water balance modeling for the two cover alternatives. Various thicknesses and permeabilities for the two basic cover types resulted in annual average infiltration rates ranging from 1.3% to 4.4% of the total annual average precipitation of 360 mm (14.1 inches) water equivalent. As shown in Figure 7, the *ET Cover* and the *Low-Permeability Cover* resulted in similar reductions in infiltration rates based on an annual average. Water balance modeling also revealed peak flux infiltration rates that were similar for the two cover types.

Other criteria considered in the selection of the cover included, for example, simplicity and cost of construction and potential long-term impacts to the cover related to such things as frost, desiccation, settlement, and animals. Based on the results of water balance modeling and other criteria, the *ET Cover* consisting of 915 mm (36 inches) of plant growth medium installed over the graded tailings surface was selected for the facility cover.

5 CONSTRUCTION

Reclamation of the Big Springs tailings facility was carried out over a 2.5-month construction period from mid-July through September 2000. Reclamation began approximately ten weeks after the seasonal supernatant pool area had completely evaporated and approximately seven weeks after the beach spray evaporation program was terminated.

Based on the results of 1998 and 1999 field tests, the technical specifications provided for the construction of access roads on 30-m (100-foot) centers within intervening in-fill sections. These access roads, underlain by geosynthetics, would be constructed using fill obtained from the tailings facility embankment and on-site borrow areas. Tailings fill materials would be placed between the access roads and compacted. While this was the prescribed method for grading the facility, the specifications allowed the contractor to select alternative grading methodologies that would meet grading objectives. In order to encourage contractor creativity and contain costs for AngloGold (Jerritt Canyon) Corp., the request for proposal sought lump sum bids for the reclamation project.

Actual grading of the facility was successfully completed with a fleet of scrapers and low ground-pressured tracked dozers. Grading activities began with the construction of a north-south access road along the center of the fill area approximately 275 m (900 feet) from the western embankment. This access road was constructed using tailings excavated, hauled, and placed by scrapers. Due to the stable nature of the underlying tailings and the compacted tailings in the access road, geosynthetics were not deemed to be necessary. From this initial access road, scrapers placed additional tailings and embankment fill materials that were generally pushed and

Figure 8. Scraper excavating tailings on western portion of facility

compacted in lifts (without underlying geosynthetics) onto the adjacent undisturbed tailings by low ground-pressure tracked dozers. Once compacted, scrapers were able to traffic on the fill, thus moving the haul route of the scrapers progressively to the east and west away from the crest of the fill. The trafficability of the placed and compacted tailings and the drained and compacted nature of the underlying tailings eliminated the need for multiple access roads and geosynthetic materials over the fill portion of the tailings facility.

The excavation of tailings on the western end of the tailings facility proceeded smoothly using scrapers (Figure 8). Although scrapers effectively operated on the tailings surface, occasionally tracked dozer support was required to push scrapers out of areas that still had locally saturated tailings. To facilitate further drying of the newly exposed tailings in the western side of the facility, the surface was ripped with a tracked dozer to a depth of approximately 610 to 915 mm (2 to 3 feet) and allowed to dry prior to excavation by scrapers.

Due to unanticipated compaction in the placed tailings, final grading of the facility included using some fill from an on-site borrow area.

The ability of scrapers to excavate and traffic on tailings without the installation of closely-spaced access roads or geosynthetics during the reclamation of the Big Springs subaerial tailings facility provided a significant cost advantage to AngloGold (Jerritt Canyon) Corp. and the construction contractor. This advantage was a direct result of a stable, drained, and dry tailings surface at the time of reclamation.

After final grading was achieved, including an overbuild for anticipated settling, a 915-mm (36-inch)-thick cover was installed using a fleet of scrapers. Cover installation included placing 610 mm (24 inches) of relatively fine-grained soils found at depth within on-site borrow areas followed by a 305-mm (12-inch) thickness of material obtained from plant growth medium salvage areas. The cover was ripped following placement and thickness verification. AngloGold (Jerritt Canyon) Corp. plans to seed the covered facility during fall 2000 using a seed mix based on its experience at the Big Springs site and other operations in the area.

6 CONCLUSIONS

A number of conclusions are presented as follows:

- Subaerial tailings deposition provides a cost-effective method for storing tailings by increasing the overall density of the tailings mass, thereby reducing the overall facility size and construction and reclamation costs.
- Subaerial tailings deposition results in a drained and consolidated tailings mass that is readily amenable to grading and reclamation.
- Field tests conducted prior to reclamation of the Big Springs tailings facility provided invaluable information with respect to the construction procedures and anticipated settlement.
- As predicted by water balance modeling, the selected *ET Cover* at the Big Springs tailings facility offers significant infiltration reductions of meteoric water. This cover consisted of 915 mm (36 inches) of plant growth medium with an established vegetation community.
- Grading of the Big Springs tailings facility using a balanced cut/fill approach consisting primarily of tailings and embankment materials was successfully completed and resulted in significantly less material requirements versus an option that would have used fill only.
- Using scrapers to excavate approximately 4.6 m (15 feet) of relatively coarse-grained tailings on the western end of the tailings facility and deposit them on the relatively fine-grained tailings on the eastern end of the tailings facility without the use of geosynthetics resulted in significant cost reductions.
- Reclamation of the Big Springs tailings facility provides an example of the cost-effective reclamation offered by subaerial tailings facilities.

REFERENCES

Gray, H., 1936. Stress distribution in elastic solids, *Proc. International Conference on Soil Mechanics.* Cambridge, Massachusetts, vol. 2, pp. 157-168.

Knight, R.B. and J.P. Haile, 1983. Sub-aerial tailings deposition, *Proc. 7th Pan-American Conference on Soil Mechanics and Foundation Engineering.* Vancouver, B.C., Canada, June: pp. 627-639.

Schroeder, P.R., T.S. Dozier, P.A. Zappi, B.M. McEnroe, J.W. Sjstrom, and R.L. Payton, 1994. *The hydrologic evaluation of landfill performance (HELP) model, engineering documentation for version 3.* Cincinnati, Ohio: U.S. Environmental Protection Agency, EPA\600\R-94\168b, September, 116 p.

Taylor, D.W., 1948. Fundmentals of Soil Mechanics. John Wiley & Sons, 700 pp.

Ulrich, B., D.R. East, and J. Gorman, 2000. *Subaerial tailing deposition – Design, construction and operation for facility closure and reclamation, Proc. Of the 7th International conference on Tailings and Mine Waste.* Fort Collins, Colorado, January: 29-37.

Tailings and Mine Waste '01 © 2001 Balkema, Rotterdam, ISBN 90 5809 182 1

Grants branch lake: A successful experiment

Donald E. Stump, Jr
Office of Surface Mining, US Department of the Interior, Pittsburg, Pa., USA

ABSTRACT: Grants Branch Lake is now a public recreation facility in Pike County, Kentucky. The site where the lake is today was a coal slurry impoundment from 1986 to 1993. Leaving a coal slurry impoundment as a permanent structure is prohibited by law, however, in this case the transformation was allowed by Federal and State regulators as an "experimental practice". Federal regulations allow the coal industry to request alternate reclamation plans as an experimental practice if specific criteria are met for public safety, land-use, and environmental protection.

The experimental practice monitoring period began in April 1998. Monitoring of Grants Branch Lake water levels, water quality, seepage through the embankment, and embankment stability continued for two years before responsibility for the lake was transferred to Pike County and State Authorities. The lake embankment is 43 meters high from the upstream toe to the spillway crest. The lake has a surface area of 8.4 hectares, contains 1,000,000 cubic meters of water, has been stocked with fish, and has a shoreline of 1,800 meters.

1 INTRODUCTION

Federal regulations do not allow coal waste impoundments to remain as permanent structures as part of the coal mine reclamation plan. However, in 1997 the Office of Surface Mining (OSM) approved a request for an to reclaim a coal waste slurry impoundment as a recreation lake in Pike County, Kentucky (Figure 1). This alternate reclamation plan was allowed as an experimental practice. As an experimental practice an inspection and monitoring period was required before the site could be considered for public use. In June 2,000 OSM determined the experimental practice to be successful and approved the Grants Branch Lake for release to the Pike County Authorities.

2 REGULATIONS

OSM was established by Public Law 95-87, the Surface Mining Control and Reclamation Act of 1977. In Section 711 (Experimental Practices) of the Act it states "In order to encourage advances in mining and reclamation practices or to allow post-mining land use for industrial, commercial, residential, or public use (including recreational facilities), the regulatory authority with approval by the Secretary may authorize departures in individual cases on an experimental basis from the environmental protection performance standards promulgated under sections 515 and 516 of this Act. Such departures may be authorized if (i) the experimental practices are potentially more or at least as environmentally protective, during and after mining operations, as those required by promulgated standards: (ii) the mining operations approved for particular land-use or other purposes are not larger or more numerous than necessary to determine the effectiveness and economic feasibility of the experimental practices; and (iii) the experimental practices do not reduce the protection afforded public health and safety below that provided by promulgated standards." It was under this authority that the request was made to revise the original coal waste slurry impoundment reclamation plan to create a recreation lake.

The experimental practice request focused on allowing departure from two of OSM's regulations (30 CFR 817.83(c)(4) and 30 CFR 817.84). In 30 CFR 817.83 (c)(4) it states that "Following final grading of the refuse pile, the coal waste shall be covered with a minimum of 4 feet (1.2 meters) of best available, nontoxic and noncombustible material, in a manner that does not impede drainage from the under drains." In 30 CFR 817.84 : Coal Mine Waste : Impounding Structures it states that "Such structures may not be retained permanently as part of the approved post mining land use."

3 EXPERIMENTAL PRACTICE

The original plan for the Grants Branch coal slurry impoundment was to use the facility from 1986 to 2005. This schedule would have allowed the entire valley to be filled to the top of the embankment and the elimination of the impoundment. However, due to changes in the local coal production the facility was no longer needed. An alternate reclamation plan was proposed to OSM in January 1995 as an experimental practice.

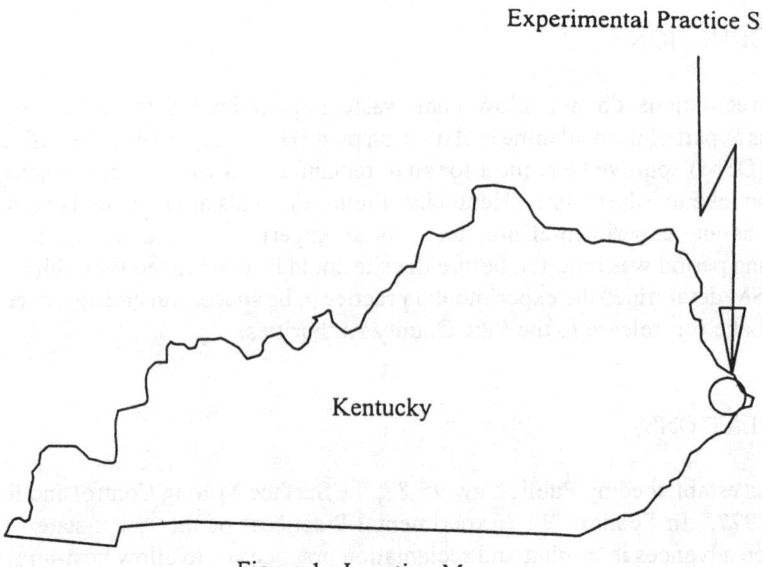

Figure 1 : Location Map

The proposal requested transforming the 66 hectare coal slurry impoundment site to a recreation facility. The facility included a lake having a surface area of approximately 8.4 hectares and containing 1,000,000 cubic meters of water with a shoreline of 1,800 meters, an entrance gate, a caretaker's residence, a maintenance building, paved roads, guardrails, picnic shelter, security lighting system, sidewalks, and fencing (Figure 2).

The proposal also included documents showing local support for the facility, population summaries to show potential users, and the future development plan by the local authorities who would be accepting responsibility for the facility upon successful completion of the experimental practice. Their development plan included construction of basketball/tennis courts, hiking trails, lavatory structure, water lines, playgrounds, volleyball courts, and a fishing pier.

The KY Department of Fish and Wildlife Resources had also proposed to participate in managing the fishery resource in the impoundment including stocking, potential habitat improvements, monitoring water quality in regard to the health of aquatic organisms, and ongoing management of fish species.

The experimental practice was reviewed by OSM, the State of Kentucky and the Mine Safety and Health Administration. Of particular concern to the regulatory agencies was insuring that the proposed facility would be safe for public use, environmentally sound and functional as a recreation facility. The proposal was approved in 1997 with the requirement that the monitoring and inspection period be at least two years before it could be considered for release and public use.

4 MONITORING AND INSPECTION

The two year monitoring and inspection required for the experimental practice covered many aspects of the facility. The monitoring and inspection period was not allowed to start until the impoundment was at its design water level (411 meters). In April 1998 the monitoring began.

The site was inspected every seven day for appearance of structural weaknesses, hazardous conditions, groundwater levels in the embankment, drainage, lake water level, underdrain discharge and embankment seepage.

Water samples were collected quarterly at the inflow into the lake, water in the lake, the discharge from the underdrain,, and the spillway outflow. The parameters analyzed were pH, acidity, alkalinity, conductivity, total iron, total manganese, sulfate, total aluminum, dissolved oxygen.

Quarterly reports were submitted to OSM and the State with lab analyses, field observations, and field work conducted. The final report summarized all the information pertaining to the experimental practice, including general water quality, uses of the lake, stocking activities, existing fish and wildlife species, success of revegetation, and recommendations.

5 RESULTS

The water level of the lake was monitored weekly and showed seasonal variations (Figure 3). The highest water levels occurred in the spring (March and April) and the lowest water levels in the late fall (November and December). The lowest water level recorded was on

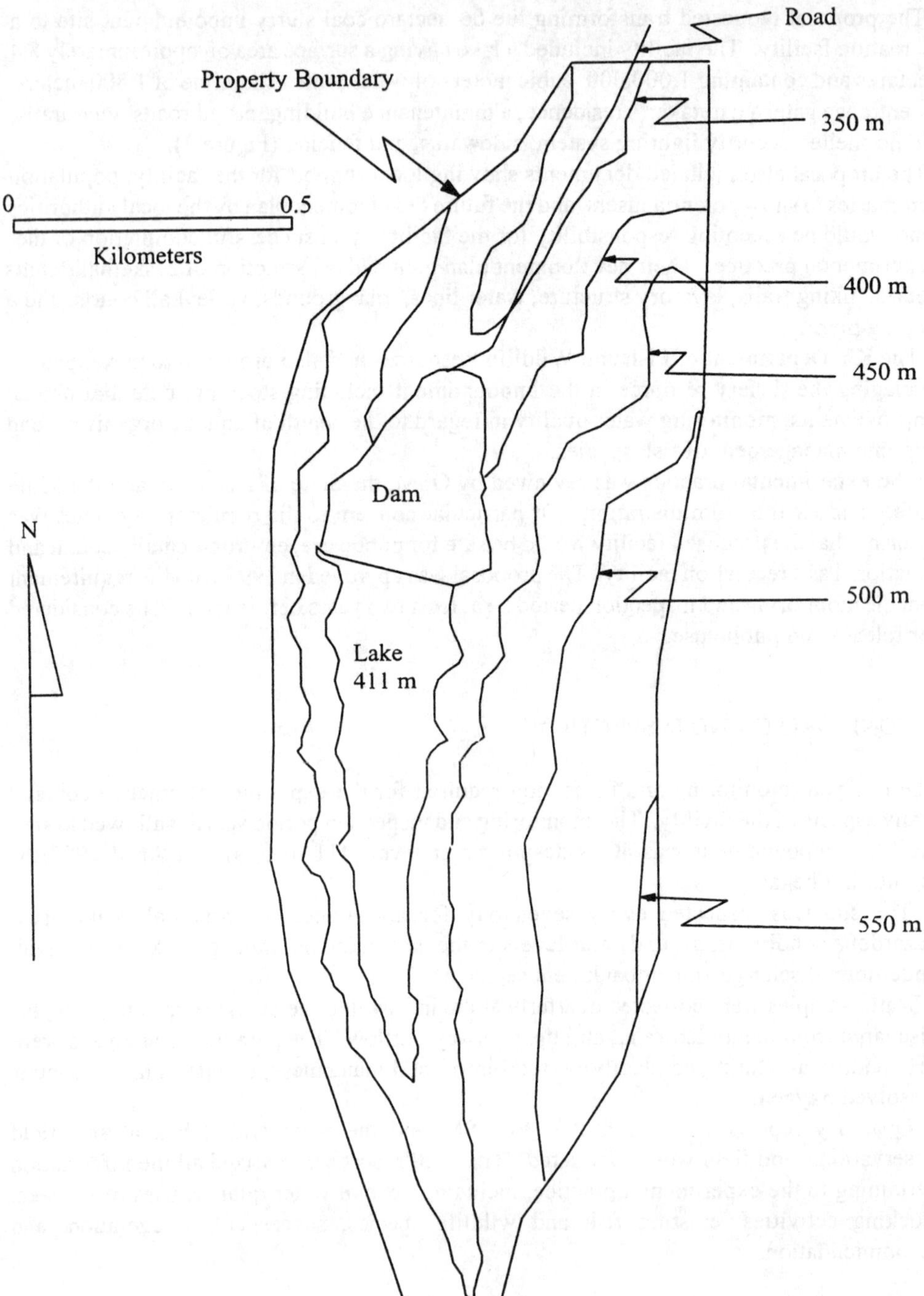

Figure 2 : Plan View

November 12, 1999 with a water elevation of 409.6 meters. This is 1.4 meters below the design water level and occurred after a very dry spring and summer. The water level increases are in response to rainfall with a water rise in the spring of 0.06 meters per day. After the peak in the spring the water levels decrease during the summer and fall at 0.006 meters per day.

Ground water levels in the embankment were measured at five piezometer locations. The measurements show the ground water levels were stable. The discharges from the underdrain and several seepages were also stable and clear.

The fish stocking of Grants Branch Lake was done by the KY Department of Fish and Wildlife Resources. In May 1999, the lake was stocked with 2,500 fingerling largemouth bass. In October 1999, the lake was also stocked with 8,400 fingerling bluegill and 1,008 fingerling channel catfish. Early indications are that the lake is developing into a fishery resource for these largemouth bass, channel catfish, and bluegill.

The water quality monitoring documented that the water is alkaline and meets all state and federal standards. The surface water quality in the lake and discharging was similar to the premining conditions documented in the original coal waste impoundment permit.

6 AWARDS

The Grants Branch Lake and Recreation Facility has already been recognized through its receiving the 1999 Governor's Environmental Excellence Award in Mining Reclamation-Eastern Kentucky; the 1999 Best Reclamation Award from the Pikeville Regional Office (this award is given jointly by the Kentucky Department of Surface Mining Reclamation and Enforcement and the Kentucky Coal Association); and, a Certificate of Honorable Mention from the Interstate Mining Compact Congress' National Mine Reclamation Award program.

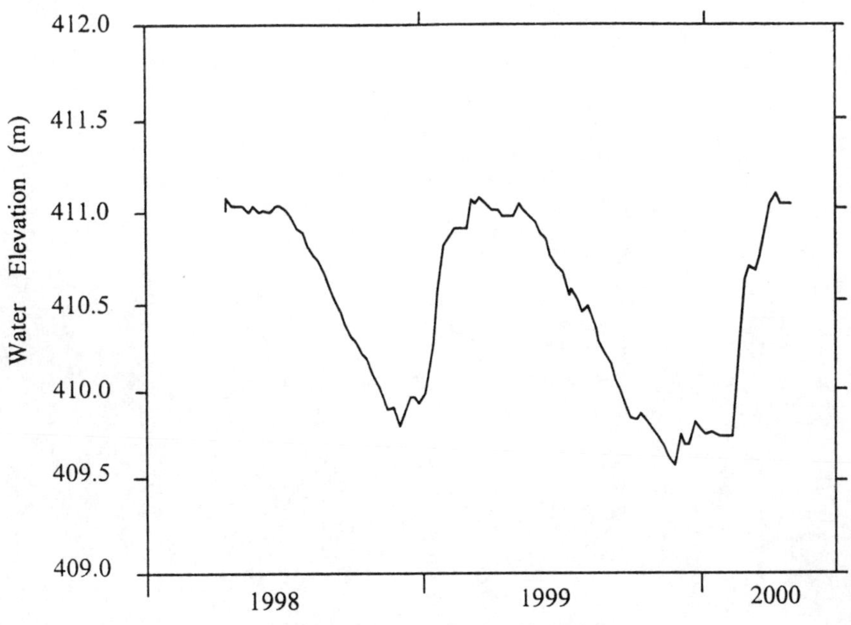

Figure 3 : Lake Water Elevations

7 FINAL DECISION

The final decision involved several regulatory agencies and was based on the success of the vegetation, the suitability of the lake for public use, the release of reclamation bonds, compliance with effluent requirements, and embankment stability. The approving federal agencies were OSM and the Mine Safety and Health Administration, and the approving state agency was the Kentucky Department for Surface Mining Reclamation and Enforcement. To insure the public's safety a requirement of the approval was that the Pike County authorities would accept the responsibilities and liabilities for the lake, and that the lake meets the requirements of the Kentucky Dam Safety Office.

8 REFERENCES

Stump, D.E.Jr. 1999. Reclamation of a Coal Mine Waste Disposal Facility to a Recreation Lake. *Proceedings of the Sixth International Conference on Tailings and Mine Waste '99.* 791-796.

Tailings and Mine Waste '01 © 2001 Balkema, Rotterdam, ISBN 90 5809 182 1

The present condition of the "Schneckenstein" uranium tailings, Germany

T.Naamoun
Institute of Applied Physics, TU Bergakademie, Freiberg, Germany

ABSTRACT: According to the SAG Wismut archive, besides processing ores from the Schneckenstein area, uranium ores from Zobes, Niederschlema, Oberschlema and the area of Thuringia were treated using physical (gravitational and radiometric) and chemical (acid and mainly soda alkaline) procedures in the uranium processing plant called "32 unit". The ore residue was discharged in special settling basins in the Schneckenstein site in the former uranium ore deposit. Thus the resulting tailings will contain material with different physical and chemical properties. One of the aims of this work was to evaluate the different soil physical parameters of the tailing material and to determine their influence on the solubility and transport of contaminants in the tailing environment. In the uppermost tailings, the soil material is found to be moderately permeable, while in the most part it is considered to be slightly permeable. Besides, the total porosity exceeds 30% in all tailings sites and is classified as very high. Therefore the water storage is also high. Moreover, the tailings material is mostly moist. This involves an important interaction between the mineral matrix and the porewaters. Furthermore, a high cation exchange capacity of the tailings material as well as a negligible transport of contaminants in most tailings environment are expected.

1 INTRODUCTION

Producing 216352 t of uranium between 1946 and 1990, the SDAG Wismut became the largest uranium producer in Europe. After the USA and Canada, the uranium production of the former GDR took the third position in the world (Barthel 1993). In the beginning of the ore exploitation, many processing plants were constructed directly near to the mine district. The residues of ore dressing were settled in special plants for the storage of tailings referred to as industrial settling basins in the surroundings of the processing plants. In the beginning of the sixty of the 20th century, many of small uranium mines as well as the corresponding processing plants were breakdown by the Wismut SDAG and the uranium ores were purified in the central processing plants in Crossen and Seelingstädt. On the basis of the lacking protection-precautions before installation of the decantation basins , these put through their potential to radionuclides and heavy metals a danger for the soil, the ground water and the air. After the unification of East- and West-Germany in 1990, the Saxon state ministry for environment and regional development had appropriate approximately 13 milliards marks for the environment problems and is working out and giving licenses for the remediation and rehabilitation of former uranium mining and milling sites (Hurst & Gläser 1998). The industrial settling basins at the Schneckenstein site is an exemplary investigation area.

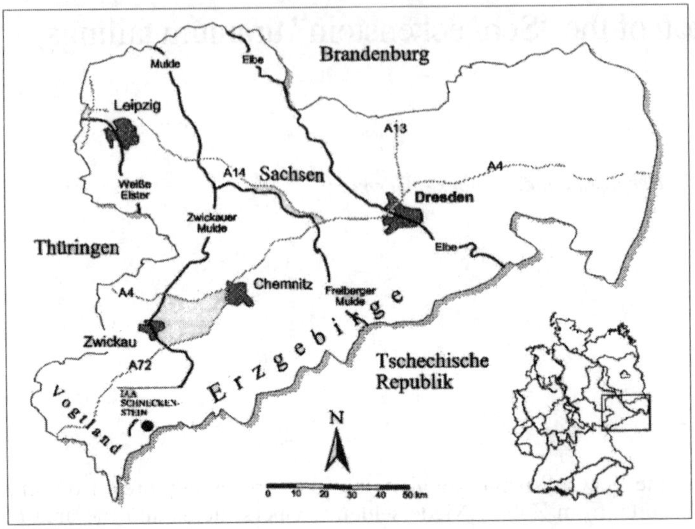

Figure 1. The area of investigation- Uranium tailings Schneckenstein.

2 A SHORT DESCRIPTION OF THE SITE

The Schneckenstein site is located in the southwest of Saxony, in the Boda valley approximately 3 km from the village of Tannenbergsthal/county of Vogtland (Figure 1). The current surface of the site is situated at an altitude of 740 to 815 m above sea level. Concerning the hydrology of the site, the uranium tailings Scheckenstein lie in the area of the watershed between the Zwickauer basin and the Eger. It belongs to the precipitation-richest of the entire Erzgebirge. Depending on the altitude, 960 to 1160 mm precipitation fall annually (I.W. 1958) and the study area receives an average annual precipitation of 1053 mm. The mean annual temperature is 5.5 °C. The calculated evaporation is of 413 mm. From the geological point of view, the area of investigation is located on the Southwest border of the Eibenstock granite. The Eibenstock granite could be considered a biotite-syenogranite and belongs to the group of the younger granites (JG1). It is medium to coarse-grained, serialporphyric and tourmaline bearing. Its intrusion is related to the upper Carboniferous series. The Eibenstock granite is covered with a weathered surface layer. Southwest of the investigation area follows the contact zone with quartz-schist.

3 EXPERIMENTAL

Four sediment cores were taken at the tailings sites by drilling four bore holes at different depths; two bore holes in each tailing respectively (figure 2). The first and the second bore hole (GWM 1/ 96; GWM 2 / 96) were drilled down to the granite foundation in Tailing 2 (IAA I). The third bore hole (RKS 1/ 96, Tailing 1(IAA II) was sunk to a depth of about eight meters,

Table 1. The location of boreholes in the tailings site

Location	Tailing I	Tailing I	Tailing II	Tailing II
Co-ordinate	45 3256 55 8723	45 3258 558727	45 3265 558703	
Depth under the deposit (m)	9.5	20.2	8	12
Filter-tube	1.00 m HDPE	1.00 m HDPE	1.00 m HDPE	1.00 m HDPE
Full-tube	9.00 m PVC	19.00 m PVC	8.00 m PVC	12.00 m PVC

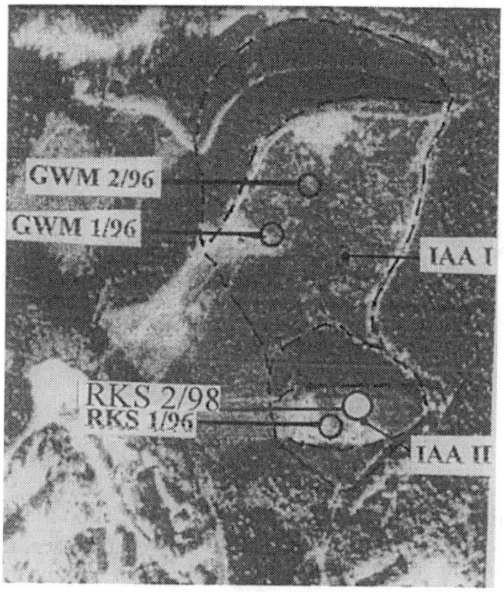

Figure 2. Location of the boreholes in the tailings sites IAA I & IAA II.

whereby the granite foundation was not reached due to technical problems. The fourth bore hole (RKS 2/ 98, Tailing1) is 12 m deep. The cores (diameter 50 mm) were cut into slices of 1 m length and transported in argon filled plastic cylinders to avoid contact with air. In the table 1: , information about field activities are summarised.

4 LABOUR ACTIVITIES (EXPERIMENTAL)

4.1 *Grain-size Analysis*

This analysis was carried out in the laboratory of soil mechanics in the institute of geotechnics according to DIN18 123. The determination of the grain-size-distribution of samples from the tailing sites was obtained by sieve and sedimentation method (wet-sieve analysis before the sedimentation). The soil material in the upper layers of the profile is a mixture of coarse and middle particles therefore it was necessary to use the wet sieve analysis for the part of sample with grain size more than 0,063mm and the hydrometer method (sedimentation) for the part of sample with grain-size less than 0,125 mm implementing the methodology described in the paragraph 6.3.2 of the Din 18 123. For other samples, it was only necessary to involve the hydrometer method because the grain-size of the soil material is less than 0,125mm. The evaluation of the analysis was conducted with the WINSIEB-program.

4.2 *Coefficient of permeability*

The results of particle-size analysis were used to determine the coefficient of permeability in different depths in the tailings sites; thus allowing the prediction of water movement in the soil. The WINSIEB program provides the calculation of this parameter after Hazen.
The permeability coefficient was obtained using the triaxial test carried out according to DIN 18 130 part 1. An undisturbed soil sample about 10 cm of length and 46 mm of diameter from each full casing was pressed with latex envelope and built-in on a basic-plate with 48 mm of diameter. The base of the sample was ventilated with three filter (1 mm, 0,63 mm and wide porous filter-paper). Then the sample was well attached on the pedestal. Between the sample-cover and the sample, three filters with the same specifics sited above were used again. By

dissolving a little quantity of Na-fluoride in the water filling the cell, the leakage of fluid was well controlled Finally an inside and outside pressure are given to the triaxial-cell. The inside-pressure was calculated using the following formula:

$P_i = i*L*0,1$ (1)

P_i - the inside pressure
i - hydraulic inclination
L - the length of the sample given in meter

The outside-pressure is 20% more than the inside pressure. In our case, the triaxial-test demand minimum two weeks for one sample.

The coefficient of permeability (Kf) was calculated according to the following formula:

$K_f = Q*L/t*F*h$ (2)

Q –flow out water volume
L _ the length of the sample given in meter
t _ interval of time given in second
F _diameter of the sample given in square meters
h _ the pressure-height

Since the coefficient of permeability (K_f) is influenced by the temperature of water, the values determined in the tests were converted to a comparison-temperature of 10°C. The calculation is based on the formula from POISEUILLE:

$K_{10} = \alpha * K_T$ (3)

α - correction-coefficient (with 20°C of test-temperature $\alpha = 0,771$)
K_T- the determined coefficient of permeability corresponding to the test temperature.

The different values of the correction coefficient of temperature are given in the following table:

Table 2. α values at different temperatures

Temperature T°C	5	10	15	20	25
α	1.158	1.000	0.874	0.771	0.686

The intermediate values can be linearly calculated.

4.3 *Water-content determination*

The water-content was computed as the following:

$\omega = (M_\omega / M_s) * 100$ (percent); (Bowles 1992) (4)

where :
M_ω - mass of water present in the soil mass.
M_s - mass of soil solids.

In order to determine the water-content in different samples, 10 g of soil material for the silt or clay material and up to 200g for sand or gravel, were taken at intervals of varying depths. The soil material is dried in an oven by a drying temperature of 105 ± 5°C to a constant mass. Due to the evaporation of water with time, the mass consequently reduces with successive weighing. Normally, several weighing are carried out to determine if a constant mass state has been obtained, but in these experiments the soil was dried for a period of 48 hours, some times more assuming that the constant mass state is reached. In order to prevent the adsorption of water from the laboratory atmosphere, the can of dry soil must be placed in a desiccator as soon as it is cool enough to handle and weigh. The testing procedure fulfilled the standards set by the German Institute for standardisation; DIN 18 121 Part 1.

4.4 *Specific gravity by capillar pyknometer method*

The specific gravity of soil solids ρ_s is defined as the mass of the solid singles-component divided by the volume V_k of the singles-component.

$\rho_s = m_t \ / \ V_k$ (g/cm^3) (5)

m_t mass of soil solids after drying at a temperature of 105°C.

Approximately 20g of well-pulverised and dried at a temperature of 105°C soil sample was weighed and then filled in a pyknometer. A quantity of distillate water was added until about 2cm above the sample-surface. In order to ventilate the sample, it was boiled for about 45 min. Subsequently the pyknometer was cooled at the room temperature and filled with distillate water up to the brim before it was closed with the plug and weighed. Finally, the temperature of the pyknometer-content ϑ was fixated.

The V_k was calculated using the following formula:

$V_k = V_p \vartheta - V'_w \vartheta$ (cm^3) (6)

where:

$V_p \vartheta$ is the pyknometer volume at the temperature ϑ

$V'_w \vartheta$ is the water volume at the temperature ϑ

The volume of the pyknometer at the temperature ϑ can be determined using the following temperature-volume-graph:

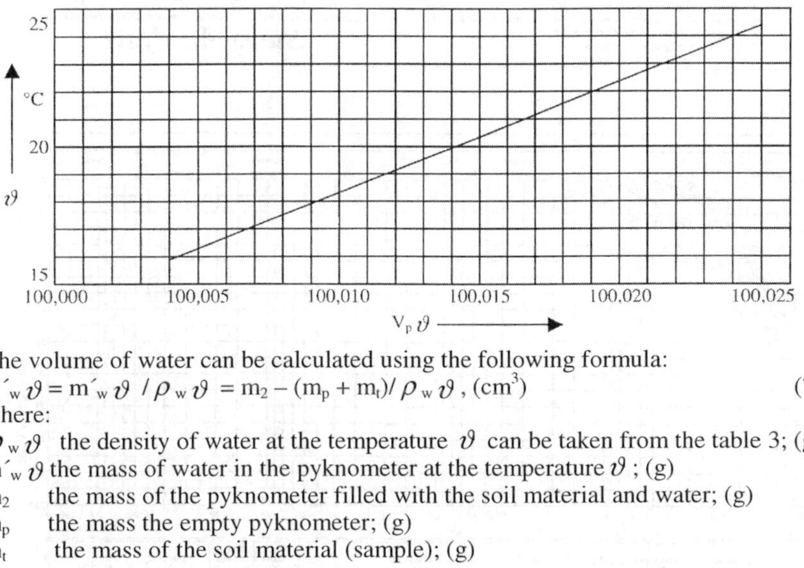

The volume of water can be calculated using the following formula:

$V'_w \vartheta = m'_w \vartheta \ / \ \rho_w \vartheta \ = m_2 - (m_p + m_t) / \rho_w \vartheta$, (cm^3) (7)

where:

$\rho_w \vartheta$ the density of water at the temperature ϑ can be taken from the table 3; (g/cm^3)

$m'_w \vartheta$ the mass of water in the pyknometer at the temperature ϑ ; (g)

m_2 the mass of the pyknometer filled with the soil material and water; (g)

m_p the mass the empty pyknometer; (g)

m_t the mass of the soil material (sample); (g)

Table 3. The water density at different temperatures

ϑ / °C	ρ / (g/cm^3)	ϑ / °C	ρ / (g/cm^3)	ϑ / °C	ρ / (g/cm^3)
15.0	0.99913	18.5	0.99853	22.0	0.99780
15.5	0.99905	19.0	0.99843	22.5	0.99768
16.0	0.99897	19.5	0.99833	23.0	0.99757
16.5	0.99888	20.0	0.99823	23,5	0.99745
17.0	0.99880	20.5	0.99813	24.0	0.99732
17.5	0.99871	21.0	0.99802	24.5	0.99720
18.0	0.99862	21.5	0.99791	25.0	0.99707

The procedure described above was accomplished according to the German institute for standardisation; DIN 18 124

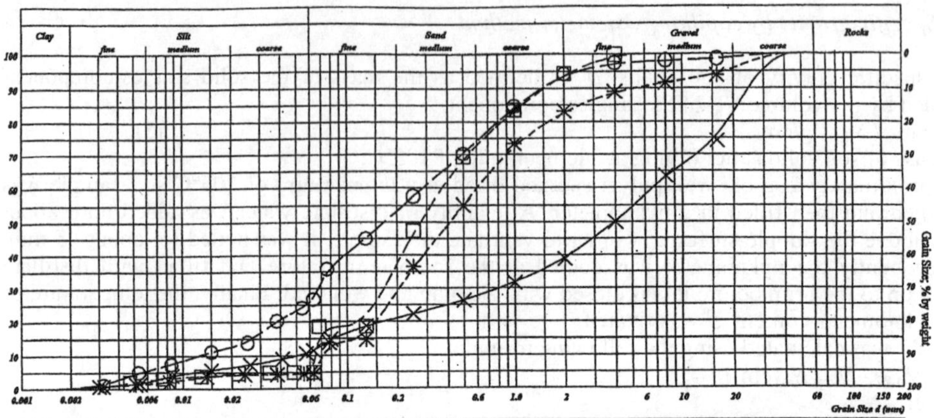

Uranium Tailings Schneckenstein	Labratory No.	Outcropp No.	Depth (m)	Line	Soil Name	Water Content (%)	Grain Density (g/cm^3)	Permeability (m/s)	VGl %	VCa %
	III/1	BPIII	0-1m	✕–✕	G,cs',ms',fs',cl'	7.13	1.77	1.9E-5	2.44	0.63
	III/2	BPIII	1-2m	⊖–⊖	S,cl	12.87	1.9	2.7E-7	4.19	2.89
Borehole N°3	III/3	BPIII	2-3m	⊟–⊟	S,fg',cl'	19.81	1.51	5.1E-5	2.42	0.49
	III/4	BPIII	3-4m	✱–✱	S,fg'	31.91	1.76	7.3E-5	6.73	3.31

Taoufik Naamoun, Dipl. Geophysicist
Institute of applied physics
TU Bergakademie Freiberg

Sieve Analysis

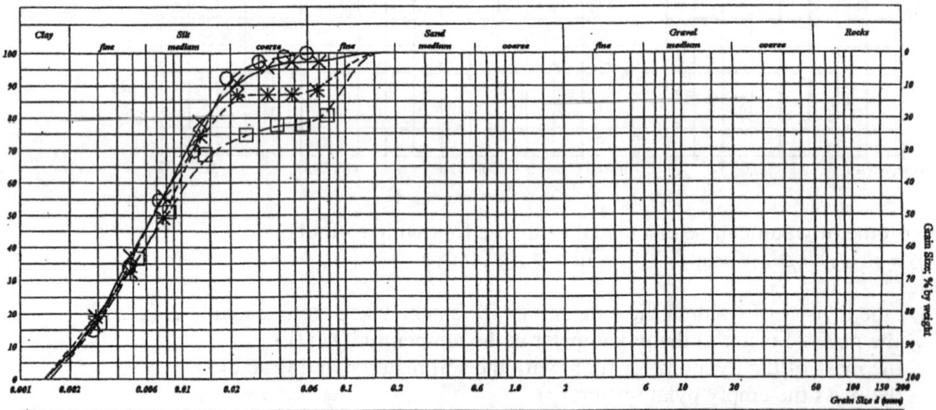

Uranium Tailings Schneckenstein	Labratory No.	Outcropp No.	Depth (m)	Line	Soil Name	Water Content (%)	Grain Density (g/cm^3)	Permeability (m/s)	VGl %	VCa %
	III/5	BPIII	4-5m	✕–✕	Cl	36.2	1.4	2.0E-8	6.68	2.70
	III/6	BPIII	5-6m	⊖–⊖	Cl	30.84	1.55	2.1E-8	4.57	1.67
Borehole N°3	III/7	BPIII	6-7m	⊟–⊟	Cl,fs	31.86	1.5	6.9E-8	5.66	1.58
	III/8	BPIII	7-8m	✱–✱	Cl,fs'	28.71	1.56	1.8E-8	5.73	3.08

Taoufik Naamoun, Dipl. Geophysicist
Institute of applied physics
TU Bergakademie Freiberg

Sieve Analysis

Figure 3. The grain-size distribution; Borehole N°3.

4.5 *Determination of soil density*

The calculation of soil density and related parameters are based on the following mathematical equations:

ρ_{wet} = mass of soil sample (m_w) / volume of the container (V_c) (8)

If the water content ω of the soil material is well known, the dry unit density of the soil material is:

$\rho_{dry} = \rho_{wet} / 1+\omega$ (9)

(the water content is determined after the DIN 18 121 part1)

The dry mass is calculated as:

$m_d = m_w / 1+\omega$ (10)

The water volume is calculated as:

$V_w = (m_w - m_d)/ \rho_w$ (11)

The dry volume is calculated as:

$V_d = m_d / \rho_s$ (12)

(ρ_s is the specific gravity determined after the DIN 18 124)

The pore volume is calculated as:

$V_p = V_c - V_d$ (13)

The total porosity is calculated as:

$n = V_p / V_c$ (14)

The saturated porosity is computed as:

$n_s = V_w / V_c$ (15)

The degree of saturation is calculated as:

$S_r = V_w / V_p$ (16)

The wet density is determined after the DIN 18 125 Part1.

A cylinder with 50 ml of volume was used to determine the soil density in each depth interval. The cylinder was filled with well pressed soil material and its weigh was determined using a scale with an error correction up to 0,1%. Each concerned parameter was determined using the corresponding formula mentioned above.

5 RESULTS AND INTERPRETATION

5.1 *Grain size distribution*

The results of particle-size analysis are given in the form of a series of fractions by weight of different size grades. These fractions are expressed as a percentage of the whole sample and are generally summed to obtain a cumulative percentage. Cumulative curves are then plotted on a semi-logarithmic paper to give a graphical representation of the particule-size distribution. The slope of the curve provides an indication of the degree of sorting (Bell 1992). Moreover, a detailed description of the labour investigations of each sample is given in special tables. Further, as in example, the grain size distribution of each examined sample from the third borehole is well illustrated in figure 3.

Besides, the soil material in the tailings sites can be described as well graded, as the value of the coefficient of uniformity (U) exceed 2 in all analysed samples especially in the uppermost tailings and the two last samples in the fourth borehole where (U) is found to be in many times more than in the other depth intervals indicating the large range of particle size in the mentioned samples in relation to the others, as the higher the coefficient of uniformity, the larger is the range of grain sizes(Bell 1992). On the other hand the coefficient of curvature (C) is found to be between 1 and 3 in most tailings confirming the same deduction. Furthermore, after the united soil classification (Wagner 1957; Anon. 1981), in the upper most tailings and the two last samples from the fourth borehole, the soil material is found to be classified as a mixture of a wide range of material types; from silt to cobbles that means that the size of grains is in a wide range between 0.006 and 200 mm while in the other part is classified as a mixture of only silt and fine sand and the size of grains is in the range between 0.006 and 0.2 mm.

The approximately uniformity of the grain sizes in the deeper depth intervals and their little sizes can be considered as a witness of the milling processing used for uranium leaching in order to increase exposure to chemical attack, and probably the alkaline method was mostly applicated as the acid leaching does not need this kind of preparation (IAEA. 1993) and this is in good agreement with the Wismut archive (Wismut 1955).

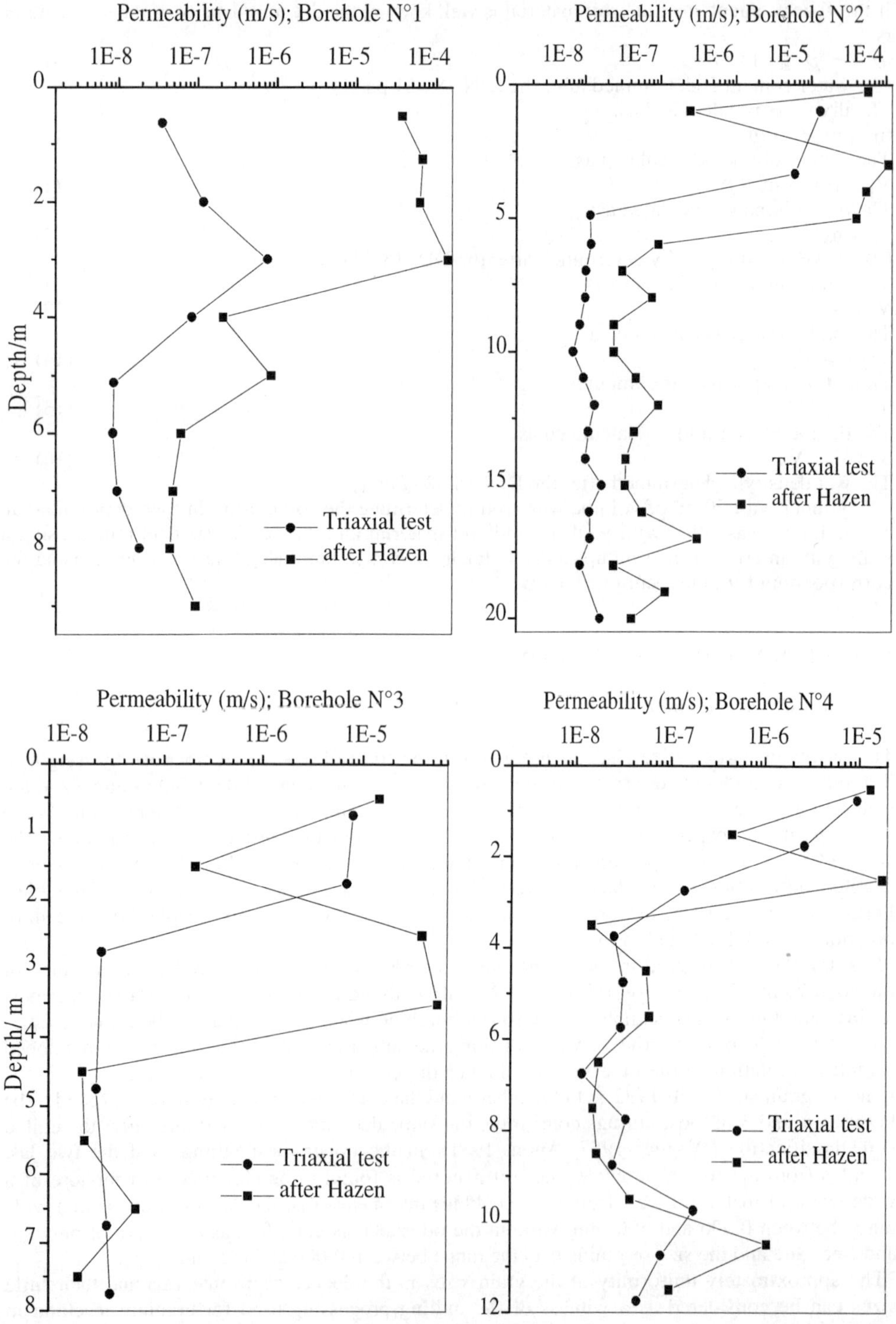

Figure 4. The coefficient of permeability by means of two methods.

From the chemical point of view, an increase of the cation exchange capacity of the tailings material is expected as the milling process increase the number of broken bonds around the edges of silica-alumina particles which are considered as a function of grain size; the smaller the particles, the larger the exchange capacity, and the irregularities in the lattice structure (lattice distortions) as well, providing a greater number of unsatisfied valences at surfaces and edges of the soil units (Degens 1965).

5.2. *The coefficient of Permeability (K_f)*

Since the K_f value after Hazen is directly proportional to the grain size of soil particles and $K_f = 116d^2_w$, where d_w is the effective diameter of grains (Kezdi 1969) and such as in the upper most tailings, the size of the soil material is found in a wide range (from silt until cobbles), the K_f values are found to be in many times higher than in the other depth intervals and its values are close to 10^{-5} m/s and after Anon. (1977), the soil material is considered as moderately permeable. For the deeper depth intervals, the K_f values are in the range between $x*10^{-8}$ and $y*10^{-7}$ except the two last intervals in the fourth borehole, and after the same classification they are considered as slightly permeable.

As demonstrated in figure 4, the K_f values became from the Triaxial test are different to those calculated by Hazen formula, and this mainly due to the heterogeneity of the tailings material and the influence of the procedure changes.

5.3 *The dry density*

The dry density is found to be less than 1.8 in all tailings and it is classified as very low after Anon. (1979).

5.4 *The grain density*

The values of the grain density are very close and exceed 2.7g/cm³ in most tailing, probably is due to the high concentration of mica or iron in the tailings material (Bowles 1992).

5.5 *The total porosity*

The total porosity seems to be dependent of the particle size distribution in the third and fourth boreholes and it increases as the grain size decreases but in the other boreholes this fact is not clear. The mentioned soil property is considered as very high (Anon. 1979) as its values is over 30 % in all tailings sites and maybe the milling process before uranium leaching carried the increase of the grains surface area ratio to their volume. On other hand, depending of the soil classification, the measured values are in a good agreement with literature (Maidment 1992).

5.6 *Degree of saturation (S_r)*

In the uppermost tailings and in the two last samples from the fourth borehole, the soil material is damp as the degree of saturation is in the range between 0.26 and 0.50. In the most other samples the tailings material is moist and S_r is in the range between 0.51-0.75 and rarely wet (Terzaghi & Peck 1967). Besides, most of the soil physical properties of the third core are illustrated in figure 5.

Besides, the change of some physical soil properties values sited above along the four cores is probably in relation with the mineralogy change of the tailings material from site to site.

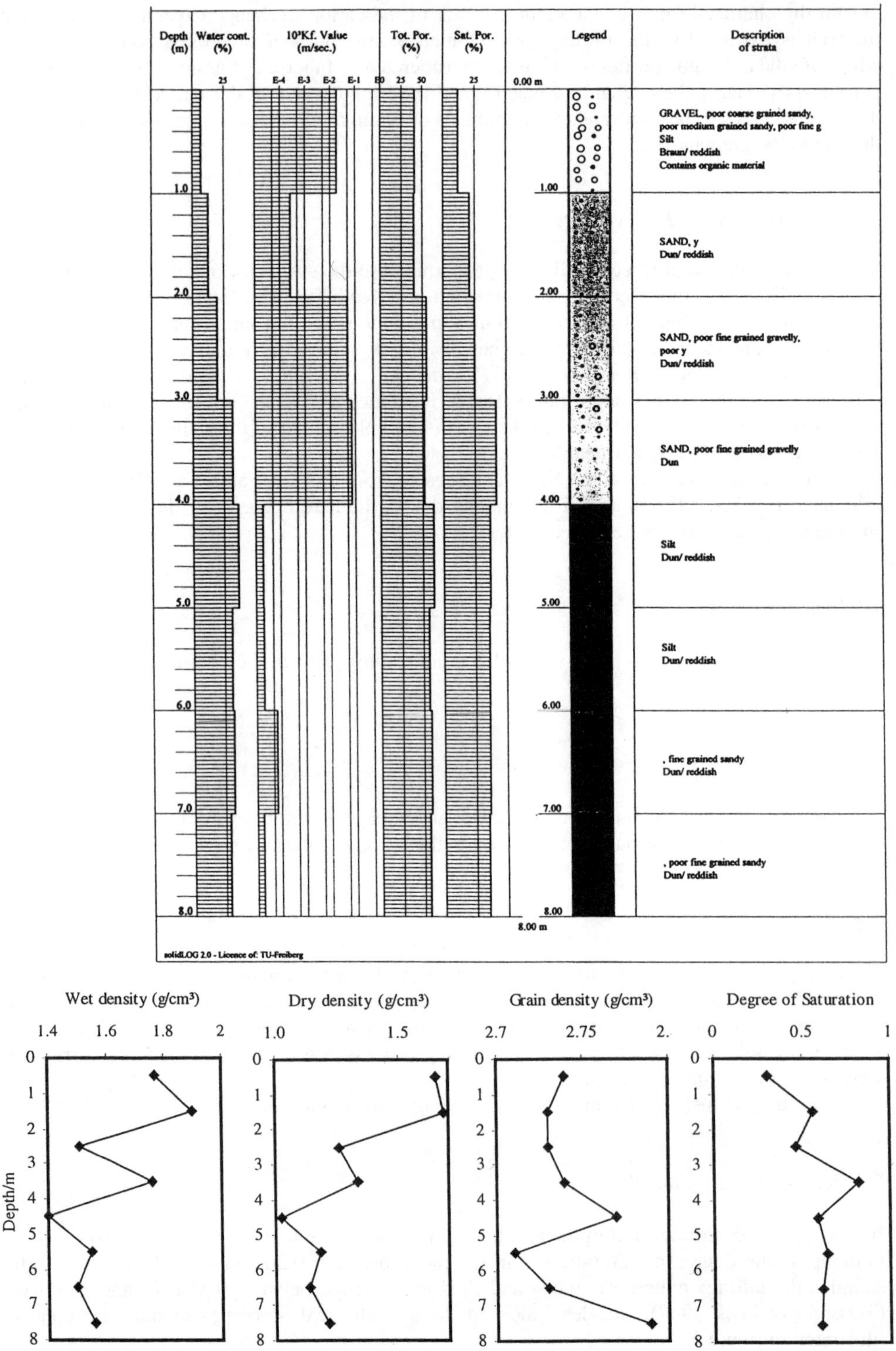

Figure 5. Different physical soil properties; Borehole N°3.

6. CONCLUSIONS

Although the uranium tailings of Schneckenstein is a considerable „reservoir“ of contaminants, as well as an important menace for the environment especially for the groundwaters. The most part of the tailings is found to be „a dead body“ for a short term monitoring, while the probable transport of contaminants is expected to be in relation to the upper part of the tailings and perhaps does not exceed a depth of the first three meters.

REFERENCES

Anon., 1977. The description of rock masses for engineering purposes, working party report, Q.J. Engg Geol., 10, p. 355-88.

Anon., 1979. Classification of rocks & soils for engineering geological mapping, part I: rock & soil materials, Bull. Inst. As. Engg. Geol., N°19, p. 364-71.

Anon., 1981. Code of practice on site investigation, BS5930, British standards institution, London.

Barthel, F.H. 1993. Die Urangewinnung auf dem Gebiet der ehemaligen DDR von 1945 bis 1990. Geologisches Jahrbuch A142. Hannover, Germany.

Bell, F.G. 1992. Engineering properties of soils and rocks, third edition, Butterworth Heinemann.

Bowles, J.E. 1992. Engineering properties of soils & their measurements, fourth edition McGraw-Hill, Inc.

Degens, E.T. 1965. Geochemistry of sediments, A brief survey, Prentice-Hall, Inc.

Hurst, S. & Gläser, K. 1998. Uranbergbausanierung-eine Herausforderung an die Fachgremien, Uranium Mining and Hydrogeology II, Proc. of the intern. Conference and workshop, Freiberg, Germany, Verlag Sven von Loga, Köln.

IAEA., 1993. Uranium extraction technology, technical reports series N°359: p. 75.

Kezdi, A. 1969. Handbuch der Bodenmechanik, seite 128, VEB Verlag Berlin.

Maidment, D.R. 1992. Handbook of hydrology, 6.9, McGraw-Hill, Inc.

Terzaghi, K. & Peck, R.B. 1967. Soil Mechanics in Engineering Practice, p. 26, Wiley.

Wagner, A.A. 1957. The use of the unified soil classification system for the bureau of reclamation, Proceedings 4th international conference of soil mechanics foundation Engg, 1, p. 125-34, London.

Wismut archiv, 1955. Razrabotka mestarazdenia uranovi rud sovietska germanskim akcionernim obchestvom Vismut za 1946-1955 godi.

Tailings and Mine Waste '01 © 2001 Balkema, Rotterdam, ISBN 90 5809 182 1

Large-scale testing of co-disposal of coarse waste rock and tailings through pumping

Å. Sundqvist & S. Sundqvist
Luossavaara-Kiirunavaara AB, Research and Development, Process Technology, Malmberget, Sweden

ABSTRACT: LKAB needs to find more space for disposal of waste rock in a near future and to ensure cost-effective transportation of waste rock. The waste rock to be disposed of amounts to some 17 Mtonnes per year in total. In 1996 a project was started to investigate future handling of waste rock. Since then, coarse particle (-60 mm) pumping and co-disposal has been tested in a large-scale test facility, with focus on energy efficiency, wear, operational reliability, disposal properties and environmental issues. Tests have shown that pumping of a mixture of tailings and waste rock may be an energy-efficient means of transportation. Furthermore, the co-disposal method seems to produce a deposit with good properties in terms of stability and landscaping possibilities. The economic feasibility of a co-disposal system based on pumping will, however, be governed also by other factors, such as wear properties of the material to be transported.

1 INTRODUCTION

Luossavaara-Kiirunavaara AB (LKAB), a Swedish iron ore mining company, needs to dispose of waste rock in new areas in Kiruna and Malmberget in a near future, and is also interested in improving the cost-effectiveness of transporting the waste rock (Sundqvist, 2000). In 1996, a project was started in Malmberget with the purpose of investigating future handling of waste rock. The transportation techniques to be investigated were trucking, belt conveying and pumping.

2 CURRENT HANDLING OF WASTE ROCK

Wet-handled fine-grained waste rock from the concentrators (tailings) is currently transported by pumping in pipelines or by gravity in flumes to tailings ponds, from which water is recirculated back to the ore treatment processes (Sundqvist, 2000).

The dry handled coarse-grained waste rock is transported by trucks to waste rock deposits. In Kiruna, the waste rock is disposed of in heaps built in stages, where each stage is approximately 30 m high. In Malmberget the waste rock is handled in a similar way, although the waste rock is also used for backfilling of old open pits. Smaller tonnages are also used for internal construction and road work, or sold for other usage.

The waste rock to be disposed of in Kiruna and Malmberget is expected to reach some 17 Mtonnes/year in the years ahead, see also Table 1.

Table 1. Expected tonnages and particle sizes of waste rock in Kiruna and Malmberget (after Sundqvist, 2000).

	Tonnage (Mtonnes/year)	**Particle size** (mm)
Kiruna		
Wet handled waste rock	1.8	0.0-2.5
Dry handled waste rock (1)	3.1	0.0-30.0
Dry handled waste rock (2)	5.1	0.0-150.0
Total	10.0	0.0-150.0
Malmberget		
Wet handled waste rock	1.7	0.0-4.0
Dry handled waste rock (1)	2.3	0.0-60.0
Dry handled waste rock (2)	3.0	30.0-120.0
Total	7.0	0.0-120.0

3 PRELIMINARY STUDY

A preliminary study involved literature reviews and discussions with various suppliers of belt conveyors and slurry pumps (Sundqvist, 2000). Site visits to coal mine operations in Australia showed that pumping of a mixture of coarse-grained waste rock and tailings was a practicable method. During the site visit, the seemingly low operational (energy) costs of pumping and the properties of the deposited material were impressive. The results of the preliminary study indicated that both pumping and belt conveying were of continued interest.

At that time, while belt conveying was considered to be a known transportation method, less experiences were documented on coarse particle pumping. Before entering into a second stage of the investigation, a decision was taken to perform loop-tests with the LKAB waste rock.

Loop-tests were carried out at Georgia Iron Works (G.I.W.) Hydraulic Institute, U.S.A., in December 1997 in a 300 mm pipeline with –60 mm waste rock from both Malmberget and –100 mm waste rock from Kiruna (Sundqvist, 2000). The loop-tests encompassed slurries made up of approximately 50% by mass of tailings and coarse waste rock, respectively, and the slurries were pumped at solids concentrations of approximately 40-60% by mass. The G.I.W. tests gave positive results with respect to the energy input requirement of pumping. Due to limitations in total time of testing, a proper evaluation of wear in pumps and pipes could not be carried out. However, the tests indicated that wear would be heavy to severe.

At this stage, LKAB decided that there was a need for tests on a large-scale basis (Sundqvist, 2000). The preliminary study had not been able to fully answer questions related to maintenance costs (wear), operational reliability or whether this application was fit for use in cold climates. There was also a need for evaluating the possibilities of the co-disposal technique for LKAB, since previously reported co-disposal operations had not involved iron ore waste materials. In early 1998, a decision was therefore taken to build a large-scale test facility for coarse particle pumping and co-disposal.

The test facility was placed in Vitåfors in the Nalojärvi area, which is currently used for tailings disposal. The plant was designed, erected and commissioned during the spring and summer, and was ready for operation in August 1998. Pumping and disposal tests were then carried out during one year (August 1998-August 1999). An internal final report on the tests was issued in the spring, 2000.

4 LARGE-SCALE TEST FACILITY

4.1 *Test facility*

The test facility consists of a tank for mixing of tailings and dry handled waste rock, a Warman 12/10 FF-AH heavy duty slurry pump with a 355 kW variable speed electric motor and a 250 mm i.d. pipeline, essentially horizontal and approximately 280 m long. A second pump is available for pumping at higher static lifting heights.

Dry waste rock from the sorting plant was transported by truck to the test site, and fed into the mixing tank via a bin, vibration feeder, belt conveyors and a protective screen (separation at 60 mm). Tailings slurry was diverted from the ordinary tailings pipeline and fed to the mixing tank via an intermediate tank, designed for diversion of requested parts of the flow into the mixing tank, the remaining portion of the flow being discarded to the tailings pond. All make-up water and gland water was taken from the adjacent tailings pond via submersible pumps.

In order to evaluate pipeline wear, the discharge pipeline was constructed with pipe sections (straight pipes and bends) from several suppliers, internally lined with different wear materials such as ceramic materials, polyurethane and combinations thereof, high chromium white cast iron, fused cast basalt, fused corundum and aluminium oxide. Ordinary steel pipes were used for comparison. Different types of slurry valves were also used and compared during the tests. Pump wear was measured for two different types of high chromium white cast irons, and for two different hydraulic designs of the impellers.

4.2 *Waste rock disposal*

The discharge end of the pipeline was located 4 m above ground level, and initially the slurry exiting the pipeline was free to flow downhill without any outer constraints. An energy dissipater in the form of a wear plate was attached to the outlet of the last pipe section. These precautions were taken in order to minimize erosion from the slurry as it hit the ground. However, a few months of operation showed signs of segregation between the finest particles and the remainder of the material. A decision was therefore taken to constrain the outflow of material, the reasons being to protect the water quality of the adjacent tailings pond, to see whether it was possible to collect more of the finest particles and also to see whether the height of the deposit could be raised without causing stability problems.

A so called seepage dam was therefore built in the mid-winter, thus limiting the growth of the deposit on the downstream side. The dam was constructed from predominantly coarse (-120 mm) waste rock from the sorting plant. The first dam was approximately 3 m high. When the dam had been almost filled up, the dam was raised another 3 m. A schematic crossection of the dam is shown in Figure 1.

5 MEASUREMENTS

The test facility is instrumented so that mass and volumetric flow rates, slurry density, pressure/differential pressure, temperature, power, pump torque and rotational speed can be measured or evaluated. All instrumentation is linked to a local control system, and data acquisition

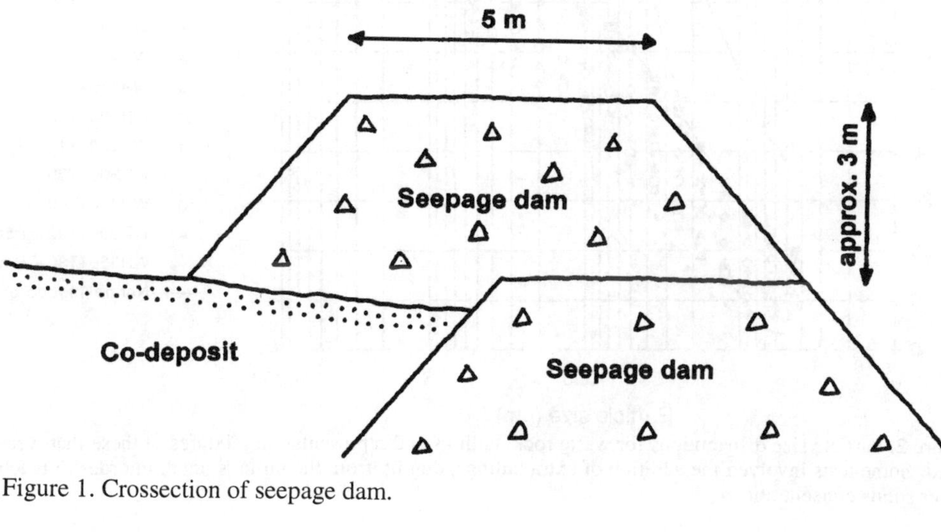

Figure 1. Crossection of seepage dam.

can be made through the ordinary computer process monitoring system. The system also enables measurement and calculation of hours of operation.

Sampling of incoming waste rock and tailings was carried out for evaluation of solids densities, particle size distributions and chemical composition. The instrumentation and sampling together has thus allowed for calcuations of incoming and discharged volumetric flow rates and slurry densities, of pipeline friction and of pump performance.

Wear was measured by weighing of pipes and bends, in combination with measurements of changes in inner diameter of the pipes and visual inspections. Wear on the pumps was measured as changes in inner diameters or construction thicknesses in locations settled beforehand.

In April 1999, a geotechnical survey of the co-deposit was performed (CPT sounding, disturbed sampling). On this occasion, standpipes were installed in the deposit, so that water levels could be monitored for the remainder of the test period. The site investigation was later supplemented with the digging of test pits in the seepage dam, and with stability calculations. Measurements of slope inclinations were carried out on several occasions, and sampling of the deposited material was made in order to establish the density of the deposited material and the variations in particle size distribution and chemical composition of the waste material with distance from the discharge point. In addition to this, load trials on the deposited material were carried out using a loader of type VOLVO L-120.

Water quality was monitored during the whole test period in order to evaluate the impact of adding dry material to the tailings slurry. A control programme set up before the tests started guided the sampling and analyses of the quality of all water coming into the test facility as well as into the tailings pond, secondary sedimentation basin and recipient, the Lina River. In addition to this, tailings slurry and make-up water entering the pump, and slurry discharged from the pump, were sampled on several occasions.

6 SOME RESULTS

6.1 *Test cases and material properties*

In total, the number of operational hours of pumping waste rock and tailings amounted to some 1700 hours. About 427 ktonnes of waste rock has been pumped, of which 325 ktonnes were dry

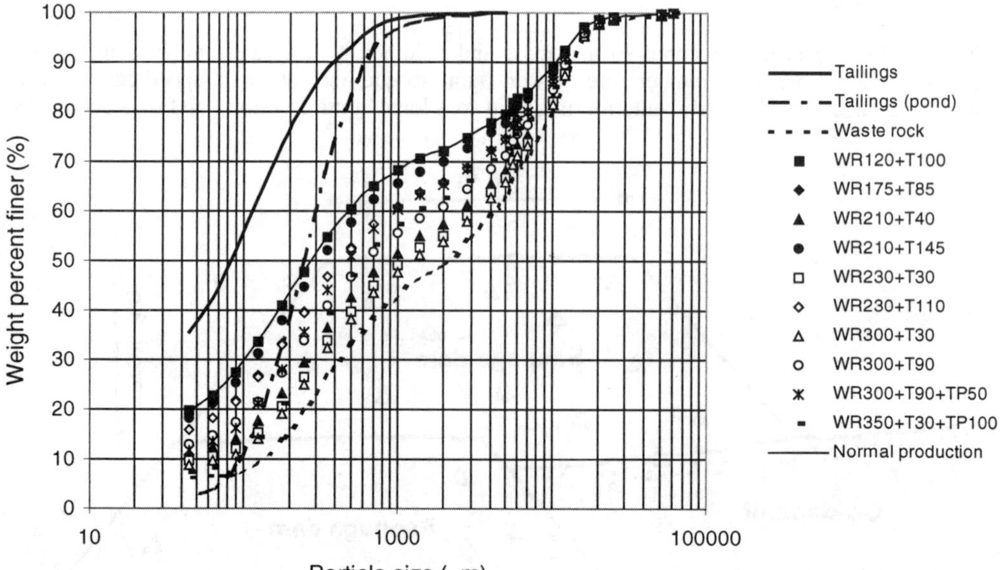

Figure 2. Particle size distributions for waste rock, tailings and representative mixtures of these that were tested. Some tests involved the addition of extra tailings, dug up from the tailings pond, in order to reach higher solids concentrations.

–60 mm waste rock and 102 ktonnes –4 mm tailings.

The solids concentration varied between 25-37 percent by mass most of the time. Some tests also involved the addition of extra tailings dug up from the tailings pond, in order to reach higher solids concentrations, >50 percent by mass. The velocity in the discharge pipeline varied between 3.7-5.3 m/s.

The solids density of the tailings was about 3200 kg/m^3, and about 3000 kg/m^3 for the coarse waste rock. Particle size distributions for the coarse waste rock and tailings are shown in Figure 2. Some sample size distributions representing the various test cases are also shown.

In Figure 2 above, the denotation "WR120 + T100" means that the slurry mixture was made up of 120 tonnes/hour of coarse waste rock and 100 tonnes/hour of tailings. The denotation "TP50" in Figure 2 above, means that extra tailings from the pond, at 50 tonnes/hour, was added to the mixture. "Normal production" denotes the particle size distribution for a case where the nominal production of –60 mm waste rock and tailings is mixed in a production-scale application.

6.2 *Operational reliability*

The operation of the test facility experienced disturbances of different kinds. A majority of these disturbances were, however, related to the auxiliary systems (feeding systems for make-up water and coarse waste rock etc) or to external factors (shortage of tailings and waste rock etc). Some 3-5% of the downtime was related to operational stops due to wear problems in the pump or pipes. Pipeline blockages were possible to avoid after final commissioning of the control system.

During the winter, operation of the pipeline took place at temperatures down to –45 °C, with the addition of make-up water of about 0 °C temperature. The transportation method as such was considered to have shown good operational reliability during the tests.

6.3 *Pumping*

Operation of the pipeline was considered to be rather safe for velocities down to about 4.0 m/s. For lower velocities, down to 3.7 m/s, the slurry appeared to be pulsating in the pipeline with possible risks for pipeline blockage. The minimum velocity required for safe and energy-efficient operation was, however, concluded to be about 4.5 m/s, which was higher than the minimum velocity estimated from the loop-test.

Measured values of pipeline friction losses and reductions in head and efficiency for the pump (due to the presence of solids) showed rather good agreement with those measured in the loop-tests at velocities ≥4.5 m/s. If a normal production of tailings (about 230 tonnes/h) and –60 mm waste rock (about 270 tonnes/h) from Malmberget is to be pumped as a mixture at 4.5 m/s in a horizontal 250 mm i.d. pipeline and at a solids concentration of approximately 44% by mass, the energy required to move the solids in the pipe (specific energy consumption) would, based on the results, be approximately 0.35 kWh/tonne, km (excluding consideration of motor and pump efficiencies).

6.4 *Co-disposal*

The successive build-up of the co-deposit, showed that the slurry stream "designed" a rounded moraine-like formation, similar to those created by melting ice during the withdrawal of the inland ice. It was possible to walk or drive on the deposit while pumping. Close to the pipeline discharge point, the waste material was found to deposit in slopes of about 6-7°. Further downstream towards the seepage dam, the slope angle successively decreased, to reach a slope similar to that of a conventional tailings deposit close to the seepage dam. The density of the deposited material was found to be higher than that of conventionally placed material, i e the use of available volume is more efficient.

The collection of finer particles close to the seepage dam retains water to a much greater extent than particles settled further upstream in the impoundment. Also, the trapping of finer particles in the coarse material of the seepage dam may cause a decrease of the internal friction of the material in the dam. In making dam stability assessments, these factors are important to consider. It was concluded, however, that if measures are taken in order to control the ground water level of the deposit, stable and high deposits may be created with this disposal method.

6.5 *Wear*

Wear was found to be rather extensive, both in pumps and pipes. Since a large portion of the pumped material, about 75% by mass, consisted of coarse waste rock, the flow pattern was partially to fully stratified, with many of the particles moving along the pipe bottom. As a result, the most extensive wear was found in the lower part of the pipes. The different liner materials gave rise to different wear results, with the most promising behaviour shown by the high chromium white cast iron. In later cost calculations, the wear life of straight pipe sections was assumed to be maximum 12000 hours, including rotating the pipes.

The pumps also showed extensive wear, and from the wear measurements the wear life of casings and impellers were estimated to be some 2500-3000 hours of operation.

6.6 *Environment*

The addition of dry handled coarse waste rock to the tailings stream means that the water is exposed to larger mineral surface areas. In underground mining, the process of blasting often gives rise to undetonated explosives (Lundkvist, 1998), the nitrogen content of which may be carried by water or adsorb to the surfaces of the ore hoisted up to the surface. Adsorbed nitrogen compounds may, as a result of the washing that takes place due to pumping, increase in the water draining out from the co-deposit. The results from water quality measurements carried out during the tests showed a significant increase of nitrogen concentration in the effluent water.

6.7 *Cost*

Cost comparisons of different trucking and pumping scenarios showed that the pumping and co-disposal alternatives were not competitive to conventional handling with trucking of dry handled waste rock and conventional disposal of tailings. In a cost comparison for Kiruna, the largest of the LKAB operations, an alternative with combined pumping/trucking showed a cost of operation and maintenance that was 26% higher compared to conventional handling. While some factors, such as energy costs, was favourable for the pumping alternative, other factors such as maintenance costs due to wear on pumps and pipes worked in favour of the trucking alternatives.

7 DISCUSSION

The tests were planned to encompass different operational cases to ensure that both normal and "extreme" operational conditions were tested in terms of particle size distribution and solids concentrations. Some of these goals were met during the tests, while some test cases could not be accomplished due to practical circumstances.

In terms of particle size distribution, the mixtures tested all represent broad particle size distributions, but with a major part of the tests with an overrepresentation of coarse waste rock material compared to a normal production scenario. In terms of solids concentration, most of the tests refer to rather moderate concentrations, while it was difficult to reach higher solids concentrations for any longer period of time for practical reasons. The loop-tests did, however, indicate possibilities of pumping up to above 60% by mass, which corresponds to volumetric solids concentrations of about 30-35% for the solids densities of iron ore waste materials.

The tests were considered to be successful in many aspects. It was shown that a mixture of coarse waste rock and tailings is pumpable, and that the energy input needed to pump the slurry met the expectations from the preliminary study and loop-tests.

The tests indicated that a coarse-particle pumping facility may be run with good operational reliability. Some additional thought must, however, be given to local circumstances which may require long-distance and/or inclined pipelines in which case possible density fluctuations in the pipeline must be able to be monitored and controlled to prevent blockages.

The co-disposal technique has shown to be a method with advantages in terms of stability and landscaping possibilities compared to traditional tailings disposal methods, even though some more development work may be necessary in order to run an efficient and safe operation.

The cost calculations performed after completed tests showed no economic advantage for the pumping and co-disposal alternative. This does not, however, exclude the method from being feasible in future or in other operations. The issue of wear in pumps and pipes requires more research and development work, in order to better predict wear lifes and to develop liner materials that are less expensive and that can stand the planned lifetime of a pumping facility. It is also of vital importance that both pumps and pipes be designed, constructed and erected in a manner that helps minimize wear. Furthermore, wear is material dependent, i e each waste material to be pumped has its own wear characteristics. Other factors that play a role in comparisons with traditional methods are investment costs (how much of the existing infrastructure needs to be altered), expected costs for decommissioning of a waste disposal site, savings potential in comparison with traditional methods (e g less need for dam raises) and the development of trucking prices.

8 CONCLUSIONS

In 1996, a project within LKAB aimed at investigating future handling of waste rock. Following a preliminary study, a decision was taken to perform tests on a large-scale basis in order to gain more knowledge related to operation and maintenance costs, operational reliability, cold climate serviceability and to the possibilities of the co-disposal technique for LKAB. In 1998, a large-scale test facility for coarse particle pumping and co-disposal was built. Pumping and disposal tests were then carried out during one year, with a total of 1700 hours of slurry pumping. The tonnage pumped amounted to some 427 ktonnes of coarse waste and tailings.

The tests were considered to be successful in many aspects. It was shown that a mixture of coarse waste rock and tailings is pumpable, and that transport was energy-efficient. Furthermore, pumping was considered to have shown good operational reliability and cold climate serviceability during the tests.

It was also concluded that, provided measures are taken in order to control the ground water level of a deposit, the co-disposal technique is a method with advantages in terms of stability and landscaping possibilities compared to traditional tailings disposal methods.

Wear was, however, found to be extensive, both in pumps and pipes, and cost calculations performed after completed tests showed no economic advantage for the pumping and co-disposal alternative. This does not, however, exclude the method from being feasible in future or in other operations, since costs and properties of the material to be pumped are site and material specific.

9 ACKNOWLEDGEMENTS

This project was partially (1/3) financed by Vattenfall, who is greatly acknowledged. Professor Anders Sellgren, Luleå University of Technology, provided valuable assistance to the various stages of the project. The Warman company supplied the pumps for the test facility, and was of assistance to the project in several ways. Thanks are also due to all companies that have supplied pipes and valves, or provided services to the project.

REFERENCES

Lundkvist, A. 1998. *The Process Water Geochemistry of the Kiirunavaara Magnetite Ore.* Licentiate Thesis 1998:08, Luleå University of Technology, ISSN: 1402-1757.

Sundqvist, S. 2000. Future disposal techniques for waste rock. In E. Forssberg & M. Thomaeus (eds.) *Bulletins from MinFo no 30 2000,* Conference in Mineral Processing, Luleå, 8-9 February, pp. 221-231 (in Swedish). English version available at http://www.lkab.com.

Tailings and Mine Waste '01 © 2001 Balkema, Rotterdam, ISBN 90 5809 182 1

Sealing of decant structure at QCM using a tailings cement grout mix

Claude Bédard, Pierre Gouslisty & John Lemieux
Journeaux, Bédard and Associates Incorporated, Dorval, Qué., Canada

ABSTRACT: Presentation of the results of a study conducted to evaluate the use of tailings incorporated in a non-shrink Portland cement grout mix to adequately seal the decant structure of the 30-meter high process water earth dam at Quebec Cartier Mines. A type I Portland cement and supplementary cementing materials (Fly Ash, Silica Fume) were tested along with the existing tailings.

A series of laboratory test and trial mixes were completed and evaluated by Journeaux, Bédard & Associates Inc. according to a series of design criteria including durability requirements, strength, density, segregation, bleeding, initial and final set, shrinkage, expansion, underwater placement, flowability, pumpability, costs, etc.

The grout mixes tested included the use of several different products including air entrainment admixture, anti-washout admixture, expansion admixture, bentonite, superplasticizer, etc.

The paper will detail the design criteria, methodology, laboratory results, different placing techniques (pressure grouting, pumping, tremie, etc.) and formwork used to seal the decant structures. The challenges and problems faced during the execution of the work will also be addressed.

1 INTRODUCTION

The gradual advancement of the tailings mine waste in the Hessé North primary settling basin at the Mont-Wright mine required that Quebec Cartier Mining (QCM) transfer the decantation point to the Hesse 4 dam and proceed with the permanent sealing of the decant structure at dam A, see figure 1.

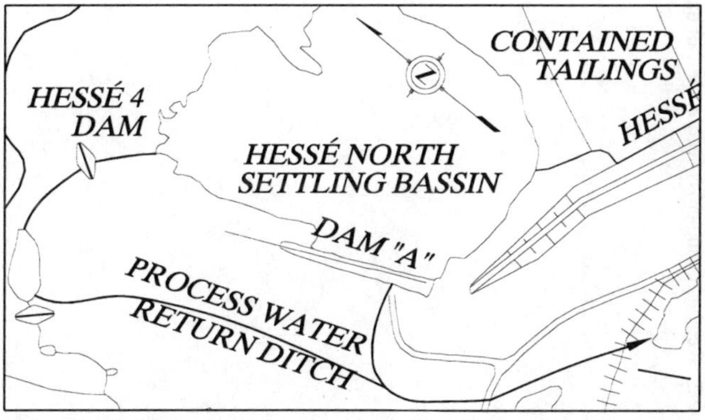

Figure 1: Sketch showing western end of tailings impoundment including: Hessé North settling basin, Dam A, Hessé 4 dam and Hessé dam.

The underwater cementing of the inlet transfer elbow between the sloping intake and the upstream decant conduit under the dam represented the first phase of this work in 1997.

The second phase followed in 1998 and involved the sealing of the decant conduit under the dam along its entire length of 185 meters. The first

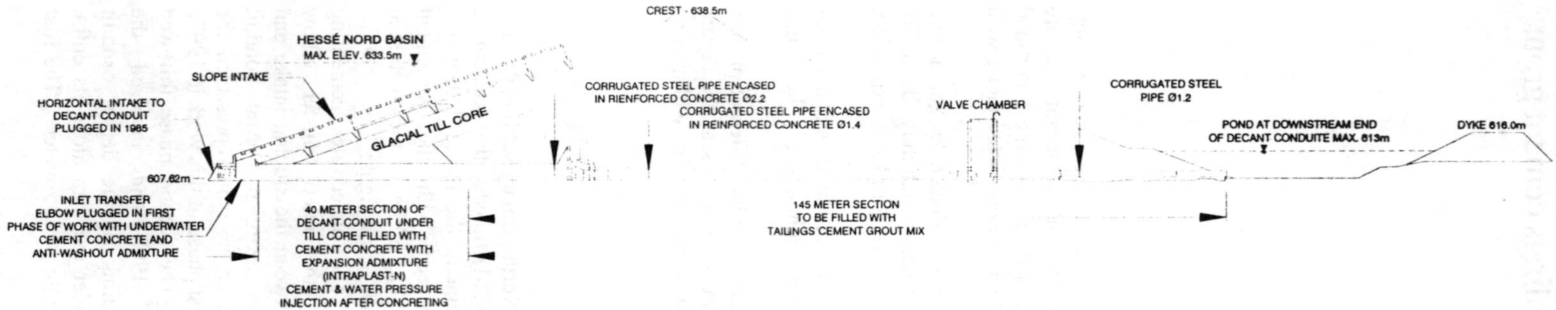

Figure 2: Sketch showing cross-section view of dam A and decant structure.

40 meters were sealed with cement concrete, while the remaining 145 meters were sealed using a specifically developed tailings grout mix. The tailings grout mix was developed and tested by Journeaux, Bédard & Associates Inc. (JBA) to help reduce the cost of sealing the bulk of the decant conduit under the dam. JBA designed and supervised all aspects of this work.

2 DESCRIPTION OF THE DECANT STRUCTURE

The decant structure of dam A consists of six (6) different principal components, including: the horizontal intake, the sloping intake, the inlet transfer elbow, the upstream decant conduit, the valve chamber, and the downstream decant conduit, see figure 2.

The simplest way to seal a decant structure is to do it under dry working conditions. In this way, it is possible to have direct control over the execution of the work and avoid using underwater divers. At the time that this work was completed the conduit was under 26 meters of red opaque process water. Draining the settling basin could not be considered as a viable option in view of the cost and time required to treat such a volume of process water was unfeasible and it was impossible to accumulate enough water in the basin to ensure operation of the mill during the winter. Structural analysis[1] of the sloping intake and the inlet transfer elbow showed that these components would collapse under the weight of the water if the sloping intake were blocked and the water drained. It would have been possible to execute the entire operation with divers however because of the opaque nature of the water, the associated cost and the control on quality desired, this option could not be considered either. However, structural analysis showed that the decant conduit, encased in reinforced concrete, could support the head of water and the following procedure was developed to allow a major portion of the work to be completed dry while reducing underwater work to a minimum.

3 LABORATORY TESTING OF TAILINGS GROUT MIXES

Approximately thirty-three (33) tailings grout mix designs were evaluated to determine the most effective and economical mixtures for the adequate sealing of sections of the decant structure.

The effects on the engineering properties of the selected supplementary cementing materials and admixtures in the tailing grout mixes were evaluated to meet certain design criteria and durability requirements. The most important engineering properties evaluated for our grout mix designs are as follows:

- Flowability
- Bleeding
- Expansion or shrinkage
- Compressive strength at 7 and 28 days
- Density

The tailings used are a uniform, medium size, angular grain sand with 5% of particles passing the No. 200 sieve, see figure 3. In general, all the mixes were done at a constant 1:1:3 ratio of water, cement and tailings. Different percentages of pre-hydrated bentonite, expansion additive (Intraplast-N), superplasticizer and air entrainment were evaluated in the grout tailings mix designs. Fly ash and silica fumes were the supplementary cementing materials tested at different percentages with the standard Portland Type I cement.

3.1 TEST RESULTS

The results of the test performed on the tailings grout mix designs are summarized and explained in the following lines.

[1] Structural analysis completed by Klohn-Crippen

3.1.1 FLOWABILITY

The standard test method for flow of grout for preplaced aggregate concrete (flow cone method) according to the ASTM C939-87 standard was used to determine the fluidity of the different mixtures. The fluidity was measured by determining the time of efflux of a specified volume of fluid hydraulic cement grout passing through a standardised flow cone, see figure 4.

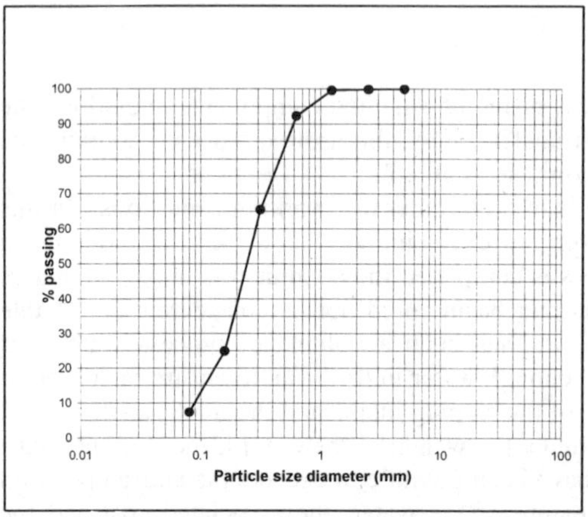

Figure 3: Typical tailings grain size distribution with a portion of finer fraction removed.

The following comments summarise some of our observations on the flowability of our tailings grout mixtures:

- With mixes done at a constant 1:1:3 ratio of water, cement and tailings and no admixtures, it was impossible to obtain a flowable mix (mixture blocked in the flow cone). The use of a superplasticizer product, which increases fluidity of cementing mixtures, was not enough to produce a more adequate flow. No time of efflux was measured as the flow cone apparatus was always blocked.
- An air entrainment admixture dramatically increased the flowability of our mixtures. Air entrainment contributed to increase consistency and workability of our mixtures, acting as a lubricant in the flow cone apparatus. The time of efflux measured with air entrainment was, in general, under 30 seconds, representing our mix design criteria for flowability.

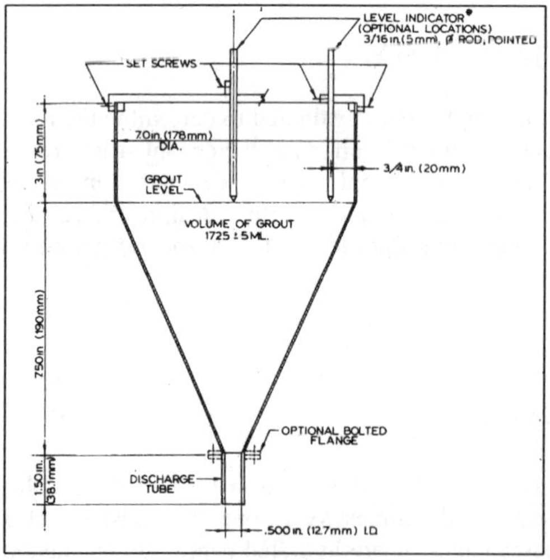

Figure 4: Typical cone apparatus.

- Having to prepare a non-shrink tailings grout to fill up sections of the decant structure, we had to use an expansion admixture of the Intraplast-N type. This admixture is an aluminum-based powder that has been found to reduce the shrinkage of grouts by producing a preset expansion of up to 15%. The Intraplast-N produces expansion by the formation of nitrogen gas. Furthermore, it is known to increase fluidity of cementing mixtures. In our study, the addition of 1% (mass of cement) of Intraplast-N was found to be adequate, in terms of costs, expansion as well as increased flowability of the mixtures, reducing the time of efflux in seconds. The effect of Intraplast-N on flowability was not as important as air entrainment.
- The addition of superplasticizer with air entrainment admixtures and Intraplast-N slightly increased the flowability of our tailings mixtures but not enough to justify the cost of using superplasticizer in our final mixture to use to seal the decant structure.
- The addition of bentonite at different percentages did not have any significant effect on the flowability of our tailings grout mixtures. Bentonite is normally used to produce grouts of high

water retention and to keep sand fillers in suspension with no segregation. Due to the high content of tailings sand used in our standard mix (three (3) times the amount of cement or water), we did not have any problems keeping our mixtures in suspension. No segregation was observed in our test mixtures even without bentonite.

3.1.2 TAILINGS GROUT BLEED

The standard test method for expansion and bleeding of freshly mixed grouts for preplaced aggregate concrete in the laboratory (ASTM C940-87) was used to determine final bleed.

The following comments summarize some of our observations on the final bleed in percentages of our test mixtures:

- Mixes with air entrainment and expansion admixtures reduce significantly the percentage of final bleed.
- No significant change of bleed was noted when bentonite was used with mixes containing expansion and air entrainment admixtures.
- A slight reduction of final bleed was noted while using supplementary cementing materials, such as silica fume or fly ash, in combination with air entrainment and expansion admixtures.

3.1.3 EXPANSION AND SHRINKAGE

The expansion and shrinkage during the initial setting of the grout was also determined using the standard test method for expansion and bleeding of freshly mixed grouts for preplaced aggregate concrete in the laboratory (ASTM C940). In a few mixes, expansion and shrinkage after initial setting (grout hardening) was also measured.

An approximate final percentage of volume change was computed in our test mixtures. Our design criteria was to obtain a non-shrink grout to adequately seal the decant structure.

The following comments summarise some of our observations:

- Tailings grout mixtures containing an expansion admixture produce a preset expansion of up to 5%, which is able to compensate for the shrinkage that occurs when the tailings grout hardens and dries. Not using an expansion admixture (Intraplast-N) produces shrinkage and would not adequately seal the decant structure.
- Bentonite, fly ash and silica fume mixed with some Intraplast-N seems to have little effect on the expansion and shrinkage of the tailings grout.
- Air entrainment by reducing bleeding contributes to a slightly reduced shrinkage in the hardened state while also contributing to the expansion in the test mixtures.

3.1.4 COMPRESSIVE STRENGTH

The standard test method for compressive strength of hydraulic cement mortars (using 2 inch or 50 mm cube specimens) was used to determine the compressive strength of our cured tailings grout specimen. In general, the tests were done after 7 and 28 days.

The design criteria was to obtain a minimum compressive strength of 15 MPa at 28 days. The following comments summarise some observations concerning the compressive strength values obtained:

- Compressive strength of mixes using over 2% of expansion admixture (Intraplast-N) showed lower values, not reaching 10 MPa at 28 days. This was due to the fact that the Intraplast-N produces small air bubbles in the grout mix, which cause the compressive strength to be reduced when it hardens.

- Compressive strength values of mixes using bentonite were lower than other mixes not containing bentonite.
- Air entrainment admixture used in higher percentages significantly reduced the compressive strength. In our final design mix, we reduced the air entrainment proportion normally used by 50% to enable us to reach our target of 15 MPa at 28 days without reducing the flowability of our tailings grout mix. In fact, we observed that only a small amount of air entrainment admixture (0.2% of the mass of cement) is needed to adequately increase the flowability, thus reducing the time of efflux in seconds.

3.2 FINAL DESIGN MIX

A few adjustments had to be done to our final design tailing grout mix to meet all our criteria.

The following grout mix proportions conformed to all our design criteria and most important had the lower costs:

Ratio in mass:	cement (type I) / water / tailing: 1.0 / 1.9 / 2.5
Admixtures:	Expansion admixture: 1% (mass cement) Intraplast-N
	Air entrainment: 0.2% mass of the cement

4 DESCRIPTION OF THE WORK

4.1 UNDERWATER

By pumping a concrete plug directly into the inlet transfer elbow and then up into the sloping intake both these components would be supported by the concrete and it was possible to allow the rest of the decant conduit to be drained.

The following paragraphs will describe the principal steps taken during the first phase of this project.

Figure 5: Photo showing pumping apparatus at downstream end of decant conduit.

<u>Construction of small dyke</u>. A small earth dyke was constructed to block water flow around the downstream end of the decant conduit and the pond formed was pumped dry, see figure 5. Once the bulkhead in place, this would allow the decant conduit to be completely drained.

Removal of accumulated silt and tailings material in the inlet transfer elbow. After fully closing the control gate in the valve chamber, divers could access the inlet transfer elbow area by diving down the sloping intake.

Installation of form assembly. Without piercing the corrugated steel pipe or anchoring into the reinforced concrete casing material it was necessary to install a form assembly to ensure that the measured volume of concrete pumped into the inlet transfer elbow was contained in the elbow and would not flow unpredictably into the decant conduit. The original horizontal intake

Figure 6: Photo taken from the crest of the dam showing form assembly.

to the inlet transfer elbow had previously been cemented in 1985, see figure 2. The form assembly, see figure 6, was tied back to this cement wall with steel rods.

Underwater concrete placement. When placing concrete underwater it has to be done under experienced supervision. The basic principals for normal concrete work in dry apply but special

Figure 7: Photo taken from the crest of the dam showing sloping intake, diver and pipeline.

points also had to be considered before pumping could begin underwater including: a minimum of 450kg/m^3 of type I cement in our design mix, a maximum water/cement ratio of 0.45 and a slump specification of 170 to 190 mm after the addition of a superplasticizer admixture. It was very important that the concrete flow without segregation having a cohesive mixture with high workability. Entrained air and anti-washout admixtures were used to obtain a cohesive mixture. Once the form assembly was installed it was possible to start pumping the concrete underwater into the inlet transfer elbow, see figure 7. The concrete mix specifications were as follows:

Compressive strength at 28 days	35MPa
Water/cement ratio	0.43
Max. aggregate nominal size	20mm
Slump (before superplasticizer)	80 +/- 20mm
Entrained air	5 to 8%
Admixtures	Air, superplasticizer and anti-washout

This material was pumped a distance of 82 meters down the 3H:1V sloping intake through a 100mm diameter steel pipe to the inlet transfer elbow. The mix had a tendency to plug up in the pipeline. To avoid this, a larger diameter pipeline, a smaller nominal diameter aggregate or a greater amount of superplasticizer (to increase workability) could have been used.

Inspection. With the plug in place the decant conduit was drained along its entire length and access could be gained from the downstream end of the decant conduit to inspect the bulkhead. After installing necessary services such as lighting and ventilation a visual inspection of the decant conduit and the bulkhead installed in phase 1 was conducted. The inspection revealed a small leak at the pipe/bulkhead interface at a constant rate of approximately 2 drops/second. Two other minor leaks were also detected coming through joints in the corrugated steel pipe at 1.5 at 12 meters downstream of the bulkhead. The pipe itself was in good condition with a thin layer of mud/silt adhering to the walls of the pipe.

4.2 DRY CONDITIONS

During the summer of 1998, one year later, work was resumed with phase 2 of the project which consisted in itself of two main steps. First, the length of the pipe passing through the till core of the dam, 40 meters, had to be completely sealed with concrete. Second, the remaining 145 meters of the decant conduit would be grouted. During the entire underground operation several security measures were taken into consideration, including a communication system, air quality surveillance program and regular water inflow measurements.

Removal of the valve assembly. First, the valve assembly was removed from the valve chamber, see figure 8.

Figure 8: Photo taken near valve chamber showing removal of valve assembly.

Cleaning of the pipe walls. With the valve components removed, it was possible to gain access to the bulkhead from the valve chamber and start cleaning the walls of the corrugated steel pipe. The mud/silt layer was removed with high pressure water jets to ensure good adhesion of the cement mixtures to the walls of the pipe. Once removed the mud/silt layer mixed with water to form a slurry and was evacuated from the bottom of the valve chamber using mobile suction pumps and a storage tank, see figure 9.

Placement of the concrete. The concrete used in this section was proportioned almost the same as for underwater concrete (see section 4.1). The only difference was that no anti-washout admixture was used and an expansion admixture such as Intraplast-N was used. This admixture has been found to reduce shrinkage of cementitious materials by producing a pre-set expansion of up to 15% in certain cases. Furthermore, it is known to increase fluidity of cementitious mixtures. Once this concrete mixture has cured for a minimum of 48 hours a 35MPa cement grout (cement and water only) was injected to a constant pressure of 300Kpa (greater than the water head pressure of 26 meters) was used to perfectly seal any voids in the upper portion of the conduit along the length of the decant conduit passing through the till core on the upstream slope of the dam. The job was done in one section of 10 meters and two sections of 15 meters each allowing inspection of each section as it was completed before pouring the next section, see figures 10 and 11.

Figure 9: Photo taken looking down on valve chamber showing removal of mud/silt residue from the bottom of the chamber.

At this stage in the project the sloping intake, the inlet transfer elbow and the section of the decant conduit passing under the till core have all been cemented with a total of 14.5 m^3 of cement grout (cement and water only) and 174 m^3 of concrete (35% of total). With this section complete it was now possible to cement the final 145 meters (330m^3 or 65% of total) of the decant conduit and the valve chamber using the much lower cost tailings grout mix designed in our laboratory and described in section 3.

Figure 10: Photo from inside the decant conduit looking upstream at the form assembly of a finished section.

5 CONCLUSION

The gradual advancement of the tailings mine waste in the Hessé North primary settling basin at the Mont-Wright mine required that QCM transfer the decantation point to the Hesse 4 dam and

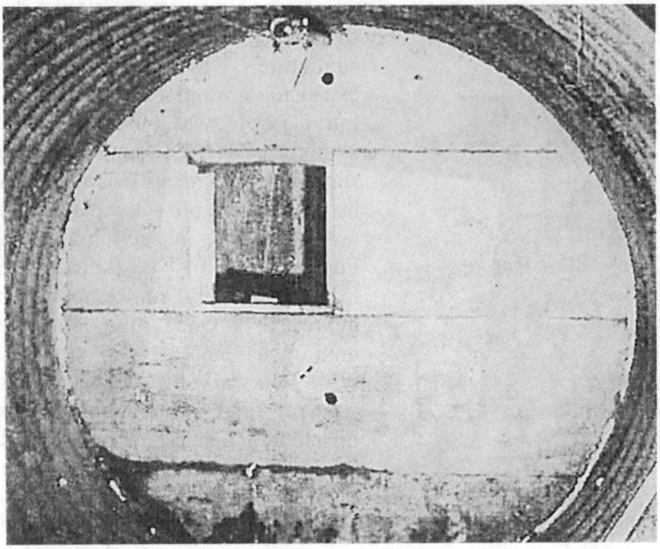

Figure 11 Photo from inside the decant conduit looking upstream at a completed section after the form assembly has been removed.

proceed with the permanent sealing of the decant structure at dam A.

The underwater cementing of the inlet transfer elbow between the sloping intake and the upstream decant conduit under the dam represented the first phase of this work in 1997. The second phase followed in 1998 and involved the sealing of the decant conduit under the dam along its entire length of 185 meters with standard cement concrete, cement grout (water and cement only) and a tailings grout mix.

The volume of cement mixtures required to completely fill the entire decant structure was measured to be 504 m^3. Sixty-five percent (65%) of which was filled with a tailings grout mix developed in our laboratory and prepared on site reducing the volume required from the local cement plant to 174 m^3 and consequently reducing the total cost of cement products for the project by almost 40%. The tailings grout mix used had a very good workability and consistency showing no signs of segregation and being easily pumped in the decant conduit.

Since the autumn of 1998, when the work was completed, the accumulated water at the outlet of the decant conduit has been drained on a yearly basis to verify for any visible water leakage seeping through the conduit. To this day no water leakage has been detected.

6 ACKNOWLEDGEMENT

The authors would like to acknowledge Daniel Lang's, from QCM, contribution in the preparation and in allowing publication of this paper.

Author index